Human Barrier Design and Lifecycle

A common source of failure in a human-dependent barrier or safety critical task is a designed-in mismatch error. The mismatch is a cognitive demand that exceeds the human capability to reliably and promptly respond to that demand given the plausible situations at that moment. Demand situations often include incomplete information, increased time pressures, and challenging environments. This book presents innovative solutions to reveal, prevent, and mitigate these and many other cognitive-type errors in barriers and safety critical tasks. The comprehensive model and methodologies also provide insight into where and to what extent these barriers and task types may be significantly underspecified and the potential consequences.

This title presents a new and comprehensive prototype design and lifecycle model specific to human-dependent barriers and safety critical tasks. Designed to supplement current practice, the model is fully underpinned by cognitive ergonomics and cognitive science. The book also presents a compelling case for why a new global consensus standard specific to human-dependent barriers is needed. Taking a novel approach, it presents its suggested basis, framing, and content. Both solutions seek to redress deficiencies in global regulations, standards, and practice. The model is guided by industry recommendations and best practice guidance and solutions from globally recognized experts. Its processes are fully explained and supported by examples, analysis, and well-researched background materials. Real-life case studies from offshore oil and gas, chemical manufacturing, transmission pipelines, and product storage provide further insight into how overt and latent design errors contributed to barrier degradation and failure and the consequence of those errors.

An essential and fascinating read for professionals, *Human Barrier Design and Lifecycle: A Cognitive Ergonomics Approach and Path Forward* will appeal to those in the fields of human factors, process and technical safety, functional safety, display and safety system design, risk management, facility engineering, and facility operations and maintenance.

Human Barrier Design and Lifecycle

A Cognitive Ergonomics Approach and Path Forward

Tom Shephard

CRC Press
Taylor & Francis Group
Boca Raton London New York

CRC Press is an imprint of the
Taylor & Francis Group, an **informa** business

Designed cover image: Top left, Valdimirovic © iStock; Top right, petroleum man © Shutterstock; Middle left, engel.ac © Shutterstock; Middle right, Avigator Fortuner © Shutterstock; Bottom left, Maksim Safaniuk © Shutterstock; Bottom right, Maksim Safaniuk © Shutterstock

First edition published 2024
by CRC Press
2385 NW Executive Center Drive, Suite 320, Boca Raton FL 33431

and by CRC Press
4 Park Square, Milton Park, Abingdon, Oxon, OX14 4RN

CRC Press is an imprint of Taylor & Francis Group, LLC

© 2024 Tom Shephard

ISBN: 9781032655949 (hbk)
ISBN: 9781032674483 (pbk)
ISBN: 9781032674476 (ebk)

DOI: 10.1201/9781032674476

Typeset in Times
by codeMantra

Contents

About the Author

Tom Shephard has 42 years of experience in global operating, engineering design, and consulting companies in the hydrocarbon and related technology industries. Tom designed and delivered safety solutions and studies to world-class global projects in O&G, refining, petrochemicals, and pipelines. Early roles included corporate technology and project leadership, HMI and safety system design and delivery, and technology and project consulting. Career spanning roles and efforts included functional safety and developing new corporate and project technology standards, tools, and project execution methods. Later roles shifted to process safety, safety and reliability studies, and department management for a world-class technical safety department. Tom's efforts to integrate cognitive ergonomics and cognitive science into human barrier design began with the Deepwater Horizon accident. He brings a seasoned practitioner's mindset, persistence, passion, and a unique blend of multi-discipline competencies to that effort.

Permissions

Figure J.1 Adapted and modified from Flin, R., O'Connor P., Crichton, M (2008), Safety at the Sharp End: A Guide to Non-Technical Skills, Figure 5.1 © Ashgate Publishing. For permission given to use and modify Figure 5.1, see CCC Marketplace License 1381872-1 dated August 1, 2023.

Guidelines for Management of Safety Critical Elements (SCEs), 3rd Ed. © January 2020 by Energy Institute. Information reproduced with the written permission of Energy Institute.

Human barriers in Barrier Management, A White Paper (2016) © by Chartered Institute of Ergonomics and Human Factors (CIEHF). Information reproduced with the written permission of CIEHF, Chief Operating Officer Tina Worthy, email dated July 31, 2023.

IEC 61511-1 (2016) © by International Electrotechnical Commission (IEC). Permission to reproduce extracts from IEC standards is granted by IEC. For permission to reproduce this standard, see CCC Marketplace License 1379024-2 dated July 31, 2023.

IOGP Report No. 460 © 2012 by the International Association of Oil & Gas Producers. Permission is given to reproduce this report in whole or in part provided that (i) the copyright of IOGP and (ii) the source are acknowledged.

ISO 11064-1:2000 © 2000 by British Standards Institution (BSI). Permission to reproduce extracts from British Standards is granted by BSI. British Standards can be obtained in PDF or hard-copy formats from BSI Knowledge: https://knowledge. bsigroup.com or by contacting BSI Customer Services for hardcopies only: Tel: +44 (0)20 8996 9001, email: cservices@bsigroup.com

Advancing Understanding of Offshore Oil and Gas Systematic Risk in the U.S. Gulf of Mexico: Current State and Safety Reforms Since the Macondo Well Deepwater Horizon Blowout © 2023 by National Academies of Science, Engineering and Medicine (downloaded from nap.nationalacademies.org website on April 7, 2023). For permission to reproduce this report, see CCC Marketplace License 9780309699778, dated July 20, 2023.

Acknowledgments

I am forever grateful for my wife's patience and support during the three long years taken to write the book and for my son's suggestion to write it.

1 Introduction

This chapter is organized as follows:

- 1.1 Intended Audience
- 1.2 Lifecycle Model Application
- 1.3 Content and Organization
- 1.4 Industry Insights and Recommendations
- 1.5 Companion Guides to This Book
- 1.6 Suggestions for Readers with Different Interests

Active human barriers are globally employed in process facilities that manufacture, process, store, or transfer hazardous materials. Like other active barriers, this type is designed to achieve a proscribed safety function within a specified safety time. Unlike other active barriers, it relies on *humans* to achieve that outcome. Active human barriers include preventive, control/recovery, and emergency response barriers, the latter being the most complex of all active barrier types.

The purpose of this book is twofold. It presents a new and comprehensive prototype design and lifecycle model for active human barriers. The model is designed to supplement current practice by addressing systemic deficiencies in cognitive ergonomics and associated areas. It also presents a compelling case as to why a new global consensus standard specific to this barrier type is required. Both approaches seek to redress acknowledged deficiencies in global current practice. The prototype processes, assessments, tools, and adaptations to existing methodologies are guided by recommendations and findings presented in industry call-to-action white papers, articles, and status reports published in response to the Deepwater Horizon accident. Examples include CIEHF (2016), IOGP (2012, 2014a,b, 2018), Johnsen et al. (2017), OESI (2016), SPE (2014), and NASEM (2023).

Industrial processes that include highly hazardous materials and activities rely on advanced control and safety systems to maintain operations within safe limits. Despite these systems, incidents and major accidents persist. These events demonstrate a continued reliance on active human barriers to save lives and prevent injuries and large-scale environmental accidents. Major accident investigations confirm existing standards and practices are failing to rid these barriers of latent design errors (cognitive-attributed) that disrupt, degrade, and disable human performance. Post-accident corrections have not adequately addressed these deficiencies. Without a fundamental change, errors of this type will persist and remain primary contributors to future incidents and major accidents.

A common source of failure in an active human barrier or safety critical task is a latent (designed-in) mismatch error. The mismatch is an unrealistic cognitive demand that exceeds the human capability to reliably respond to that demand. Demand situations often include increased time pressures and difficult environmental conditions.

DOI: 10.1201/9781032674476-1

A primary design goal for the prototype model is to reliably reveal the mismatch (and other error types), and then prevent, eliminate, or mitigate its source or effects. Example cognitive processes, attributes, limitations, and behaviors addressed in the model include the following:

- **Automatic cognitive process** functioning, capabilities, performance, attributes, behaviors, and tendencies (also referred to by the terms *System 1, subconscious,* and *unconscious*).
- **Conscious cognitive process** functioning, capabilities, capacity limits, performance, influencers, attributes, behaviors, and tendencies (also referred to by the terms *System 2* and *working memory*).
- Wide variations in **cognitive processing speeds** (the abovementioned processes).
- **Interactions between these cognitive processes** (asynchronous and situationally dependent) that determine which process is the final arbiter that forms and drives perceptions, decisions, and actions.
- **Short-term memory** (STM) storage capacity and retention duration limitations, both readily degraded by a wide range of internal and external factors. (STM is essential to working memory and conscious cognitive processes.)
- The nature and limitations of **attention** and **attention management**.
- The nature and functioning of **long-term memory** (LTM). LTM provides instant access to safety critical skills and knowledge.
- **Mental model** (MM) development, content, accuracy, stability, and retention. MMs provide safety critical, *in-the-head* knowledge.
- **Human visual system** capabilities and limitations.
- Human limitations in **tracking clock time** reliably.
- The conditions and mechanisms that lead to **skill fade** and **drift**.
- **Non-technical skills** essential to multi-person barriers, which include shared situation awareness, leadership, communication, coordination, and trust.
- **Non-rational biases and behaviors** including confirmation bias, loss aversion bias, task-switch errors, bounded rationality, and many others.

As a prototype, this first-of-kind lifecycle model is presented as a white paper process. A goal in its development is not to provide "the" answer. Instead, it presents the widest possible range of processes, methods, and tools that identify what may be missing in current practice and demonstrate what can be achieved using new and modified methods. The information is sufficient to evaluate the model for its potential efficacy, validity, level of effort, and benefits. Source references identify the basis for each approach and the source of key material. They include industry standards and guidance documents, professional society publications, books from globally known experts, accident investigation reports, and peer-reviewed research articles. Supporting information is also provided to perform a cost-benefit analysis or deploy the model in a demonstration project.

The model uses flowcharts to identify lifecycle phases, sequencing, and the processes and activities included in each phase. Processes are described, explained, and supported by rich sources of guidance, analysis, and the topic-based appendices. The model exceeds what is possible using common human-centered design standards and practices. It develops new requirements and information that are essential to a cognitive-focused approach. Information captured in data tables and identified documents ensures the information is discoverable, readily accessible to all users, and fully traceable through the entire lifecycle. Inline notes provide examples, explanations, solution options, guidance from globally recognized experts, and insights from major accidents. Finally, the model provides adaptations and extensions to existing methods providing improved support for a cognitive-centric approach.

The second stated purpose is to delineate the need for a new global consensus standard for active human barriers. A dedicated chapter presents a compelling case as to why the standard is needed and why it is the essential next step toward reducing and preventing future incidents and major accidents. Achieving this end requires preventing or minimizing cognitive-attributed errors that lie within existing and new barrier designs and lifecycle activities. The prototype model suggests options to do so by identifying attributes and features that should be considered when framing and developing the new standard. As a demonstration, the model includes attributes and elements adopted and adapted from an existing globally accepted and widely employed standard for a different active barrier type, namely, IEC 61511. Examples include its lifecycle phasing, key documents essential to a lifecycle approach, and other selected concepts and processes. By example, the prototype model provides contextually rich and previously unavailable guidance to those attempting to define the scope, constructs, framing, and topics to include in a new global standard for active human barriers.

This book contributes to the overall knowledge and understanding of active human barriers. Core constructs and processes are equally applicable to safety critical tasks, other barrier types, and other industries and process sectors. The appendices (roughly half the book) address key topics such as team design, human cognition basics, performance influencing factors, verification, validation, training, procedures, remote barrier support, procurement, and project execution. Appendix I presents a new cognitive assessment process derived from the latest science on human cognition. Information from the Deepwater Horizon accident provides contextual insight into the active and active human barrier successes and failures. New visual representations present complex information and concepts in easily understood forms.

This is an expansive work that required three years to write and a decade to research and develop the base concepts and ideas. I believe readers will discover a wide range of fascinating, new, and insightful information, ideas, and perspectives. It is the book I wished I had had to support my 40-year career. I hope readers find it to be equally helpful and valuable.

1.1 INTENDED AUDIENCE

The intended audiences include the following:

- Owner/operators
- Engineering, procurement, and construction contractors
- Industry and professional organizations
- Regulatory agencies
- Academics and researchers

- Accident investigators
- Other Industries and sectors

For process facility **owner/operators**, the book presents a first-of-kind, cognitive ergonomics–focused design and lifecycle model that proposes an approach to improve the reliability and performance of active human barrier and safety critical tasks. The model shows the new information developed by these more advanced process methods and tools. It provides the means to examine existing standards and practices, identify vital information they do not create or use, and assess the potential safety consequences of that missing or unused information. The model also shows where and how company policies provide input to and affect barriers and safety critical tasks. It shows how model processes identify and mitigate hidden barrier deficiencies that develop in the operate and maintain phase. The capital project execution guidance suggests approaches to integrate, manage, and share new barrier requirements with other disciplines and organizations, and when and why owner/operator participation is recommended. New visualizations, process flowcharts, technical information, and contextually rich examples may markedly advance and enrich the reader's understanding of human cognition, cognitive ergonomics, active human barriers, and safety critical tasks. Cross-reference tables and the topic-based appendices enhance the book's usability and value as an essential technical and project execution resource.

For **engineering, procurement, and construction contractors**, the lifecycle model includes guidance on project planning and execution. It provides supporting information to develop plans, budgets, schedules, and personnel requirements. Guidance is also provided to support procurement, interface planning, information management, and other project activities. The detailed process descriptions and notes provide insight into the scope, nature, and timing of technical and project information exchanges, coordination, and interdependencies with other technical disciplines and organizations.

For **industry and professional organizations**, the book identifies the need and presents the case for developing a new global standard for active human barriers. The provided information suggests a path forward for addressing its potential scope, framing, organization, and content. Analysis and reviews identify deficiencies in current practice and why they are a major source of risk. The unique challenge of human cognition is discussed to understand the lack of progress in cognitive ergonomics from a historical, technical, and industry perspective. The challenges discussed include the need to gather, assess, and deploy this information in forms that can be applied to projects and operating facilities. Further, the new information, processes, and methods require new expertise and competencies.

For **regulatory agencies**, the book provides insight into areas that are not adequately guided by regulations. As discussed in the Introduction, a recent assessment of the oil and gas industry (U.S. Gulf of Mexico) identified active human barriers and safety critical tasks as a main source of systemic risk. Without further change, deficiencies in both will continue to contribute to incidents and major accidents. The absence of a dedicated consensus standard for active human barriers and a broad understanding and application of cognitive ergonomics should both be sources of concern.

For **academics and researchers**, the book demonstrates the many possible applications of research findings from the fields of cognitive psychology, neuropsychology, behavioral psychology, and human factors. The prototype model shows how this information may be used to improve the safety of industrial facilities. It may also provide insight for future research. Broadly, the model and materials suggest ways to frame and present research to attract and reach a process industry audience. On the contribution side, the safety of process facilities can be enhanced by contributing to the knowledge, tools, and methods used to progress a cognitive ergonomics approach. As an example, the information in the cognitive assessment tables (Appendix I.3) reflects the latest research findings. Another is the development of guidance tools for correctly applying published information on human memory (e.g., the information in Radvansky, 2021).

For **accident investigators**, the book compiles and integrates information from the Deepwater Horizon accident investigation reports, live hearings, authoritative books, and post-event interviews. The detailed timeline shows the failed/degraded active human barriers (and related information) that were causal contributors to the accident chain. If the goal were to identify the root cause of these failures, the accident report findings and recommendations did not identify the absence of a dedicated industry standard for active human barriers or the inadequate application of cognitive ergonomics as root causes. Further, the NASEM (2023) report implies a cause-and-effect relationship between the omissions in the recommendations and the current state. If true, this may justify a concerted effort to understand this finding and determine how it should be used to guide future investigations. This book provides information, analysis, and insights that may provide important contributions toward that effort.

For **other industries and sectors**, many of the included processes, methodologies, and tools may be readily adapted and applied to safety critical tasks and functions in those industries and sectors.

1.2 LIFECYCLE MODEL APPLICATION

The prototype lifecycle model applies to all active human barriers. Most processes may also be applied to safety critical tasks. Example barrier types include the following:

- Preventive barriers that prevent the occurrence of a defined hazardous event or condition.
- Mitigation barriers designed to control and recover from a hazardous event once it occurs.
- Emergency response barriers that limit the effects of the hazardous event once it occurs.

Note: Some of the lifecycle model terminology, assessments, and examples may be more common to the energy sector, for example, oil and gas drilling and production, and petroleum and petrochemical manufacturing, processing, storage, distribution, and pipelines. However, at its core the model is task-based. As such, it includes industry-agnostic processes, methodologies, and tools that should be applicable and adaptable to any industry that has safety critical tasks or active human barriers. Example industries may include healthcare, cyber security, rail, marine, non-hydrocarbon pipeline, pharmaceutical, aviation, and nuclear industries.

1.3 CONTENT AND ORGANIZATION

Chapter 2 defines the terms and definitions, and presents the foundational concepts, constructs, and framing applied in the prototype lifecycle model.

Chapter 3 presents the *conceptual* and *preliminary design phase* processes, methods, analyses, and activities (lifecycle phases A and B).

Chapter 4 presents the *detailed engineering and design phase* processes, methods, analyses, and activities (part of lifecycle phase C).

Chapter 5 presents the *implementation phase* processes, methods, analyses, and activities (part of lifecycle phase C).

Chapter 6 presents the *construction, installation, and commissioning phase* processes, methods, analyses, and activities (lifecycle phase D).

Chapter 7 defines the *operate and maintain, modification, and decommissioning phase* processes, methods, analyses, and activities (lifecycle phases E, F, and G).

Chapter 8 presents the *case for developing a global consensus standard* for active human barriers. It includes the background, justification, and suggested path forward to achieve this end.

Chapter 9 presents *additional observations, findings, and insights* that are incidental to the development of this book. As often occurs with leading-edge developmental efforts, the prototype model and supporting information provide unexpected insights and new and deeper understandings. One section discusses the historical hurdles that have challenged and will continue to challenge progress in cognitive ergonomics and the actions needed to overcome those hurdles. Others discuss closing the work-as-imagined/work-as-done gap, the unique challenges when executing a cognitive ergonomics methodology, different views of workspace design, and the limits of human reliability and performance. Yet others discuss the need for new terms and language, the DWH accident investigations and reports, and a powerful new skill the reader may develop when the prototype model and supporting materials are understood and internalized. Examining an existing standard or practice through this new lens may markedly improve the ability to quickly identify gaps and errors in these documents (cognitive-attributed) and understand why and how those gaps and errors may lead to barrier system or safety critical task degradation and failure.

Appendix A provides selective project execution input to *project resourcing and staffing plans, scheduling, interface management, procurement, information management, and quality processes* through each lifecycle phase. It also provides insight into the potential interactions, timing, and coordination requirements (interdependencies) with other technical disciplines and organizations.

Appendix B provides the suggested table of contents for three essential documents included in the lifecycle model, namely, the *barrier safety management plan, safety requirements specification*, and *remote barrier support design basis*. It addresses several *technical topics*, including why humans look but do not see, display salience, and why time is a primary adversary in a barrier system. Section B.6 presents the "CX" component-level sub-process used in several detailed design phase processes. A different section provides cross-reference tables that enhance the book's usability as a technical, process, and project resource. A closing section provides guidance on selecting and developing task *support (job) aids.*

Appendix C, dedicated to barrier *team design* topics, provides technical information and application guidance on the adopted team situation awareness model, implicit and explicit coordination, and trust. (Team design topics specific to non-technical skills are individually discussed in many sections.)

Appendix D addresses the adopted *performance standards* (PSs) that support the verification and validation processes discussed in Appendix G. It reviews the commonalities and differences between several published practice and guidance standards. It highlights the importance of "flowing-down" barrier and barrier element PS requirements to barrier-dependent external support systems and external protective barriers. It also presents the adopted PS types for physical elements (design and operations), human elements, and the barrier system (operations-based). A closing section reviews and comments on the "human performance standard" proposed in CHIEF (2016).

Appendix E provides *procurement* guidance (what, how, when, and who) for integrating barrier requirements into a purchase order or contract for barrier-dependent equipment, systems, packages, and buildings. Example guidance includes requisition preparation and reviews, order kick-off, design reviews, assessments, inspections, testing, verification, and validation. Useful tables suggest methods for integrating requirements into the typical procurement package. It presents common execution challenges and example approaches to address each challenge. One suggested solution responds to procurement activities that occur very early in the project cycle, for example, a long lead or early need requisition or contract. Another offers suggestions for coordinating quality activities with those having primary responsibility for a requisition package or contract.

Appendix F provides the *information on human cognition* that underpins the prototype model processes, methods, and tools. It includes an insightful, compelling, and first-of-kind table that presents and compares the unique nature, capabilities, and limitations in the two vastly different automatic and cognitive processes common to all humans. It presents specific cognitive issues to consider and address in the barrier design and lifecycle, such as attention, working memory, change blindness, bounded rationality, and common non-rational biases and behaviors. Other sections provide information on the inherent limitations in the human visual system, the cognitive effects of acute stress, and example methods to minimize or manage acute stress and its effects. A closing section provides information to better understand mental models and long-term memory type, content, retention, recall, and change over time.

Appendix G defines the suggested *design reviews, inspections, testing, verification, and validation* activities employed in the prototype model. A unique visualization shows their typical or proposed timing in each lifecycle phase. One section provides example execution guidance (who, when, how, competency, and independence requirements), and the example forms of examination and tangible evidence suggested for each activity. Another section assesses ISO 11064-1 (2000), EI (2020b), and IEC 61511-1 (2016) for commonalities, differences, and differing perspectives, information that guided the approach adopted for the model. The less familiar topic of validation is discussed to understand its application when progressing a cognitive ergonomics approach.

Appendix H presents the two adopted *performance influencing factor (PIF) assessments* performed in the prototype model. One section evaluates PIFs and performance shaping factors, referenced in several widely known industry standards and guidelines, by analyzing and assigning them to one of four categories. Those in three of the categories are more thoroughly addressed by dedicated lifecycle model processes and therefore excluded from the PIF assessments. Those in the remaining category are included in the PIF assessment for barrier system tasks and task phases. A new PIF is added (societal influencer) that is a frequent contributor to major accidents, though seldom considered in current practice and risk assessments. Working environment, the focus of the second assessment, applies to barrier elements having a human-system interface.

Appendix I presents a new first-principles-based *cognitive assessment process*. The prototype process examines task phase activities and interactions in a manner that reveals a cognitive demand that cannot be reliably met within the specified time given the plausible situations and environments. The included tools provide a range of solution options (physical, human, or organizational) that may be effective in preventing or eliminating the identified mismatch or its consequences.

Appendix J demonstrates the relationship between *company programs and policies*, barrier tasks, *competency, training, procedures, drift*, and *skill fade*. It provides the basis and guidance for procedures, training, staffing, and competency requirements and processes employed in the prototype model. Other sections explain and provide essential information on skill fade and drift, and methods to *improve human response time performance.*

Appendix K presents an expanded view of a *workspace* as adopted and employed in the prototype model. The guidance and design information include workspace descriptions, boundaries, conventions, and tools. Defined workspace types are physical, communication, human, and information. Extensions to existing physical workspace design practices are suggested to improve support for a cognitive ergonomics approach. Example physical workspaces include a backlit switch, VDU-based display, control console, control room, and engineered egress/escape route.

Appendix L provides guidance on *barrier system tasks performed from a remote location* (e.g., from a Remote Operations Center). It guides the development and integration of the numerous new requirements into the lifecycle model. A suggested design basis document, described in Appendix B, captures these new requirements, and constitutes a valuable tool for guiding and coordinating the many technical disciplines and organizations engaged in the various lifecycle processes. Potential opportunities and risks are reviewed and discussed.

Appendix M provides an extensive compilation of information from the *Deepwater Horizon accident*. It reviews the active and active human barriers that contributed to the accident scope, escalation, and outcomes. A minute-to-minute timeline identifies key barriers, events, and actions taken, beginning with the failed cement barrier. It presents information on each active and active human barrier that was an element in the causal chain. Barrier information includes its safety function, prior events and history, health status, activation (if/when), and whether the safety function was achieved. This material provides important insight into the likely and plausible cognitive challenges presented to the crew and emergency responders. (This information influenced many of the processes included in the prototype model.)

1.4 INDUSTRY INSIGHTS AND RECOMMENDATIONS

The following are statements and recommendations from industry call-to-action articles, white papers, accident investigation reports, and post-accident status reports published in response to the Deepwater Horizon accident. This book is a direct response to those recommendations. Additional excerpts are inserted throughout the book to show where and how recommendations are addressed in the lifecycle model.

According to the CIEHF (2016), Human Factors in Barrier Management – White Paper:

> From Table 1 (p. 19), the following are "Characteristics of good human barrier elements." The expectations on human performance should be realistic and include the following, "a) identify the situation that needs action; b) knowing or being able to work out what needs to be done; c) being able to do it in the time available, with the resources and equipment available, and under the likely conditions; d) having some means of knowing that the action has the intended effect."
>
> *(Abstract)*

> The intentions and expectations of human performance that are implicit in the decision to rely on people as part of the barrier system are rarely made explicit.
>
> *(p. 39)*

> The performance needed to deliver the required functionality should be capable of being described clearly: i) what state or events would initiate the performance, ii) what task(s) are involved in carrying out the function, and iii) when the function has been achieved.
>
> *(p. 44)*

From the International Association of Oil and Gas Producers (IOGP 2012):

> IOGP's "Human factors Sub-Committee believes the improved understanding and management of the cognitive issues that underpin the assessment of risk and safety critical decision-making could make a significant contribution to further reducing the potential for the occurrence of incidents."
>
> *(p. 1)*

> The report discusses four themes: situation awareness, cognitive bias in decision-making, interpersonal behavior, [and] awareness and understanding of safety critical tasks. The first three are concerned with non-technical human skills, while the fourth is concerned with organizational preparedness for critical operations.
>
> *(p. 1)*

When the cognitive dimension of incidents is properly investigated, there is often a significant discrepancy between what the operator thought was the state of the world, what was happening, or how an equipment or a process would have behaved and what the actual state of the world was, or how the system did behave.

(p. 4)

Cognitive bias is fundamental to human cognition and can have both positive and negative effects... Negative effects can include a distorted interpretation of the state of the world, poor assessment of objective risk, and poor decision-making.

(p. 6)

Operational decision makers need information, not data. And they need the right level of detail, in the right format and in the right place and time. Designing equipment, displays and working practices that recognize the important of SA is not trivial and can require significant expertise.

(p. 4, SA = situation awareness)

IOGP recognizes the "potential for insufficient awareness and understanding of the psychological complexity of safety critical tasks...members should work towards adopting practices to identify and understand safety critical human tasks." As the objective, IOGP "members should be able to demonstrate that safety critical human barriers will actually work and the risk of human unreliability in performing them is ALARP."

(p. 13, ALARP = As Low As Reasonably Practicable)

In Missing Focus on Human Factors – Organizational and Cognitive Ergonomics – in the Safety Management for the Petroleum Industry, Johnsen et al. (2017) state:

A study of the petroleum industry and the safety authorities in Norway

revealed an immature focus and organization of Human Factors. Expertise on organizational ergonomics and cognitive ergonomics are missing from companies and safety authorities and are poorly prioritized during development.

(Abstract)

...there seems to be imprecise definitions of HF, poor focus initially in projects and poor structure of the HF work.

(p. 401)

The missing focus on cognitive and organizational ergonomics (the prerequisite physical ergonomics) may create weaknesses and holes in defenses/barriers.... These issues have not been properly addressed in new versions of the NORSOK standard planned to be published in 2017.

(p. 405)

HMI design for presentation of safety critical information and alarm design must be improved to sustain situational awareness and sensemaking.

(p. 404)

Latent errors are related to designers, high-level decision makers and managers, where the adverse consequences may lie dormant within the system for a long time and only becoming evident when combining with other factors to breach the system's defenses.

(p. 403)

From the SPE (2014) Technical Report, The Human Factor: Process Safety and Culture:

Incident investigations often identify deficiencies in the design or implementation of the interface between people and technology as contributing to the loss of reliable human performance. This is sometimes referred to as "design-induced human error.

(p. 5)

...sometimes decisions have regrettable consequences. This has to do with cognitive limits and biases in human decision-making, physiological factors that affect an individual's ability to make decisions... and the organizational conditions under which decisions are made. Decisions are often required under the stress of time-pressure, with competing goals and with inadequate, insufficient, or uncertain understanding and unknown consequences.

(p. 11)

Use a mental models approach to IT systems design: Design IT systems to reflect the optimal mental model that is taught rather than training being bent around the way systems work.

(p. 20)

According to OESI (2016) in Human Factors and Ergonomics in Offshore Drilling and Production: The Implications for Drilling Safety:

...technical standards often lack design features necessary to incorporate human factors and ergonomics considerations.

(p. 10)

...poor HMI design can negatively impact workers' situation awareness and ultimately their performance; thus inadequate HMI design is a major risk factor for offshore O&G.

(p. 10)

In the offshore O&G environment, many of the current interfaces for daily and emergency tasks have not been specifically designed to facilitate and support human performance.

(p. 12)

According to the root cause analysis of several incidents, inaccurate procedure use contributed to the highest number of incidents in high-risk environments. Findings suggest that workers did not completely trust their procedures and checklists. This distrust was based on the worker not knowing if or when an update had occurred.

(p. 14)

Stress can reduce working memory capacity as well as attention. As a result, stress has been found to significantly reduce SA.

(p. 18)

Incident analysis and investigative reports revealed the cause of the incident was not related to poor decision-making, but rather the absence of SA and poor mental models." "Studies have reported finding the loss of SA can result from something as simple as inattention and is also a function of experience and training.

(p. 21)

From a study of 332 offshore drilling incidents, "135 were related to SA, and more than 40% of drilling activity incidents were associated with inadequate SA."

(p. 21)

From a recent study, "These analyses highlight how an inaccurate mental model and resulting expectations can impact successive cycles of SA, influencing the interpretation of cues and how the situation is anticipated to develop."

(p. 22)

Confusion occurs when a worker misinterprets the observed behavior of the operating system in light of their mental model of the system.

(p. 24)

Judgement and the decision-making process are higher order processes than memory and attention for effective problem solving under high workload conditions. This means that one cannot make good judgment and decisions without remembering essential information or attending to critical aspects of the environment.

(p. 37)

The National Academy of Sciences (NASEM 2023) provides a current-state assessment of the U.S. offshore O&G industry (Gulf of Mexico). The context is the industry and regulatory response to recommendations in the Deepwater Horizon accident investigations and reports.

Even today, neither regulation nor industry standards sufficiently reflect modern terminology, thinking, and practice about barriers...

(p. 124)

There is little industry or regulatory guidance available for contingent barriers, which the committee views as a main source of industry systemic risk…. Contingent barriers are those that require active human interaction or intervention to prevent the accidental release of hydrocarbons.

(p. 124)

Recommended Practice 75 and SEMS regulations based on it lack guidance on (a) contingent barrier management that depends on human judgement and intervention, and (b) application of human factors to safety critical procedures.

(p. 4)

Many companies did understand and include human factors in design and operation to protect against human error, but it was an early level of understanding and a narrow scope relative to today's concepts of human factors and HSI.

(p. 117)

Both the current (Third) and new RP 75 editions (Fourth) still refer to human factors only generally and do not mention contingent barrier management. Hence, BSEE regulations and audits are in the same situation.

(p. 118)

From risk element eight, "…the industry in general still puts too much emphasis on a 'fix the worker' mentality if there are problems or errors when workers are engaged in barrier management…"

(p. 118)

From conclusion 4-1, "…there has been little improvement in testing for and demonstrating worker competence in safety critical tasks."

(p. 135)

1.5 COMPANION GUIDES TO THIS BOOK

The following resources are suggested companion guides to the material in this book. (Consider the latest edition of each resource as they become available.)

- *Applied Attention Theory*, by C. D., Wickens, J. S. McCarly, R. S. Gutzwiller, CRC Press (2023)
- *Cognitive Basis for Human Reliability Analysis* (NUREG-2114), by Office of Nuclear Regulatory Research (2016)
- *Designing for Situation Awareness: An Approach to User-Centered Design*, by M.R. Endsley and D.G. Jones, CRC Press (2012)
- *Engineering Psychology and Human Performance, Pearson Education Inc.,* by C.D. Wickens, J.G. Hollands, S. Banbury, R. Parasuraman, Pearson Education Inc. (2013)

- *Guidelines for Management of Safety Critical Elements (SCEs),* by Energy Institute (2020b)
- *Handbook of Cognitive Task Design,* Ed. Eric Hollnagel, CRC Press (2003)
- *Handbook of Distributed Team Cognition, Contemporary Research, Models, Methodologies, and Measures in Distributed Team Cognition,* Ed. M. McNeese, E. Salas, M.R. Endsley, CRC Press (2021)
- *Human Error,* J. Reason, Cambridge Press (1990)
- *Human Factors Handbook for Process Plant Operations, Improving Safety and Systems Performance,* Center for Chemical Process Safety, John Wiley, and Sons Inc. (2022)
- *Human Factors in the Design and Evaluation of Central Control Room Operations,* by N.A. Stanton, P. Salmon, D. Jenkins, and G. Walker, CRC Press (2010)
- *Human Memory,* by G.A. Radvansky, Routledge (2021)
- *Mental Model Matrix: Implications for System Design and Training,* by J. Borders, G. Klein, and R. Besuijen, *Journal of Cognitive Engineering and Decision-making* (2024)
- *Safety at the Sharp End: A Guide to Non-Technical Skills,* by R. Flin, P. Connor, P., and M. Crichton. Ashgate Publishing (2008)
- Situation Awareness Misconceptions and Misunderstandings, by M.R. Endsley, *Journal of Cognitive Engineering and Decision-making,* 9(1), 4–32 (2015)
- *Situation Awareness for Emergency Response,* by R. Gasbury, PennWell Corp. (2013)
- *Snapshots of the Mind,* by G. Klein, MIT Press (2022)
- *Teams that Work: The Seven Drivers of Team Effectiveness,* by S. Tannenbaum and E. Salas, Oxford University Press (2021)
- *Thinking, Fast and Slow,* by D. Kahneman, Farrar, Straus and Giroux (2011)
- *Understanding Mental Models,* by Rob Fisher, Kindle Direct Publishing (2022)
- *User-Centered Requirements Engineering: Theory and Practice,* by A. Sutcliffe, Springer (2002)

1.6   SUGGESTIONS FOR READERS WITH DIFFERENT INTERESTS

This is an atypical book given its varied objectives, unique scope, comprehensiveness, innovations, insightful analysis and examples, and its many new tools and resources. Readers will have varied reasons and interests for buying and using this book. The following suggests options to accommodate those varied reasons and interests:

Why Should I Read this Book? Most readers should find the answers in the Introduction to this book. If more is needed, see Chapter 8 explaining the current deficiencies in current practice and the consequences of those deficiencies. (This book suggests solutions to address each of these deficiencies.)

Exploring Comprehensive Lifecycle Models. This reader may want to begin with a brief overview of the prototype lifecycle. To do so, review the figures listed in Section 1.6.1. It may also be helpful to briefly review Appendices C and F to gain awareness of the cognitive underpinnings of each model lifecycle. From there, proceed to Chapters 2–7.

New to Cognitive Ergonomics or Active Human Barrier Design. For readers in the initial stages of their journey into cognitive ergonomics, active human barrier, or safety critical task design, consider reading Chapter 1 (in full) and then Chapter 9. The latter provides a wide range of topics that were incidental findings, insights, and perspectives gained while developing the prototype model and supporting materials.

Looking for Reasons to Consider or Apply the Lifecycle Model. This reader may want to start with Introduction Section 1.4 followed by Chapter 8. The latter reviews the current state of industry standards and practice and explains why incidents and major accidents may continue to happen if the necessary changes are not made. See Section 9.3 for additional potential benefits.

Need Information on a Specific Topic. The book provides a wide range of rich information that applies to active human barriers and safety critical tasks. The index and the barrier data cross-reference tables in Appendix B.7 may be helpful starting points. Another may be one of the many topic-specific appendices that provide technical information, new tools, technical and project execution guidance, or useful excerpts from industry standards and reputable resources. Some analyze current practice when defining the basis for selecting, adapting, or developing a process, method, or tool used in the lifecycle model. If looking for information on a specific topic, the reader may choose to review the model processes that address that topic. (Most appendices provide cross-references that identify the topic-specific processes. The expansive index is another helpful starting point.) The lifecycle model explains each process and identifies its input and output information. Many processes include examples, insights, explanations, and additional information. Taken together, this material should contribute to developing a broad and deep understanding of cognitive ergonomics and cognitive science and their application to active human barriers and safety critical tasks.

1.6.1 An Overview of the Lifecycle Model Processes and Activities

To gain a quick overview of the prototype lifecycle model processes, assessments, and activities, refer to the following figures:

- Figure 3.1. Active Human Barrier Lifecycle Process Overview (Chapter 3).
- Figure 3.3. Preliminary Design Phase Overview (Chapter 3).
- Figure 4.1. Detailed Design and Engineering Phase Overview (Chapter 4).
- Figure 5.1. Implementation Phase Overview: Procurement, Fabrication, and Human and Organizational Elements (Chapter 5).
- Figure 6.1. Construction, Installation, and Commissioning Phase Overview (Chapter 6).
- Figure 7.1. Operate and Maintain, Modify, and Decommissioning Phase Overview (Chapter 7).

1.6.2   AN OVERVIEW OF THE DOCUMENTS AND INFORMATION CREATED IN THE LIFECYCLE MODEL

For listings of the documents and information created by the prototype lifecycle model, refer to the following tables.

Concept (Lifecycle Phase A)

- Table 3.1. Concept Phase: Technical and Project Documents.

Preliminary Design (Lifecycle Phase B)

- Table 3.2. Preliminary Design: Barrier and Task Requirements Tables.
- Table 3.3. Preliminary Design: Project and Technical Documents.
- Table 3.4. Preliminary Design: Studies and Assessments.

Detailed Design and Implementation (Lifecycle Phase C)

- Table 4.1. Detailed Design: Barrier and Task Requirements Tables.
- Table 4.2. Detailed Design: Technical, Procurement, and Project Documents.
- Table 4.3. Detailed Design: Studies and Assessments.
- Table 5.1. Implementation: Technical, Procurement, and Project Documents.
- Table 5.2. Implementation: Studies and Assessments.

Construct, Install, and Commissioning (Lifecycle Phase D)

- Table 6.1. Construct, Install, and Commission Phase: Technical, Procurement, and Project Documents.
- Table 6.2. Construct, Install, and Commission Phase: Assessments.

Operate and Maintain (Lifecycle Phase E)

- Table 7.1. Operate and Maintain Phase: Technical and Project Documents.
- Table 7.2. Operate and Maintain Phase: Assessments.

1.6.3   KEY LIFECYCLE MODEL DOCUMENTS

If interested in the unique and essential documents developed and deployed in the prototype lifecycle model, refer to the following appendices for the suggested table of contents:

- Appendix B.2. Barrier Safety Management Plan.
- Appendix B.3. Safety Requirements Specification.
- Appendix B.4. Remote Barrier Support Design Basis.

REFERENCES

Borders, J, Klein, G., Besuijen, R. (2024), Mental model matrix: Implications for system design and training, *Journal of Cognitive Engineering and Decision-making*. sagepub.com

CCPS (2022), *Human Factors Handbook for Process Plant Operations, Improving Safety and Systems Performance*, New York: John Wiley & Sons Inc., Center for Chemical Process Safety (CCPS)

CIEHF (2016), Human barriers in Barrier Management, a white paper by the Chartered Institute of Ergonomics and Human Factors, 12/2016, CIEHF

EI (2020b, January), *Guidelines for Management of Safety Critical Elements (SCE)*, 3rd Ed, London: Energy Institute

Endsley, M.R. (2015), Situation awareness misconceptions and misunderstandings, *Journal of Cognitive Engineering and Decision-Making*, 9(1), 4–32

Endsley, M.R. (2021), *Handbook of Distributed Cognition: Contemporary Research Models, Methodologies, and Measures in Distributed Team Cognition*, 1st Ed. McNeese, M., Salas, E., Endsley, M.R. (Eds.). Boca Raton, FL: CRC Press

Endsley, M.R., Jones, D.G. (2012), *Designing for Situation Awareness: An Approach to User-Centered Design*, 2nd Ed, CRC Press

Fisher, R. (2022), *Understanding Mental Models*, Kindle Direct Publishing

Flin, R., O'Connor P., Crichton, M. (2008), *Safety at the Sharp End: A Guide to Non-Technical Skills*, Ashgate Publishing

Gasbury, R.B. (2013), *Situation Awareness for Emergency Response*, PennWell Corporation

Hollnagel, E. (2003), *Handbook of Cognitive Task Design*, Ed. Hollnagel, E., Mahwah, NJ: Lawrence Erlbaum Associates Inc. (Reprinted by CRC Press, 2010)

IEC 61511-1 (2016), *Functional Safety – Safety Instrumented Systems for the Process Industry Sector – Part 1: Framework, Definitions, System, Hardware and Application Programming Requirements*, 2nd Ed, International Electrotechnical Commission

IOGP (2012), Cognitive issues associated with process safety and environmental incidents, London: International Association of Oil and Gas Producers, IOGP Report No 460, 7/2012

IOGP (2014a), Crew resource management for well operations team, London: International Association of Oil and Gas Producers, IOGP Report No 501, 4/2014

IOGP (2014b), Guidelines for implementing crew resource management training, London: International Association of Oil and Gas Producers, IOGP Report No 501, 12/2014

IOGP (2018), Introducing behavior markers of non-technical skills in oil and gas operations, International Association of Oil and Gas Producers, IOGP Report No 503

ISO 11064-1:2000, Ergonomic design of control centres – Part 1: Principles for the design of control centres, International Organization for Standardization, 1st Ed, 2000–12–15

Johnsen, S.O., Kilskar, S.S., Fossum, K.R. (2017), Missing focus on human factors – organizational and cognitive ergonomics – in the safety management for the petroleum industry, *Journal of Risk and Reliability*, 231(4), 400–410

Kahneman, D. (2011), *Thinking, Fast and Slow, Farrar, Straus, and Giroux* (paperback edition)

Klein, G.A. (2022), *Snapshots of the Mind*, MIT Press

NASEM (2023), Advancing Understanding of Offshore Oil and Gas Systematic Risk in the U.S. Gulf of Mexico: Current State and Safety Reforms Since the Macondo Well Deepwater Horizon Blowout (2023), National Academies of Science, Engineering and Medicine, Prepublication Copy (Downloaded from nap.nationalacademies.org website on April 7, 2023)

NUREG (2016), *Cognitive Basis for Human Reliability Analysis*, NUREG-2114, Whaley, A.M., Xing, J., Boring, R.L., Hendrickson, S.M.L., Joe, J.C., LeBlanc, K.L., Morrow, S.L., Office of Nuclear Regulatory Research, U.S. Nuclear Regulatory Commission, Washington, DC

OESI (2016), Human factors and ergonomics in offshore drilling and production: The implications for drilling safety, Ocean Energy Safety Institute, 12/2016

Radvansky, G.A. (2021), *Human Memory*, 4th Ed, Routledge

Reason, J. (1990), *Human Error*, Cambridge: Cambridge University Press

SPE (2014), The human factor; process safety and culture, SPE Technical Report, Society of Petroleum Engineers, March 2014

Stanton, N.A., Salmon, P.M., Walker, G.H., Jenkins, D.P. (2010), *Human Factors in the Design and Evolution of Central Control Room Operations*, CRC Press

Sutcliffe, A. (2002), *User-Centred Requirements Engineering, Theory and Practice*, 1st Ed, Springer

Tannenbaum, S., Salas, E. (2021), *Teams That Work, the Seven Drivers of Team Effectiveness*, Oxford University Press

Wickens, C.D., Hollands, J.G, Banbury, S., Parasuraman, R. (2013), *Engineering Psychology and Human Performance*, 4th Ed, Pearson Education Inc.

Wickens, C.D., McCarly, J.S., Gutzwiller, R.S. (2023), *Applied Attention Theory*, 2nd Ed, CRC Press

2 Terms, Definitions, and Model Constructs

This chapter is organized as follows:

- 2.1 Terms, Definitions, and Abbreviations
- 2.2 Defining Barriers as Tasks
- 2.3 Frame Tasks in the Form of the Detect, Decide, and Act Model
- 2.4 Integrate Situation Awareness into the Detect, Decide, and Act Model
- 2.5 Barrier Elements: Physical, Human, and Organizational

2.1 TERMS, DEFINITIONS, AND ABBREVIATIONS

This section introduces the terms and definitions used in this book. At the time of publication, a global consensus standard that defines commonly used terms for active human barriers is not available.

Preventive, control/recovery, and mitigation barriers provide pre-determined responses to a major accident event (MAE) caused by defined hazards. Customary practice employs a defense-in-depth approach that implements multiple barriers of diverse types to address the possibility that one or more barriers fail for unforeseen reasons:

- A **preventive barrier** prevents the occurrence of an MAE.
- A **mitigation barrier** responds to the occurrence of an MAE. This barrier type includes control/recovery and emergency response barriers.
- A **control/recovery barrier** limits the scale, intensity, and duration of an MAE.
- An **emergency response barrier** limits the potential consequences of an MAE, i.e., limits its potential effects on personnel or the environment.

Barriers are also classified as "active" or "passive."

- An **active barrier** performs the required safety function only upon detection of a pre-defined condition or state. An active barrier includes preventive and mitigation types that take the form of an active human barrier or a fully automatic safety instrumented function.
- A **passive barrier** is continuously available to perform its barrier function, for example, a blast wall or fireproofing. Upon exposure to the hazardous condition, the barrier provides a defined protective function for a defined period or duration.

DOI: 10.1201/9781032674476-2

2.1.1 Definitions

Active Human Barrier – An active barrier that relies on a human to perform a barrier/task phase function, i.e., the function provided by the task detect, decide, and act phases. (This barrier type includes the "Active Hardware + Human Barrier" defined in CCPS (2018, p. xv). Emergency response barriers are active human barriers.)

ALARP – "As Low as Reasonably Practicable—a term used to describe a target level for reducing risk that would implement risk reducing measures unless the costs of the risk reduction in time, trouble or money are grossly disproportionate to the benefit" (CCPS 2018, p. xv).

Alternative Display Type – A more advanced HMI display design that translates raw data into a more advanced information-rich form that more directly supports a specified comprehension (situation awareness level 2 or SA-2) or projection (situation awareness level 3 or SA-3) requirement. The purpose for using this display type may be to significantly reduce a challenging training/competency requirement, or the elapsed time needed to achieve that requirement under operational conditions. (As an example, see the note "Alternative Display Type" in detailed design step C5-1c.)

Barrier – "A control measure or grouping of control elements that on its own can prevent a threat developing into a top event (prevention barrier) or can mitigate the consequences of a top event once it has occurred (mitigation barrier)" (CCPS 2018, p. xv).

Barrier Function – The safe state achieved when all barrier tasks and phases are completed and executed as intended. The barrier/task function is realized by the collective and coordinated action of each element in the barrier system.

Barrier Phase – The uniquely different detect, decide, and act stages (activities) that comprise a task and collectively achieve the task goal and function. Each stage is referred to as a phase.

Barrier Response Time – For an active human barrier, the actual elapsed time period that begins with barrier activation and ends when its specified safety function and safe state is fully achieved.

Barrier Safety Time (BST) – For an active human barrier, the specified (maximum) elapsed time period that begins with barrier activation and ends when its specified safety function and safe state is fully achieved. (The BST should be less than, but never exceed, the PST.)

Barrier System – "A barrier system is a system that has been designed and implemented to perform one or more barrier functions. A barrier system describes how a barrier function is realized or executed" (Sklet 2006). In the lifecycle model, the barrier system boundary encompasses all tasks required to achieve the barrier function. (Barrier-dependent external resources, external support systems, and external protective barriers lie outside this boundary.)

Common Cause Failure – "Failure of more than one device, function or system due to the same cause" (CCPS 2015, p. xx).

Common Mode Failure – "A specific type of common cause failure in which the failure of one or more device, function or system occurs due to the same cause, and failure of the devices occurs at the same time" (CCPS 2015, p. xx).

Component – A discrete object in a barrier system. A local display, HMI display element, and a fire hose are example physical elements that may be included in the barrier system. Likewise, a procedure, support aid, or purposely designed training drill are example organizational elements.

Design Accident Load – "Accident load/action used as a basis for design" (NORSOK 2021, cl. 3.6).

Direct-Use PE – A barrier system physical element (PE) to which a barrier-assigned person has a direct physical or sensory interaction that is critical to achieving a task goal or barrier safety function. Direct-use PEs include those used to perform an act phase activity (e.g., a fire hose) or present/display information that must be detected to support a detect phase requirement (e.g., view a passive or active visual indicator, hear an audible display, or communicate via a hand-held radio or a hands-free headset).

Drift – A barrier system deviation that progresses slowly over time. Examples include changes in procedure usage and in-the-head knowledge, skills, behaviors, goals, and beliefs. Timely detection requires periodic monitoring using appropriate detection methodologies.

Element (Barrier or Task) – A barrier system component or object that may be physical, human, or organizational.

Engineered Area – A protected physical workspace or area required to achieve the barrier function or safe state. Examples include an egress/escape route or safe haven building, room, or delineated outdoor space.

External Protective Barrier – An external passive or active barrier that provides a defined protective function to the barrier system or its dependent external support systems, protective barriers, or an additional resource. (The protection is provided for a specified period or duration.)

External Support System (ESS) – Equipment, systems, and utility services on which the barrier system depends to maintain its uninterrupted capabilities and functioning. Examples include environmental control, technical, and utility systems and services.

Final Element – A barrier system element that affects the controlled process in a way that directly achieves the proscribed barrier/task function and safe state. A process isolation valve is an example where the function is to stop the flow from a vessel. When applied to a task, the final element may also be an HE action taken to achieve the specified task goal, i.e., a conveyed instruction or command.

HMI Display – A purposely developed, configured, or programmed VDU-based display that presents or provides user entry access to one or more HMI display elements.

HMI Display Element – A single display object such as an indication or control entry object customized/configured to meet a specified user-interface requirement.

Human Element (HE) – A person assigned to perform one or more of the barrier system tasks required to achieve the barrier function. (The term does not apply to others who perform barrier system maintenance or provide other services or blunt-end activities.)

Independence – "The condition that no significant common mode of failure exists that would degrade two or more barriers simultaneously in an incident pathway" (CCPS 2018, p. xvii).

Independent Protective Layer (IPL) – "A device, system, or action that is capable of preventing a scenario from proceeding to the undesired consequence without being adversely affected by the initiating event or by the action of any other protection layer associated with the scenario" (CCPS 2015, p. xxi).

In-Place PE – A physical element (e.g., a technical system or workspace) that must be in place to enable, realize, or maintain a barrier function or capability. For example, a required HMI display element is realized through the use and application of the (in-place) VDU and technical system that gathers external input information and presents that information at a display element. The software configuration that creates and presents the element is part of the barrier system.

Knowledge-Based Competency – A competency type achieved when a required technical, procedural, or task execution knowledge is adequately learned and retained in long-term memory, and can be promptly recalled when needed.

Muster area – A "designated area where personnel report when required to do so" (ISO 13702:2015, cl. 3.1.34).

New Skill and Knowledge – A task-required skill or knowledge requirement (a task within the barrier system) that is not one of those required by standard operating procedures (SOPs). (This statement assumes all skill and knowledge competencies required by the applicable SOPs are already achieved and verified. Those not achieved or verified become pending requirements in the competency management system.)

Organizational Element – Barrier system elements that are not physical or human, for example, procedures, training modules, staffing plans, barrier rosters, and competency management systems.

Physical Element – Any barrier system element that is physical in nature, for example, a technical system, paint marking, hand-held device, or an engineered area or workspace.

Process Safety Time – "The time period between a failure occurring in the process, or its control system, and the occurrence of the consequence of concern" (CCPS 2015, p. xxii).

Protected Facility – The facility where the barrier system provides its protective function. The term is used to differentiate this facility from a Remote Operations Center from which a remote barrier support task is performed.

Prototype Lifecycle Model – All processes, assessments, and activities that fully encompass and describe the complete barrier lifecycle as described in Chapters 3–7. (Variations may include "prototype model," "lifecycle model," or "model.")

Remote Barrier Support – A barrier system task performed from a remote operations center. (Depending on how and the task is defined, it may or might not be essential to achieving the barrier function and safe state.)

Remote Operations Center – A generic term that refers to a remote located room or facility where a remote barrier support task is performed. (A ROC is physically distanced from the protected facility and therefore not physically affected by a hazard condition that may occur at that facility.)

Safe Haven – A purposely engineered indoor or outdoor area that provides temporary protection to occupants from a defined hazard or hazard effects for a specified period. (The term is used interchangeably with "temporary refuge.")

Safety Instrumented Function – "A safety function allocated to a Safety Instrumented System (SIS) with a Safety Integrity Level (SIL) necessary to achieve the required risk reduction for an identified scenario of concern" (CCPS 2015, p. xxiii). (SIFs are commonly implemented as fully automated functions achieved by a technical system. However, by design, some SIFs rely on a human to achieve its safety function. SIFs of this type are active human barriers that meet SIL and possibly other requirements if the design is guided by IEC 61511-1 (2016) or other standards.)

Skill-Based Competency – The competency type achieved when a required cognitive or physical action or activity sequence can be accurately and automatically performed while placing little to no demand on working memory. (Skills are automatic cognitive processes.)

Skill Fade – A progressive decrement in a required skill that occurs as the time period between use and refresher training increases. The degradation rate and time period vary by skill type. Many biomechanical skills, for example, riding a bike, tend to degrade slowly over a longer period of time. A complex skill, for example, those common to complex multi-step procedures or teamworking environments, may begin to degrade within months of non-use.

Support PE – Physical element (e.g., equipment) that provides a required support function during the HE performance of an assigned barrier task. The element type is typically worn or carried. Examples include a flashlight, smoke hood, stretcher, hand-held gas detection meter, and firefighting gear and clothing.

Target Task Safety Time – The time (duration) allocated to start and complete a task (all activities) and achieve the task goal and (if applicable) task safe state.

Task Phase Safety Time – The time (duration) allocated to start and complete a task phase activity.

Temporary Refuge – See "safe haven".

Validation – The definition employed in the prototype lifecycle model is from ISO 11064-1 (2000). See Appendix G.1.2 for that definition.

Verification – The definition employed in the prototype lifecycle model is from ISO 11064-1 (2000). See Appendix G.1.1 for that definition.

2.1.2 ABBREVIATIONS

AHB – Active Human Barrier
ALARP – As Low as Reasonably Practicable
AR – Act Phase Response
AVL – Approved Vendor List
BPCS – Basic Process Control System
BSMP – Barrier Safety Management Plan
BST – Barrier Safety Time
CCR – Central Control Room
CMS – Competency Management System
CP – Control Console or Panel
DAL – Design Accident Load
DDA – Detect, Decide, and Act

EL – Environmental Load
EPC – Engineering, Procurement, Construction
ESD – Emergency Shutdown
FSA – Functional Safety Assessment
HAZOP – Hazard and Operability Study
HE – Human Element
HMI – Human Machine Interface
HSE – Health and Safety Executive (UK)
HVAC – Heating, Ventilation, Air Conditioning
ICB – Incident Command Board
ICC – Incident Command Center
IPL – Independent Protection Layer. An active human barrier is an IPL if it is designated as such in a LOPA or equivalent risk assessment process.
LOPA – Layer of Protection Analysis
OE – Organizational Element
PDMS – Procedure Development and Management System
PE – Physical Element
PIF – Performance Influencing Factor
PPE – Personal Protective Equipment
PSF – Performance Shaping Factor
PST – Process Safety Time
RA – Risk assessment
RBS – Remote Barrier Support (Functions may be included in the term *Integrated Operations*)
ROC – Remote Operations Center
SCBA – Self-Contained Breathing Apparatus
SIF – Safety Instrumented Function
SIS – Safety Instrumented System
SPOF – Single Point of Failure
SRS – Safety Requirements Specification
TDMS – Training Development and Management System
TPST – Target Phase Safety Time
TTST – Target Task Safety Time

2.2 DEFINING BARRIERS AS TASKS

An active human barrier comprises one or more tasks assigned to one or more humans. Figure 2.1 indicates a barrier with tasks assigned to three people, one assigned as barrier leader. The barrier function is achieved when all tasks are correctly performed, coordinated, and completed within the specified barrier safety time.

From the IOGP Report 460: Cognitive Issues Associated with Process Safety and Environmental Incidents (IOGP 2012):

> A 'safety critical human task' is an activity that has to be performed by one or more people and that is relied on to develop, implement or maintain a safety barrier.

(p. 12)

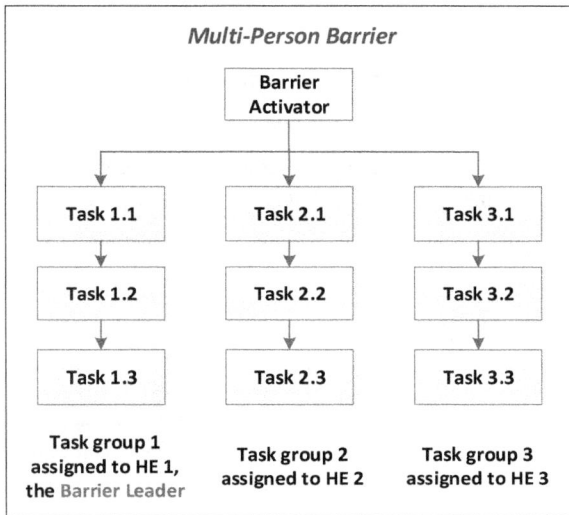

FIGURE 2.1   Multi-Person, Multi-Task Barrier.

There also seems to be an insufficient understanding of the demands that safety critical tasks can make on human performance, what is needed to support the required level of performance, and the ways in which human performance could fail in undertaking the tasks, or the inherent unreliability associated with the tasks.

(p. 12)

Members should work toward adopting being able to satisfy themselves that safety critical human barriers will actually work and the risk of human unreliability in performing them is effectively managed and reduced.... Members should work toward adopting practices to identify and understand safety critical human tasks.

The UK Health and Safety Executive document, Assessment Principles for Offshore Safety Cases (APOSC), defines salient regulatory principles. "Principle 8, The major accident risk evaluation should take account of human factors" (HSE 2021). **Example clauses within this principle include the following:**

Safety and environmental critical tasks should be analysed to demonstrate that task performance could be delivered to the specified performance when required. This demonstration should draw on recognized good practice in human factors.

(cl. 45)

Human performance problems should be systematically evaluated. This should involve evaluating the feasibility of tasks, identifying control measures, and providing an input to the design of procedures and personnel training, and of the interfaces between personnel and plant. The depth of the analysis should be appropriate to the severity of the consequences of failure of the task.

(cl. 47)

TABLE 2.1

Comparison between Active Human Barrier and Safety Critical Tasks

	Active Human Barrier			
Barrier Type	**Preventive**	**Control/Recovery (Mitigation)**	**Emergency Response (Mitigation)**	**Safety Critical Task**
Occurrence	Unplanned	Unplanned	Unplanned	Planned
How many HEs (assigned persons)	1 (typical)	1[a] (typical)	2 or more (typical)	Varies
Tasks needed to achieve the barrier function	1 (typical)	1 or more[a] (typical)	2 or more (typical)	Varies
Active barrier?	Yes	Yes	Yes	No[b]
Workload demand	Typically, manageable	Depends on the number of simultaneously active barriers	Situation and peak workloads may exceed HE capacity for periods of time	Assumed manageable
Specified safety time (BST)	Yes	Yes	Yes	Generally, no

[a] The number of unique tasks depends on how the tasks are framed and defined by the task analysis team.

[b] Active only when scheduled.

A barrier that requires two or more people (common to emergency response barriers) introduces a new set of design challenges. It increases the range of skills (nontechnical skills) that each person needs to interact, communicate, and coordinate in ways that can reliably achieve the barrier function within the specified safety time. The barrier leader requires additional skills like leadership and team monitoring.

Active human barriers and safety critical tasks are similar in that they both perform safety critical functions and rely on one or more humans to achieve the desired safety function. Table 2.1 compares and identifies a few key differences between the two. The activation of an active human barrier is unplanned and therefore a surprise. In contrast, a safety critical task (e.g., maintenance or simultaneous operations – SIMOPs) is planned. Time pressure is not as acute, and the work is not performed when a major accident event is in progress.

Active barriers must achieve the barrier function within a specified period to prevent the hazardous event from occurring or mitigate the undesired consequences thereof. Achieving the function within that time is subject to other performance requirements including equipment design and reliability, the appropriateness of individual HE actions, and the quality and timing of the barrier team interactions and coordination. (Similar performance requirements may also apply to a safety critical task.)

2.3 FRAME TASKS IN THE FORM OF THE DETECT, DECIDE, AND ACT MODEL

Tasks are a compilation of cognitive and physical activities. According to CIEHF (2016, p. 20),

Active barriers must have detect-decide-act functionality – i.e., they must have one or more elements that allow them to:

- …Detect the condition that is expected to initiate performance of the barrier function.
- …Decide what action needs to be taken, and;
- …Take the necessary action.

The ability to define and assess these activities requires that the task be expanded into a form that supports the requirements definition and design process. The model adopted in this book is the detect, decide, and act (DDA) model of the Centre for Chemical Process Safety (CCPS) as shown in Figure 2.2.

For reference, each activity in Figure 2.2 is referred to as a "phase." The detect phase may include an alarm or a different indicator type to activate the barrier/task.

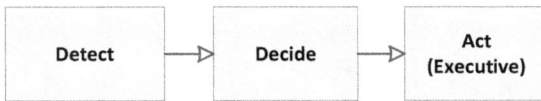

FIGURE 2.2 Activity Phases in an Active Barrier Task.

Source: **Modified and adapted from CCPS (2018, Figure 2–7).**

2.4 INTEGRATE SITUATION AWARENESS INTO THE DETECT, DECIDE, AND ACT MODEL

The call-to-action white papers referenced in the introduction affirm the need to improve the understanding and application of situation awareness (SA) to the barrier system definition and design process.

OESI (2016, p. 23) notes, "SA has been acknowledged as the basis for good decision-making within complex systems, including the O&G industry where poor performance can lead to devastating results."

According to IOGP (2012. p. 2), "SA… must be understood and applied at an adequate level of technical depth. SA also needs to be understood and managed at both the individual and team levels."

The referenced white papers provide little guidance on how to implement SA. However, SA models and applications have been successfully used in other high-risk, high-consequence industries for several decades. The dominant and most widely recognized and referenced model is Dr. Mica Endsley's three-stage SA model (Endsley 1995, pp. 34–37), which is adopted for the purposes of this book. It proposes three stages of situation awareness:

- **Perception (SA-1)** refers to the acquisition of information that is perceivable and available to our senses.

- **Comprehension (SA-2)** is the product of combining the SA-1 information with one's stored knowledge and experience to develop an understanding (mental picture) of what the information means.
- **Projection (SA-3)** is the capability to project and anticipate what may happen in the near term based on how the SA-1 information changes. This capability requires extensive domain expertise and knowledge (procedural and technical).

Figure 2.3 shows the approach (adopted in this book) for integrating Endsley's model into the DDA model and thus into the barrier design process. This construct is similar to those used in other publications and guidance documents, which are noted in Table 2.2.

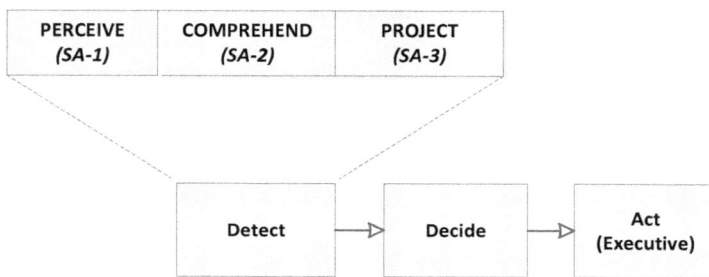

FIGURE 2.3   Integration of the Situation Awareness Model into the Detect, Decide, and Act (DDA) Model.

Source: **Modified and adapted from Figure 2.1, Endsley and Jones (2012). DDA is the active human barrier model in Figure 2.3.**

TABLE 2.2
Comparison of Task Models

Reference	Model Phases
SINTEF (2011, Figure 5.1): CRIOP	Observation/Identification, Interpretation, Planning/Choice, Action/Execution
IFE (2022, Table 20) Petro-HRA Guideline, Vol. 1	Detect, Diagnose, Decide on Actions, Execute Actions
NUREG (2016, Figure 2–6)	Detecting and Noticing, Understanding and Sensemaking, Decision-making, Action
EI (2020a, p. 28)	Detect, Diagnose, Decide, Activate

Note: The referenced situation awareness model does not define the cognitive underpinnings that achieve the realization of each level. The mapping of each level to those underpinnings (e.g., attention, mental models, and long-term memories) is detailed in later chapters.

2.5 BARRIER ELEMENTS: PHYSICAL, HUMAN, AND ORGANIZATIONAL

The final construct needed to support the design process is to name and frame the barrier elements that together achieve the barrier function and safe state. These elements (some or all) comprise each task phase. According to CHIEF (2016):

Barriers should be considered as barrier systems: i.e., in nearly all cases, for the barriers to perform as expected, a combination of elements needs to perform their individual functions in a coordinated manner. (p. 43)

Failure of any barrier or barrier element to perform its function, or to be identified as being unlikely or incapable of performing its function when demanded, should therefore be treated as a significant event. (p. 44)

As indicated in Figure 2.4, the terms and framing adopted for the lifecycle model barrier elements are physical, human, and organizational. As framed and defined, they encompass every component and element required to complete the barrier system.

2.5.1 DISCUSSION

One of the more significant step-outs in the lifecycle model is its adoption of the framing and definition for the three barrier elements physical, human, and organizational. At least one regulatory regime (e.g., PSA 2017) and some industry documents (e.g., CIEHF 2016, SINTEF 2016, and CCPS 2018) use the alternate terms technical, organizational, and operational. PSA (2017, p. 1) defines the terms as follows:

Technical elements are "Equipment and systems involved in the realization of a barrier function." Organizational elements are "Personnel with defined roles and functions and specific competence involved in the realization of a barrier function." Operational elements are "The actions or activities which personnel must perform in order to realise a barrier function."

The *technical* and *organization element* definitions are similar to the *physical* and *human* element definitions adopted in the lifecycle model and defined in Section 2.1.

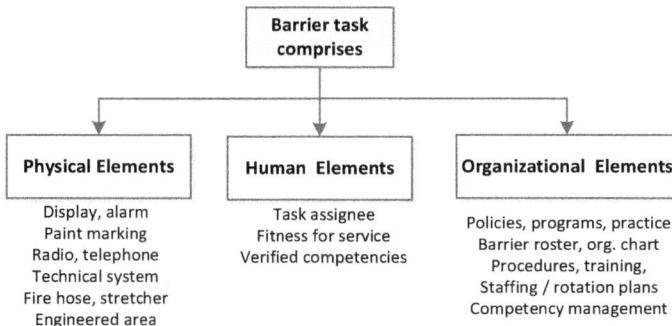

```
              ┌─────────────────┐
              │  Barrier task   │
              │    comprises    │
              └─────────────────┘
```

Physical Elements	Human Elements	Organizational Elements
Display, alarm	Task assignee	Policies, programs, practice
Paint marking	Fitness for service	Barrier roster, org. chart
Radio, telephone	Verified competencies	Procedures, training,
Technical system		Staffing / rotation plans
Fire hose, stretcher		Competency management
Engineered area		

FIGURE 2.4 Barrier/Task Elements.

Note: **Every element must function and perform as designed to achieve the task goal/barrier function and safety state.**

PSA's definition of an *operational* element is not similar to the lifecycle model term *organizational* element. The PSA definition is activity-based rather than a more common use of the term, for example, one that encompasses procedures, training, and competency management. Readily apparent in the lifecycle model presented in Chapters 3–7, the adopted terms, physical, human, and organizational elements, logically and elegantly support and guide all phases and activities in the lifecycle model. An equivalent model using the PSA definitions may introduce a few challenges.

REFERENCES

CCPS (2015), *Guidelines for Initiating Events and Independent Protection Layers in Layer of Protection Analysis*, New York: John Wiley & Sons Inc., Center for Chemical Process Safety (CCPS)

CCPS (2018), *Bow Ties in Risk Management: A Concept Book for Process Safety*, Hoboken, NJ: John Wiley & Sons Inc., Center for Chemical Process Safety (CCPS)

CIEHF (2016), Human barriers in Barrier Management, a white paper by the Chartered Institute of Ergonomics and Human Factors, 12/2016, CIEHF

EI (2020a, January), *Guidance on Human Factors Safety Critical Task Analysis*, 2nd Ed, London: Energy Institute

Endsley, M.R. (1995), Toward a theory of situational awareness in dynamic systems, *Human Factors*, 37(1), 32–64

Endsley, M.R., Jones, D.G. (2012), *Designing for Situation Awareness: An Approach to User-Centered Design*, 2nd Ed, CRC Press

IEC 61511-1 (2016), *Functional Safety – Safety Instrumented Systems for the Process Industry Sector – Part 1: Framework, Definitions, System, Hardware and Application Programming Requirements*, 2nd Ed, International Electrotechnical Commission

HSE (2021), Assessment Principles for Offshore Safety Cases (APOSC), UK Health and Safety Executive

IFE (2022), The Petro-HRA Guideline, Rev. 1, Vol. 1, IFE/E-2022/001, ISBN 978–82–7017–937-4, Institute for Energy Technology

IOGP (2012), Cognitive issues associated with process safety and environmental incidents, London: International Association of Oil and Gas Producers, 110 Report No 460, 7/2012

ISO 11064-1:2000 (2000), Ergonomic design of control centres – Part 1: Principles for the design of control centres, International Organization for Standardization, 1st Ed, 2000–12–15

ISO 13702:2015 (2015), Petroleum and natural gas industries, Control and mitigation of fires and explosions on offshore production installations – Requirements and guideline, International Organization for Standardization, 2nd Ed, 2015-08

NORSOK (2021), Technical Safety, NORSOK S-001:2020+AC:2021 (en), Standards Norway

NUREG (2016), *Cognitive Basis for Human Reliability Analysis*, NUREG-2114, Whaley, A.M., Xing, J., Boring, R.L., Hendrickson, S.M.L., Joe, J.C., LeBlanc, K.L., Morrow, S.L., Office of Nuclear Regulatory Research, U.S. Nuclear Regulatory Commission, Washington DC

OESI (2016), Human factors and ergonomics in offshore drilling and production: The implications for drilling safety, Ocean Energy Safety Institute, 2016-12

PSA (2017), Principles for Barrier Management in the Petroleum Industry, Barrier Memorandum 2017, Petroleum Safety Authority, Norway

SINTEF (2011), CRIOP: A scenario method for crisis intervention and operability analysis, SINTEF Technology and Society, Report SINTEF A4312, 2011-03-07

SINTEF (2016), Report: Guidance for barrier management in the petroleum industry, SINTEF Technology and Society, Report SINTEF A27623, 2016-09-23

Sklet, S. (2006), Safety barriers: definition, classification and performance, *Journal of Loss Prevention in the Process Industries*, 19, 494–506

3 Conceptual and Preliminary Design (Model Phases A and B)

This chapter is organized as follows:

- 3.1 Conceptual Design (Phase A)
- 3.2 Preliminary Design (Phase B, Front End Engineering Design)

3.1 CONCEPTUAL DESIGN (PHASE A)

Figure 3.1 provides an overview of the active human barrier lifecycle model presented in this book.

Chapter 2 defines an active human barrier as a barrier system comprised of one or more tasks. In the following processes, each task is further assessed to reveal the detect, decide, and act phase requirements that establish the basis for its design. For each task phase, this process continues by identifying, selecting, and specifying the human, physical, and organizational elements needed to achieve the task and barrier requirements. These processes apply to every task that makes up the barrier system. (Many of these processes may also be applied to tasks that lie outside the barrier system boundary. Examples include barrier-dependent, safety critical maintenance and support tasks.) In all cases, the scope of these processes is intended to be limited to only those elements that are specific to active human barriers.

The prototype lifecycle model begins with the conceptual design phase (Process A). The suggested activities include the following:

- Execution and organizational plans and planning for the new activities performed in the preliminary design phase.
- Develop the active human barrier philosophy.

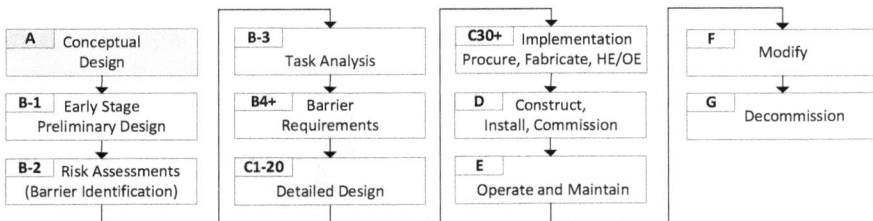

A Conceptual Design	B-3 Task Analysis	C30+ Implementation Procure, Fabricate, HE/OE	F Modify
B-1 Early Stage Preliminary Design	B4+ Barrier Requirements	D Construct, Install, Commission	G Decommission
B-2 Risk Assessments (Barrier Identification)	C1-20 Detailed Design	E Operate and Maintain	

FIGURE 3.1 Active Human Barrier Lifecycle Process Overview.

DOI: 10.1201/9781032674476-3

TABLE 3.1

Concept Phase: Technical and Project Documents

Active human barrier philosophy[a]

Barrier safety management plan[a] (See Appendix B.2 for suggested content.)

Plan – Preliminary Design Phase Scope and Execution Plan[a]

Facility Basis of Design: asset type, capacity, etc. Selected processes and technologies.

 Expectations for emergency response, remote barrier support/Remote Operations Center, etc.

Regulatory/statutory requirements

Industry and project standard and practice documents

Facility plot plans and layout

Conceptual General Arrangement Drawings – buildings, rooms, etc.

Process flow diagrams

Philosophy (technical disciplines) – process safety, fire and gas, emergency response,

 instrument and control systems, electrical system, facilities, etc.

Philosophy – operations and maintenance

Studies and assessments, for example, concept phase hazard and risk studies

[a] Suggested new documents specific to active human barriers.

- Develop the barrier safety management plan. (See Appendix B.2 for content suggestions.)

Table 3.1 lists these and other documents commonly developed in this phase given current practice for the larger capital projects.

This process suggests developing three documents specific to active human barriers, namely, a barrier safety management plan (BSMP), an active human barrier philosophy, and preliminary design phase execution plan for active human barriers. For the BSMP, see Appendix B.2 for the suggested content. It provides high-level guidance that applies throughout the barrier's lifecycle. The active human barrier philosophy establishes the base active human barrier requirements, constraints, and concepts. The planning document(s) define the active human barrier scope, deliverables, budget, schedule, and who performs this work. These documents provide essential input to guide the preliminary design phase activities.

3.2 PRELIMINARY DESIGN (PHASE B, FRONT END ENGINEERING DESIGN)

The following sections present the suggested preliminary design and assessment processes highlighted in Figure 3.2. As noted in Section 3.1, the constructs, framing, and methodologies described in Chapter 2 apply to every barrier system task addressed in this phase. Each process step in this phase is described and supported by examples and supporting information. Selective execution guidance is also provided, such as the suggested participants to include in a task analysis workshop. (As noted earlier, many of these steps equally apply to safety critical tasks.) Table A.1 (Appendix A.2.3) identifies the suggested participants in select activities.

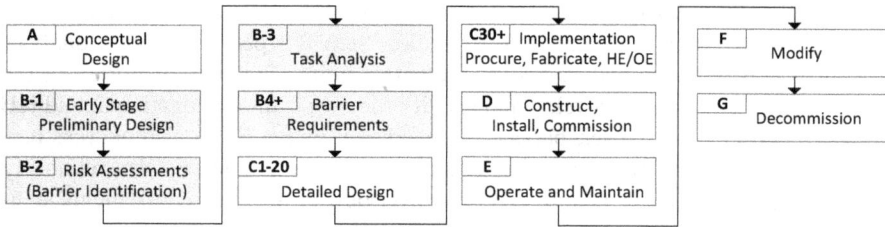

FIGURE 3.2 Preliminary Design: Early Phase Activities.

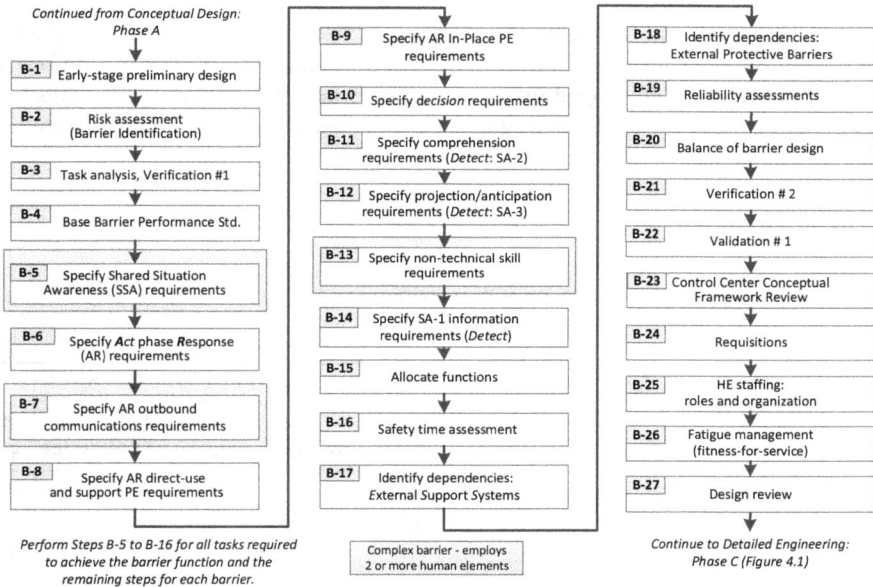

FIGURE 3.3 Preliminary Design Phase Overview.

Note: Current practice employs mature and familiar methodologies, some of which are purposely aligned to regional regulatory regimes. In most cases, the prototype lifecycle model is intended to supplement (not replace) existing methodologies. Current practice often stops the barrier decomposition process at the task level. As will become clear, latent design errors (cognitive mismatch errors) tend to be revealed only when the task is further decomposed to its task phase and phase elements. Because of the wide variations in methods employed worldwide, the reader may choose to evaluate existing practice to determine if it can match the increased comprehensiveness, new information, and possible results that appear achievable with these new processes and methods. If not, the prototype lifecycle model may provide a more advanced starting point to modify those practices.

Figure 3.3 summarizes the preliminary design phase steps and assessment processes. Note the shading used to indicate steps that apply to barriers that employ two or more HEs. Step B-1 develops the design and documents that are essential inputs to the Step B-2 risk assessment (RA) activities. Step B-3 performs a task analysis to identify the tasks needed to realize the active human barriers identified in the RAs. The remaining steps develop and specify the physical, organizational, and HE requirements for each barrier, barrier task, and task phase. Steps also address barrier dependencies, staffing, and other organizational areas, and perform reliability assessments and phase verifications. (For an overview of the task requirements identified in this phase, see Figure B.1 in Appendix B.3.1.)

3.2.1 DOCUMENT AND INFORMATION INPUT AND OUTPUTS

The document outputs from conceptual design (Table 3.1) are examples of inputs to the preliminary design phase. Tables 3.2–3.4 summarize the requirements, information, and documents developed in this phase. This new information contributes to a common technical design basis to guide the associated work from other disciplines and organizations.

TABLE 3.2
Preliminary Design: Barrier and Task Requirements Tables

		Table Info			
		Direct-Use PE		New Skills and Knowledge	See Step
Table Number	Table Description	SA-1 Info	Act Phase Response		
3.5	Barrier Origination Requirements	Yes	—	—	B2-x
3.6	Task Origination Requirements	Yes	—	—	B3-1
3.7	Barrier Performance Standards: Base Requirements	—	—	—	B-4
3.8	Shared Situation Awareness Requirements	Yes	—	Yes	B-5
3.9	Act Phase Response Physical Element Requirements	—	Yes		B-6
3.11	Communication Requirements	Yes	Yes	Yes	B-7
3.12	Act Phase: Direct-Use PE Requirements	Yes	Yes	Yes	B-8
3.13	Act Phase Response: Support PE Requirements	Yes	Yes	Yes	B-8
3.14	Detect and Act Phase: In-Place PE: Engineered Area Requirements	—	—	—	B-9
3.15	Detect and Act Phase: In-Place PE: Building Requirements	—	—	—	B-9
3.16	Detect and Act Phase: In-Place PE: Technical System Requirements	Yes	—	—	B-9

(Continued)

TABLE 3.2
(Continued)

Table Number	Table Description	Table Info: Direct-Use PE: SA-1 Info	Table Info: Direct-Use PE: Act Phase Response	Table Info: New Skills and Knowledge	See Step
3.17	Decide Phase Requirements	Yes	—	Yes	B-10
3.18	Detect Phase: SA-2 Comprehension Requirements	Yes	—	Yes	B-11
3.19	Detect Phase: SA-3 Projection Requirements	Yes	—	Yes	B-12
3.20	NTS Requirements: Teamworking	—	—	Yes	B-13
3.22	NTS Requirements: Leadership	—	—	Yes	B-13
3.24	NTS Requirements: Monitor and Manage Acute Stress	—	—	Yes	B-13
3.25	Detect Phase: SA-1 Information Requirements	Yes	—	Yes	B-14
3.27	Barrier, Task, and Phase Safety Times	—	—	—	B-16
3.28	External Support System (ESS) Requirements	Yes	—	—	B-17
3.29	External Protective Barrier Requirements	Yes	—	—	B-18
3.30	SPOF Reliability Assessment	Yes	—	—	B-19
3.31	Shared Element Reliability Assessment	Yes	—	—	B-19
3.32	Performance Standards	—	—	—	B-20
3.33	Additional Resource Requirements	—	—	—	B-20
3.34	Fatigue Monitoring and Management Requirements	—	—	—	B-26

The *New Skills & Knowledge* field in Table 3.2 identifies tables that define new skill or knowledge requirements, i.e., requirements beyond the standard expected skills and knowledge. This provides input into personnel selection, procedures, training requirements, etc.

Note: See Section 2.1 for the definition of the term "new skill or knowledge." An accurate understanding of the definition is needed to correctly perform lifecycle model activities.

Table 3.3 identifies documents and technical and requisition input information developed during this process phase. Table 3.4 lists the assessments performed in this process phase.

TABLE 3.3
Preliminary Design: Project and Technical Documents

Documents	New, Update, or Input	See Step	Refer
Documents from Table 3.1	Update	B-1	
Active human barrier design basis	**New**	B-1	
Remote barrier support design basis (if applicable)	**New**	B-1	See App. B.4
Risk studies – scope and execution plan	Input	B-1	
Task analysis – scope and execution plan	**New**	B-1	
Facility plot plans and layout drawings	Input	B-1	
Process and instrument diagrams (P&IDs)	Input	B-1	
Instrument index	Input	B-1	
Design basis and design guidelines (other disciplines)	Input	B-1	
Data sheets	Input	B-1, B-20	
Task analysis report	**New**	B3-1	
Base barrier performance standards	**New**	B-4	
Barrier block diagrams	**New**	B-15	
Reliability study reports	**New**	B-19	
Safety requirements specification	**New**	B-20	See App. B.3
Safety equipment list	Input	B-20	
Functional design: cause-and-effect charts, logic diagrams, functional narratives, and specifications	Input	B-20	
Specifications: direct-use and support components, technical systems, buildings, control consoles and panels, engineered areas, external support systems, external protective barriers	Input	B-20	
Performance standards: barrier-specific or barrier-dependent, component, technical system, packaged equipment system, building, engineered area, external support system, external protective barrier	Input/**New**	B-20	See App. D.3
Block diagrams: technical system, external support system (barrier interfaces)	Input	B-20	
Location and layout drawings: buildings, rooms, and walk-in enclosures (In-Place PE)	Input	B-20	
Location and layout drawings: engineered area (In-Place PE)	Input	B-20	
Location and layout drawings: control consoles and panel (In-Place PE)	Input	B-20	

(Continued)

TABLE 3.3

(Continued)

Documents	New, Update, or Input	See Step	Refer
Location drawings: direct-use PE, stored Support PE	Input	B-20	
Requisition input: scope of work, technical package, vendor data requirements, inspection, and test requirements	Input	B-24	See App. E
Barrier, HE rosters, and organization chart	**New**/Input	B-25	

TABLE 3.4

Preliminary Design: Studies and Assessments

Table Number	Documents	See Step
—	Risk Studies: HAZOP, LOPA, etc.	B2-x
—	Task analysis	B3-1
—	Verification #1	B3-3
3.30	SPOF Reliability Assessment	B-19
3.31	Shared Element Reliability Assessment	B-19
—	Verification #2	B-21
—	Validation #1	B-22
3.34	Fatigue Assessment	B-26
—	Design Review	B-27

3.2.2 Step B-1, Early-Stage Preliminary Design

The early-stage activities in step B-1 develop the documents and information that are essential inputs to the risk assessments and other disciplines. This period may cover the first half or more of the preliminary design phase duration. Table 3.3 identifies the suggested early-stage documents. Example documents include an active human barrier design basis and, if applicable, the remote barrier support (RBS) design basis. See Appendix L for further information on remote barrier support provided from a remote location. Appendix B.4 suggests the content and topics to address in the RBS design basis.

This step should also complete the planning for the task analysis in step B3-1, which may include the following:

- Scheduling the event, for example, the date, duration, and location.
- Identify, acquire/contract the analysis facilitator, and scribe.
- Develop the analysis method, scope, terms of reference, result report/content, etc. See Appendix A.2 for general guidance and the suggested task analysis participants.
- Identify the required participants and confirm availability.

3.2.3   STEP B-2, RISK ASSESSMENTS AND BARRIER IDENTIFICATION

3.2.3.1   Steps B2-1 to B2-4, Risk Assessments

The studies and activities in this step identify barriers that may be an active human type. Figure 3.4 gives examples of risk assessment (RA) processes that may be performed in steps B2-1, B2-2, and B2-3x.

Note: Steps B2-1, B2-2, and B2-3x are existing processes. The information from step B-1 provides input to these processes as applicable.

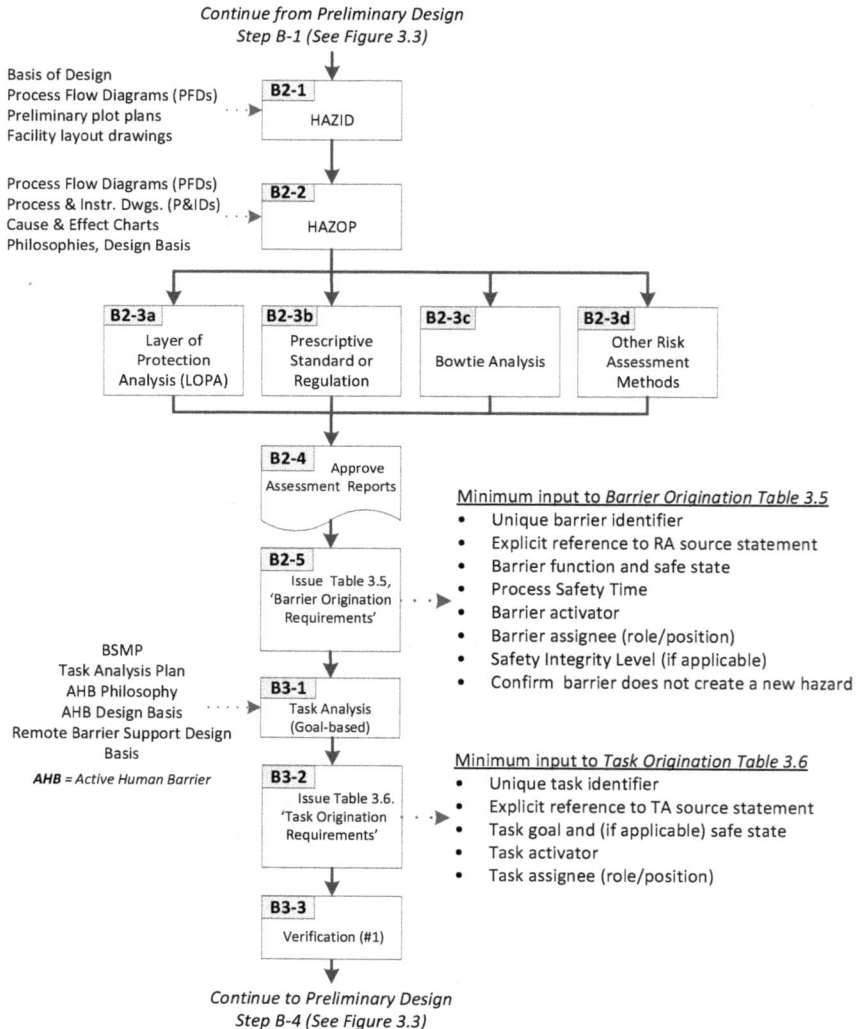

FIGURE 3.4   Preliminary Design Step B-3: Risk Assessments and Barrier Identification.

Note: The risk assessments should capture all the information indicated in Figure 3.4 as "Minimum Input Information to Barrier Origination Table 3.5." Failure to do so can delay work and contribute to errors in this information. Late identification that a proposed barrier creates a new hazard (i.e., the proposed barrier is not viable) can also delay work and contribute to later rework with its associated cost and potential schedule impact.

Step B2-4 addresses several additional considerations when approving and issuing the risk assessment reports.

Note: Preferably, the risk assessment reports should be issued and approved before beginning the task analysis. Errors in the reports may lead to errors in the task analysis.

3.2.3.2   Step B2-5, Capture Barrier Requirements in the Barrier Origination Table

Step B2-5 migrates the active human barrier information from the risk assessment reports to Table 3.5, the Barrier Origination Requirements. Populate all applicable fields. (The barriers in this table are those addressed in the lifecycle model.)

Note: When populating Table 3.5, identify the HEs/roles assigned to each barrier using a unique ID or identifier.

Note: Active human barriers assessed in a LOPA study may include preventive barriers. An example is a proscribed manual operation action taken in response to the barrier activator. The operator is expected to promptly detect the activator, make timely decisions, and perform the required manual response actions that achieve the barrier function within the barrier safety time. The manual action may be as simple as selecting a manual process shutdown pushbutton. Other barrier types may be more complex; for example, they may require specific expertise to detect the activator condition, require unique knowledge, invoke many possible decisions, or require one or more complex action responses. These topics are addressed in the lifecycle model.

3.2.4   Step B-3, Task Analysis, Data Capture, and Verification #1

3.2.4.1   B3-1, Task Analysis

Step B3-1 is the task analysis (TA) that defines the tasks comprising each barrier. Suggested inputs to this process are noted in Figure 3.4. Tasks are the basis for defining the barrier elements and HE actions needed to achieve the barrier function and safe state within the specified safety time. See Section 2.2 for additional background, discussion, and guidance.

> *A particular risk is that the inputs to a barrier analysis are not realistic and properly informed about operational realities. The operator needs to be enabled to contribute fully to the process through training and preparation. It is also essential that any analysis session has an adequate task analysis as an input....*

(CIEHF 2016, p. 33)

TABLE 3.5
Barrier Origination Requirements

Barrier ID	Barrier Title or Descr.	Source Reference	Barrier Activator (SA-1 Info)	Safety Function	Safe State	Process Safety Time (PST)	Human Element (HE) ID	Safety Integrity Level (SIL)	Creates a New Hazard?	Mode of Operation
Enter barrier ID	Enter barrier title or description	Enter reference to barrier origination source (from RA)	Enter barrier activator (SA-1 info item, e.g., alert for an unsafe state or condition)	Enter barrier safety function	Enter barrier safe state	Enter PST (minutes)	Enter HE/role ID *Single task barriers only*	LOPA designated barriers only	Enter Y/N	Enter all modes that apply to this barrier, e.g., start-up, shutdown, normal, upset, emergency

There are several options for decomposing a barrier into individual tasks, of which two are discussed here. The first and more familiar approach is a hierarchical task analysis (HTA). For examples and guidance, see Kirwan and Ainsworth (1992), Shepherd (2001), NUREG (2020, Appendix B.2.1 – Function and Task Analysis), and IFE (2022, pp. 28–34). The second is the goal-directed task analysis (GDTA) presented in Endsley and Jones (2012, Ch. 5).

There are pros and cons to each approach. However, regardless of the selected approach, the defined barrier tasks must reliably achieve the barrier function within the specified safety time. Each task should be assigned a valid task goal and encompass the most cognitively demanding activities within that task. The GDTA process seems well suited to both objectives. The task goal guides the task decisions. The task goal and decision requirements help to identify the SA-1 information needed to support and guide task decisions and act phase response actions. This approach may have an intuitive advantage, though additional steps may be needed to define the act phase response actions required within each task. The HTA is not inherently a cognitive-aware process. Once defined, the group may need additional steps to verify the task goals are appropriate and encompass the most demanding cognitive activities.

Note: Multi-person barriers introduce additional requirements for timely and efficient HE-to-HE interactions and coordination. The terms non-technical skills (NTS) and team situation awareness (TSA) provide the frame to define these requirements. Both are addressed in the lifecycle model. Also see Appendix C for background.

Note: See Appendix L for additional requirements that apply to a remote barrier support task performed from a remote location.

For the suggested HTA participants, see Table A.1 in Appendix A.2.3. Here, participation by senior operations experts is essential, as mistakes in this step can lead to potentially profound and costly design errors. According to CIEHF (2016, p. 39),

> There is often a lack of awareness of the difference between 'work-as-imagined' and 'work-as-done.' 'Work-as-imagined' reflects an idealized, office-based view of how task and processes are to be performed without recognising the many situational factors established work practices, practical difficulties, uncertainties, completing goals and stresses – that exists in reality at the front line. 'Work-as done' captures the reality of how work is actually done, including the compromises and adaptations made with carrying out tasks under real-world constraints and pressures.

A gap between work-as-imagined (WAI) and work-as-done (WAD) can indicate a potential barrier design error or other conditions that can lead to barrier failure. Participation by operations experts that have extensive active human barrier experience is essential for achieving a key design goal to minimize or prevent this gap.

TABLE 3.6
Task Origination Requirements

Barrier ID	Task ID	Task Title or Description	Source Reference	Task Activator (SA-1 Info)	Task Goal	Task HE ID	Task Function or Safe State
Enter barrier ID	Enter Task 1 ID	Enter task title or description	Enter reference to task origination source (from TA)	Enter task activator (SA-1 info item that activates this task)	Enter task goal	Enter ID for HE assigned to this task	Define expected function/result/ safe state at task completion

3.2.4.2 B3-2, Capture Task Requirements in the Task Origination Table

Step B3-2 migrates the active human barrier information from the TA study and report to Table 3.6, the Task Origination Table. The results of the HTA are captured in Table 3.6.

In Table 3.6, uniquely identify each barrier task and each person (or role) assigned to each task. Assign these IDs before commencing step B3-3, as this information is needed for later processes.

3.2.4.3 B3-3, Verification #1

Verification #1 is a selective verification of the risk assessment and task analysis processes, such as performance against published guidance requirements, attended by the appropriate personnel, and so on. It also verifies the accurate selection and migration of the information to Tables 3.5 and 3.6. For requirements, see Appendix G.5, and Tables G.5 (processes) and G.6 (execution plan). The discussion in Appendix G.5 provides further background on this less common process.

3.2.5 Step B-4, Barrier Performance Standards: Base Requirements

This step develops a base-level performance standard that applies to all active human barriers. Using the guidance from the active human barrier design basis, develop a performance standard that applies to all active human barriers identified in the risk assessments. Record the requirements in *Table 3.7.*

Note: See Appendix D for background on performance standards. See Appendix L for the unique requirements that apply to a remote barrier support task. Also see Table 9 (App. B.7.1) for sources of additional information.

3.2.6 Step B-5, Shared Situation Awareness Requirements (Barrier Team)

This step, applicable to barriers with two or more HEs, defines the minimum and specific shared situation awareness (SSA) requirements needed to achieve the barrier

TABLE 3.7

Barrier Performance Standards: Base Requirements

Application of Requirements (examples)	Applicable Codes and Standards	Functionality	Response Performance	Capacity	Reliability Availability	Survivability
Enter requirement application: barrier system, barrier elements (PE, HE, OE), external support systems, external protective barriers	Enter all that apply	Enter base guidance requirements	Enter base requirements	Enter base requirements	Enter base requirements	Enter base requirements

function and safe state within the specified safety time. The barrier-assigned HEs do not need to know and understand everything. Instead, they need a minimum and common shared understanding and time sense that achieves coordinated team functioning and maintains a shared focus on common barrier goals. SSA is a shared understanding of the current situation that enables HEs to implement a coordinated response to a change and respond to the anticipated needs of others. See Appendix C.1.2 for additional information, including the unique SA-1 information requirements that may apply to a remote-located HE assigned to perform a remote barrier support task. The activities in this step include defining and entering the SSA requirements and supporting information into *Table 3.8*.

For Endsley and Jones (2012, p. 196), SSA is "the degree to which team members have the same SA on shared SA requirements." For additional information, see Endsley and Jones (2012, Ch. 11).

Hollnagel (2003, Ch. 31 pp. 758–759) suggests,

It is also hypothesized that shared mental models have a positive impact on more generally formulated teamwork processes. The basic idea is that when team members have a common understanding of the team's goal and each other's roles, responsibilities, and task, teamwork is facilitated. Thorough knowledge of the teammate's tasks (presumably well-structured and organized in a mental model) is needed to be able to monitor each other's performance, provide each other constructive feedback, and back each other up.

Note: See Tannenbaum and Salas (2021, Ch. 8) for an expanded discussion on SSA (e.g., what it is, how to achieve it).

Note: Achieving SSA relies on communications and timely access to other sources of shared information. Steps B-7 (communication exchanges) and B-14 (SA-1 sources and access locations) further develop the information conveyance requirements.

TABLE 3.8
Shared Situation Awareness Requirements

Barrier ID	SSA Req. Task ID	SSA Req. ID	SSA Req.	Required SA-1 Info	SSA Req. Function	SA-1 Info Form	SA-1 Info Timing	Support Aid Type	Support Aid Function	New Skills & Knowledge
Enter barrier ID	Enter task ID 1	Enter unique SSA ID #	Enter SSA requirement One per row *See examples.*	Enter the required SA-1 info item *See examples.*	Enter the function of purpose of the SSA requirement. *See examples.*	Enter form: *See examples.*	Enter timing: *Event-based*, e.g., task start or stop *Periodic update*, e.g., intervals	Enter aid type (entered in process C-6)	Enter aid function or purpose	Enter task-specific requirement: • Skill • Knowledge

Note: See Appendix L for the unique requirements that apply to a remote barrier support task.

Populate all table fields guided by the information in the table and the following guidance. Provide a unique ID for each SSA requirement.

SSA Requirement identifies the need for a shared understanding between all HEs and cases where the SSA is only needed between select HEs, for example:

- Which barriers are active?
- Priorities (exigencies) and constraints (missing, declining, or limited resource)
- Time needed to complete one's own task or tasks performed by others
- Threats/contributors that may cause barrier degradation or failure
- Threat escalation

Required SA-1 Information is the information needed to achieve the SSA requirement, for example:

- Notice of barrier activation
- Instruction to start, hold, or halt a barrier task or response action
- Notice of a missing crew member and the last seen location
- Location of a moving HE (e.g., a search and rescue, fire, or medical response person or team)
- Update on the success/failure of fire control and suppression barriers
- Remaining time before the loss of an external support system or external protective barrier
- Remaining time available to control and recover from the hazard that activated the barrier.

Examples of the **SSA Requirement Function**:

- Coordinate one's own actions with other HEs, such as timing and pace.
- A global request, notification, or status information from the barrier leader that changes team actions or response.
- Coordinate action to support search and rescue operation.
- Changes to an HE's task (start, stop, or hold) that affects other team members.
- Coordinate one's task with that of another HE.
- Input to an SA-3 assessment (projection, anticipation, timing).

SA-1 Info Form refers to how the SA-1 information is presented to the intended HE receiver:

- Verbal information presented over a public address system
- Visual information provided on a shared device, shared HMI display, or VDU-based display wall
- Visual and verbal information conveyed in a face-to-face exchange

- Visual information presented on signs or painted markings on an egress/escape route
- Audio, visual, tactile, or olfactory information presented at an incident scene

New Skills and Knowledge include the following:

- *Knowledge* – the SA-1 information they need and information they may have that another HE needs
- *Knowledge* – general knowledge of tasks assigned to others and how those tasks interact with one's own tasks
- *Knowledge* – procedural knowledge on how to maintain one's own shared SA, communicate information, etc.
- *Skill* – automatic (unconscious) use of communication equipment and protocols
- *Skill* – automatic (unconscious) use of other tools and methods employed to maintain shared SA, such as displays or audio/visual information available to all or multiple HEs
- *Skill* – automatic (unconscious) updating, checking, and maintaining one's own shared SA

The **Support Aid** fields are addressed in the detailed design phase process C-6.

Note: Communications may be initiated as a "push." For example, the barrier leader provides the minimum necessary and timely information that others need to maintain their minimum SSA. Communications may also be initiated as a "pull"; for example, an HE requests information when needed to maintain or confirm one's own SSA.

Note: See Appendix K.5 for a discussion of the information workspace, including information latency and retention, among others.

3.2.7   Step B-6, Act Phase Response and Required Physical Elements

This step defines and specifies the act phase responses (AR) required to achieve the barrier/task safety function and safe state. For this, enter information into Table 3.9. A barrier or task may have several act phase responses or response steps. Example response actions are as follows:

- Pressing an emergency shutdown pushbutton in the field or at a control room console
- Controlling a remote-controlled fire or foam monitor

TABLE 3.9

Act Phase Response Physical Element Requirements

Barrier ID	Task ID	Act Response (AR) ID	Act Phase Response	Direct-Use PE		Support PE	
				PE ID	Use Location		*See Table 3.9 extension*
Enter barrier ID	Enter task ID	Enter unique AR ID For a communication response, enter ID from Table 3.11.	Describe required HE response action	Enter PE ID One per row *See Note*	Enter all that apply, e.g., control room, electrical building, process area, indoor/ outdoor escape route, "anywhere in facility."	Enter PE ID, all that apply *See Note*	

	In-Place PE			
	Engineered Area (EA)	Building or Room	Technical System (TS)	Act Phase Safety Time (Minutes)
See Table 3.9 extension	Enter EA ID or description, all that apply *See Note*	Enter building or room ID, all that apply, unique ID *See Note*	Enter TS ID, all that apply *See Note*	Estimated time to complete action response *once initiated.*

Note: If available, use an existing equipment identifier that may reside in a facility master equipment list.

- Recovering an injured person using a hand-carried stretcher
- Updating an Incident Command Board by hand-marking the information on this board

This step introduces the following new terms. (See Chapter 2 for formal definitions.)

- **Direct-Use PE**: a physical element (PE) that HE must directly access or use. Access/use may be sensory (visual, audible, etc.) or physical (held, touched, accessed entry target on an HMI display, etc.).
- **Support PE**: physical equipment and devices that protect or support the HE during their performance of a task.
- **In-Place PE**: a PE that must be in place to enable, support, or provide an interface to direct-use PE. Example In-Place PEs include engineered areas (egress/escape route or safe haven), rooms, buildings, technical systems, or a fire water supply and distribution systems.

Populate all table fields guided by the information in the table and the following guidance. For each response action performed in this task (one per line), enter a unique response ID, and the HE performed response action.

Direct-Use physical element examples:

- Command or data entry target on an HMI display
- Hand-held radio or telephone
- Fire hose or monitor
- Lifeboat
- A manually activated physical emergency shutdown (ESD) pushbutton mounted on a control console or field panel
- Stretcher or medical device
- Safe haven (see Note)

Note: Classify the safe haven as a direct-use physical element if the barrier safe state is achieved by personnel transiting to and occupying this area.

Support PE examples:

- Self-contained breathing apparatus (SCBA)
- Smoke hood
- Flashlight
- Life jacket
- Hand-held flammable or toxic gas detector

In-Place PE: Engineered Area examples:

- Safe Haven – A protected indoor or outdoor muster/rally area
- Safe Haven – A protected lifeboat embarkation area
- Egress/escape route located within a building or outdoors

In-Place PE: Building and building-enclosed room or area

- Central Control Room
- Incident Command Center (may be a dedicate room or a designated space located in a control room)
- Emergency response station, for example, a designated area in a building
- Living quarters (may be a muster point/rally area)
- Walk-in enclosures (the type often provided with large, packaged equipment systems or provided as a local control or electrical equipment building)
- Remote Operations Center (See Appendix L for background on the ROC and remote barrier support.)

In-Place PE: Technical System examples:

- Facility monitoring and control system
- Facility general alarm system
- Control console or panel
- Safety instrumented system
- Building Heating, Ventilation, and Air Conditioning (HVAC) system

- Packaged Equipment Control System (supplied with packaged equipment system)
- Fire suppression system, such as a sprinkler, foam, deluge, or water mist system
- Uninterruptible Power Supply (UPS) System
- Lifeboat davit/launching system
- Public address/general alarm system (PA/GA)
- Telephone system
- CCTV system

Example, Offshore O&G muster barrier (simplified hypothetical).

Consider the application of the above process to an offshore O&G muster barrier. Table 3.10 indicates the possible act phase responses to tasks assigned to barrier HEs and non-essential personnel:

- All personnel safely muster to their designated safe locations or emergency response stations.
- Upon arrival at the safe location, they report to the person assigned to record the status of all persons on board (POB) and identify possible missing persons.

For this barrier, every person at the facility has an assigned role. The assigned roles for this example are as follows:

- *Non-Essential Personnel (NEP)* – Personnel not assigned to the Emergency Response Team or ERT. (The remaining personnel are assigned to the ERT.)
- *Barrier Leader (BL)* – As this is a multi-person barrier, this person is assigned the barrier leadership position.
- *Muster Leader (ML)* – Assigned to a designated muster station to record and report personnel reporting to this station. (MLs may be assigned to the primary and secondary muster stations.)
- *Scribe* – Assigned to gather and record information onto the Incident Command Board, and request status updates.
- *Radioperson* – Assigned to radio and telephone communications, and manage information to/from the scribe and BL.
- *Control Room Operator (CRO)* – Brings the process to a safe sate. Monitors the operational and safety status of the facility and essential systems.
- *Fire Team (FT).*

Note: The task analysis defines the tasks assigned to the NEPs and HEs recorded in Table 3.6 (steps B3-1 and B3-2). Table 3.10 indicates the complexity of this barrier type and the many possible implementation challenges.

Note: Some organizations use personnel-carried RFID cards and passive RFID tracking systems to reduce the time and effort to complete a full roll call, an activity needed to progress and complete the muster process.

TABLE 3.10
Example Action Response Table – Offshore Muster Barrier (Simplified)

Barrier/Task Name: *Muster Barrier*   Barrier/Task Number: *TBD*

Act Phase Response PE Requirements

Action Response ID	Action Response	NEP and Barrier Team Personnel	Direct-Use PE — Required PE	Direct-Use PE — Use Location	Required Support Physical Elements	In-Place PE — Technical System	In-Place PE — Room	In-Place PE — Engr'd Area	Specified Barrier Safety Time (Minutes)
Enter unique ID for each act phase response	On alarm, safely transit to assigned primary muster station or ER station.	Barrier leader (BL) / Muster leader (ML) / Non-essential personnel (NEP) / Scribe / Radioperson (RP) / Control room operator (CRO) / Fire team (FT)	General alarm horns/beacons. Display indicators along route (passive, active)	Egress/escape route	Std. PPE; On route locations: life jacket, smoke hood, etc.	Public alarm/public address system; Monitoring, control, fire and gas systems. External support systems (lighting, power, etc.)		Egress/escape route	10 minutes to complete all muster actions
	Go to secondary if blocked								
	Report in accounted/missing POB	**NEP: Report to ML.** **Pri. ML:** Check in. Report POB to RP/command team **Sec. ML:** Check in. Report POB to RP/command team	See Support PE / Emergency radios, telephone / POB checklist or system	Primary and secondary muster areas	*See above*			Primary/secondary muster area	
	Report to ICC	**BL:** Check in. Monitor ICB. Request status. Identify missing persons. **Radioperson:** Check-in. Receive/request POB reports. Communicate results to Scribe.	Emergency radios, telephone, IT systems, weather systems, SCADA displays	ICC	*See above*	Emergency radio, telephone, monitoring, control, and IT systems	ICC		
	Receive POB reports from MLs and others	**Scribe:** Check-in. Record POB status on ICB (accounted, missing, time). Communicate results to BL.	POB checklist / Incident Command Board		*See above*	External Support Systems (lighting, power, etc.) / HVAC (indoor muster area)			
	Report to CCR. Report in. Monitor/maintain facility state.	**Control room operator:** Check in. Monitor/maintain facility state.	SCADA/HMI displays. Fire and emergency panels. Emergency radio, telephone	CCR	*See above*		CCR		(See step B-16 and Table 3.27 for safety times.)
	Report to ERT	**Fire Team.** Report in. Don gear. Await instructions.	Emergency radio, telephone, other gear/equipment	ERS, TBD	Fire gear/clothing, air packs		ERS		

Note: POB – persons on board; CCR – Central Control Room; ERS – Emergency Response Station; ICB – Incident Command Board; ICC – Incident Command Center; HVAC – Heating, Ventilation, and Air Conditioning

3.2.8   STEP B-7, ACT PHASE: OUTBOUND COMMUNICATIONS

This step defines barrier requirements for outbound communications. Barriers with two or more HEs require information exchanges between HEs. The barrier leader conveys status information and instructions to coordinate actions. Others convey status feedback information or requests. This step defines the act phase response actions that convey SA-1 information. Enter the information into *Table 3.11*.

Note: The communication requirements identified in this step are those required for explicit coordination and to assure that barrier-assigned HEs have the timely and sufficiently complete information needed to progress assigned tasks and maintain team coordination, cohesion, and alignment to common shared goals and situation awareness. See Appendix C.3.2 for a discussion on implicit and explication coordination. See Appendix K.3 for further information on the communication workspace.

Table 3.11 identifies the estimated timing for required communications. Each communication should be uniquely identified, and the following information defined or specified:

- Sender and all intended receiver(s)
- Message function and purpose: convey instruction, coordinate actions, etc.
- Message form: real-time (two-way) communication, email, voice communication using a public address system (one-way), etc.
- Sender location
- Receiver(s) location(s)
- Estimated message frequency, timing, and duration
- Medium form: voice, visual, conference call, text, etc.
- Medium system (PE): telephone, radio, public address, video conferencing, etc.

Populate all table fields guided by the information in the table and the following guidance.
New Skills & Knowledge field: Examples of **Skills**:

- Timely, efficient, and effective communication is used in different team and ambient environments and situations.
- Automatically and accurately use the protocol and vernacular defined in barrier procedures.
- Listening skills, including tone of voice, background sounds.
- Communicating as or to non-native language speakers.

New Skills & Knowledge field: Examples of **Knowledge**:

- Understanding the correct use, functioning, and limitations of a communication system. Limitations of equipment, such as gaps in radio coverage and usability in different ambient environments
- Procedure guidance on protocols, vernacular, timing, etc.

TABLE 3.11
Communication Requirements

See Table 3.11 Extension

Barrier ID	Task ID	Comms ID	Message function/purpose	Message Sender			Message Receiver	
				Sent Msg. Content (SA-1 info to receiver)	Sender Loc.	New Skills & Knowledge	Target Msg. Rec'r	Rec'r Loc. (s)
Enter barrier ID	Enter task ID	Enter unique ID for Comms req.	**Barrier Leader Commands** Activate, request update, hold, halt **Other HE** commands: Request **All:** Provide status/info to coordinate actions	Enter conveyed SA-1 information *Enter all SA-1 info needed to support this function and conveyed in this message.*	Enter all possible locations: – Building: CCR, ICC – ER station – Equip. room, – Process area – Machine space – Hull	Examples: Learn/use standards for Msg. protocols and format Speaking tone, pace, body language	Enter HE ID for target rec'r *One per row* Next Rec'r ID for this Msg.	Identify rec'r loc.(s) *List all that apply*

Communications Detail

Msg. Timing	Msg. Duration	Msg. Form	Medium PE
Enter – Periodic – On request – Barrier/task activation – Ad hoc	Enter maximum target message duration (seconds)	Enter – In-Person – Text – Verbal – Live video: 1-way, 2-way – HMI display	Enter medium PE type: – email – Phone – Radio – Public alarm/public address (PAGA) – Video conferencing

Message Receiver

New Skills & Knowledge
Examples: Learn/use standards for Msg. protocols and format. Listening skills

See Table 3.11 Extension

Note: CCR – Central Control Room; ICC – Incident Command Center; ER –Emergency Response.

Note: The sender and intended receiver(s) must fully attend to (focus on) the exchange to prevent incorrect, incomplete, or inappropriate message conveyance and receipt. During the exchange, the sender and receiver(s) may not be available for other activities; i.e., this is a sustained vigilance activity. As such, the exchange should be limited to the most efficient and effective exchange in the least amount of time. Overly lengthy exchange durations (time penalty) or a message error (conveyance or receipt) can contribute to a degraded or failed barrier/task.

3.2.9 STEP B-8, ACT PHASE: DIRECT-USE AND SUPPORT PHYSICAL ELEMENT REQUIREMENTS

This step further defines requirements for direct-use and support physical elements identified in Table 3.9 (step B-6). Enter requirements into Tables 3.12 and 3.13, respectively.

Note: The HE directly interfaces to and interacts with the direct-use physical element. Later processes perform physical and cognitive ergonomics and performance influencing assessments to address the human-system integration aspects attributed to the interface and situational use.

TABLE 3.12

Act Phase Response: Direct-Use PE Requirements

Direct-Use PE ID	PE Description	Barrier ID	Task ID	Response Action ID	PE Function	PE Use Duration (Minutes)	SA-1 Feedback from PE	
Enter ID from Table 3.9	Enter PE description	Enter barrier ID 1	Enter task ID 1	ID from Table 3.9 (See Note) For this barrier/task	Enter PE function	Enter PE use duration	Identify requirement for DU PE status feedback Also identify if monitoring is Local, Remote, or both	*See Table 3.12 Extension*

	New Skills & Knowledge	PE Usage Hazards	Support Aid		Required In-Place PE?
			Type	Function	
See Table 3.12 Extension	Enter required new skill, knowledge	Enter potential use hazard to user if any	Enter aid type (Entered in process C-3)	Enter aid function or purpose	*Enter ID for In-Place PE listed in Table 3.9*

Note: If the action is to convey information, enter the Comms ID from Table 3.11. Refer to Table 3.11 for other applicable fields.

TABLE 3.13

Act Phase Response: Support PE Requirements

Support PE ID	Barrier ID	Task ID	Response Action ID	PE Function	PE Use Duration (Minutes)	SA-1 Feedback from PE	Special Skills & Knowledge	PE Usage Hazards	Use Locations
Enter Support PE ID *One row for each unique ID*	Enter barrier ID	Enter task ID	Enter ID from Table 3.9 For this barrier/task.	Enter Support PE function (How used)	Enter PE duration	Identify req. for Support PE status feedback. Also identify if monitoring is *Local*, *Remote*, or both.	Enter new skill, knowledge	Enter use hazard with each PE user, if any	Enter all locations where PE used, worn, carried.

Populate all table fields guided by the information in the table and the following guidance.

PE Function examples:

- Activate emergency shutdown of the south facility
- Active foam release at tank 541

PE Use Duration examples:

- One-time discrete use (pushbutton)
- Ad hoc, variable-use duration within the task or barrier activation period
- Continuous use for the full duration of the task or barrier activation

SA-1 Feedback from AR Physical Element (Direct-Use):

- Identify requirements for real-time SA-1 feedback (if any) from an act phase response initiated or performed using a direct-use physical element.
- Identify if this required monitoring is by the physical element user, by a remotely located HE, or by both.

Example SA-1 feedback information:

- The monitored information may be visual, for example, locally viewed, or viewed remotely via a CCTV system.
- A technical system monitors the status of a manually activated ESD pushbutton and provides visual feedback using the pushbutton backlighting. This feedback may be designed to confirm receipt of the activation request, the requested ESD action is initiated, or requested ESD action achieved the required safe state. The pushbutton status and confirmation feedback may be mimicked on an HMI display to notify others assigned to monitoring this information.

Note: Entering "Remote" monitoring in this field identifies a new task, i.e., add the task to Table 3.6 for design and development.

New Skills and Knowledge – Enter new skills and knowledge (if any) required to correctly and safely use the direct-use physical element to perform the action response.

Direct-Use Physical Element Usage Hazard examples:

- Lose control of a hand-held fire hose (uncontrolled movement of heavy object)
- Drop a stretcher during transfer of an injured person
- Drop a direct-use object that may be heavy, awkward to use, etc. (Dropped object hazard.)

The **Support Aid** fields are addressed in the detailed design phase process C-3.

In-Place PE – Enter the unique ID or descriptor for the required In-Place PE, as identified in Table 3.9.

Note: The HE directly interfaces with Support PE. Later processes perform physical and cognitive ergonomics and performance influencing assessments to address the human-system integration aspects of the design.

The following provides guidance on the field entries in Table 3.13:
Support PE Function examples:

- Self-contained breathing apparatus required to provide safe air when entering areas that may have reduced air quality (e.g., smoke), low O_2, or toxic or damaging airborne agents.
- A smoke hood limits smoke inhalation when transiting in response to a muster/rally activation alarm.
- Flashlight to aid access to visual information if area lighting fails or is inadequate.

Support PE Use Duration examples:

- Full period of barrier activation
- Limited to the duration of the task that required the use of Support PE
- Other times or periods

SA-1 Feedback from Support PE:

- Identify requirements for real-time SA-1 feedback (if any) from Support PE.
- Identify if this required monitoring is by the physical element user, remotely located HE, or both.

Example SA-1 feedback information:

- Information that provides real-time feedback on the operational state of the Support PE, for example, the remaining air in a SCBA. (Monitoring this feedback may be safety critical and frequent or sustained vigilance task.)

Note: Entering "Remote" monitoring in this field identifies a new task. Enter the task in Table 3.6 for design and development.

New Skills and Knowledge – identify new skills and knowledge (if any) required to use specialized Support PE correctly and safely.
Support PE Usage Hazard examples:

- Failure of a self-contained breathing apparatus places the HE user at risk.
- Incorrect use/application of a hand-held flammable or toxic gas detection meter
- Incorrect use of a smoke hood

3.2.10   STEP B-9, ACT (AND DETECT) PHASE: IN-PLACE PE REQUIREMENTS

This step further defines the requirements for In-Place PE recorded in Table 3.9 (step B-6) or in Table 3.25 (Step B-14). Enter the expanded In-Place PE requirements in Tables 3.14 (engineered area), 3.15 (buildings), and 3.16 (technical systems).

Note: This table is common to the detect and act phases. Repeat step B-9 to capture the In-Place PE requirements (those within the EA) that apply to detect phase PE (if any).

Populate all table fields guided by the information in the table and the following guidance.
Engineered Area ID – Enter the unique ID for this engineered area and brief description. (This engineered area contains one or more barrier components or elements.)
Barrier ID – Enter the ID for the barrier that relies on this engineered area.
Task ID – Enter the ID for the task that relies on this engineered area.
Enter the **Act Phase Response (AR) ID** that identified an EA requirement (from Table 3.9). (When applied to the detect phase, enter the **Detect Phase SA-1 ID** that identified an EA requirement from Table 3.25.)
Enter **Area Protective Function** – Identify the EA function in the defined detect or act phase activity. The area provides temporary protection from the hazard that activated the barrier or other stated hazards, conditions, or situations. Examples are as follows:

- A muster/rally area (safe haven) that provides temporary protection to occupants from the hazard that activated the barrier or from its effects. *The area may be designed as an indoor or outdoor location.*
- The escape route in a building or process area provides temporary and defined protection to transiting personnel from the identified hazard or hazard effect.

Endurance Time – Enter the period the engineered area must provide or support the specified function in the presence of a defined condition or situation. (The endurance time may be the period of barrier activation, the barrier safety time, or a different period or interval.)
Support PE in this Area examples:

- Firefighting gear and self-contained breathing apparatus worn by the fire team responders.
- Other worn or carried equipment, clothing, or gear, for example, life vests and water emersions suits.

Note: Identify bulky Support PE that may require changing the area configuration or dimensions to accommodate equipment. Consider dimensions for airlocks, stair landings, and doorways.

TABLE 3.14

Detect and Act Phase: In-Place PE – Engineered Area Requirements

Engr'd Area ID	Barrier ID	Task ID	Enter AR or Detect Phase ID	Area Protective Function	Endurance Time (Min)	HEs in this Area	Support PE in this Area	Max. Occupants	Min. Dim. of Area
Enter unique ID for EA and descr. (*From Table 3.9 or 3.25*)	Enter barrier ID 1 (From Table 3.5)	Enter task ID 1 (From Table 3.6)	Enter AR ID from Table 3.9 or detect ID from Table 3.25	State HE protective function provided by this space, if any.	Enter min. endurance time	Enter all HE IDs in this space for this barrier/ task	Enter all Support PE expected/ used in this area	Enter maximum occupants possible in this area (design basis)	Enter estimated area and dimensions (L, W, H)

Maximum Occupants in this area. Examples are as follows:

- A lifeboat embarkation area sized to accommodate the maximum number of lifeboat occupants.
- The maximum number of persons expected in a building entry/exit airlock or a stair landing. For example, consider the maximum number and positioning of persons carrying a stretcher through an airlock. The airlock size should accommodate all persons and the stretcher with both doors simultaneously closed.

Minimum Dimensions for this Area. Examples are as follows:

- A lifeboat embarkation area may be sized to accommodate the maximum number of persons (noted above) while wearing the "Required PE in this Area" identified in the table.
- An egress/escape route section must be sized to accommodate continuous two-way traffic or some other peak loading in that area
- Consider the varying conditions and situations that may affect the maximum feasible number of people in the area.

Note: The barrier-task activity that places the most onerous demands on an engineered area (e.g., maximum occupants or minimum sizing requirements) provides the basis for the design of that area.

Note: This table is common to the detect and act phases. Repeat step B-9 to capture the In-Place PE requirements (those within the EA) that apply to detect phase PE (if any).

The following provides guidance on the field entries in Table 3.15.
Building ID – Enter the barrier ID for this building and a brief description. (This building contains/protects one or more barrier components or elements.)
Barrier ID – Enter the ID for the barrier that relies on this building.
Task ID – Enter the ID for the task that relies on this building.
Enter the **Act Response Phase (AR) ID** that identified the requirements for this building from Table 3.9. (When applied to the detect phase, enter the **Detect Phase SA-1 ID** that identified an EA requirement from Table 3.25.) Building examples are as follows:

- If the indicated area is a room within a building, enter the room and building IDs and brief descriptions.
- If the indicated area is a walk-in enclosure (building), provide the name of the enclosure equipment.
- If this is an emergency response station located within a building or room, enter the ER station ID and a brief description.

Building Function – Identifies the primary barrier functions provided by this building in support of the identified detect or act phase activity. Examples are as follows:

- Protects personnel or equipment from the hazard that activated the barrier or other specified conditions.

TABLE 3.15

Detect and Act Phase: In-Place PE – Building Requirements

Building ID	Barrier ID	Task ID	Enter AR or Detect Phase ID	Building Function	Endurance Time (Min)	HEs in this space	Required PE in this Building	Max. Occupants	Min. room dim.
Enter unique ID for building, and description (*From Table 3.9 or 3.25*)	Enter barrier ID 1 (*From Table 3.5*)	Enter task ID 1 (*From Table 3.6*)	Enter AR ID from Table 3.9 or Detect ID from Table 3.25	State function provided by this space, if any.	Enter minimum endurance time	Enter all HE IDs in this space for this barrier/task	Enter all direct-Use and Support PEs used in this space for this task	Enter maximum occupants possible in this space (design basis)	Enter estimated area/dimensions (L, W, H)

- Includes a room, engineered area, or other physical space required by a barrier function or activity.
- Provides an ambient environment suitable to HEs, equipment, and barrier activities performed in this building, such as lighting, heat/humidity control, or reduction of airborne irritants that may degrade personnel performance or damage equipment.

Endurance Time – Enter the period the building must provide or support the specified function in the presence of a defined condition or situation. (The endurance time may be the period during barrier activation, the barrier safety time, or a different period or interval.)

Required PE in this Building. Examples are as follows:

- Rooms in which barrier functions or elements take place or reside, such as a control room or Incident Command Center.
- Control console or a panel that provides access to direct-use PE, such as an ESD pushbutton or HMI display.
- Storage areas for barrier-required equipment, such as worn or carried Direct-Use and Support PE.

Minimum Room Dimensions (if applicable) – The information in this table provides a subset of the information needed to guide the building sizing, configuration, and layout. Considerations may include equipment within the room, the required spacing (minimum or maximum) between personnel or between personnel and accessed equipment, and to accommodate the normal and off-normal of personnel with the space.

Note: This table is common to the detect and act phases. Repeat step B-9 to capture the In-Place PE requirements (those within the EA) that apply to detect phase PE (if any).

The following provides guidance on the field entries in Table 3.16.
Technical System IDs – Enter the unique ID for this technical system and a brief description. (This technical system contains one or more barrier components or elements.)
Barrier ID – Enter the ID for the barrier that relies on this technical system.
Task ID – Enter the ID for the task that relies on this technical system.
Enter the **Act Phase Response (AR) ID** that identifies the requirements for this technical system, from Table 3.9. (When applied to the detect phase, enter the **Detect Phase SA-1 ID** that identified an EA requirement from Table 3.25.)
Technical System Function. Enter the barrier function that resides in or is performed by this technical system.
Use Condition – Enter all conditions in which this technical system must continue to perform or support the barrier-dependent function. Conditions may be environmental or the effects of the hazard that activated the barrier.
Endurance Time – Enter the period during which the barrier function that resides in or is performed by the technical system must continue to perform in the presence

TABLE 3.16

Detect and Act Phase: In-Place PE Technical System Requirements

Technical System ID	Barrier ID	Task ID	Enter AR or Detect Phase ID	Technical System Function	Use Condition	Endurance Time	SA-1 System Status	Other requirements?
Enter unique ID for technical system and description (*From Table 3.9 or 3.25*)	Enter barrier ID (From Table 3.5)	Enter task ID (From Table 3.6)	Enter AR ID from Table 3.9 or Detect ID from Table 3.25	State function provided by this system	Enter all conditions in which the system must function	Enter period that systems must reliably provide the stated function	Enter requirement for SA-1 status feedback on system status	Enter additional technical system requirements if any

of a defined condition or situation. (The endurance time may be the period of barrier activation, the barrier safety time, or a different period or interval.)

SA-1 Status (this technical system) – Provide SA-1 status feedback to alert HEs of a failure or fault of a technical system that degrades or causes the failure of a barrier component or function that resides within or relies on this technical system.

Other Requirements: Provide other information that may be important to understand how, what, and when the technical system contributes to the barrier function.

3.2.11 Step B-10, Decide Phase Requirements

This step defines every decision required to guide each barrier/task function and safe state. Once identified, define all SA-1 information required to make each decision. Enter the decisions, decision type, and additional fields into Table 3.17.

Populate all table fields guided by the information in the table and the following guidance.

Decision: identify all required decisions. See the discussion below for guidance. Decision types:

- **Skill-based:** the decision type eventually becomes intuitive and automatic. (A skill-based decision is appropriate and reliable if the situation does not change and is accurately perceived and understood.)
- **Rule-based:** the decision is based on the requirements in a procedure. The cognitive challenge is to recall the correct procedure. This invokes a conscious process to evaluate the validity of the procedure recalled from memory. (The need to rely on memory may be offset by a support aid that automatically and reliably presents the correct procedure. The operator may use the aid if trusted, convenient, and takes minimal effort to use.)
- **Recognition-Primed Decision Model (RPD)** – RPD requires a high level of experience and expertise. This type of decision-making is commonly used by those charged with making high-risk, high-consequence decisions, especially when it occurs under time pressure. The approach relies on its experienced and skilled use, sufficiently accurate mental models, and the quality and accuracy of the mental simulations performed to understand the plausible outcomes and recognize when the outcome appears adequate (a satisficing approach). (For additional information on mental models and RPD, see Appendices F.5 and I.3.5, respectively.)
- **Knowledge-based**: Knowledge-based decisions are more prone to error. Reason (2008, pp. 45–46) theorizes that "Errors at this level arise from resource limitations ('bounded rationality') and incomplete or incorrect knowledge." To further explain, "At the KB level…mistakes result from changes in the world that have neither been prepared for or anticipated" (Reason 1990, p. 61). A procedure is not available to address a novel situation that was unforeseen. Conscious processes are required to make this type of decision, an approach that is known to be less reliable and more error-prone (Reason 2008, pp. 54–55).

TABLE 3.17

Decide Phase Requirements

Barrier ID	Task ID	Decision ID	Decision	Type	SA-1 Info Required	New Skills & Knowledge	Support Aid	
							Type	Function
Enter barrier ID 1	Enter task ID	Enter unique decision ID	**Enter** decision	Enter: - Skill-based - Rule-based - Knowledge-based - RPD	Define all SA-1 information needed to guide this decision	Enter new skills or knowledge	Enter aid type (Entered in process C4-1c)	Enter aid function or purpose

Note: Decision-making can be negatively affected by skill fade and drift over time. For information on both topics, see Appendix J.7. For information on the potential effects of non-rational biases and bounded rationality, see Appendices F.2.5 and F.2.6.

Support Aid: These fields are addressed in the detailed design phase process C4-1c. (See this step and Appendix B.8 for detail.)

The barrier leader assigned to an emergency response barrier or other complex barrier types should demonstrate the competency of RPD decision-making. (For more information, see Flin et al. (2008, pp. 48–51, 57–58), Klein (1993), and Appendix I.3.5.) If a knowledge-based decision is a possibility, this warrants an assessment to determine if and why this type of decision is needed; i.e., the barrier/task may be infeasible. As appropriate, consider the information in Tables 3.2–3.4 as possible input to this step.

3.2.11.1   Considerations and Examples

"The situations people find themselves in can also influence the quality of their decision-making. Time pressure, poor information presentation, ambiguity of information and conflicting goals can lead to poor decisions" (SPE 2014, p. 12).

Barrier decisions are often a primary contributor to barrier failure, and often tend to be cognitively demanding. Cognitive demand (workload) and the time needed to complete the decision-making process can increase if:

- The SA-1 input information changes rapidly
- The barrier requires numerous and/or complex decisions
- Goal conflicts exist (Sträter 2005, p. 51; Woods et al. 2010, p. 88; Gasbury 2013, Ch. 14).
- A complex decision environment exists. For example, a multi-task situation with concurrent tasks and a high workload or a challenging work environment (e.g., many interruptions and distractions) that can lead to attention tunneling and task-switch errors may increase complexity. (See the background and elaboration in Appendix F.)

Other example sources of decision errors (discussed in Appendix F) include the following:

- Too much time, which may make a pending safety critical decision easier to forget.
- Too little time may increase reliance on the automatic cognitive processes, which have no awareness of risk and tend to be impulsive.

A late decision may result in a failure to achieve the safe state within the specified barrier safety time or target task safety time. (Step B-16 further defines the target phase safety time available for decision-making.)

Note: Refer to Flin et al. (1996) for valuable insight into the decision-making environment (emergency barriers) for an offshore O&G facility. Much of the focus is on the barrier leader (Offshore Installation Manager).

Furthermore, the time needed to make decisions may increase exponentially as the number of decisions increases. (See **Hick's Law:** Response time $= \text{Log}_2 (n+1)$, where "n" is the number of decisions. This applies to binary-type decisions (Hick, 1952).

Note: In current practice, the full range of necessary and plausible decisions are seldom identified, evaluated, or addressed in the barrier design process. (Some are implied only.) Consider the following. With many active human barriers, the operating company chooses to insert a human into the barrier to perform a function that, as perceived, cannot be reliably performed by a fully automated barrier residing in a safety instrumented system. The common expectation is the operator's experience, knowledge, and judgment will reduce the number of unintended barrier activations, i.e., nuisance trips. This expectation is often encompassed within an organizational practice or denoted by the terms "Good Process Practice" or "Well-Control." In these cases, the operator is expected to make production versus safety judgments. These types of expectations are implied requirements often not documented or integrated into the barrier design process. With all such requirements, this creates a potential entry point for hidden design errors that place the barrier/task at risk.

With the possible exception of emergency response barriers, active human barriers provide a degree of flexibility; i.e., the operator can choose when to complete the action response within the available barrier safety time. This offers flexibility but can place the barrier at risk. The following are plausible decisions that occur with active human barriers:

1. What is the required action response for this barrier or task?
2. Is the barrier/task activator signal valid?
3. Do I have enough information to act?
4. Do I initiate the barrier response now or wait?

Item 2 could be approached from a production or a safety decision. Procedures and training programs should clearly and explicitly state the expected mindset. *Item 4* introduces a special set of cognitive issues that may be unknown to barrier designers. Consider the following situation: the barrier activator alarm is detected, and the operator chooses to delay the action response for one of many reasons. This decision sets up the following possible scenarios. (The following conversation is an early venture into the cognitive information included in Appendix F.)

1. *Reliance on prospective memory:* The need to remember a future action relies on prospective memory, a known human weakness. A future action is stored in working memory (short-term memory), which has a limited storage capacity. In addition, the information in the short-term memory can fade (is forgotten) if not periodically refreshed. This can occur in as little as 20–30 seconds (Endsley and Jones 2012, p. 33).
2. *Reliance on tracking clock time internally:* Humans have a limited and often unreliable ability to internally track clock time to varying degrees of accuracy.

This ability often degrades with excessive workload, stress, fatigue, and distractions. It can also degrade when a person experiences challenging and complex situations at work; for example, one or more barriers activate unexpectedly or several barriers are simultaneously active. For these reasons, the barrier design should provide externally driven, clock-time alerts and tools to notify and guide time-sensitive actions. Failure to do so creates the need to *watch the clock*, an unreliable and unreasonable expectation. (Cognitively, losing track of time is a hidden deficiency that often leads to barrier failure.)

Note: A design that unnecessarily relies on prospective memory or internal tracking of clock time is a latent design error. Both are common contributors to barrier failure when the barrier function is not performed (forgotten) or not performed at the correct time (occurs too late or too early).

Consider: Consider the case of an active human barrier (preventive type) identified in a LOPA and assigned a risk reduction (RR) factor of 10. To increase the likelihood that this RR can be approached or realized, consider adding features to improve the barrier system reliability. For example, provide a timer that starts timing when the barrier activator first occurs. Generate a suitably salient operator alert that sounds when the timer reaches x minutes from the BST AND the required operator action is not yet initiated. This alarm is a means to capture and direct the operator's attention to the pending (incomplete) action. Example considerations when selecting the alert time may include the following:

- The time needed for an operator to disengage from other activities.
- The time needed to return to this activity and understand the current state and required action.
- The time needed to initiate the response action and achieve the safe state before the BST is reached.
- The addition of a safety margin commensurate to the plausible uncertainties in the above considerations.

  If the operator does not respond within a predefined interval after the above alert activation, consider automatically initiating the action response.

  If adding the alert, the design process should consider the potential for unintended effects. For example, can this alert inappropriately re-direct attention from a higher priority pending action? Do the selected alert tones adequately differentiate between the two and appropriately and reliably guide the operator?

3. During the deferred action period, what happens if additional demands occur? Here, another high (or higher) priority alarm could activate, or a shutdown pre-alarm may alert the operator of a pending process shutdown that will occur if prompt action is not taken. These new events introduce one or more increasingly complex decisions, i.e., which issue should be attended

to first? This places the original barrier at risk because attention and work-ing memory/short-term memory (WM/STM) are limited resources, and the time-pressured situation increases the likelihood that task-switch errors, non-rational biases, and other behaviors may influence or drive these decisions.

4. Another consequence of deferring a required action is that the operator may believe this act "frees" cognitive capability to address issues that may seem to be more immediate. Factually, the deferred action continues to consume WM/STM; holding the pending action requires attention and STM to remind oneself of the future action. The same may happen if a sec-ond action is deferred. The load effect is additive and continues to consume the limited STM capacity. As another consideration, STM capacity may be reduced in response to stress, excessive workload, lack of sleep, and fear. Information in STM may fade (forgotten) or misremembered. When the cognitive workload approaches the WM/STM limits, the HE behavior may revert to attention tunneling, where one's attention is purposely limited to one item and does so without regard for other tasks that may be of higher priority.

Note: Humans are often driven by behaviors and hidden, non-rational biases that can lead to non-optimal and potentially problematic behaviors and outcomes. For further information, see Appendices F.2.3 (task-switch error), F.2.5 (cognitive ease and non-rational biases), and F.2.6 (bounded rationality/keyhole effect).

What additional design questions that can affect barrier reliability should be considered?

- What are acceptable reasons to defer a barrier response action once the activator has triggered?
- What is an acceptable practice on how long the operator can wait to perform the response action(s)? Is it acceptable to do so at the last minute as a matter of practice?
- If multiple demands occur at the same time, what guidance and training are provided on which priorities to address first? Does the training include drills under time pressure?
- Should the barrier safety time be set to one-half (½) the process safety time, as is commonly done for fully automated safety instrumented functions?

3.2.12   STEP B-11, DETECT PHASE: COMPREHENSION (SA-2) REQUIREMENTS

This step defines the minimum understanding and comprehension requirements needed to correctly guide each decision and action response. The information is entered into Table 3.18. Also define the SA-1 information needed to realize each requirement.

As appropriate, consider the documents (applicable information) listed in Tables 3.2–3.4 as input to consider in this step.

TABLE 3.18
Detect Phase: SA-2 Comprehension Requirements

| Barrier ID | Task ID | SA-2 Comprehension | | | Supported Decision/Act Ph. Requirements | | | New Skills & Knowledge | Support Aid | |
		ID	Requirement	Type and Description	ID	Description	SA-1 Info Required		Aid Type	Function
Enter barrier ID	Enter task ID	Enter unique SA-2 Req. ID	Enter comprehension requirement See examples.	Enter type: – *Technical* – *Procedural* – *Execution* – *Situational* – *Hazard,* and Description	Enter ID of supported act or decide phase req.	Enter description of supported requirement	Enter Req'd SA-1 Info (If any)	Enter new skill or knowledge required	Enter aid type (Enter in process C5-1)	Enter aid function or purpose

Populate all table fields guided by the information in the table and the following. **New Skills and Knowledge** examples:

- Identify any new skills needed to comprehend the presented and accessed information. Skills may guide the required and timely access to the SA-1 information needed to achieve the comprehension requirement.
- Identify the barrier-/task-specific knowledge (procedural, technical, and execution) required to achieve the stated comprehension given the availability of the listed SA-1 information.

Comprehension Requirement, Type and Description, SA-1 Info Required examples:

Comprehension is achieved by correctly understanding the meaning of the accessed/perceived SA-1 information and its relationship to decisions and actions. The breadth and depth of a comprehension requirement depends on the process and hazard type, barrier safety function and safe state, and nature and source of the SA-1 information. The following examples may provide insight into the thought processes needed to reveal and confirm comprehension requirements.

The **Support Aid** fields are addressed in the detailed design phase process C5-1c.

Example 1

Facility: Process Plant. **Barrier safety function:** Activate (press) the process safety shutdown pushbutton on activation of the High (HH) level alarm on tank A. The barrier is defined as a single task assigned to and performed by the Control Room Operator.

- Understand the correct response action when the HH level alarm activates (i.e., the barrier trigger alarm) and barrier safety time (BST).
- How long ago did the alarm activate? How much time remains to complete the action response within the PST?
- What is the priority of this barrier relative to other barriers and safety critical alarms?

Example 2

Facility: Offshore production platform. **Barrier safety function:** When the general alarm activates, non-essential personnel (NEPs) and HEs use the designated egress/escape routes to transit to their assigned muster stations or emergency response stations and do so safely and promptly. (NEPs do not have roles in these barriers.) Required NEP and HE comprehension includes knowledge of the barrier procedures and how to respond to situations and location conditions that may interfere with these activities. This barrier is typically supported by weekly, all-hands drills whereby each person is expected to transit to their assigned muster or emergency response stations promptly and safely. The repetitive training allows each person to gain experience with the transit routes and movements, safe and potentially unsafe actions (e.g., an unauthorized shortcut), and other muster barrier tasks, and do so automatically in a manner that achieves the specified barrier response times and performance. Example awareness and comprehension needed to guide individual decisions and actions may include the following:

- The general alarm is activated. What is my response?
- What and where are my assigned primary and secondary muster stations or emergency response stations?
- Is sufficient time available to reach the preferred muster area?
- Is my current route passable? Conditions and considerations may include a route blocked by debris or damage, dense smoke prevents seeing what lies ahead, a detected threat is present (fire or an audible/visible high pressure gas leak), or an area toxic gas alarm beacon is active (lit).
- If the route is blocked or the designated muster area or emergency response station is not available, what are the remaining options? Which of those are viable and reachable?

*Case Study, Deepwater Horizon Accident: The muster alarm (muster barrier) was manually activated **after** the following occurred: well blowout in progress, several explosions, widespread fires, injuries and fatalities, the loss of all emergency lighting, and destruction of building sections and escape routes. This contributed to crew confusion and unsafe deviations from barrier procedures. For additional details, see Appendix M, Note 11. For insight into the actions, mental state, and the many challenges presented to mustering personnel, see Skogdalen et al. (2011).*

Example 3

Facility: An offshore production platform. **Situation:** A non-recoverable catastrophic event places personnel at acute risk; the facility will soon be uninhabitable. **Barrier function and safe state:** Abandon the facility using the primary evacuation method (if feasible) or one of the alternate escape methods (life raft, ladder to sea, etc.) and promptly move away to a safe distance from the facility. A **barrier task is** assigned to the barrier leader (i.e., the Offshore Installation Manager): *Assess conditions and decide if they warrant activation of the abandon alarm (e.g., activate the abandon facility alarm).*

- Do current conditions warrant abandoning the facility?
- How much time do I have to make this decision?
- How much time is needed to achieve the specified safe state, such as evacuating and moving personnel far enough away from the facility?
- Where are personnel currently located? Is everyone accounted for?
- How many injured people do we need to move?
- What is the status of the lifeboats?
- Is everyone wearing the required PPE and gear?
- What is the weather and sea state? How does that affect the evacuation process and personnel survivability?
- Once evacuated, how long before rescue? By whom?
- Who is nearby that can assist with rescue operations? How far away are they, i.e., what is the earliest arrival time?

- What hazards are attributed to the abandon process, such as the potential for delays, injuries, or fatalities?
- What notifications are needed? Examples may include notifications to the home office (emergency response center), a regulatory agency (major environmental spill in progress), or an external resource contacted to provide support to this operation, such as a local ship in the area or contracted helicopter support services.

3.2.12.1   Importance of Mental Models and Long-Term Memories

The view of IOGP (2012, p. 4),

> All three levels of SA involve significant cognitive complexity and rely heavily on what psychologists often refer to as the operator's 'mental model.' A mental model captures the operators understanding of how a system operates and how it behaves. We only really learn about the importance and limitations of mental models when major accidents happen.

They "are probably one of the single most important concepts in cognitive engineering." Mental models (MM) are long-term memory structures and content. MMs "are the mechanisms whereby humans are able to generate descriptions of system purpose and form, explanations of system functioning and observed system states, and predictions of future system states" (Rouse and Morris, 1985, p. 351).

Comprehension relies on having adequate mental models (unique to each person) that encompass the SA-1 information, where it resides (or how to access), what it means for the assigned task, and conditions that may impede that effort (under time pressure or the information source is obstructed by smoke). The required MM is the product of having the right experience (e.g., depth, duration, and applicability to the barrier) and knowledge (procedural, technical, and activity/task execution). The product of combining the MM with the SA-1 information should provide the needed understanding and comprehension. See Appendix F.5 for additional information on MMs and other forms of long-term memory.

3.2.13   STEP B-12, DETECT PHASE: PROJECTION/ANTICIPATION REQUIREMENTS

This step defines the minimum requirements (capability) to project or anticipate near-term, barrier-specific events or changes to conditions. Enter the SA-3 requirement into Table 3.19. Also, identify and populate the table with any new SA-1 information needed to realize each requirement.

Because time is often the single resource that places the greatest demand on HEs, it is important to recognize it as an essential aspect of SA. Endsley and Jones (2012, p. 19) explain,

> An often critical part of SA is understanding how much time is available until some event occurs or some action must be taken. The dynamic aspect of real-world situations is another temporal aspect of SA. An understanding of the rate at which information changes allows for the projection of future situations.

TABLE 3.19
Detect Phase: SA-3 Projection Requirements

| | | SA-3 Project/Anticipate Requirements | | | Supported Decision/Act Ph. Requirements | | | New Skills & Knowledge | Support Aid | |
									Aid Type	Function
Barrier ID	Task ID	ID	Requirement	Type and Description	ID	Description	SA-1 Info Required			
Enter barrier ID	Enter task ID	Enter unique SA-3 Req. ID	Enter project/ anticipate requirement	Enter type: *See examples*	Enter ID of supported act or decide phase req.	Enter description of supported requirement	Enter Req'd SA-1 Info (If any)	Enter new skills or knowledge required	Enter aid type (***Entered in process C5-1***) *One per row*	Enter aid function or purpose

Key Point: This SA-3 capability develops when advanced knowledge and extensive experience are gained by a person who becomes an expert in that domain. An inherent limitation, humans need time to perceive information, determine what it means, and develop a course of action. The pace of an emergency can easily exceed the human ability to match that pace and respond in kind. A HE having the required SA-3 competency automatically and frequently looks ahead to see what can happen and when it might happen. This ability to project and anticipate allows the HE to become aware and begin to plan a course of action based on what might happen. Without this capability, incidents and emergency scenarios can greatly outpace an HE's ability to respond at the rate needed to prevent escalation, achieve a rapid control/recovery, or mitigate the potential consequences of the hazard.

Note: The competency needed to meet an "expert-level" requirement may limit the range of personnel who can qualify for the identified task (role). The broad and deep experience, training, and knowledge (domain-specific) required to become a domain expert may require 10 years or more working in that domain. This applies to complex barriers, for example, the expertise needed by an emergency response barrier leader. A person who is an expert in chemical separation processes does not necessarily have the required competency if reassigned to a catalytic cracker or a cryogenic unit. (This may be a common misperception when considering moving senior operators between units.)

Populate all table fields guided by the information in the table and the following guidance.

Project/Anticipate Requirements and description examples:

- A capability to project how and how quickly conditions may escalate. An example of a toxic or flammable gas release/leak: anticipate the potential cloud size, concentration, location, and direction of movement.
- A capability to anticipate knock-on effects that may occur when a barrier action response is initiated.
- Anticipate workload spikes that can occur during emergency operations.

Project/Anticipate Type examples:

- Anticipate the effects of an action, for example, it activates an emergency shutdown function, sounds the general alarm, or activates the search and rescue barrier.
- Estimated timeline for an escalating hazardous event.
- Estimated time for the individual HE (or other on a multi-person barrier) to start or complete an action.
- Estimated timing on the potential loss of an additional resource, external support system, or external protective barrier.
- Remaining time available to provide medical aid or move an injured person in response to an encroaching hazard.
- Anticipate how an active hazard may affect personnel, for example, their response to fear, panic, focus, rational decision-making.

- Anticipate the direct and knock-on consequences of a failed or failing barrier element, external protective barrier, technical system, or external support system (e.g., a fuel system or an electrical power distribution and supply system).

SA-1 Info Required – Identify and enter the SA-1 information needed to support the SA-3 requirement.
New Skills and Knowledge (expert-level) examples:

- Advanced experience (skill-based) in the barrier-specific requirement.
- Advanced knowledge of the hazard, i.e., nature, potential for escalation, types of potential escalation, and knock-on effects on other systems and process units. This may include advanced knowledge of aspects such as organizational and regulatory regime procedures, process chemistry, fire chemistry and dynamics, and effectiveness of fire suppression agents under different use conditions.

Note: SA-3 requirements can only be met by an HE with expertise developed over many years. This limits the personnel who have the necessary competency to fulfill this role. See Appendix J for additional information on experts and expertise. See Appendix B.5.3 for a discussion of the challenges of time as a constrained resource.

A Look Ahead: the detailed design process C-5 further examines the SA-3 requirements. As mentioned, personnel with the required expertise may be limited. This process considers other options to meet the intent of the requirement, such as providing a projection or anticipatory display or tool as a support aid. For further on this topic, see Wickens et al. (2013, pp. 214–218).

The **Support Aid** fields are addressed in the detailed design phase process C5-1c. (The "alternative display type" is introduced as a support aid in the form of an HMI display element.)

3.2.14 Step B-13, Non-Technical Skills

This step defines the specific non-technical skills (NTS) required to maintain timely and effective barrier team functioning, interactions, and performance. Like the detect, decide, and act phase requirements, NTS are also required to achieve barrier/task safety functions and safe state within the specified safety time.

Flin et al. (2008, p. 1) identify the following NTS, described as "the cognitive, social, and personal resource skills that complement technical skills, and contribute to safe and efficient task performance."

1. Teamworking – *defined in this step*
2. Leadership – *defined in this step*
3. Managing stress – *defined in this step*

4. Situation awareness (see steps B-5, B-11, B-12, and B-14)
5. Communications (see steps B-7 and B-14)
6. Decision-making (see step B-10)
7. Coping with fatigue (see step B-26)

Multi-HE barriers are more complex because of the added communications and interdependencies between HEs. Many mitigation barriers and most or all emergency response barriers require two or more HEs. This expands the importance of NTS. From Johnsen et al. (2017), "There is a lack of non-technical skills such as communications and decision-making."

Also important is that

> If an incident occurs, the first minutes of the response are critical to escalation prevention and to the successful conclusion of the event.... Before personnel can go forward for formal assessment in emergency management, they first require training in handling emergencies at the scene and an appraisal of their capabilities under duress. Emergency management also requires specific qualities and skills, which are essentially different from those demanded by daily routine.
>
> *(OPITO 2014)*

Thus,

> Developing proper technical and nontechnical competencies are a critical part of assuring operational safety. Both are necessary, but neither alone is sufficient. In the operational safety context, key nontechnical competencies typically include situational awareness, leadership, teamwork, communication, decision-making, risk awareness, etc.
>
> *(SPE 2014, p. 14)*

Case Study: BP Thunderhorse Offshore O&G Facility Riser Failure. Though not a barrier, Crichton et al. (2005) provided insight and useful information on the non-technical skills needed to stand up an ad hoc team charged with responding to a sudden, unforeseen, and novel hazardous event. Thunderhorse was a newly constructed floating, offshore O&G drilling-production facility that was progressing through its initial start-up operations. A vertical riser (pipe) connected this facility to pipelines (flowlines) and production wells on the seafloor. The riser suddenly failed (severed) roughly at its midpoint causing a loss of containment event. It was initially unknown if the falling riser damaged critical equipment and pipelines on the seafloor. As an unforeseen event, there were no plans in place to guide the emergency response. An ad hoc team was formed from operations, technical specialists, and other disciplines and organizations to determine and execute a path forward. Leadership, communication, and teamwork were needed to develop these individuals into a functioning team that could effectively and more quickly respond to the incident. Members in this new multi-discipline, interdependent, and diversely located team also needed non-technical skills to effectively interact and function toward common objectives and priorities. An example communication issue occurred when technical specialists attempted to convey complex technical information to non-specialists. The use of unfamiliar terms, language, and technical knowledge impeded communications. The team needed to learn how they could exchange information and ideas in complex and unfamiliar areas. Team leaders discovered they needed different decision-making skills and abilities/methods for achieving and maintaining team shared understanding, objectives, and priorities.

3.2.14.1 B13-1, Teamworking

This step defines the teamworking requirements. Flin et al. (2008, p. 94) identify the following teamworking behaviors and capabilities:

- Coordinating with others
- Supporting others
- Solving conflicts
- Exchanging information

Teamworking is required within the barrier team and may also apply to HE interactions with additional resources (step B20-4) and others (e.g., NEPs). See Appendix C.4 for additional information to consider, such as the effect of trust on team behavior. The suggested process to populate *Table 3.20* is as follows:

1. See Table A.1 in Appendix A.2.3 for the suggested participants.
2. Review Appendix C and the pertinent barrier documents and tables generated to date, including task tables, standard operating procedures, philosophy, and design basis documents.
3. Identify teamworking capability requirements in each barrier task (if any). If the capability is task-phase–specific, also enter the phase (detect, decide, act). Enter the following information in Table 3.20:

- **Barrier** and **Task ID**.
- **Task Phase**, if applicable.
- **Teamworking Capability ID** and **Requirement**.

TABLE 3.20
Non-Technical Skills Requirements: Teamworking

Barrier ID	Task ID	Teamworking Capability		Task Phase	New Skills & Knowledge	Recommended Solution	
		ID	Requirement			Type	Solution
Enter barrier ID	Enter task ID	Enter capability ID	Enter capability: – Coordination – Support others – Solve conflicts – Exchange info	Enter task phase, if applicable: – Detect – Decide – Act	Enter required new skills and knowledge	Enter type: – OE – HE *See Table 3.21 for examples*	Enter the proposed implementation solution *See Table 3.21 for examples*

4. Repeat step 3 to identify all required capability requirements.
5. For each capability requirement, identify a recommended implementation approach. Table 3.21 provides examples, although other solutions are possible. The solution may reflect the owner/operator's philosophy and standardized approach. Under the **Recommended Solution** header, enter the recommended or proposed **Type** and **Solution**.
6. Repeat step 5 so that every capability requirement has a recommended solution, when warranted.
7. For each capability requirement and solution, identify if it requires new skills or knowledge. If so, enter that information in Table 3.20 as well.

3.2.14.2 B13-2, Leadership

This step identifies and defines the required leadership capabilities from the barrier leader and other HEs. To define these requirements, follow the applicable steps in B13-1 (teamworking) to develop and populate *Table 3.22* using the applicable information from *Table 3.23*.

Flin et al. (2008, p. 130) identify the following leadership behaviors and capabilities:

- **Use authority and assertiveness** – "Create a proper challenge and response atmosphere, by balancing assertiveness and team member anticipation and being prepared to take decisive action if required by the situation. The leader must know when to apply his or her authority to achieve the safe task completion" (Flin et al. 2008, p. 132).
- **Maintain standards** – The leader monitors the team's application of procedures, timely task completion, communication discipline, and so on, and intervenes when a deviation from standards may place the barrier at risk.
- **Plan and prioritize** – The leader develops and conveys plans and priorities as a tool to guide HEs in a manner that achieves task and barrier functions at the required level of performance. Plans may include selective delegation authority.
- **Manage workload and resources** – The leader understands his/her own and the team's capabilities (e.g., workload, skills) and how to manage and apply these resources to achieve the barrier function and required performance. This necessitates an understanding of the type and timing of workload demands and peaks that may be placed on each HE and managing peaks that may exceed individual capacity or capability. Management includes monitoring for stress and fatigue.

Flin et al. (2008, p. 144) provide additional guidance on leadership characteristics and competencies. Furthermore, see Appendix C.4 for additional information to consider, such as the effect of trust on team behavior.

Also important is that "Under stressful conditions charaterised by time pressure, risk, dynamic conditions, high information load and uncertainty, team performance has been linked to the leader's effectiveness" (Flin et al. 2008, pp. 142). "Leaders of emergency

TABLE 3.21

Team Working Non-Technical Skills Examples

Examples of Teamworking Implementation Solutions

Teamworking Capability	Select Solution (Element Type)	Define Solution	Example Methods to Verify Capability to Meet Requirement
	OE – procedures	Define coordination requirements: coordinate with who for what purpose, timing, action	Test knowledge of procedures
	OE – team-based training, drills, and exercises	Develop and perform drills designed to develop implicit coordination (see Appendix C.3 for background)	Monitor for reductions in explicit coordination, e.g., communication frequency and duration while maintaining or improving team coordination and performance
Coordinate actions with other HEs (For reference, see Flin et al. 2008, p. 98)	OE – staffing/organizational chart	Maintain consistent team staffing (A stable team make-up contributes to higher team performance.)	Verify barrier roster changes are within defined metrics
	OE – procedures recommend HEs to monitor radio communications	Define the use of passive radio monitoring to monitor the activities of others while reducing the need for explicit coordination (e.g., communication)	Observe/verify behavior
	PE – employ shared displays to maintain team awareness	Specify use of shared displays that provide information that guides team activities while reducing the need for explicit coordination (e.g., communication)	Observe/verify behavior
Support other HEs (For reference, see Flin et al. 2008, p. 96)	OE – procedures	Define support requirements and expected supportive behaviors	Test knowledge of procedures
	OE – team-based training, drills, exercises	Learn/gain experience on/with support actions in a dynamic environment that support team functioning and performance	Observe/verify behavior
	HE – staffing	Maintain consistent team staffing (a stable team make-up contributes to higher team performance)	Verify barrier roster changes are within defined metrics

(Continued)

TABLE 3.21
(Continued)

Examples of Teamworking Implementation Solutions

Teamworking Capability	Select Solution (Element Type)	Define Solution	Example Methods to Verify Capability to Meet Requirement
Solve conflicts (For reference, see Flin et al. 2008, pp. 96–97)	OE – procedures	Define expectations for conflict resolution and mutual cooperation.	Observe/verify behavior Tabletop exercises
	OE – team-based training, drills, exercises	Learn/gain experience on/with support actions in a dynamic environment that supports team functioning and performance	Observe/verify behavior
	OE – staffing/organizational chart	Maintain consistent team staffing (A stable team make-up contributes to higher team performance.)	Verify that barrier roster changes are within defined metrics
	OE – procedures	Define information exchange requirements (exchange information, timing, HE to-from) and the methods used to achieve that exchange, e.g., communications (see step B-7), technical systems (e.g., displays)	Test knowledge of procedures. Observe/verify behavior
	OE – team-based training, drills, exercises	Experience with information exchanges that are essential to team functioning and performance in dynamic environments	Observe/verify behavior
Exchanging Information (For reference, see Flin et al. 2008, pp. 97–98)	OE – staffing/organizational chart	Maintain consistent team staffing (a stable team tends to require fewer and briefer exchanges.)	Verify barrier roster changes are within defined metrics
	PE – employ shared radio systems to maintain team awareness	Specify use of shared radio channel (monitoring) to support coordinated actions	Observe/verify behavior
	PE – employ shared displays to maintain team awareness	Specify use of shared displays to exchange information (see Appendix C, Meta SA for one possible application/purpose)	Timing and frequency of display access and apparent information acquired and applied

TABLE 3.22
NTS Requirements: Leadership

Barrier ID	Task ID	Leadership Capability			Task Phase	New Skills & Knowledge	Recommended Solution	
		ID	Requirement				Type	Solution
Enter barrier ID	Enter task ID	Enter capability ID	Enter capability: – Use of authority – Standards – Plan/prioritize – Manage workload and resources		Enter task phase, if applicable: – Detect – Decide – Act	Enter required new skills and knowledge	Enter type: – OE – HE *See Table 3.23 for examples*	Enter the proposed implementation solution *See Table 3.23 for examples*

TABLE 3.23

Leadership Capability Requirements and Solution Examples

Examples of Leadership Implementation Solutions

Leadership Skills/ Knowledge	Select Solution (Element Type)	Describe Solution	Example Methods to Verify Capability to Meet Requirement
Use of authority and assertiveness (For reference, see Flin et al. 2008, p. 132)	OE – procedures	Define the leader's responsibilities and authorities to direct team, but also maintain a work environment that supports open feedback	Competency assessment to verify knowledge of procedures
	OE – team-based training, drills, exercises	Use training drills and exercises to practice and demonstrate the leader's use of authority and maintain environment for team feedback. *This may also apply to other HEs.*	On defined criteria, assess the use of authority and the team environment that use creates
	OE – industry-certified training and assessment program	Enroll barrier leader in an industry-provided program specific to this competency requirement (e.g., OPITO 2014) *This may also apply to other HEs.*	Program assessment metrics
	OE – staffing/hiring	Select/hire personnel with the required leadership capabilities	Employ personality and leadership test when selecting barrier leaders
Providing and maintaining standards (For reference, see Flin et al. 2008, p. 132)	OE – procedures	Define leader responsibilities for maintaining team use of accepted practice and procedures, e.g., radio communication discipline and protocols.	Competency assessment to verify knowledge of procedures
	OE – industry-certified training and assessment program	Enroll barrier leader in an industry-provided program specific to this competency requirement (e.g., OPITO 2014) *This may also apply to other HEs.*	Program assessment metrics
	OE – team-based training, drills, exercises	Use training drills and exercises to practice and demonstrate supervision and monitoring ability to maintain team alignment to accepted practice and procedures	Observe/verify behavior against expectations. Review feedback from other HEs.

Planning and prioritizing (For reference, see Flin et al. 2008, p. 133)		
OE – procedures	Define the leader's role and responsibility for prioritizing, planning, and adjusting team activities to achieve priorities and barrier functions	Competency assessment to verify knowledge of procedures
OE – team-based training, drills, exercises	Use team training, drills, and exercises to practice and demonstrate the ability to appropriately prioritize and plan response to changing feedback results, observed behaviors and performance, changing conditions and time pressures	Assess the timeliness and appropriateness of the expected planning and prioritization responses.
OE – industry-certified training and assessment program	Enroll barrier leader in an industry-provided program specific to this competency requirement (e.g., OPITO 2014) *This may also apply to other HEs.*	Assess the timeliness and appropriateness of the expected planning and prioritization responses.
OE – staffing/hiring	Select/hire personnel with the required aptitude and leadership capabilities	Employ aptitude and leadership tests when selecting barrier leaders
Manage workload (own and others) and resources (For reference, see Flin et al. 2008, pp. 133–134)		
OE – procedures	When not adequately defined in HE requirements, procedures define the barrier leader's expectations and actions	Competency assessment to verify knowledge of procedures
OE – team-based training, drills, exercises	Use team training, drills, and exercises that require the barrier leader to monitor team for potential work overload that can lead to mistakes or degrade barrier response. In response, observe the leader's attempt to adjust resources accordingly where feasible.	Observe/assess leader monitoring for workload and assess the adjustments made to account for peaks
OE – team-based training, drills, exercises	Use team training, drills, and exercises that require the barrier leader to develop and demonstrate behaviors and capability for managing resources	An expert monitors and assesses the barrier leader's actions and performance against requirements
OE – industry-certified training and assessment program	Enroll barrier leader in an industry-provided program specific to this competency requirement (e.g., OPITO 2014) *This may also apply to other HEs.*	Program assessment metrics
OE – staffing/ organizational chart	Select/hire personnel with inherent leadership attributes and behaviors.	Employ personality and leadership test when selecting barrier leaders

response teams must be able to change leadership style in response to a fast-moving situation. That is, they must have the ability to switch to a more directive style rather than consultative in response to situational demands" (Flin et al. 2008, pp. 142).

3.2.14.3 B13-3, Monitor and Manage Acute Stress

This step defines the "managing and coping with acute stress" capabilities HEs should have and demonstrate. To define these requirements, follow the applicable steps in B13-1 (teamworking) to populate *Table 3.24*. For step 5, see Appendix F.4 for suggested guidance and methods to achieve the stated requirements.

Flin et al. (2008, p. 158) identify the following capabilities to manage and cope with acute stress:

- Recognize and monitor acute stress (in self and others)
- Manage and cope with acute stress (self)

3.2.15 Step B-14, Detect Phase: SA-1 Information Requirements

Table 3.2 identifies the documents that define SA-1 information requirements. This step progresses the SA-1 design by confirming its requirement source, access locations, presented form, and the In-Place PE needed to capture or acquire and present that information. Enter this information into *Table 3.25*.

Note: See Appendix L for the unique requirements that apply to a remote barrier support task.

The following are the suggested activities in this step:

1. Review the applicable tables and documents listed in Tables 3.1–3.3 for SA-1 information requirements (barrier required).

TABLE 3.24

NTS Requirements: Monitor and Manage Acute Stress

Barrier ID	Task ID	Monitor/Manage Capability		Task Phase	New Skills & Knowledge	Recommended Solution	
		ID	Requirement			Type	Solution
Enter barrier ID	Enter task ID	Enter capability ID	Enter capability: – Recognize and monitor acute stress in self and others – Manage acute stress in oneself	Enter task phase, if applicable: – Detect – Decide – Act	Enter required new skills and knowledge	Enter type: OE, HE	Enter the proposed implementation solution *See Appendix F.4 for examples.*

TABLE 3.25
Detect Phase: SA-1 Information Requirements

See Table 3.25 Extension

SA-1 Info	Requirement Source		Target HE/non-HE	SA-1 Info Function	Access Location(s)	SA-1 Info ID (Presented Form ID)	ID of SA-1 Info Source
	Barrier ID	Task ID					
Enter SA-1 Info. Req.	Enter barrier ID	Enter task ID	Enter target HE/non-HE ID *See examples*	Enter Info function *See examples*	Enter all access locations *See examples*	Enter unique ID for tagged SA-1 info element	Enter ID of originating SA-1 source *See examples*

Table 3.25 Extension

Presented Form	Detect Senses	New Skills & Knowledge	Support Aid		Required In-Place PE?
			Type	Function	
Enter form See Table 3.26 and entry detail below	Enter sensory input type *See examples*	Enter required new skills and knowledge	Enter support aid type *(Assessed in process C8-1c)*	Enter aid function or purpose	Enter ID for In-Place PE

2. From the review from step 1, enter the following information into Table 3.25:
   - **SA-1 Info**. Enter the information as it is described/identified in the source table or document.

     Note: If the SA-1 information is conveyed in an in-person or device-enabled communication, include the information from Table 3.11 or add a note that refers to that table for that information.
   - **Barrier ID**: Enter barrier requiring the SA-1 information.
   - **Task ID**: Enter task requiring the SA-1 information.
   - **Target HE/non-HE**: Enter the ID for the HE or non-HE ID requiring the SA-1 information.
   - **SA-1 Info Function**: Enter the function of the SA-1 information. (Refer to Table K.1 (Appendix K.6) for the standard function types.)
   - **Access Locations**: Enter all locations where the HE may require access to the SA-1 information. Example access locations include the following:
     - Process area
     - Central Control Room
     - Egress/escape route to Muster Station A (north route)
     - Process area (north side of exchanger B-500, electrical building A)
     - Muster/rally area A, north side
     - Emergency response station A-ES-7
     - Incident Command Center (e.g., information on a hand-written whiteboard)
     - Electrical room
     - Remote Operations Center (supports remote barrier support function)

   Note: Table 3.25 identifies distinct SA-1 information items that may have many users, support different functions, and require access from various locations. This affects design. The presented form may be guided by the number of HE users, their locations, the situation of use, and how that information is used to support the barrier/task function. This information may affect the design of a communication exchange or an HMI display and its supporting or enabling technical system.

3. **SA-1 Info ID (Presented Form ID)**: If the SA-1 info is a tagged item, enter the unique tag. Variants of this tag may be required if the information is presented in several forms and locations.
4. **SA-1 Info ID – Info Source**: Enter the originating source of the SA-1 information, for example, instrument transmitter, CCTV camera, person, or incident scene.
5. **Presented Form**: Enter the form of the presented information guided by Table 3.26.
   a. Enter "non-VDU" or "VDU-based"
   b. Enter the presented display type, for example, "Passive indicator," "Live Video," or "HMI display"
   c. Enter the form detail, for example, "Windsock," "CCTV Source," or "Third Party HMI Display"

TABLE 3.26

Categories: Presented SA-1 Information Forms

Non-VDU	VDU-Based
Passive Indicator Windsock, signage, physical labels, painted markings (e.g., use of reflective paint to denote boundary or warning), mechanical device (e.g., valve position indicator)	**VDU – Video Display** Live or recorded video from CCTV, weather, radar, and satellite TV systems
	VDU – Video-Based Communications, e.g., live video conferencing system
Simple Active Device Integral readout indicator mounted on a local active instrument, lamps, beacons, or strobe, use of backlighting to indicate state of a pushbutton or multi-position switch, alarm horn (different tones), area speaker (recorded or live voice messaging	**VDU – COTS Display** Email, standard network-based, office suite displays, e.g., electronic call-up of stored procedures, spreadsheets, reports, or video files. Other VDU-based display systems that provide non-configurable options.
Public Alarm/Public Address Notification System Area- or facility-wide alarm notification. The alarm form may be audible (e.g., a horn or siren) or visual (e.g., color-coded beacons). May be combined with a live or recorded message announced over area speakers.	
Communications (device-enabled) Telephone, radio, conference calls (without video)	**VDU – Engineered Display** Third-Party HMI Displays: Vendor options may limit the range and number of changes to add new information to an existing display, add a new display function (e.g., alarm), or minor limited changes to display presentation or function
Communications (face-to-face) Spoken, language, words used, tone, urgency, or fear in voice. Visual image of person: body language, appearance, location, position, visible fear, signs of injury.	
Incident scene Visual indication of damage to equipment or loss of containment Visual indication of fire, smoke, electrical arcing Visual (person): body language, location, injury, appearance, hand gestures, facial expressions (fear/anxiety, anger) Weather: rain, fog, ice, icy surfaces, waves Auditory (person): verbal, tone, urgency, fear Auditory: ambient explosion, gas release, jet fire Olfactory: smell gas, oil, ozone, etc. Heat/cold: fire, cryogenic release, weather-related Location of personnel, signs of injury, proximity to danger	HMI Displays: Fully configurable (customized) displays purposely developed to provide SA-1 information in a form, format, and presentation that directly supports the function required by the SA-1 information (e.g., directly supports) and SA-2 comprehension or decision support requirement. HMI displays include displays (and display systems) that provide a full or partial range of customization.
Other Sources Hand-written on incident command board. (This is not included as a passive indicator because of the complexity and options available in how the information is arranged and presented. As such, it is not a "simple" display.)	

Note: COTS – Commercial Off-the-Shelf.

The categories in Table 3.26 are normalized and consistently employed throughout the lifecycle model. The "presented form" entered in Table 3.25 is the direct-use physical element (or object) that is the actual form the target user is expected to view or access to acquire this information. (The term "direct-use" is formally defined in Section 2.1 and first introduced in step B-6.) Examples of sensed information are visual access to a gauge or HMI display and visual/audible access from an in-person communication exchange.

Detect Senses: Identifies the HE sensory mode to detect the presented information. Examples are as follows:

- Visual, for example, text, graphical, color-coded lamps, body language, observation of location conditions
- Auditory, for example, horn tones or ambient sounds
- Audible, for example, voice
- Tactile, for example, vibration intensity, type, potential source, or cold vapor on skin
- Smell, for example, hydrogen sulfide, electrical arcing, or smoke
- Taste

Note: The same SA-1 information may be presented in different sensory forms to improve detectability and achieve mode redundancy, which would enable addressing environmental interferences and sensory channel saturation.

New Skills and Knowledge: Identify specialized skills and knowledge (if any) required to use, access, or gain access to this display.

The **Support Aid** fields are addressed in the detailed design phase processes C7-2c and C8-1c.

Required In-Place PE, formally defined in Section 2.1 and first introduced in step B-6: enter the physical element that must be in place to make the SA-1 information available and accessible to the HE user (e.g., a technical system or lighting provided as an external support system). See Tables 3.9, 3.12, 3.16, and 3.28 for example sources of this information.

Case Study, HOSL Buncefield Petrol Tank Overfill and Fire (COMAH 2011, pp. 9–14, 16–19, 31): on the day of the accident, a failure in the tank gauging system (a frequent occurrence) caused the loss of the real-time tank level data and level alarms the control room supervisors had come to rely on. In addition, the supervisors were unaware the audible alarm and a shutdown function triggered by the independent tank level switch alarm were disabled. Both were assessed to be primary contributors to the tank overfill, fire, and spill accident. The supervisors also did not have access to real-time data from the incoming pipeline. At a time when the tank was approaching its fill limit, without notice the pipeline company increased the pipeline flow rate to the tank from 550 m³/hr to 900 m³/hr. This reduced the time when the tank would reach and exceed its maximum capacity. Another factor that increased the confusion in the control room, a deficient shift handover did not provide the new shift with a clear understanding of which pipelines were flowing to each tank. (For further information on both failures, see Chapter 7, processes E2-5 and E4-3).

(Increasingly common in large depots, control room operators are supported by a large overview display that shows every pipeline, tank, equipment lineup, and all critical, real-time process and alarm data. Displays of this type more effectively support the operators' need to maintain a complete view of all major facility operations, event changes, and alarms.)

Case Study Background. *The Hertfordshire Oil Storage Ltd (HOSL) located in Buncefield, UK, experienced a catastrophic explosion and fire when a storage tank overflowed 250,000 liters of petrol. In the windless night, a flammable vapor cloud formed that encompassed the overflowing tank, a nearby kerosine tank, and a car park. Believed ignited by a starting firewater pump, the cloud ignition resulted in an unusually destructive explosion. From COMAH (2011, p. 11), the "ensuing fire, the largest seen in peacetime UK, engulfed over 20 tanks on the HOSL and adjacent sites and burnt for several days...Fuel and firefighting chemicals flowed from the leaking bunds down drain and 'soakaways', both on and offsite." Forty people were injured (no fatalities).*

Under normal operations, the depot received batched fuel deliveries through three pipelines, two controlled by other companies. In response to an ever-increasing workload over many years, supervisors became increasingly reliant on system-generated alarms to guide tank filling operations. An automatic tank gauging system (ATG) captured and displayed real-time tank level data and generated tank level alarms at different tank levels. Control room supervisors shared a single ATG workstation and monitor that was limited to displaying data one tank at time. The system failed 14 times in the months that preceded the accident. When that occurred, the tank level data no longer updated (froze). The failure effectively disabled level alarms that relied on the level data updates. The control room did not have real-time access to the current flow rate for two of the pipelines. Getting updated information required a call to the pipeline company. The tank filling rate changed when the pipeline company changed the incoming pipeline flow rate, occasionally without notice. The level was also affected if the tank was simultaneously discharging product to the depot's tank truck loading facility. This could occur at the same time the tank was receiving product through a pipeline. Supervisors used a small, home-type alarm clock to track pipeline product interfaces and, on occasion, to provide an audible reminder to check the tank level.

Two barriers were in place to prevent a tank overfill and an overflow loss of containment event. 1) The supervisors monitored the tank level data and took the necessary actions needed to prevent a level exceedance. 2) Tanks were also supplied with an independent level switch. A level that reached the level switch setpoint activated a control room alarm and initiated a shutdown function that stopped all inflow to the tank. Both barriers failed the night of the accident.

3.2.16   Step B-15, Barrier/Task Function Allocation

This step performs a functional allocation process that selects, allocates, and assigns barrier tasks to personnel (HE's or roles), and functions within each task to an HE or a physical element (e.g., technical systems). The process develops the barrier roster showing the allocation of tasks to HEs. The process creates and updates a functional block diagram that guides and supports the process and provides a visual representation of key task activities and interfaces. Specifically, the diagram shows the assigned HE (or role), key task functions, each HE-PE interface (identifies the direct-use PE, use, and context), and the In-Place PEs that directly connect to and enable the direct-use PE.

The allocation objectives for this process are as follows:

1. Multiple-HE barriers: Logically distribute barrier tasks between HEs. Considerations may include allocating (grouping) tasks based on work type, location, or areas of personnel expertise. (This process may also be guided by owner/operator practice standards and/or regulatory requirements.)
2. Allocate task functions to limit the potential for excessive HE workloads (base and peak loads) that contribute to barrier failure. (For background on the effects of excessive workload, see the index references for the term "workload").
3. Allocate task functions to the HE or PE(s) considering the capacities and capabilities inherent to that element, such as functionality, response performance, and reliability.
4. With Item 3, consider the potential need (or opportunity) to introduce flexibility to respond to a wide range of plausible or unforeseen situations. (Element selection may be guided by a need to introduce a resilience element in the task design. A HE may be able to make dynamic adjustments and responses to conditions and situations that were not considered in the design.)

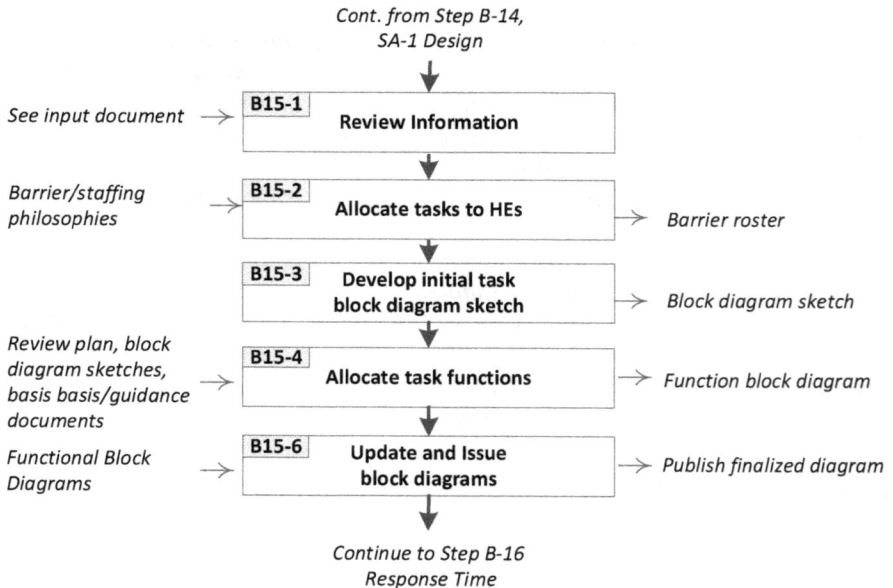

FIGURE 3.5   Preliminary Design Step B-15: Function Allocation Process.

5. Using prior experience from the owner/operator organization, the allocation process should be aware of and consider any HE-HE and HE-PE interface issues that are known problems in existing barriers.

Note: Prior steps focus on barrier elements and components. This step expands the design focus that views a barrier task as a system of interconnected elements. The diagram created by this process (Figure 3.5) provides an important visual tool that helps the user understand the nature and complexity of the task that may be driven by the number of task locations and HE interfaces (interface types and locations).

Note: Previous design steps are a form of progressive functional analysis. This step provides a way to step back and address the functional design process from a broader systems perspective. It presents one of many demonstrations of how the prototype lifecycle model aligns to the nine key principles in ISO-11064-1 (2000). In this case, it demonstrates alignment with Principle 3 (improvements using an iterative design process).

3.2.16.1   B15-1, Review Information

Review the following information for requirements, constraints, and considerations that apply to this process.

- Allocation objectives
- Philosophy documents from the conceptual design phase
- Design basis documents developed in step B-1
- Applicable recommendations from the risk assessments in step B-2
- Task analysis and report from step B3-1
- Barrier and task requirements developed in steps B4 to B-14.

These references include information on allocation decisions made to date. In such cases, this process should verify those allocations are appropriate and accepted.

3.2.16.2   B15-2, Allocate Barrier Task to HE Roles

Guided by the information from step B15-1, this step allocates barrier tasks to HE roles. One or more barrier tasks may be assigned to a named role. The product of this step is an initial barrier roster that identifies all barrier roles. (This work is reviewed and, if necessary, updated in step B-25.)

3.2.16.3   B15-3, Develop Barrier/Task Function Block Diagram (Sketch)

This step develops a draft sketch that shows the items identified in the prior step, element interfaces, and physical locations. It provides a "best guess" functional allocation given the input information. Depending on the barrier, the sketch may be effectively developed by a single discipline with inputs from others as required. Figure 3.6 is an example of this diagram.

Note: The form and level of detail in the sketch may be guided by the active human barrier philosophy and design basis documents. Keeping the end in mind, the diagram content, terminology, presentation conventions, and level of detail should be

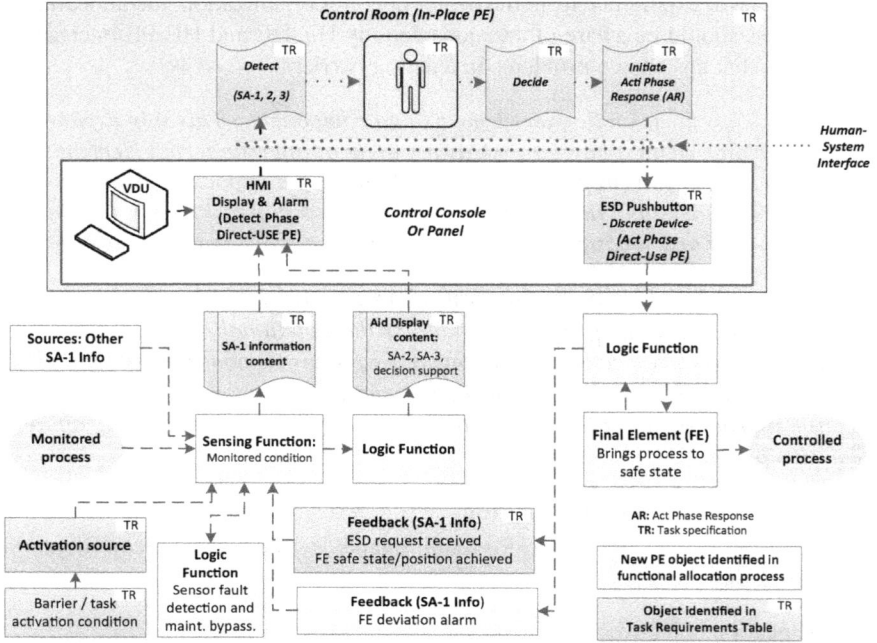

FIGURE 3.6 Example: Barrier/Task Function Block Diagram; TR, Task Requirements Tables.

guided by how and by whom the information will be used in later design processes and other barrier lifecycle phases.

Note: The block diagrams from a prior project may be a useful starting point. With many barriers, the diagrams may share a high degree of similarity between projects.

3.2.16.4   B15-4, Allocate Task Functions

Facilitated by the block diagram developed in step B15-3, this step allocates every task function to an HE or a named PE. The diagram should be corrected (marked/redlined) during the process if a required function is missing or not correctly represented. The suggested product of this step is a set of finalized barrier/task block diagrams that accurately reflect the collective and mutually agreed input from operations specialists and others who may be responsible for the selection and design of elements indicated in the diagram.

The following are example questions to consider as this process progresses:

- Does the diagram correctly identify and show all required functions?
- Are the allocations consistent with published and approved philosophy and design basis documents?

- Do the indicated direct-use physical elements adequately support the task assignee (HE) performance of the task?
- Are the required In-Place physical elements (e.g., technical system) identified and shown correctly?
- Are interfaces correctly shown?
- Can the indicated elements (human and physical), configuration, allocations, and interfaces/interactions reliably achieve the indicated function within the expected safety/response time?

The suggested process may be highly structured, i.e., guided by an experienced process facilitator. Alternatively, the process can be managed by the discipline responsible for the development and management of the document. It may also be useful to develop and approve a Functional Allocation Review Process plan before proceeding with this review. The example contents of the plan may include the following:

- Draft block diagram sketches and other documents proving inputs to this process
- Required and supporting participants (See Appendix A.2.3, Table A.1 for the suggested participants in this process.)
- Assessment planning and execution documents (The planning documents developed in step B-1 and guidance in Appendix A.2.)
- Assessment methodology and guidance documents (For example, see ISO 11064-1: 2000, Section 7.3 for guidance on the process of allocating functions to humans and/or machines for guidance.)
- Requirements for finalizing and publishing the diagram and allocation process report

Note: This process may create recommendations that change prior documents, like a task phase requirement. Processes should be in place to manage and process these recommendations. Though this is not intended to be a design validation process, recommendations of this type may result.

3.2.16.5 B15-5, Update Block Diagrams for Review and Issue
This step updates the block diagrams to include the approved input from step B15-4. The updated block diagrams should proceed through the normal review and approval stages before the final issue.

3.2.17 STEP B-16, SAFETY TIME, BARRIER, TASK, AND PHASE

This step evaluates and specifies the barrier safety time (BST) and allocates this time to barrier tasks and task phases. Figure 3.7 summarizes the activities within this step.

3.2.17.1 B16-1, Barrier, Task, and Task Phase Safety Times
This step defines the design basis barrier safety time (BST). The risk assessments that identified the barrier (step B-2) evaluated and identified its process safety time (PST), as recorded in Table 3.5. The PST is an upper BST limit. Table B.11 (Appendix B.7.4)

Cont. from Step B-15,
Functional Allocation

Table 3.5 $\rightarrow$

| B16-1 | **Specify Barrier Safety Time (BST)** | $\rightarrow$ *Table 3.27 (BST)* |

| B16-2 | **Specify Target Task Safety Time (TTST)** | $\rightarrow$ *Table 3.27 (TTST)* |

Table 3.27 (BST)
Table B.11 (Appendix B.7.4) $\rightarrow$

| B16-3 | **Specify Target Task Phase Safety Times (TPST)** | $\rightarrow$ *Table 3.27 (TPST)* |

Table 3.27 (BST, TTST)
Table B.11 (Appendix B.7.4) $\rightarrow$

| B16-4 | **Review safety times** |

Table 3.27 (BST, TTST, TPST) $\rightarrow$

Continue to Step B-17,
Identify Dependency - ESS

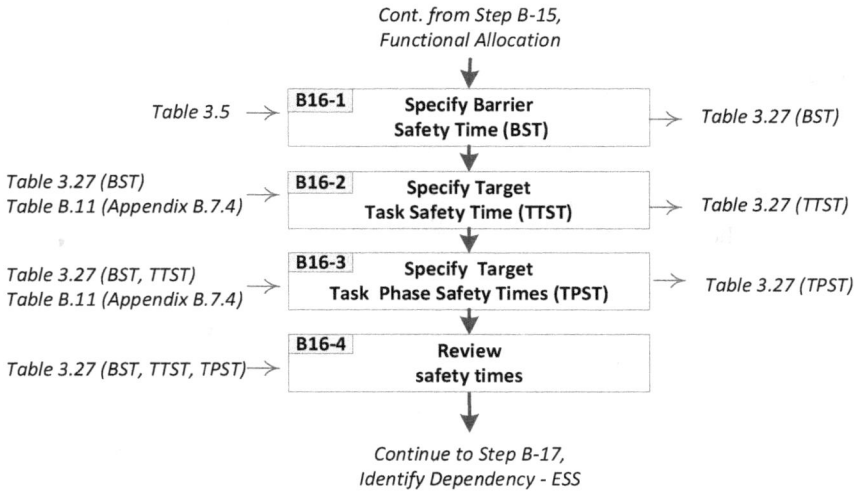

FIGURE 3.7   Preliminary Design Step B-16: Define Barrier, Task, and Phase Safety Times.

summarizes the documents that may provide additional input and guidance to this and the remaining steps. Record the barrier ID and the specified BST in *Table 3.27*. The following should also be considered:

1. An emergency response barrier often warrants a time-of-the-essence response, i.e., select the shortest feasible safety times. (This contributes to a time contingency budget that should be included to account for unknown events.)
2. The hazard event that activated a barrier is an unplanned event. Responders need time to shift their attention and focus to this new event. Further, the time needed to begin the barrier response and complete all required actions may be affected by the local conditions and unique situations presented to each responder.
3. The HE may be performing a task that is already in progress and has an equal or greater priority to the newly activated barrier or barrier task. Situationally, the HE may be missing or is no longer available (e.g., of an injury or fatality caused by the hazard), or a sudden illness or departure from the company.
4. With fully automated safety instrumented functions (SIFs) guided by IEC 61508 and IEC 61511-1, the recommended practice specifies that the safety time be half the process safety time. This provides a margin that accounts for potential inaccuracies in the PST and physical element performance variations.
5. The specified BST should be achieved reliably and repeatably. Considerations must also be given to plausible situations where the HE is assigned to two or more barriers that may be active at the same time.
6. The available safety time may be constrained by the endurance time of a barrier-dependent external protective barrier or the continued functioning of an external support system or "additional resource."

TABLE 3.27
Barrier, Task, and Phase Safety Times

Barrier		Task		Detect Phase				Act Phase		
ID	Barrier ST (BST)	Task ID	Target Task ST (TTST)	Sense/ Notice	Detect	Comprehend and Project / anticipate (SA-2, SA-3)	Decide Phase	Initiate Act Response	Achieve Safe State	Sum of Phase STs
Enter barrier ID	Enter barrier safety time	Enter task ID	Enter TTST (minutes)	Enter T_s Time (minutes)	Enter T_d Time (minutes)	Enter T_{cp} Time (minutes)	Enter T_{dec} Time (min.)	Enter T_{ai} Time (minutes)	Enter T_{ar} Time (minutes)	Sum of $T_s + T_d + T_{cp} + T_{dec} + T_{ai} + T_{ar}$

Note: An incident that triggers an emergency response (ER) barrier, such as a fire or explosion, warrants a prompt HE response to begin actions to gain control of the event and limit its threat to personnel, the environment, and the facility. A failure to do so enables new hazard escalation pathways that may increase at an exponential rate. Both can rapidly increase the danger and risk. Now, consider the HE response. The hazard event is unexpected and may even trigger a startle response. (See Figure F.1 in Appendix F.1.) The HE needs time to disengage from what he/she was doing and begin to access information and comprehend what it means and what is happening. The barrier system design and the known human limits on cognitive capacity and throughput constrain how quickly this occurs.

Humans are also limited in their ability to accurately track and understand events that change in a non-linear way or at an exponential rate (Reason 1990, pp. 92–93). It affects the person's (HE) response and responsiveness. Design options to overcome these challenges may include adding proactive response actions, support aids, pre-loaded plans, and planning, and exploiting SA-3 capabilities (e.g., the ability to anticipate and project how and how quickly the situation may evolve). Another option takes a time-of-the-essence (TOE) approach that seeks to implement and achieve the shortest possible barrier/task/phase response times, irrespective of the available barrier safety times. The approach seeks to create a time contingency budget that allows for unplanned-for and unforeseen events that create new workload demands. Example events include the simultaneous activation of several barriers (same HE) or an unexpected hazard escalation that requires diverting efforts to or from other barriers.

A different situation may be the result from an unsafe deviation in a procedure that dramatically reduces the actual process safety time from the PST identified in the risk assessment. (Refer to the Deepwater Horizon accident information in Appendix M. A barrier procedure specified directing the well flow from a major kick to a divert overboard line. Instead, it was manually lined up to a vessel that had a vent line directed toward the drill floor and possible ignition sources. The procedure deviation dramatically reduced the time interval from when the kick was first detected to the development of a massive flammable gas cloud that encompassed one or more ignition sources.)

Note: Existing processes may fail to assess and specify a BST. An active barrier not designed according to a specified safety time should not be claimed to be or classified as a "barrier" in a Hazard and Operability Study (HAZOP) or shown as such on a bowtie diagram. A safety function of this type is at most a safeguard.

3.2.17.2   B16-2, Target Task Safety Time

This step allocates the barrier safety time (BST) to its individual tasks. The term used here is target task safety time (TTST). Barrier tasks may be executed sequentially, in parallel, or as a mix of both. Allocate the BST period to the longest task string. For guidance, consider the suggested act phase safety time recorded in Table 3.9 and other documents listed in Table B.11. Then, record the task IDs and specified TTSTs in Table 3.27.

Note: This step may identify a task or a series of connected (interdependent) tasks that cannot be completed within the BST. This may require a return to earlier processes to seek a path forward that reduces the time, which may include changing the BST value, a PE/HE function allocation, or the distribution of barrier tasks among several HEs.

3.2.17.3   B16-3, Task Phase Safety Times

In this step, the target task safety time (TTST) is distributed (allocated) to the phases within each task. Each allocated time is referred by the term task phase safety time or *TPST*.

From CHIEF (2016, p. 51),

> The standard for successful performance of the barrier should be defined. For example, criteria could be:
>
> > Time to detect an event or situation expected to trigger the barrier.
> > Time to initiate an intervention.
> > Time to complete an intervention.

To determine the TPST, it may be helpful to start with phases that have known or obvious time constraints. This varies by barrier type. In practice, the decide phase often takes longer that envisaged in the design process. A task may have two or more SA-1 detection/perception actions (the task activator and other SA-1 information) or act phase responses that (most likely) must occur consecutively/sequentially. In this case, the allocated time must be sufficient to reliably complete both phase activities within the assign phase time. It may be prudent to consider the most onerous (and plausible) cases and situations when assessing and allocating the phase time. Regardless of case, only a single time value is entered for each phase. On completion, enter each phase time value into Table 3.27.

1. T_s: maximum time allocated to sense or notice the unsafe condition or task activator

   If the function relies on an instrument sensor/alarm to detect and notify the HE about the unsafe condition, T_s may be a few seconds. However, the time may be considerably longer if an HE performs the detect function. Input from several disciplines may be needed to assess this time.

2. T_d: maximum allocated time for the HE to detect and become fully aware of the notification/barrier activator.

   Case Study, Deepwater Horizon Accident: *Several active human barriers failed because a kick was not detected for 49 minutes. A decision to employ simultaneous operations (SIMOPs) to save time severely degraded the detection process. (A kick is an uncontrolled well flow.) Had a barrier or detect phase safety time been established, perhaps that might have triggered a change management process to assess the effect of the proposed SIMOPs on the kick detection response performance and reliability.*

3. T_{cp}: maximum allocated time for the HE to comprehend the notification meaning (SA-2) and anticipate what may happen next (SA-3) when determining the active phase response plan.

   See Tables 3.18 and 3.19 for the SA-2 and SA-3 requirements.

4. T_{dec}: maximum allocated time to complete all required decisions and plan the response action.

   See Table 3.17 for the decide phase requirements. In practice, this period often consumes much barrier BST. (See considerations and examples in step B-10 for background.)

5. T_{ai}: maximum allocated time to *initiate* the selected act phase response plan. The time starts at completion of the decide phase, T_{dec}.

   See Table 3.9 for the act phase response requirements.

6. T_{ar}: maximum allocated time to achieve the barrier/task safe state. See Table 3.9 for the act phase response physical element requirements.

***Case Study: PG&E Natural Gas Transmission Pipeline Rupture and Fire** (NTSB 2011, pp. x–xii, 1, 14–19, 56–57, 92, 97–102). PG&E did not install automatic or remote-operated pipeline valves that could quickly close and isolate a pipeline section that experienced a rupture or a leak. Their absence was a primary contributor to PG&E's 95-minute delay in the emergency response. Field personnel required 56 minutes to travel to and manually close the upstream and downstream valves once the control room gave instructions to do so. The rupture-supplied fire created a 183-meter fire impingement zone that prevented firefighters from reaching the rupture and initiating containment operations. The delay contributed to the breadth and scope of the fire damage. The fires destroyed or damaged 108 homes and 75 vehicles, and burned an area park. The firefighting efforts achieved a 75% containment at ~ten hours after the rupture. The firefighting effort continued for several days.*

***Case Study Background:** Pacific Gas & Electric (PG&E) owned and operated a 30" (0.76 meter) diameter natural gas transmission pipeline that transited underground through a residential neighborhood. An uncontrolled pressure increase ruptured a defective seam weld that existed when the segment was originally installed in the 1950s. PG&E did not employ the required integrity assessment and monitoring methods, so the defect remained unknown and undetected. The accident caused 8 fatalities and 58 injuries (10 serious, 48 minor). The 50-hour emergency response included 600 firefighting and emergency response and 325 law enforcement personnel.*

Note: If a safety margin is added, determine where to add the margin. Document the basis and justification for the decision.

3.2.17.4   B16-4, Review Safety Times

This step performs a cursory review of the safety times (BST, TTST, and TPST) recorded in Table 3.27. The suggested approach relies on an expert-level operations specialist to perform this step. The form of the review is one of inspection. Review the allocated times to confirm they appear reasonable and provide sufficient time to reliably achieve the task and phase activities within the specified times. When a safety time does not seem reasonable, changes may be needed.

- Revise the time allocations (BST, TTST, or TPST).
- Revise the function allocations (step B-15). Seek options to reduce workload or improve HE performance.
  - Change the function allocations between HE and physical elements (step B-15)
  - Change the allocation of tasks among the barrier HEs to better balance the workload (cognitive and physical)
  - Add HEs to reduce individual HE workload
- If the above options are not achievable, return to step B3-1 (task analysis) to:
  - Redefine tasks in a way that simplifies the task to reduce workload or activity time
- If all such options fail, the barrier may be unfeasible as defined. This may require a return to step B-2 (risk assessment).

Note: At this stage, the phase safety time targets are design assumptions that are further evaluated and confirmed in the detailed design phase. These assessments provide an early indication of a possible unfeasible barrier safety time or other potential barrier design challenges.

3.2.18   Step B-17, Dependencies: External Support Systems

This step identifies barrier-driven requirements that apply to barrier-dependent external support systems (ESSs) such as lighting, power, and HVAC. The requirements are design inputs provided to those responsible for designing these systems. They may be inputs to the ESS design (e.g., sizing and application), specifications, or performance standards. This step may identify cases where an ESS cannot meet barrier requirements, a condition that triggers additional design activities.

Consider: Many barriers may rely on the same ESS and therefore are mutually affected by an ESS failure; i.e., the barriers are not independent. Furthermore, in some cases, the reliability and continued barrier functioning cannot exceed that achieved by the ESS. As such, the design rigor and performance standards applied to the ESS may warrant exceeding those applied to the supported barrier.

The suggested steps in this process are as follows:

1. Review the barrier requirements information to identify barrier components and elements that depend on one or more external support systems. ESSs are commonly required to support or enable barrier components and In-Place PEs such as technical systems, engineered areas, and buildings. See Table B.19 in Appendix B.7.11 for a listing of example input documents in this step.

Note: Figure D.2 (Appendix D.2.2) may clarify why barrier requirements should be applied to an ESS and other systems on which the ESS may depend.

TABLE 3.28
External Support System (ESS) Requirements

ESS Info		Supported Barrier Information				ESS Requirements			
ESS ID	System	Barrier ID	Task ID	PE ID	ESS Support Function	Demand Period (minutes)	Demand Capacity (minutes)	Performance Requirements	SA-1 Info to Monitor
Enter ID for ESS *See Note*	Enter ESS type/ descr.	Enter IDs for supported barrier	Enter IDs for supported task	Enter ID for the PE supported by the ESS	Enter ESS function in supported PE *See examples*	Enter demand period *See examples*	Enter capacity duration/ period *See examples*	Enter requirements *See examples*	Identify SA-1 info to monitor

Note: If available, use an existing equipment identifier that may reside in a facility master equipment list.

2. Based on the abovementioned findings, identify and record the information in Table 3.28. Populate all table fields guided by the information in the table and the following guidance.

ESS ID: Enter the external support system ID (see note for Table 3.28)
System: Enter the ESS system type/descriptor:

- Lighting and lighting systems
- Electrical generation and distribution systems
- Battery-backed uninterruptible power system
- Heating, Ventilation, and Air Conditioning (HVAC) systems
- Other environmental control systems, for example, systems to limit ingress of corrosives and toxic gases, eliminate smoke within a building, etc.
- Other utility systems such as instrument air, nitrogen, plant air, and hydraulic power.

Barrier/Task ID: Enter ID for the supported barrier/task
Supported PE ID: Enter the ID for the barrier physical element supported by this ESS. Examples are as follows:

- Passive SA-1 information source (lighting required to see a non-powered sign or paint markings)
- Control building HVAC system (maintains habitable environment in HE work area)
- Control valve (e.g., instrument air provides motive power to stroke valve)
- Instrument sensor (e.g., electrical power for remote powered instrument)

Support Function examples:

- Power required to maintain barrier component/system functionality
- HVAC maintains main control room habitability in support of the HE functions in this room
- Maintain lighting to detect passive SA-1 information sources such as paint markings and signage
- Maintain lighting to support safe HE transits over an egress/escape route (engineered area)
- Provide motive power to powered barrier element, for example, a control valve
- Maintain power to In-Place technical systems such as the radio system, fire, and gas detection system

Demand Period examples:

- Continuous (always available)
- Duration of barrier activation beginning with barrier activation and ending when the barrier achieves the required safe state
- Only when the identified barrier (or barrier task) is active
- Defined period of operation, or only when in defined modes of operation
- Other periods (define)

Demand Capacity examples:

- Sized, sourced, and designed to provide uninterrupted service for the demand period
- Sized to above-noted capacity plus other specified load demands and plausible worst-case (specified) situations and conditions

Performance Requirements examples:

- Survivability requirements, for example, must survive the event that triggered the barrier without interruption to the ESS support provided
- System reliability that equals or exceeds the reliability and integrity requirements.

Note: Much of the above may overlap with other processes, generate considerable information, and require significant effort. This is not the intent. Rather, the objective is to identify the most onerous demands on an ESS to ensure they are addressed in its design. Alternative methods may achieve the same result with less effort such as using automated condition checks embedded within a computer-based database application.

3. Based on the captured information, identify the most onerous performance requirements. Confirm these requirements against the specified ESS design (e.g., procurement specifications, data sheets) to confirm it meets or exceeds the barrier requirements. If a requirement is not met, provide that information to those responsible for the ESS design and confirm if changes can be made to meet the barrier requirement.

Note: If the requirement cannot be met, changes to the barrier design may be necessary. For example, the ESS may be unable to meet the desired demand period. If so, this may require a change to the barrier or task safety time, which requires a return to an earlier design process. If the scenario risk (frequency x likelihood) is very low/rare, the path forward may be an ALARP assessment to determine if the design is accepted as is.

4. The operational status of a barrier-dependent ESS should be monitored. Identify the SA-1 information to be monitored and the monitoring frequency. (Step B20-5 defines how this information is used.)

Case Study, Deepwater Horizon Accident: *An explosion at or near engine room #3 destroyed critical components in the emergency power distribution system. This event contributed to the loss of all emergency power to all firewater pumps (loss of the firefighting barrier), emergency lighting (degraded emergency response barriers, e.g., muster and abandon barriers), and the facility positioning thrusters (loss of station-keeping barrier). The loss of this external support system (and failure of the backup generator to start) disabled key systems needed to control and mitigate the effects of the blowout and fires and placed personnel at risk.*

Case Study, PG&E Natural Gas Transmission Pipeline Rupture and Fire *(NTSB 2011, pp. 12–14). Firefighters arrived at the rupture site within 2 minutes. Located*

~300 meters from the rupture area, personnel at a local fire station saw the fire and immediately responded. Eleven (11) minutes later, the firefighters discovered the area hydrants were dry. The rupture and explosion damaged the underground water supply line to the hydrants. Additional time was then required to deploy 300 to 600 meters of large diameter hose supplied for other hydrants. Sixteen (16) minutes later, two water tenders were called to provide additional water. The response delay could have been much longer were it not for the exceptional response from the local fire station company. (For background information on the accident, see Chapter 3, step B16-3.)

3.2.19 STEP B-18, DEPENDENCIES: EXTERNAL PROTECTIVE BARRIERS

This step identifies requirements that apply to barrier-dependent external protective barriers (EBs). EBs may be passive, such as a fire wall or fireproofing, or active, for example, a fire detection and suppression system. The requirements are design inputs provided to those assigned EB design responsibility for integration into its design. They become considerations or direct inputs to the EB design (e.g., protective function endurance time), specifications, and performance standards. This step may identify cases where an EB cannot meet one or more barrier requirements, a condition that triggers additional design activities.

This step is like the ESS design in step B-17. As such, the guidance and comments from that step also apply to EBs.

The suggested steps for this process are as follows:

1. Follow B-17 step 1 as it applies to the EB definition and requirements. See Table B.20 in Appendix B.7.12 for a list of input documents specific to EBs.
2. Based on the abovementioned findings, identify and record the information in Table 3.29 (all fields except **SA-1 Info to Monitor**).

Populate all table fields guided by the information in the table and the following guidance.
EB ID: Enter the external protective barrier ID (see Table 3.29, Notes 1 and 2)
EB Type: Enter the EB type (passive, active) and description

- Passive barrier: Examples, a fire wall or blast wall, fireproofing, and fire-resistant doors
- Active barrier: Examples, a building or area fire detection and suppression system, toxic or flammable gas detection that initiates a building shutdown that includes an HVAC trip (stop fans) and closes the HVAC fresh air inlet/fire dampers.

Barrier and Task ID: Enter ID for the supported barrier/task
Protected PE or HE: Enter the ID/descriptor for the protected receptor, for example, for a human (HE) or an In-Place physical element: an engineered area, building, or a technical system. (Technical systems include external support systems.)
Protective Function: identify the required protective function given the identified hazard. Examples:

- Human element receptor: Limit threat from fire, heat, toxic gas, explosive overpressure, etc.
- Technical system receptor: Limit potential for critical damage to technical system equipment, cabling, etc.

TABLE 3.29

External Protective Barrier (EB) Requirements

EB Info			Supported Barrier Information				EB Requirements		
EB ID	EB Type	Barrier ID	Task ID	Protected PE or HE	Protective Function	Protected Period (minutes)	Identify Hazard	Performance Requirements	SA-1 Info to Monitor
Enter ID for EB *See Note 1*	Enter EB type: – Passive – Active	Enter IDs for protected barrier	Enter IDs for protected task	Enter ID(s) for the protected PE or HE	Enter the protective function *See examples*	Enter protected period/ endurance time *See examples*	Enter hazard for which the EB provides protection *See examples*	Enter requirements *See examples*	Identify SA-1 info to monitor

Note 1: If available, use the unique equipment identifier that may reside in a facility master equipment list.

Note 2: Use this table to identify an EB that provides barrier-required protection to an ESS listed in Table 3.28.

- External support system receptor: See "Technical System" above.
- Building receptor: Damage to building structure or equipment (or personnel) located within the building.

Protective Period/Endurance Time: identify the required protective period or endurance time. (The EB provides its protective function when the identified hazard is present.) Examples:

- Duration of barrier activation
- Duration of an identified barrier task
- Continuous period covering the activation period that includes several simultaneous or sequential barrier activation periods (see Note below).
- A different period

Note: As a consideration, the external protective barrier should be designed to provide its protective function for a period equal to the barrier safety time. However, this period may be longer in some cases. Consider the example in Figure 3.8. *Activating the muster barrier causes personnel to muster to a location protected by the EB. Non-essential personnel remain in this location, while attempts are made to control and recover from the event that activated the muster. If the control and recovery barriers fail, additional time is needed to complete an abandon facility barrier. Continuous protection from the EB is required to protect the mustered personnel through this series of sequential barrier activations.*

Identify Hazard: Identify the hazard for which the EB provides its protective function. Where important, clarify hazard specifics such as fire type or location. Examples:

- Fire (pool, jet, other)
- Explosion, associated potential shrapnel
- Toxic or flammable gas release
- Cryogenic fluid release

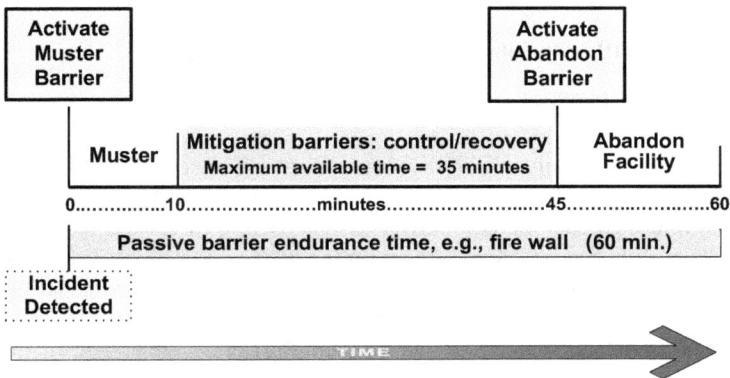

FIGURE 3.8 Dependence on External Protective Barrier Endurance Time (Offshore Example).

Performance Requirements examples:

- Survivability requirements. Examples: a requirement to survive the event that triggered the barrier without interruption to its provided protective function.
- Minimum capability to provide the specified level of EB protection to the identified physical or human receptors.

3. From the stated barrier protection required from each EB, identify the most onerous requirements such as endurance time and the efficacy of the provided protection. Confirm those requirements against the EB design, design specifications, performance standards, and data sheets. If a requirement is not met, provide the information to those assigned EB design responsibilities for integration into its design.

Note: If the requirement cannot be met, changes to the barrier design may be required. For example, it may require a change to the barrier safety time, which may then require a return to a prior design process to address the change. If this is an infrequent or less likely case, the path forward may be an ALARP assessment that determines whether the design is acceptable as is.

4. To the extent possible and practicable, the operational status of an external protective barrier should be monitored as a task within the barrier system. Identify the SA-1 information that should be monitored. For a passive EB, this may be limited to a simple countdown timer that shows the remaining time the EB can continue to provide protection (a time defined in its design documents.) For an active EB, the task may be to monitor an EB system alarm (or alarms) that identifies when its protective function degrades, fails, or approaching an imminent risk of failing.

3.2.20   Step B-19, Preliminary Reliability Assessments

This step performs two high-level reliability assessments:

- Single Point of Failure (SPOF) Assessment – Objective: Identify barrier system elements that, upon failure, cause the failure of the barrier function. Reviewed elements include barrier-assigned HEs; direct-use, support, and In-Place PE; other elements shown on the barrier/task block diagrams; and barrier dependencies including external support systems, external protective barriers, and additional resources.
- Assess Elements Shared (SE) with Other Barriers – Objective: Identify if any above-assessed elements that are shared by two or more barriers AND the failure of that element cause the simultaneous (or imminent) failure of those barriers. (This is an example of a common mode failure.)

Both assessments identify corrective actions to address the identified SPOF or SE effects. The following defines the failure criteria basis.

Failure Criteria: The failure or loss of a non-redundant physical or human element ensures the barrier cannot achieve its safety function and safe state within the barrier safety time.

See Appendix A.2 for suggested planning and execution activities to support and guide these activities. Table A.1 (Appendix A.2.3) identifies the suggested assessment participants. Appendix L.2.2 provides basic guidance on remote barrier support, such as adding new tasks, personnel, facilities, and additional resources.

Note: Existing methodologies (e.g., risk-based) may also be suitable to replace one or both assessments.

3.2.20.1 B19-1, Single Point of Failure (SPOF) Assessment

The suggested SPOF assessment process is as follows:

1. Identify barrier system elements to consider in the SPOF assessment process. Potential SPOF elements include the following:

   - Barrier-assigned HEs
   - Direct-Use, Support, or In-Place PEs
   - Barrier-dependent elements (e.g., an external support system (ESS), protective barriers (EB), or an additional resource identified in step B20-4.)

   Documents that may provide input and additional guidance to this process include the following:

   - Active human barrier philosophy, design basis, or like applicable documents listed in Table 3.1
   - Task requirements tables (see Table 3.2)
   - Barrier functional block diagrams (see step B-15)
   - Other documents (see Tables 3.3 and 3.4)

   Example SPOF sources are as follows:

   - Human element: Loss of the HE (single HE barrier), loss of the barrier leader (multiple-HE barrier), etc.
   - Physical element: A failure of or inability to access a direct-use physical element, the loss of an external support system (e.g., electrical power), or an external protective barrier that provides protection to an occupied safe haven.

2. Review each element from step 1 and identify those that meet the *Failure Criteria* statement provided in the step introduction. Enter each element that conforms to this criterion into *Table 3.30*. Enter the **Element ID**, **Element Type**, **Barrier ID**, and **Barrier Title**. For each entry, evaluate and enter (all that apply) the **Failure Scenario** and **Failure Consequence**.

TABLE 3.30

Reliability Assessment: Single Point of Failure

Element ID	Element Type	Element Description	Barrier ID	Barrier Title	Failure Scenario	Failure Consequence	Corrective Action	Status
Enter element ID	Enter PE HE	Enter description for this element	Enter ID of barrier employing this element	Enter barrier title/ descriptor	Enter SPOF scenario	Enter consequence of element failure *Enter all that apply to this scenario*	Enter recommended corrective action	– ALARP – Implement – Defer to detailed design process C-1

3. Based on the failure scenarios and consequences, assess whether the design should be revised to prevent or mitigate the failure or its consequences. If so, evaluate and select the recommended **Corrective Action**. If not, proceed to the next SPOF. Example corrective actions:

   - Accept the design as is, i.e., ALARP.
   - Employ redundancy, for example, add a fully qualified HE backup to the SPOF HE role/position or employ a redundancy solution for a PE.
   - Increase barrier system fault tolerance.
   - Select a physical element that meets an increased reliability/integrity standard, for example, employ components and subsystem designs that achieve a specified Safety Integrity Level of 1 or higher. (See Marszal and Scharp (2002), Safety Integrity Level Selection, Chapter 10.)
   - Enhance the maintenance program (frequency, type) or competency of maintenance personnel
   - Increase the performance standard, inspection, testing, or verification requirements. (See Appendices D.3, G.3, G.4, and G.5 for guidance.)

     Enter the recommended **Corrective Actions** in Table 3.30. In the **Status** field: enter ALARP, if so determined in this step. If not, leave this field blank for a follow-up in Step B20-6.

3.2.20.2   B19-2, Shared Element (SE) Assessment

The suggested SE assessment process:

1. From the list of identified SPOF elements recorded in Table 3.30, identify those that shared with two or more active human barriers. In Table 3.31, enter the **Shared Element ID**, **Element Type**, and **Element Descr.**, and the **Barrier ID** and **Barrier Name** for all barriers that share this element.
2. Evaluate each element for its effect on all barriers that share and rely on this element. Identify those that meet the *Failure Criteria*. For those that do, enter "Yes" in the **SPOF?** field. Otherwise, enter "No" and proceed to the next potential SE element for review.
3. If "Yes" is recorded in the **SPOF?** field, evaluate and enter the **Failure Scenario** and **Failure Consequences** in Table 3.31. (If one or both duplicate the entries in Table 3.30, add a reference to those entries.)
4. Based on the entered scenario and consequences, evaluate and select (recommend) the corrective action(s) that prevent or mitigate the SE failure or minimize its consequences. The solution may be one of those noted in B19-1 (step 3) or other options. Enter the recommended **Corrective Action(s)** in Table 3.31. In the **Status** field, enter ALARP if that is the determination. If not, leave the field blank for follow-up in Step B20-6.

TABLE 3.31

Reliability Assessment: Shared Elements

Shared Element ID	Element Type	Element Descr.	Barrier ID	Barrier Title	SPOF?	Failure Scenario	Failure Consequence	Corrective Action	Status
Enter ID of Shared Element	Enter shared element type: – *PE*, – *OE*, – *HE*	Enter description for this element	Enter ID of barrier using this element	Enter barrier title/descriptor	Is this a SPOF for this barrier? Enter – *Yes* – *No*	Enter the failure scenario	Enter consequence of element failure. *Enter all that apply to this scenario*	Enter recommended corrective action(s)	Enter: – ALARP – Implement – Defer to detailed design process C-1

3.2.21   Step B-20, Balance of Barrier Design

This step addresses preliminary design phase activities not covered in other steps. Activities included in this step are as follows:

- B20-1, Provide Barrier Requirements and Design Information to Others
- B20-2, Review Documents Developed by Others
- B20-3, Develop New Documents
- B20-4, Additional Resources
- B20-5, New Monitoring Tasks
- B20-6, Address Recommendations from Assessments (Reliability, Fatigue Management)
- B20-7, Remote Barrier Support and Remote Operations Center

Note: The source tables in Appendix B.7 identify information developed in the concept and preliminary design phase. Refer to these documents when gathering requirements and design information used in the activities in step B-20.

3.2.21.1   B20-1, Provide Barrier Requirements and Design Information to Others

This step compiles barrier requirements and design information and provides that information to others for integration into their designs and design activities. Here, "Others" are those disciplines and organizations with primary responsibility for developing design and procurement documents for equipment, facilities, and systems that may contain elements in a barrier system. "Others" also refers to those responsible for external support systems, external protective barriers, and additional resources. See the applicable input sources for this information. (See the source tables in Appendix B.7).

Note: Planning, communication, and coordination are needed to reach mutual agreement with those who receive this information. The scope of the agreement may include expectations for integrating the barrier requirements into their work processes and designs, and the delivery form and timing of the provided information. See Appendix A.8 for information on the unique challenges common to these activities.

Example inputs and requirements may include the following:

- Applicable regulations, industry standards, and guidance practice standards
- Direct-use physical element required to initiate or perform act phase response actions
- Component selection requirements
- Sizing and capacity requirements
- Performance requirements such as functionality, reliability, availability, and survivability

- Location and layout requirements
- System interface design
- Display design requirements: non-VDU- and VDU-based displays
- Others

Example document types are as follows:

- Performance standards
- Specifications regarding function, system design, procurement, etc.
- Data sheets
- Safety equipment lists
- Location and layout drawings
- System block diagrams

The documents and activities listed above apply to In-Place PE:

- Engineered areas
- Buildings, rooms, walk-in shelters
- Technical systems

Example technical systems may include the following:

- Control consoles and panels
- Non-VDU- and VDU-based displays
- Packaged equipment system or PES (possible PES functions in the barrier system: direct-use, In-Place PE, and external support system)
- General alarm/public address system
- CCTV systems
- Radio, telephone, and video conferencing system
- Basic IT infrastructure including email, the Internet, and file storage systems
- Safety instrumented systems
- Fire and gas detection systems
- Fire suppression systems such as water mist, deluge, and sprinkler systems
- HVAC

The documents and activities mentioned above may also apply to:

- External support systems (see B-17 for examples)
- External protective barriers (see B-18 for examples)
- Additional resources (from step B20-4)

3.2.21.2   B20-2, Review Documents Developed by Others

As a follow-up to step B20-1, this step reviews the documents developed by others to confirm that the barrier requirements are correctly integrated and included in these documents.

3.2.21.3 B20-3, Develop New Documents

This step develops the documents identified in Table 3.3 identified as *New* or *Input/ New*. The safety requirements specification and performance standards are examples of "new" documents.

Note: The source tables in Appendix B.7 summarize the potential input sources of information. The updates to this document occur at different lifecycle stages that reflect when the information is developed, approved for use, or approved for construction.

3.2.21.3.1 B20-3.1, Safety Requirements Specification

This step develops the safety requirements specification (SRS). Appendix B.3 summarizes the suggested content to include in this document. This content should be developed to the extent possible (based on the information available at this stage) and added to the SRS.

For a capital project, a single SRS may be appropriate to address the active human barriers to the project.

Note: As with IEC-61511-1 (2016), the SRS is essential to managing the barrier over its full lifecycle. Active human barriers are a collection of many diverse and dispersed elements. Maintaining awareness of this system of parts (human, organizational, and physical) and their function in the barrier system remains a challenge over the barrier lifecycle. The SRS is the primary document serving this function. As such, it is the basis for evaluating the effects of a proposed change such as to components and systems, staffing or maintenance policy, operating procedures, training modules, or competency requirements.

3.2.21.3.2 B20-3.2, Performance Standards

This step identifies the required performance standards (PS) and their type, content, function, and development timing. (See Appendix D.3 for guidance and detail.) The following are the steps in this activity. Enter all the information in *Table 3.32*. Refer to the guidance provided in the table and the following.

1. Identify the barrier-required performance standards. Enter the **PS ID/ Reference**.
2. For each entered PS, determine and enter the assigned responsible party and planned (or proposed) development timing. Under the **Execution header**, enter the **Phase Developed** and **Assigned To** files.
3. For each entered PS, enter its **Type**, and the **Applies To**, **Element Identifier (ID)**, and **Function/Purpose** fields. See Appendix D.3 for background and additional guidance.
4. For PS assigned to others, follow the applicable guidance in step B20-1.

Note: As applicable, the prototype lifecycle model assumes a design-focused performance standard is developed for use in the step B-21 verification process.

TABLE 3.32
Performance Standards

		Identify Requirements			Execution	
PS ID/ Reference	**Type**	**Applies To**	**Element Identifier (ID)**	**Function/Purpose**	**Phase developed**	**Assigned to:**
Enter unique ID, title, or descriptor	Enter type: Design Operational	Enter element type: – Barrier – HE role – Barrier component – In-Place PE: technical system, building, engineered area, packaged equipment, etc. – Dependent system: external support system, external protective barrier, additional resource	Enter element ID or descriptor	Enter – Verification – Validation	Enter timing: – Preliminary design – Detailed design (process C-13) – Implementation (phase D) – Other	Enter the assigned responsible party (technical discipline or organization) *One per row*

3.2.21.4   B20-4, Additional Resources

This step identifies the requirements for the barrier-required resources not addressed in other steps. (This type of resource affects the barrier function, performance, and reliability. Remote barrier support is not an additional resource.) Examples of *additional resources* include the following:

- Third-party contracted communication resources needed to enable remote barrier support from a remote location (e.g., satellite, fiber-optic networks)
- Third-party Internet provider
- Third-party fire and emergency response such as the fire brigade and emergency rescue
- Third-party logistics and transportation resources such as helicopters and specialized emergency vehicles or equipment
- Third-party medical emergency resources like telemedicine services

See Appendix L for requirements unique to a remote barrier support task performed from a remote location.

The following are the suggested activities. Record the information in Table 3.33.

1. Enter the **Barrier ID**, **Task ID**, and **Resource ID** and description for each required resource.
2. Enter the resource **Function** in the barrier system.
3. Enter the required information to achieve this function and performance, for example, **Response Time** and **Capacity/Capability**.
4. Enter the known or proposed resource **Provider** and the current acquisition/reservation **Status**, for example, signed annual contract and existing agreements.
5. If known, enter a new **Training Req.** (if any) that may be required to use this resource. (Status is entered in a later phase.)
6. Identify and enter any potential **Hazard Exposure** that can affect the resource function or performance, such as the hazard that activated the barrier, severe weather, unplanned traffic delays, and cyber events.
7. Define the **Resource Verification** status and verification method.

Note: The remote barrier support design basis may identify additional resources needed to enable the RBS functions from a remote location.

Case Study, DuPont La Porte (Texas) Toxic Chemical Release. *When an accident occurred at the DuPont La Porte plant, the emergency response plan relied on an external, shared emergency response resource. According to CSB (2019):*

On November 15, 2014, approximately 24,000 pounds of highly toxic methyl mercaptan was released from an insecticide production[b] unit (Lannate® Unit) at the E. I. du Pont de Nemours and Company (DuPont) chemical manufacturing facility in La Porte, Texas.[c] The release killed three operators and a shift supervisor inside a manufacturing building.[d] They died from a combination of asphyxia and acute exposure (by inhalation) to methyl mercaptan.

(p. 7)

TABLE 3.33
Additional Resource Requirements

Barrier ID	Task ID	Resource					
		ID & Describe	Function	Response Time (Minutes)	Capacity/Capability	Provider	Status
Enter barrier ID	Enter task ID	Enter resource ID and description	Enter resource function in barrier system	Enter minimum resource response time	Enter minimum resource capacity/capability *Each function*	Enter resource provider name	Enter resource acquisition/reservation status *Each provider*

See *Table 3.33* extension

Table 3.33 Extension

Hazard Exposure	Training		Resource Verification
	Training Req.	Status	
Enter potential hazards and hazard exposures *All that apply*	Enter resource training req. *(if any)*	Enter training status	Enter verification status and method

As with most shifts, the Lannate® area process coordinator on the night of the incident was the Shift Supervisor. The Shift Supervisor in this case, however, was a victim of the toxic chemical release, and DuPont La Porte had not designated an on-site backup process coordinator for this shift.[a] As a result, these important functions were either performed in a disorganized manner by other operations personnel or never performed at all.

(p. 34)

At about 3:50 am, the Board Operator called for the plant emergency response team (ERT)[a] to respond to the manufacturing building, stating over the intercom system, 'We need rescue people, and there's people missing'.

(p. 20)

....To the ERT, a request for 'rescue' meant specifically high-angle or confined-space rescue. Therefore, the ERT responded to the request for help by gathering only technical rescue gear (e.g., harnesses and ropes), not knowing that there was a major toxic chemical release.

(pp. 20–21)

At about 4:15 am, ERT members assigned an operator (Operator 7) to bring them SCBAs because the ERT mini-pumper truck holding SCBAs could not respond to the scene.... Operator 7 went alone to retrieve these SCBAs and unknowingly walked into the path of the methyl mercaptan being released from the manufacturing building (Figure 3.2).[a] The hot zone was not clearly identified or communicated to plant personnel. As a result, when Operator 7 went to retrieve the SCBAs, she was unaware that she was entering a potentially hazardous area.

(p. 23)

Beginning at 5:08 am, three external firefighter groups arrived on-site, and at 5:15 am, the ERT conducted its second entry into the manufacturing building.

(p. 24)

3.2.21.5   B20-5, New Monitoring Tasks

Several steps in the preliminary design process may add a new monitoring task, such as adding a new task to monitor the status of safety equipment worn by others (support PE) or monitoring the operational status of In-Place PE, ESS, EB, or an additional resource. Because the task is identified after completing the risk assessment and task analysis, the approach warrants a return to step B-3. Add the task to Table 3.6 and progress its design through all applicable preliminary design steps.

3.2.21.6   B20-6, Address Recommendations from Assessments (Reliability, Fatigue Management)

The SPOF and shared element reliability assessments in step B-19 may recommend corrective actions (CA), as recorded in Tables 3.30 and 3.31. The fatigue management assessment in step B-26 may recommend corrective actions, as recorded in *Table 3.34.*

The design considerations (e.g., entry point in the preliminary design process) and effects (effort, schedule, cost) depend on the nature of the CA. Initial decisions may include the timing of when the CA should be further evaluated and if/when it should be implemented. Enter the planned timing to address each CA in the abovementioned tables. Example options include the following:

1. Implement the CA in the preliminary design process. (Enter "Implement" in the **Status** field.)
2. Defer and address the CA in the detailed design process, C-1. (Enter "Defer to (C-1)" in the **Status** field.)
3. Accept the design as ALARP. (Enter "ALARP" in the **Status** field.)

Note: If option 1 is selected, the path forward depends on the nature of the recommendation. For example, a recommendation that adds a new task will likely require a return to earlier steps that define the new task requirements and design. If it modifies a task phase requirement, add the new requirement to the respective table and progress the design from that step forward. Option 2 defers the action to the detailed design phase. The detailed design process C-1 updates the preliminary design. Addressing corrective actions (CAs) at that time may be less disruptive if the preliminary design schedule is highly compressed. It may be prudent to identify open items in the detailed design phase, request for proposal package.

3.2.21.7   B20-7, Remote Barrier Support and Remote Operations Center

If the barrier system employs remote barrier support, see Appendix L for example design activities.

3.2.22   STEP B-21, VERIFICATION #2

Step B-21 performs the verification #2 activities suggested in Appendix G.5. Tables G.5 and G.6 provide the suggested examination form, tangible evidence, and execution plan to guide and perform this activity.

3.2.23   STEP B-22, VALIDATION #1

Step B-22 performs the validation #1 activities suggested in Appendix G.6. Tables G.7 and G.8 provide the suggested examination form, tangible evidence, and execution plan to guide and perform this activity.

3.2.24   STEP B-23, CONTROL CENTER CONCEPTUAL DESIGN FRAMEWORK (EVALUATION AND INPUT)

This step performs the applicable activities in ISO 11064-1 (2000, Clause 8). Applicable activities from this clause are limited to active human barrier defined requirements such as an identified control center that supports barrier activities.

In the prototype lifecycle model, a control center may be a central control room, a secondary/backup control room, an incident command center, or the remote operations center discussed in Appendix L. Control centers included in the scope of the model are limited to those that directly support the defined barrier activities or functions. See Appendix L for the unique requirements that apply to a remote operations center from which a remote barrier support task is performed.

3.2.25   STEP B-24, REQUISITIONS AND CONTRACTS

This step gathers, assembles, and provides barrier system information that should be added to the applicable draft procurement documents and packages. It also includes a review of the draft procurement/contract package prior to issue and the returned proposal. (The procured equipment contains one or more barrier system elements, or is a barrier-dependent ESS, EB, or additional resource.)

For guidance on these activities, see the following:

1. Appendix E.1, Input to Request for Proposal/Quotation
2. Appendix E.2, Proposal Review

Note: A project may require early procurement activities to ensure the timely delivery of equipment that has a lengthy or early delivery time. This may also occur if the project needs accurate cost, schedule, or technical information to support budgeting and planning, and provide technical information needed to progress critical path facility and technical design activities.

3.2.26   STEP B-25, HE STAFF – ROLES AND ORGANIZATION

This step compiles barrier HE requirements and provides that information to those assigned personnel assignment and management responsibilities. The suggested steps are as follows:

1. Gather and review requirements and information on the HEs assigned to barriers and their role in the barrier function. Potential inputs to this activity include the following:
   - Active human barrier philosophy, design basis
   - Barrier safety management plan
   - Active human barrier philosophy and design basis
   - Remote barrier support design basis
   - Staff policies and practice
   - Tables 3.5 and 3.6, Barrier and Task Base Requirements
   - Task analysis and report (step B3-1)
   - Barrier/Task Functional Block Diagrams and Roles (step B-15)
   - Table 3.11, Communication Requirements (Multiple-HE Barriers)
   - Table 3.34, Fatigue Monitoring and Management Requirements
   - Other documents that may include applicable recommendations:
   - Table 3.30. Reliability Assessment: Single Point of Failure.

- Table 3.31. Reliability Assessment: Shared Elements.
- Table 3.33. Additional Resources.
- Standard operating procedures
- Other documents containing key assumptions and information

2. Based on these documents, identify the requirements:
   - Identify HE roles, functions, and activities; interaction with other HEs and others; competency requirements, etc.
   - Physical locations where activities are performed
   - Design assumptions such as personnel rotations, shifts, turnover, and availability
   - Requirements affecting maintenance and barrier support personnel
   - Other work assigned to HEs that may lead to excessive workload peaks or conflicts

3. Roster: From the above, confirm the roles identified in prior steps. Update the roster to show:
   a. Final HE roles/positions for each barrier
   b. HE required backups to identified roles, including expectations on backup competency, availability, call-out response time, and events that may cause both to be unavailable
   c. HE roles providing remote barrier support from remote locations
   a. Other permanent roles defined in the barrier safety management plan

4. Organization chart: The chart provides a visual representation of the roster information such as barrier roles and required backups, and relationships between HEs. Examples of the information included are as follows:
   a. Pertinent information from the roster, i.e., information needed by the likely organization chart user
   b. Relationships between HE roles, such as the relationship between the barrier lead and other HEs on the protected facility, or RBS HEs providing support from a remote location
   c. HE relationship to additional resources (if any)

Note: The number of personnel, their locations, in-person communication require-
ments, and similar considerations are critical inputs to the design activities per-
formed by other disciplines and organizations. The information can affect the sizing,
layout, and provisioning of the control console, rooms, buildings, and engineered
areas. Changes in these areas, if only first discovered in the detailed design phase,
can have a detrimental effect on project cost and schedule. Staffing changes may
also contribute to project capital and operating costs or commercial benefits (e.g.,
rate of return). In addition, HE roles may present competency challenges to address
in personnel selection, training plans, and so on.

Note: See Appendix L for the unique requirements that apply to a remote barrier
support task.

3.2.27 Step B-26, Fatigue Management (Fitness for Service)

This step identifies a maximum fatigue target that applies to barrier HEs. The target fatigue level is a function of sleep including timing and duration. The step evaluates and identifies plausible situations when a maximum fatigue target is exceeded and recommends corrective action to prevent or mitigate this exceedance or source thereof.

The metrics, methods, and guidance assumed in this process are defined in IOGP (2014, Report 492). This report (p. 3) defines fatigue (mental fatigue) as a "progressive decline in alertness and performance caused by insufficient quality or quantity of sleep, excessive wakefulness or the body's daily circadian rhythm."

Note: IOGP (2019d) is a later document. Refer to this document for additional information and possible updates to this step.

The activities in this step are as follows:

1. Identify a maximum fatigue limit. Enter the target in Table 3.34.
2. For each HE, identify and enter the barrier ID for barriers assigned to that HE.
3. Review the available information. For each HE, identify every plausible exceedance situation that can lead to extended periods of limited sleep opportunities and the estimated duration of that exceedance. Record both requirements in Table 3.34.
   Consider the following sources of information, as applicable.
   • Barrier safety management plan
   • Active human barrier, design basis

TABLE 3.34
Fatigue Monitoring and Management Requirements

Maximum fatigue limit: *enter limit (rating)*

HE ID	Barrier ID	Exceedance Situation	Exceedance Duration (Hours)	Corrective Action (CA)	Status
Enter HE ID	Enter barrier ID	Enter situation(s) that may cause a fatigue limit exceedance	Enter estimated exceedance duration (may be a range)	Enter recommended corrective action if any *See Table 3.35 for examples*	Enter – Identified recommended corrective action (CA) – CA approved – CA implementation complete – Defer action to detailed design process C-1

- Owner/operator staffing policies
- Standard operating procedures
- Typical work activities assigned to job classes/positions that include the HE
- Table 3.6 (HE roles)
- Roster and organization chart from step B-25. (If known, the roster may include the HE names.)
- Performance standards (see Table 3.7 and other documents developed in step B20-3)
- Applicable recommendations from the step B-19 reliability studies (see Tables 3.30 and 3.31)
- Historical records on staffing peaks and gaps, and causes such as vacation, leave, turnover, training, and event-based personnel demands (assignments to project team)
- Historical records on unplanned personnel events including sickness/injury and employee turnover
- Process operations (historical norms) and workload change, including start-up/shutdown, change feedstock, simultaneous operations, plant upset, and emergency events
- Historical success developing, maintaining, and managing staff competency to meet well-defined competency requirements
- Known/planned staffing events such as the work shift, rotation, vacation, and reassignment

4. For each exceedance, assess and recommend corrective actions to prevent or minimize the exceedance duration and/or frequency. See Table 3.35 for examples of solutions. Record the recommended actions in Table 3.34. (Recommendations from this step are addressed in step B20-6.)
5. Repeat the above steps for each HE.

3.2.27.1 Additional Considerations and Background

Fatigue caused by insufficient sleep increases the likelihood of forgetting a step in a procedure. It also reduces the capability to comprehend information (input to decisions) or develop appropriate plans (CCPS 2022, p. 164).

Note: Poor health (physical and mental) and chronic stress can have effects like sleep-attributed fatigue. The prototype lifecycle model assumes the owner/operator has base policies and practices that monitor and address basic health and chronic stress. Operate and maintain steps E2-4 and E5-4 suggest processes to monitor and minimize conditions that contribute to sleep-related fatigue. For a more complete discussion on this topic, see CCPS (2022, Ch. 15), Edmonds (2016, Ch. 22), and Flin et al. (2008, Ch. 8).

3.2.28 Step B-27, Design Review

This step performs the end-of-phase design review suggested in Appendix G.

TABLE 3.35

Example Corrective Actions for Fatigue Exceedance

Objective (Example)	Corrective Actions (Examples)	How Achieved (Examples)
Manage HE staff and backup resources to prevent or limit fatigue exceedance events	Modify schedules and rotation policies or practice	OE – modify personnel scheduling and rotational practice to limit overtime requirements, periods between shifts that may be less than 24 hrs., etc.
	Modify organizational assignments	Develop and assign a verified competent backup to each HE position.
	Reduce staffing gaps/ mismatches through improved staff management	OE – increase frequency of staff planning and scheduling reviews and improve tools to track and monitor planned and unplanned (historical metrics) events that contribute to excessive overtime and personnel shortages.
	Increase management awareness and support	OE – increase management awareness of the risk attributed to insufficient sleep and gain management support for this unique barrier requirement.
	No corrective action required, i.e., ALARP	

REFERENCES

CCPS (2022), *Human Factors Handbook for Process Plant Operations, Improving Safety and Systems Performance*, New York: John Wiley & Sons Inc., Center for Chemical Process Safety (CCPS)

CIEHF (2016), Human barriers in Barrier Management, a white paper by the Chartered Institute of Ergonomics and Human Factors, 12/2016, CIEHF

COMAH (2011), "Buncefield: Why Did it Happen?", COMAH Competent Authority. Downloaded August 23, 2023 from https://www.hse.gov.uk/comah/buncefield/index. htm. Follow web link: '*Buncefield: Why Did It Happen? (PDF)*'

Crichton, M.T., Lauche, K., Flin, R. (2005), Incident command skills in the management of an oil industry drilling incident: A case study, *Journal of Contingencies and Crisis Management*, 13(3), 116–128

CSB (2019), Toxic chemical release at the DuPont La Porte chemical plant, Report no 2015-0I-I-TX, U.S. Chemical Safety and Hazardous Investigation Board

Edmonds, J., (2016), *Human Factors in the Chemical and Process Industries, Making it Work in Practice*, Elsevier

Endsley, M.R., Jones, D.G. (2012), *Designing for Situation Awareness: An Approach to User-Centered Design*, 2nd Ed, CRC Press

Flin, R., O'Connor P., Crichton, M., Slaven, G., Stewart, K. (1996). Emergency decision making in the offshore oil and gas industry, *Human Factors* 38(2) 262–277

Flin, R., O'Connor P., Crichton, M. (2008), *Safety at the Sharp End: A Guide to Non-Technical Skills*, Ashgate Publishing

Gasbury, R.B. (2103), *Situation Awareness for Emergency Response*, PennWell Corporation

Hick, W.E. (1952), On the rate of gain of information, *Quarterly Journal of Experimental Psychology*, 4(1), 11–26

Hollnagel, E. (2003), *Handbook of Cognitive Task Design*, Ed. Hollnagel, E., Mahwah, NJ: Lawrence Erlbaunm Associates Inc. (Reprinted by CRC Press, 2010)

IEC 61511-1 (2016), *Functional Safety – Safety Instrumented Systems for the Process Industry Sector – Part 1: Framework, Definitions, System, Hardware and Application Programming Requirements*, 2nd Ed, International Electrotechnical Commission

IFE (2022), *The Petro-HRA Guideline*, Rev. 1, Vol. 1, IFE/E-2022/001, ISBN 978-82-7017-937-4, Institute for Energy Technology

IOGP (2012), Cognitive issues associated with process safety and environmental incidents, London: International Association of Oil and Gas Producers, IOGP Report No 460, 7/2012

IOGP (2014), Assessing risks from operator fatigue, guidance document for the oil and gas industry, London: IPIECA and International Association of Oil and Gas Producer, IOGP Report No 492

IOGP (2018), Introducing behavior markers of non-technical skills in oil and gas operations, International Association of Oil and Gas Producers, IOGP Report No 503

IOGP (2019d), Managing fatigue in the workplace, a guide for the oil and gas industry, London: IPIECA and International Association of Oil and Gas Producer, IOGP Report No 626

ISO 11064-1:2000, Ergonomic design of control centres – Part 1: Principles for the design of control centres, International Organization for Standardization, 1st Ed, 2000-12-15

Johnsen, S.O., Kilskar, S.S., Fossum, K.R. (2017), Missing focus on human factors – organizational and cognitive ergonomics – in the safety management for the petroleum industry, *Journal of Risk and Reliability*, 231(4), 400–410

Klein, G.A. (1993), A recognition-primed decision (RPD) model of rapid decision-making. In G.A. Klein, J. Orasuanu, R. Calderwood, & C.E. Zsambok (Eds.) *Decision-Making in Action: Models and Methods* (pp. 138–147). Westport, CT: Albex Publishing

Marszal, E., Scharpf, E., (2002), *Safety Integrity Level Selection, The Instrumentation, Systems and Automation Society.*

NTSB (2011), Pacific Gas and Electric Company Natural Gas Transmission Pipeline Rupture and Fire, San Bruno California, September 9, 2010, National Transportation Safety Board, Pipeline Accident Report NTSB/PAR-11/01. Washington, DC

NUREG (2020), Human-Systems Interface Design Review Guidelines, NUREG-0700, O'Hara, J.M., Fleger, S., Rev. 3, Office of Nuclear Regulatory Research, U.S. Nuclear Regulatory Commission, Washington, DC

OPITO (2014), Major Emergency Management Initial Response Training, Revision 1, OPITO Standard Code 7228, OPITO, March 13, 2014

Reason, J. (1990), *Human Error*, Cambridge: Cambridge University Press

Reason, J. (2008), *The Human Contribution, Unsafe Acts, Accidents and Heroic Recoveries*, Ashgate Publishing Ltd.

Rouse, W.B., Morris, N.M. (1985), On looking into the black box: Prospects and limits in the search for mental models, *Psychological Bulletin*, 100, 349–363 (Report no 85-2)

Shepherd, A. (2001), *Hierarchical Task Analysis*, CRC Press

SPE (2014), The human factor; process safety and culture, SPE Technical Report, Society of Petroleum Engineers, March 2014

Sträter, O. (2005), *Cognition and Safety: An Integrated Approach to Systems Design and Assessment*, 1st Ed, Ashgate Publishing Ltd.

Tannenbaum, S., Salas, E. (2021), *Teams That Work, the Seven Drivers of Team Effectiveness*, Oxford University Press

Wickens, C.D., Hollands, J.G, Banbury, S., Parasuraman, R. (2013), *Engineering Psychology and Human Performance*, 4th Ed, Pearson Education Inc.

Woods, D.D., Dekker, S., Cook, R., Johannsen, L., Sarter, N. (2010), *Behind Human Error*, 2nd Ed, Ashgate Publishing

4 Detailed Design, Engineering, and Requisitions (Model Phase C)

Figure 4.1 gives an overview of the suggested design, engineering, and procurement processes to complete the detailed design and engineering phase of the barrier life-cycle. In all cases, the scope of these processes is intended to be limited to only those elements that are specific to active human barriers. For inputs to this phase, see Tables 3.2–3.4 for documents and information developed in the preliminary design phase. *Tables 4.1–4.3* summarize the documents and information developed in this phase. Table A.2 (Appendix A.2.3) identifies the suggested participants in select activities.

Figure 4.1 identifies four assessment sub-processes performed for each barrier task:

- *Sub-process CX* design activities that apply to barrier system components that are direct-use and Support PE. Example activities include component selection, design, identifying new skills, knowledge, and procedure requirements, and develop input to procurement processes. (The CX sub-process is defined in Appendix B.6.)
- *Sub-process CP* performs a component-level working environment assessment. The assessment, defined in Appendix H, is embedded into the CX sub-process.
- *Sub-process CI* performs a new cognitive assessment and mitigation process applied at the task phase level. (See Appendix I.2 for detail.)
- *Sub-process CS* confirms the previously defined task phase safety time assessments.
- *Sub-process CF* performs a functional assessment for barrier-dependent control consoles and panels, rooms, and engineered areas.

Note: The CX process does not replace existing practice and design activities that select, design, and implement these components. Instead, the CX process is intended to supplement existing practice by adding this new set of activities focused on the task phase and human-system integration.

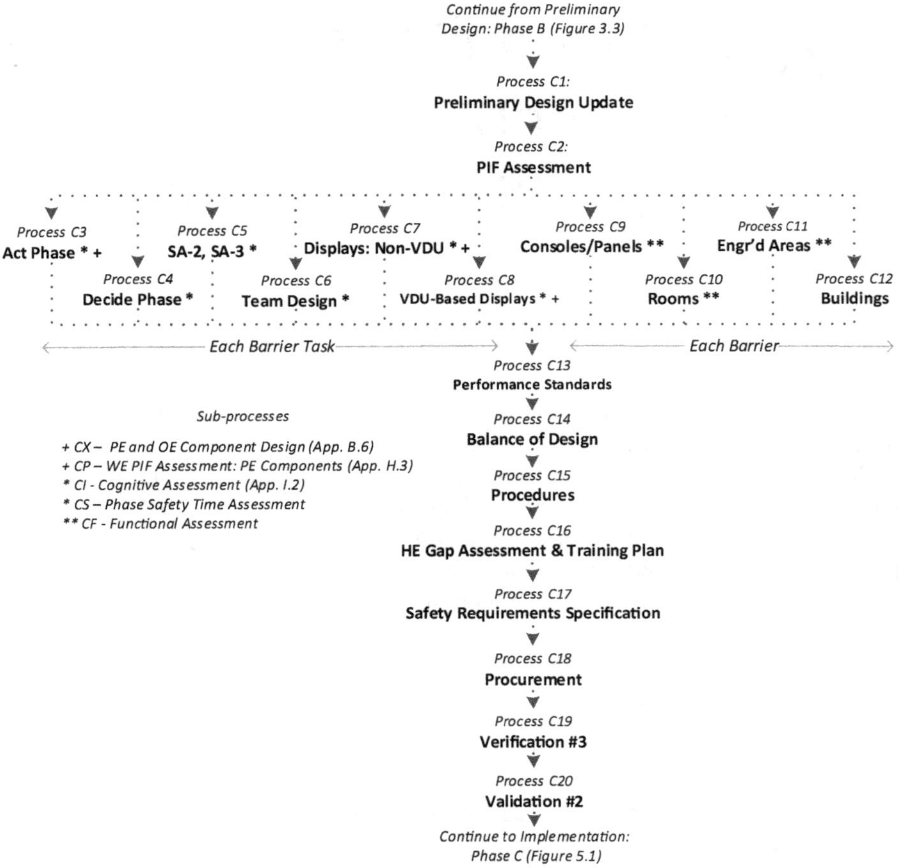

Continue from Preliminary
Design: Phase B (Figure 3.3)

Process C1:
Preliminary Design Update

Process C2:
PIF Assessment

Process C3	Process C5	Process C7	Process C9	Process C11
Act Phase * +	**SA-2, SA-3 ***	**Displays: Non-VDU * +**	**Consoles/Panels ****	**Engr'd Areas ****
Process C4	Process C6	Process C8	Process C10	Process C12
Decide Phase *	**Team Design ***	**VDU-Based Displays * +**	**Rooms ****	**Buildings**

←———————— Each Barrier Task ————————→   ←———— Each Barrier ————→

Process C13
Performance Standards

Sub-processes

Process C14
Balance of Design

+ CX – PE and OE Component Design (App. B.6)
+ CP – WE PIF Assessment: PE Components (App. H.3)
* CI - Cognitive Assessment (App. I.2)
* CS – Phase Safety Time Assessment
** CF - Functional Assessment

Process C15
Procedures

Process C16
HE Gap Assessment & Training Plan

Process C17
Safety Requirements Specification

Process C18
Procurement

Process C19
Verification #3

Process C20
Validation #2

Continue to Implementation:
Phase C (Figure 5.1)

FIGURE 4.1   Detailed Design and Engineering Phase Overview.

The following tables summarize the assessment, design, and procurement documents developed in this lifecycle phase.

- Table 4.1. Detailed Design: Barrier and Task Requirements Tables
- Table 4.2. Detailed Design: Technical, Procurement, and Project Documents
- Table 4.3. Detailed Design: Studies and Assessments

Note: The conceptual and preliminary design phase documents provided as input to this phase are summarized in Tables 3.2–3.4. *Many of these documents may be updated in process C-1, preliminary design update.*

4.1   PROCESS C-1, UPDATE PRELIMINARY PHASE DESIGN

Process C-1 re-performs preliminary design phase activities.

TABLE 4.1
Detailed Design: Barrier and Task Requirements Tables

Table Number	Table Description	Table Info			
		Direct-Use PE		New Skills & Knowledge	See Process/ Step
		SA-1 Info	Act Phase Response		
3.2	Updated preliminary design phase documents listed in Table 3.2	Update			C-1
4.6	Remote Monitoring of Act Phase Response and Support PE	Yes	Yes	Yes	C3-8
4.9	Design: SA-1 Incident Scene Information	—	—	—	C7-7
4.10	Design: Incident Command Board	—	—	—	C7-8
4.12	HMI Display Elements: Detect Phase	Yes	—	—	C8-4
4.13	HMI Display Elements: Support Aids and Alternative Display Types	Yes	Yes	—	C8-4
4.14	HMI Display Element: Act Phase (Controls)	Yes	Yes	—	C8-4
4.15	Integrated HMI Displays	Yes	Yes	—	C8-5
4.16	CCTV Image Elements and PE Requirements	Yes	—	—	C8-7
4.30	Maintenance Procedures (see Note)	—	—	Yes	C15-2b

TABLE 4.2
Detailed Design: Technical, Procurement, and Project Documents

Table	Documents	New, Update, or Input	Process/Step
3.3	Updated preliminary design phase documents listed in table	Update	C-1
4.5	Operating Procedures – Barriers/Tasks	New	C3-3
4.17	Off-Console VDU Displays	New	C8-8
4.19	Control Console or Panel (CP): Barrier Systems and Support Elements	New	C9-1b
4.20	Rooms: Barrier Systems and Support Elements	New	C10-1b
4.21	External Protective Barriers	New	C10-1b
4.22	Engineered Area: Barrier System and Support Elements	New	C11-1b

(Continued)

TABLE 4.2

(Continued)

Table	Documents	New, Update, or Input	Process/Step
4.23	Buildings: Barrier Systems and Support Elements	New	C12-1a
4.30	Maintenance Procedures	New	C15-3
4.31	Procedure Development Plan	New	C15-3
4.32	HE Gap Assessment and Training Plan: Knowledge (Technical, Procedure, Execution)	New	C16-2
4.33	HE Gap Assessment and Training Plan: Skills (PE Usage, Execution)	New	C16-2
4.34	HE Gap Assessment and Training Plan: Skills (Leadership, Teamwork, Stress Management)	New	C16-2
B.5	Operating Procedures: Direct-Use Components	New	C-3 to C-8 (CX-0.2, App. B.6)
B.6	Maintenance Procedures: Direct-Use Components	New	ditto
B.7	Training Requirements: Direct-Use Components (Operate and Maintain)	New	ditto
—	Design requirements: technical systems (In-Place)	Input	C-9 to C-12
—	Design requirements: external support systems	Input	ditto
—	Design requirements: external protective barriers	Input	ditto
—	Requisition input: components, technical systems, panels/control consoles, packaged equipment, buildings	Input/ **New**	C-3 to C-8 (CX-0.3, App. B.6), C-18

TABLE 4.3
Detailed Design: Studies and Assessments

Table	Documents	See Process/Step	Refer
3.4	Updated preliminary design phase documents listed in table	C-1	—
H.2	PIF Assessment – Barrier, Task, and Workspace Activities	C-2	App. H.2
H.3	PIF Assessment – Components (Working Environment)	C-3 to C-8, (CX-0.1e, App. B.6)	App. B.6, H.3

(Continued)

TABLE 4.3
(Continued)

Table	Documents	See Process/Step	Refer
I.1	Cognitive Assessment	C-3 to C-8	App. I.2
4.18	Functional Analysis	C-9 to C-12	—
—	Verification #3	C-19	App. G.5
—	Validation #2	C-20	App. G.6

Note: In large capital projects, design and planning activities often continue into the interim period between the end of the preliminary design phase and start of the detailed design phase. New information is also commonly provided at the start of this phase. Process C-1 integrates this new input into the preliminary design documents and assessments.

Consider the activities that occur early in this phase. The product of this early work can significantly affect the design activities performed by many disciplines and other organizations, for example:

- Updates to risk studies (affects barriers and tasks).
- Updates to the barrier safety management plan, active human barrier design basis, etc.
- Updates to process block diagrams, process flow diagrams, and process and instrument diagrams.
- Recommendations for risk and design assessments.
- Changes or clarification to the project scope or execution plan. This may include changes to the project schedule, contract strategies, regulatory approach, employed industry or owner/operator standards, assignment of barrier design elements and activities to different disciplines or organizations, procurement planning and strategy, or the selected facility certification authorities.
- Updates to the applicable regulatory regime, or industry and company standards.
- Early start required to progress critical path procurement activities.

4.2 PROCESS C-2, ASSESSMENT OF PERFORMANCE INFLUENCING FACTORS

This process performs the assessment of performance influencing factors (PIFs) described in Appendix H.2.

Note: The integration of the approved recommendations should occur in the affected processes.

+

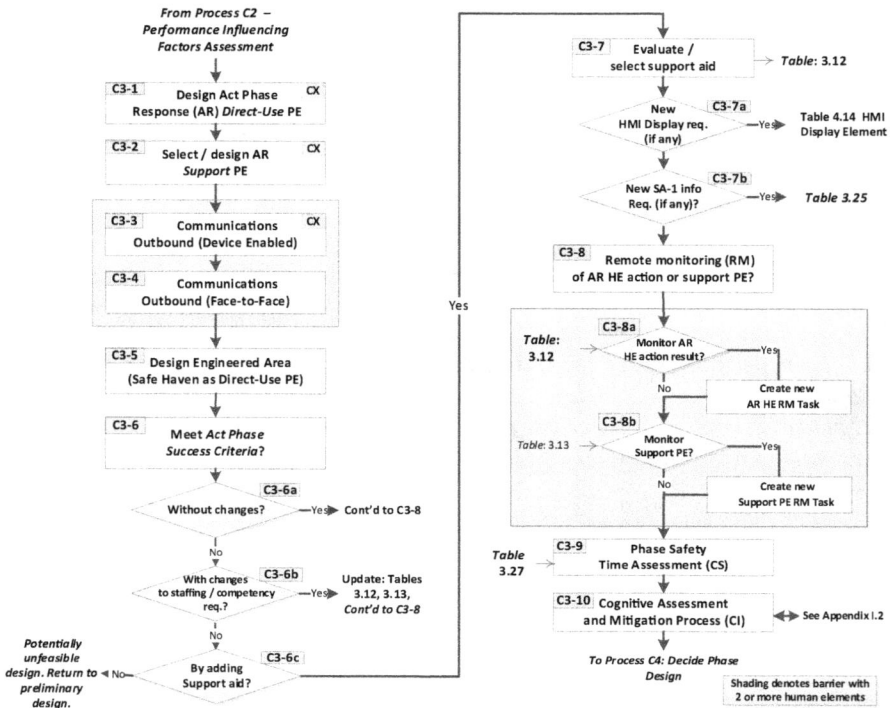

FIGURE 4.2 Detailed Design Process C-3: Act Phase Design. Perform this Process for Every Barrier Task.

4.3 PROCESS C-3, ACT PHASE

Figure 4.2 provides an overview of the act phase design process and steps. See preliminary design steps B-6 through B-9 for background, references, and discussions.

Act Phase Success Criteria: The act phase function and safe state can be reliably achieved within the specified phase safety times when the HE is under time pressure and exposed to designed-for situations and working environment conditions. The criteria are met in all modes of operations identified in Table 3.5 – Barrier Origination Requirements.

Common inputs to processes include the following:

- Barrier safety management plan
- Active human barrier design basis, remote barrier support design basis
- Safety requirements specification
- Barrier roster and organization chart
- Table 3.5. Barrier Origination Requirements (defines safety function, safe state, and operating modes)

- Table 3.6. Task Origination Requirements (defines task goal and safe state)
- Table 3.7. Barrier Performance Standard: Base Requirements
- Table 3.9. Act Phase Response Physical Element Requirements
- Table 3.11. Communication Requirements
- Table 3.12. Act Phase Response: Direct-Use PE Requirements
- Table 3.13. Act Phase Response: Support PE Requirements
- Table 3.16. Detect and Act Phase: In-Place PE Technical System Requirements
- Table 3.27. Barrier, Task, and Phase Safety Times
- Table 3.28. External Support System (ESS) Requirements
- Table 3.30. Reliability Assessment: Single Point of Failure
- Table 3.31. Reliability Assessment: Shared Elements
- Table 4.1. Detailed Design: Barrier and Task Requirements (applicable information)
- Table 4.2. Detailed Design: Technical, Procurement, and Project Documents (applicable information)
- Table 4.3. Detailed Design: Studies and Assessments (applicable information)
- Standard operating procedures
- Barrier block diagrams (developed in preliminary phase, step B-15)
- Specifications (e.g., components, technical systems, packaged equipment systems)
- Data sheets (e.g., components, technical systems, packaged equipment systems)
- HMI Display Design Guidelines and Style Guides, NUREG (2020), ISO-11064-5 (2008), etc.
- Location and layout drawings (e.g., process areas, engineered areas, rooms, control consoles and panels)
- Appendix C (team design information)
- Appendix F (foundational cognitive information)
- Appendix J (OE and HE planning and development)
- Appendix K (workspace design guidance)

Phase-specific inputs:

- Table 3.9. Act Phase Response Physical Element Requirements
- Table 3.11. Communication Requirements
- Table 3.12. Act Phase Response: Direct-Use PE Requirements
- Table 3.13. Act Phase Response: Support PE Requirements

4.3.1   STEP C3-1, ACT PHASE RESPONSE DIRECT-USE PE

This step completes the detailed design and engineering of the direct-use physical element employed in the act phase response. Table 3.12 and the barrier function block diagram identify these PEs.

The applied design process depends on the type of PE. If the direct-use device is a commercial off-the-shelf (COTS) component, use the CX sub-process in Appendix B.6 to select and progress its design. If the physical element is an engineered item, see *Table 4.4* for the suggested design process. The direct-use physical element functionality and design provide design input into the In-Place physical element design. Update Table 3.12 and the block diagram with new information (if any).

TABLE 4.4
Design Processes: Act Phase –Direct-Use PE

| | Direct-Use PE: Act Phase Response | | | | Apply Design Process | |
| | | | | | In-Place PE | |
PE Type	Examples	Function	Use Location	Direct-Use or Support PE	Tech. System	Engr'd Area
Simple active device	Switch, hand valve, lever, pushbutton	Activates safe state function via SIS	Control room		C14-2	—
			Process area			
			Engr'd Area			
Equipment (COTS)	Stretcher	Recover injured person	Indoor	CX		C-11 (e.g., safe haven, escape route)
			Process area	(see App. B.6)	—	—
Engineered PE	Firewater (FW) hose	Fire suppression	Many		C14-2 (FW system)	—
Engineered PE	Lifeboat	Physical means to abandon facility	Open sea, Engineered area, e.g., boarding area	CX, C14-2	C14-2 (e.g., lifeboat launch system)	C-11 (e.g., safe haven, escape route)
Hand-held comms. device	Telephone or radio handset	Convey: – Requests – Commands – Status	Process area	CX	C14-2	—
			Room	(see App. B.6)		
			ER station			
			Engr'd Area			
HMI display	Command entry target	Initiate AR via target on HMI display	Control room	C8-4	C14-2	—
			Process area panel		C14-2	—
Engineered area (safe haven)	Outdoor muster area	Protect occupants from hazard	Outdoor muster area	CX		C-11
	Indoor muster area		Indoor muster area	*C-11*		C-11

Support PE

					Apply Design Process	
Equipment (COTS)	Smoke hood	Temporary protection from smoke effects	Egress/escape routes	CX (see App. B.6)	—	—
Equipment (COTS)	Self-contained breathing apparatus	Entry into low O2 or toxic air environments *(no remote monitoring)*	Process area, buildings, etc.		—	—
		Add remote monitoring		CX, C3-8	C14-2	—

Example: Refer to the functional block diagram example in Figure 3.6 *(step B15-3). In this example, the HE's act phase response uses an emergency shutdown (ESD) pushbutton to initiate an ESD. A technical system receives the pushbutton signal and initiates actions that achieve the safe state. (The final element that achieves the safe state, e.g., a control valve, is not a direct-use physical element because there is no direct HE interaction.) Another example is using a hand-directed fire hose to control the spread of fire or cool adjacent equipment. The HE handles and directs a fire hose in the manner that achieves the required task result (e.g., the barrier or task safe state). The firewater system (e.g., water source, pumps, piping) provides an In-Place physical element needed to enable the barrier function. Another example, an HE uses a telephone handset (direct-use PE) to initiate a response that communicates the required information or a request. The telephone system and infrastructure (the In-Place PE) enable the communication.*

4.3.2 Step C3-2, Support PE Selection and Design

This step progresses the detailed design activities for the Support PEs listed in Table 3.13. The Support PE may be a COTS component or an engineered item. Table 4.4 provides the suggested design approach given the physical element type, for example, a COTS or an engineered item. As a last step, update Table 3.13 with the new information (if any).

A Support PE may rely on an In-Place PE (e.g., technical system) to enable a remote monitoring task. The Support PE function and performance requirements provide design input to the In-Place physical element design.

4.3.3 Step C3-3, Communications (Device-Enabled)

This step further evaluates the device-enabled communication exchanges recorded in Table 3.11 and progresses its design (PE and HE). The process evaluates the conveyed message, form, and media for appropriateness and its effect on the sender and receiver(s). It designs the direct-use physical element and provides design input to the In-Place physical element as applicable. The communication requirements are then addressed in the procedures

Figure 4.3 is an example of device-enabled communication system.

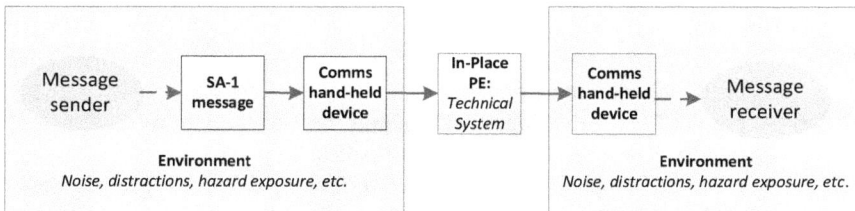

FIGURE 4.3 Example of Device-Enabled Communication.

The following are steps in the C3-3 process:

1. Select a communication requirement in Table 3.11 for evaluation.
2. Review the following inputs as they pertain to the selected communications:
   - Applicable input documents listed in the introduction
   - Communication equipment specifications
   - Area noise studies
   - Ergonomics studies, for example, of the control room
3. Evaluate the *conveyed information* against the above inputs, and the guidance and considerations in Appendices C.1 and C.3. Confirm the conveyed information is necessary and appropriate to the stated function. If not, determine whether a change is required. An example change may be one that modifies or deletes the communication requirement. (If warranted, employ the change management process.)
4. Evaluate the *message form and media* against the guidance in the standard operating procedures and the following:
   - Appendix K.3 – Communications workspace design
   - Appendix K.5 – Information workspace design
        If the proposed message form and media are appropriate, proceed to step 5. If not, determine the change needed. Examples may include a change to the proposed message form or proposed media. Update the revised information in Table 3.11 and other documents affected by the change. (If warranted, employ the change management process.)
   - This step should consider the intended receiver(s) of the information to confirm the conveyance method is optimal for this communication. See Appendix K.3 for example considerations and guidance.
   - This step should consider the additional attributes of the conveyed information to determine its effect (if any) on the selected method of conveyance. See Appendix K.5 for example considerations and guidance.
5. Evaluate the exchange effect on the *message sender and receiver.*

Note: Communication is a sustained (captive) attention task for the message sender and receiver(s) that extends for the duration of the exchange. The exchange requires accurate and timely message delivery (sender). For receivers, it requires an accurate perception and understanding of the message, and the need to remember the information for later recall, a short-term memory demand.

Consider the information from step 4 and the following:
- Appendix F.2 – Attention, working, and short-term memory
- Appendix K.4 – Human element workspace
     If the exchange does not negatively affect the sender or receiver, then proceed to step 6 (e.g., a negative effect may be when the communication timing interferes with a higher priority barrier task or activity). Otherwise, define solutions to eliminate or mitigate the undesirable

effects and confirm that the solution does not introduce a new deficiency. Update Table 3.11 and other affected documents with the approved change. (If warranted, employ the change management process.)

6. Using the information from the previous steps, proceed with the detailed design of the communication equipment (direct-use components) using the CX process described in Appendix B.6.

7. If a new procedure is required to address this communication requirement, identify the requirements in Table 4.5. Populate all **Procedure Requirement** fields, and, if known, the **Barrier ID** and **Task ID** fields. (The procedure may apply to other barriers and tasks.) Enter Table 3.11 into the **Source Ref.** field if it is the basis for the procedure addition. If the provider is the facility owner/operator, enter O/o in the **Provider** field. If O/o is not the provider, leave the **Constraint on Use** field blank (The **Support Aid** field is addressed in step C3-6c. Other fields under the **Procedure Detail** header are defined in implementation phase process, C32-2b.)

8. Repeat the steps above for the next communication requirement in Table 3.11.

4.3.4   Step C3-4, Communications (In-Person)

This step evaluates and develops the in-person communication requirements recorded in Table 3.11. (Table 3.11 identifies the "Msg. Form" and "Media Type" as "In-Person.") Perform all steps in C3-3 above, except for step 6.

4.3.5   Step C3-5, Design of the Engineered Area Identified as Direct-Use PE

This step addresses a unique case where an engineered area is the direct-use PE. In this case, use process C-11 to design the engineered area.

Note: For an offshore O&G production facility, the designated muster area (safe haven) is an example where an engineered area is the direct-use PE. The safe state is achieved when assigned personnel occupy a safe haven that provides temporary protection from hazards.

4.3.6   Step C3-6, Meets Act Phase Success Criteria?

See the ***Act Phase Success Criteria*** and input information listed in the C-3 introduction. In this step, guided by an operations expert, review and assess each HE action, and all plausible interactions with the direct-use and Support PE identified in Table 3.9. Consider the following positive and negative (and like) effects:

- Current and specified competencies for the proposed HE
- Nature of the protected process, and the hazard that activated the barrier and its potential consequences
- Inherent nature and functioning of the barrier or task
- Target task and task phase safety times (Table 3.27)

TABLE 4.5

Operating Procedures – Barriers/Tasks

Procedure Requirement				Procedure Detail					
Procedure									
Source Ref. ID	ID	Provider	Type/Use	Content	Format/Type	Constraints on Use	Support Aid	Task ID	Barrier ID
Enter source reference or Table	Enter unique ID for this proc.	Enter – O/o – Vendor – Contractor – Existing – Other	Enter – Start-up – Normal – Abnormal – Shutdown One per row	Enter required content	Enter procedure format or type (from standard templates)	Enter considerations that may affect type, format, or level of detail	Enter – Yes – No Yes means a support aid is required	Enter task ID. All that apply	Enter barrier ID

- Recommendations from reliability studies (preliminary design, step B-19; and the detailed design phase process, C-1 update) and performance influencing factor assessment (process C-2)
- The likelihood that other barriers may be activated at the same time (e.g., excessive workload)
- Design basis environmental and workspace conditions that may impede or interrupt this activity

4.3.6.1   C3-6a, Achievable Without Changes?

Confirm whether the stated success criteria can be reliably achieved without changes to the current design. If yes, proceed to step C3-8. If not, proceed to step C3-6b for further evaluation.

4.3.6.2   C3-6b, Achievable with OE or HE Changes?

1. Assess and determine whether the success criteria can be met with an HE or OE change. (For examples, see Appendices J.6, K.3, K.4, and K.5, and the applicable solutions in Table I.13 (Appendix I.3.6).)
2. If achievable with an HE or OE change, add the change to the affected design documents, then proceed to step C3-8. If not, proceed to step C3-6c. Example HE or OE changes:
   - Add a new skill or knowledge requirement to Table 3.11, 3.12, or 3.13, as applicable.
   - Add a new operating procedure requirement to Table 4.5.
   - Increase the number of HEs or change distribution of barrier tasks between the barrier HEs. (Indicate the proposed change in the barrier roster and organization chart.)

     (These types of changes may require a return to the preliminary design phase to fully address the change. Use of the change management process may be warranted.)

     Note: This step raises an issue pertaining to the owner/operator philosophy. Should the order of preference be an OE change, HE change, or addition of a new physical element? Each approach has its pros and cons. This topic may be one to address in the active human barrier design philosophy or design basis.

4.3.6.3   C3-6c, Achievable with Support Aid (PE)?

Assess and determine whether the success criteria can be met by adding a support aid. (For example support aid types and options, see Appendix B.8.)

If the success criteria can be met with the identified aid(s):

- In Table 3.12 (direct-use PE), add the support aid (type and function) and the new skill or knowledge needed (if any) to understand and correctly use the aid.
- Proceed to step C3-7.

If the success criteria cannot be met with a support aid, the barrier or task may be unfeasible. A return to the preliminary design phase may be needed.

4.3.7 Step C3-7, Specify Support Aid Requirements

This step further specifies and progresses the design of the support aid identified in step C3-6c.

- If the aid is an element on an HMI display (added information or alternative display type), proceed to step C3-7a.
- If the aid is a COTS device, follow the CX sub-process steps in Appendix B.6.
- If the aid is a checklist or similar device, define and develop the checklist.
- If the support aid is an engineered component (other than an HMI display element), follow the design process most suited to the aid type. Consider providing a *Support Aid Functional Specification* that may include the following:
  - A more detailed definition of the presented information and form of information.
  - Identify new additional SA-1 input information (if any) needed as input to the support aid.
  - Specify the location where the support aid information must be available for use or access by the target HE.

4.3.7.1 C3-7a, Define Custom HMI Display Requirements (If Applicable)

If the support aid takes the form of an HMI display element, follow the process in C8-4b – VDU-Based Support Aids and Alternative Display Types.

4.3.7.2 C3-7b, Additional SA-1 Information Required

If the support aid requires new SA-1 information, add the information to Table 3.25 – Detect Phase: SA-1 Information Requirements.

Note: The follow-up design of this new SA-1 information occurs in process C-7 or C-8.

4.3.8 Step C3-8, Remote Monitoring Function

Remote monitoring (RM) requirements and tasks identified in the preliminary design phase or process C-1 should be integrated into the normal task design processes. This step applies to an RM requirement added in the detailed design phase. RM functions (examples) may take two forms.

- The result of an act phase HE response action is monitored by someone other than the HE who initiated/performed the response action.
- Support PE (operational status) is monitored by someone other than the Support PE user.

The RM task may be a recommendation from one of the assessments, for example, a means to improve function reliability or resilience. Alternatively, the task may be required if the physical element cannot be reliably monitored by the HE user. For background on information attributes to consider, see Appendix K.5 and Table K.7.

Note: The information types in Table K.7 (Appendix K.5) are normalized and employed throughout the lifecycle model.

4.3.8.1   C3-8a, Remote Monitoring Act Phase HE Response Action

Suggested steps in this process are as follows:

1. If a new RM task is identified in the detailed design phase, several possible path forward options may include the following:
   a. Consider a return to the step B2-5 risk assessment to confirm the RM task is viable and necessary. If so, add it to the Task Origination Table 3.6. Enter the RM act phase response and associated information and fields in Table 3.9 (preliminary design, step B-6). As applicable, enter the appropriate in Direct-Use PE Table 3.12 (preliminary design, step B-8). From there, progress the design through the remaining applicable preliminary and detailed design phase steps and processes.
   b. Progress the task definition and design from this step forward guided by the information in Table 4.6.

   Note: Both of the above options may warrant use of the change management process.

2. With either option in step 1, define, develop, and enter the information into all Table 4.6 fields (when known).

*Under the **RM Task** header, enter:*
   ID: Assign ID for this new RM task.
   Task Goal: Identify the RM task goal.
   Task HE: Identify the HE(s) assigned to this task (one per row).
   *Under the **Detect Function** header, enter:*
   SA-1 Info: Identify the monitored SA-1 information.
   Monitored **Info Form or Source:**

- Observed Status: Real-time, direct observation, for example, monitored via CCTV display.
- HMI Display: Presents real-time information from the monitored HE action or Support PE status and provides visual or audible alerts that should trigger and act phase response.
- Other forms or sources.

Monitors **From Where:** Identify the physical location from where the RM task is performed, such as a Central Control Room or Remote Operations Center.

TABLE 4.6
Remote Monitoring of Act Phase Response and Support PE

Barrier/Task Name

	RM Task			Detect Function						Action Response		
ID	Task Goal	Task HE	SA-1 Info	Info Form	From Where	Activator Event	Decide	HE Response Action		Safe State	Task Safety Time (BST)	Consequence
Enter task ID	Enter RM task goal One goal per line *See example*	Enter ID/role for HE assigned to this task	Enter SA-1 Info	Enter Info form *See example*	Enter RM location(s) *See example*	Enter activator event or condition that requires action response *See example*	Enter Req'd decision (s)	Enter action response to a detected trigger event or condition *See example*		Define expected result/ safe state	Enter task safety time for this task.[a]	Enter consequence of delayed or no response

[a] *The RM task is itself an active human barrier. As such, it will have its own barrier safety time (BST).*

Activator/Event: Define the detected SA-1 condition or event that activates the RM task.

*In the **Decide** field,* define all required task decisions.

Under the ***Action Response*** header, enter:

- **Response Action:** Define the required act phase response.
- **Safe State:** Define the required safe state or condition achieved by the response action.
- **Consequence:** Identify the potential consequence(s) if the RM task fails or is not achieved within the stated safety time.

4.3.8.2　C3-8b, Remote Monitoring of Support PE

Apply the steps in C3-8a to progress the design of the Support PE remote monitoring (RM) task.

To monitor Support PE, the RM goal may be to monitor or identify a pending safety limit or failure and promptly notify the user of the condition. This may be a time-sensitive, safety critical task. For example, a remotely located HE is assigned the task of monitoring the remaining air or air pressure in a self-contained breathing apparatus. The equipment is worn by a different HE who is actively performing a task in a potentially hazardous and time-pressured environment. When or before a procedure-specified minimum air pressure limit is reached, the RM HE notifies the Support PE user of the alert condition and the procedure-recommended response. This type of task may add new direct-use and In-Place PE requirements (e.g., an HMI display (with alert) or technical system). This new function may add new skill and knowledge requirements (required competencies) that apply to the RM task assignee and Support PE user.

4.3.9　Step C3-9, Confirm Act Phase Safety Time

This step confirms the act phase response actions can reliably and repeatedly achieve the specified safe state within the allocated phase safety time (T_{ai} plus T_{ar} from Table 3.27) and under all defined situations, conditions, and modes of operation.

Separately assess the physical element and HE phase safety time contributions. The suggested steps are as follows:

4.3.9.1　Physical Elements

1. Assess the phase response contribution from each physical element employed (directly and indirectly) to achieve the specified safe state. (For direct-use components, this assessment is performed by the CX sub-process in Appendix B.6).
2. If the net (collective) response does not meet the phase safety time requirements, identify changes to meet or exceed these requirements. Example solutions may include the following:
   - Select different components or change its implementation (performed in CX sub-process in Appendix B.6).
   - Improve the response time in the supporting In-Place PE (e.g., technical system).

3. If the considered changes in step 2 cannot meet requirements, consider the following options:
   - Return to the safety time assessment in the preliminary design in step B-16. Determine whether changes to the task and phase allocations can meet requirements without creating similar issues in other phases or tasks.
   - If permitted by the documents listed in the introduction, change the barrier safety time (BST) to a higher value, though still below the process safety time (PST) recorded in Table 3.5. (Design documents such as the design basis and SRS may require a minimum safety margin in the assigned BST.)
4. If the requirement cannot be met by the above changes, a return to the risk assessment process may be needed to seek a deeper assessment of the barrier PST. If the PST does not significantly change, the barrier as currently defined may be unfeasible, which may require a return to the risk assessment process to seek alternative barriers.

4.3.9.2   Human Elements

5. Evaluate the response safety time contributions from the HE assigned to perform response actions. Perform an assessment to confirm that the actions can be reliably achieved within the stated times while under time pressure and exposure to the stated conditions and environment (physical and task).
6. If the individual response does not meet requirements, identify the changes needed to meet those requirements. Examples of changes include the following:
   - See Appendix J.6 for possible methods to improve human performance (reduce human response and task execution time).
   - Reduce the time allocated to the physical element response and adding that time to the HE response. (This may increase the performance demands placed on PE.)
7. If the options in step 6 do not achieve the required response, consider those defined in step 3 above.
8. If the required safety time is not met by the above options, see step 4 above for suggested actions.

Note: Use of the change management process may be warranted for changes of this type.

4.3.10   STEP C3-10, PERFORM THE COGNITIVE ASSESSMENT AND MITIGATION PROCESS

This step performs the cognitive assessment process defined in Appendix I.2 and the Evaluation Input Guidance Table I.6.

4.4   PROCESS C-4, DECIDE PHASE

Figure 4.4 provides an overview of the decide phase detailed design and engineering process. See preliminary design step B-10 for the background, references, and discussions.

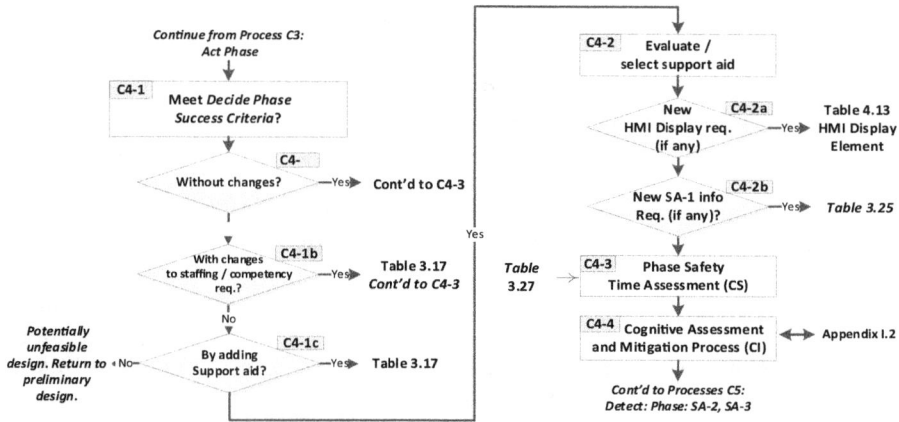

FIGURE 4.4   Detailed Design Process C-4: Decide Phase. Perform this Process for Every Barrier Task.

Decide Phase Success Criteria: *The decide phase requirements (functions and performance) can be reliably achieved within the specified phase safety times when the HE is under time pressure and exposed to designed-for situations and working environment conditions. The criteria are met in all modes of operations identified in* Table 7 – *Barrier Origination Requirements.*

Inputs to this process:
- See the "common inputs" listed in the process C-3 introduction.
- Table 3.17 – Decide Phase Requirements.

4.4.1   Step C4-1, Meets Decide Phase Success Criteria?

See the ***Decide Phase Success Criteria*** and input information listed in the C-4 introduction. In this step, guided by an operations expert, review and assess each decision requirement in Table 3.17. Consider the following positive and negative (and like) effects listed in the act phase process step C3-6.

4.4.1.1   C4-1a, Achievable Without Changes?
Confirm whether the stated success criteria can be reliably achieved without changes to the current design. If yes, proceed to step C4-3. If not, proceed to step C4-1b.

4.4.1.2   C4-1b, Achievable with OE or HE Changes?
1. Follow the process in C3-6b step 1. *Exception*: For the reference to Appendix I.3, refer to Table I.12 in Section I.3.5.
2. Follow the process in C3-6b step 2 with the following exceptions:
   - In Table 3.17, add the new skill or knowledge requirement.
   - In the roster and organization chart, identify the proposed HE change.
3. If the success criteria are met by the above change(s), proceed to step C4-3. If not, proceed to step C4-1c.

4.4.1.3   C4-1c, Achievable with Support Aid (PE)?

Follow the guidance in act phase step C3-6c for this step with the following exception: In Table 3.17, add the support aid (type and function) and the new skills or knowledge needed (if any) to understand and correctly use the aid. Proceed to step C4-2.

If the success criteria cannot be met with a support aid, the barrier or task may be unfeasible. A return to the preliminary design phase may be needed.

Note: A wide range of decision aids are available to support the varied task-unique situations. An example aid may be a list of the available act phase options (situation-dependent). Others may provide guidance on a pending decision to initiate a task response action now or wait, or what task to attend to next, or the response options when faced with a time-of-the-essence situation and the available information is incomplete or conflicting.

4.4.2   Step C4-2, Specify Support Aid Requirements

Follow the guidance in act phase step C3-7 for this step.

4.4.2.1   C4-2a, Define Custom HMI Display Requirements (If Applicable)

If the support aid takes the form of an HMI display element, follow the process in C8-4b – VDU-Based Support Aids and Alternative Display Types.

4.4.2.2   C4-2b, Additional SA-1 Information Required

Follow the guidance in act phase step C3-7b for this step.

4.4.3   Step C4-3, Confirm Phase Safety Time

This step evaluates the decide phase decisions to confirm whether all decisions can be reliably completed within the allocated time, T_{dec} (Table 3.27), for all modes of operation and plausible situations and conditions that can affect the phase safety time (see Table 3.5 for the barrier modes of operation).

Follow the guidance in the act phase step C3-9 for this step. If the design provides no PEs to support this phase, limit the assessment to the HE only.

4.4.4   Step C4-4, Perform the Cognitive Assessment
and Mitigation Process

This step performs the cognitive assessment and mitigation process defined in Appendix I.2 and the Evaluation Input Guidance Table I.5.

4.5   PROCESS C-5, DETECT PHASE: SA-2, SA-3

Figure 4.5 provides an overview of the detect phase (SA-2 and SA-3) detailed design and engineering process. See preliminary design steps B-11 and B-12 for the background, references, and discussions.

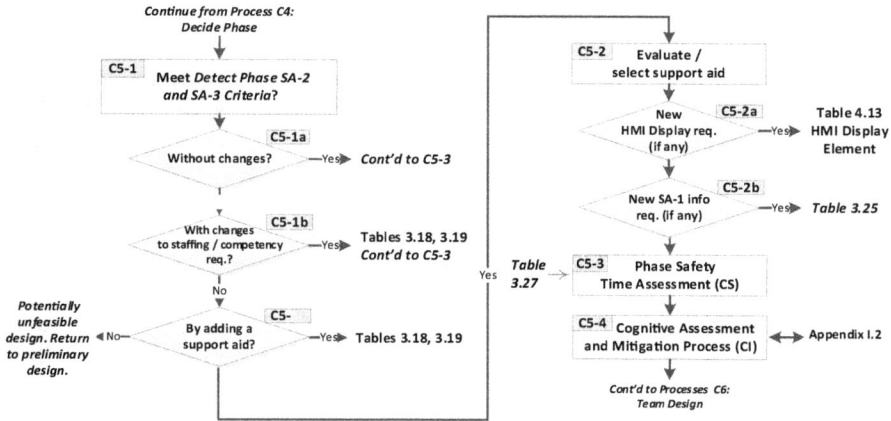

FIGURE 4.5   Detailed Design Process C-5: Detect Phase SA-2, SA-3. Perform this Process for Every Barrier Task.

Inputs to this process:
- See the "common inputs" listed in the process C-3 introduction.
- Table 3.18. Detect Phase: SA-2 Comprehension Requirements.
- Table 3.19. Detect Phase: SA-3 Projection Requirements.

Detect Phase SA-2 and SA-3 Success Criteria: *The detect phase SA-2 (comprehension) and SA-3 (projection) requirements (functions and performance) can be reliably achieved within the specified phase safety times when the HE is under time pressure and exposed to designed-for situations and working environment conditions. The criteria are met in all modes of operations identified in* Table 3.5 *– Barrier Origination Requirements.*

4.5.1   STEP C5-1, MEETS DETECT PHASE SA-2 AND SA-3 SUCCESS CRITERIA?

See the ***Detect Phase SA-2 and SA-Success Criteria*** and input information listed in the C-5 introduction. In this step, guided by an operations expert, review and assess each SA-2 and SA-3 requirement in Tables 3.18 and 3.19, respectively. Consider the following positive and negative (and like) effects listed in the act phase process step C3-6.

4.5.1.1   C5-1a, Achievable Without Changes?

Confirm whether the stated success criteria can be reliably achieved without changes to the current design. If yes, proceed to step C5-2. If not, proceed to step C5-1b.

4.5.1.2   C5-1b, Achievable with OE or HE Changes?

1. Follow the process in C3-6b step 1. *Exception*: For the reference to Appendix I.3, refer to Tables I.10 and I.11 in sections I.3.3 and I.3.4, respectively.
2. Follow the process in C3-6b step 2 with the following exceptions:
   - In Tables 3.18 and 3.19, add the new skill or knowledge requirement.
   - In the roster and organization chart, identify the proposed HE change.

3. If the success criteria are met by the above change(s), proceed to step C5-3. If not, proceed to step C5-1c.

Consider: An unusually high competency requirement in terms of experience, expertise, skill, or knowledge may be unrealistic or not aligned with an operational objective to minimize changes to existing programs, for example, in staffing and staff selection.

4.5.1.3   C5-1c, Achievable with a PE Support Aid?

Follow the guidance in act phase step C3-6c for this step with the following exception:

- To Tables 3.18 and 3.19, add the support aid (type and function) and the new skills or knowledge needed (if any) to understand and correctly use the aid. Proceed to step C5-2.

If the success criteria cannot be met with a support aid, the barrier or task may be unfeasible. A return to the preliminary design phase may be needed.

Note: SA-2 and SA-3 support aids differ from those provided for decisions. A SA-2 support aid may provide a checklist or flowchart that contributes to understanding a current situation given the available information. A SA-3 support aid may be a checklist that shows plausible event escalation paths or the consequences of a degrading external support system or external protective barrier.

Alternative display type. A support aid may take the form of an alternative HMI display type that instantly conveys meaning and understanding without requiring advanced knowledge. (Step C8-4b progresses the design of this display type.) Consider the example of a barrier activated by flammable gas detection alarms. The detector measurements and alarms are often misunderstood, contributing to reduced or delayed barrier response and reliability. The presented indication may be a percentage of the lower explosive limit (% LEL) or an inferred indication of the actual hazard, i.e., the effect if the gas ignites. The ignition effect depends on the gas make-up, cloud size and concentration, gas release rate and location (congested, enclosed space, non-classified area), and environmental conditions such as wind direction and dilution rate. Unsurprisingly, active human barriers that rely on flammable gas detection tend to be problematic. An option to improve reliability and performance may utilize the findings from the fire and blast study to create a display that directly conveys the meaning of the gas alarm activations. A physical arrangement display shows the objects of interest (key receptors) within the monitored area. As alarms activate, color-coding indicates the threat level (heat and overpressure) to each receptor.

Additional thoughts on the abovementioned example: A final thought on the alternative display-type example follows. Operators seldom have full knowledge of the fire and gas studies and the base case scenarios used to design the facility and guide detector coverage (how many and where). Procedures and training should make the HE aware of the consequences when one or more gas detectors in the monitored area fail or are removed for maintenance. A further problem with barriers activated by flammable gas detection may occur if an HE has repeated experiences of gas release without ignition. Over time, this may reduce the HE's perceived risk regarding this barrier type, and also change the HE's mental model accordingly, i.e., a source of drift (unsafe direction) that is undetected. This is a drift-attributed example and plausible justification for periodic re-training, competency assessment, and reinforcement.

4.5.2   Step C5-2, Specify Support Aid Requirements

Follow the guidance in act phase step C3-7 for this step. Consider the appropriate application of Meta and Compatible SA discussed in Appendix C.

4.5.2.1   C5-2a, Define Custom HMI Display Requirements (If Applicable)

If the support aid takes the form of an HMI display element, follow the process in C8-4b – VDU-Based Support Aids and Alternative Display Types.

4.5.2.2   C5-2b, Additional SA-1 Information Required

Follow the guidance in act phase step C3-7b for this step.

4.5.3   Step C5-3, Confirm Phase Safety Time

This step evaluates the detect phase SA-2 and SA-3 elements and functions to confirm the phase can be reliably completed within the allocated time: T_{cp} (Table 3.27) for all modes of operation and plausible situations and conditions that can affect the phase safety time. (See Table 3.5 for the barrier modes of operation.)

Follow the guidance in the act phase step C3-9 for this step. If the design provides no PEs to support this phase, limit the assessment to the HE only.

4.5.4   Step C5-4, Perform the Cognitive Assessment
and Mitigation Process

This step performs the cognitive assessment and mitigation process defined in Appendix I.2 and the Evaluation Input Guidance Tables I.3 and I.4.

4.6   PROCESS C-6, TEAM DESIGN: SHARED
SITUATION AWARENESS

Figure 4.6 provides an overview of the team design, shared situation awareness (SSA) detailed design, and engineering process. See the preliminary design steps B-5 and B-13 for the background, references, and discussions.

Inputs to this process:
- See the "common inputs" listed in the process C-3 introduction.
- Table 3.8. Shared Situation Awareness Requirements.
- Table 3.20. NTS Requirements: Teamworking.
- Table 3.22. NTS Requirements: Leadership.
- Table 3.24. NTS Requirements: Monitor and Manage Acute Stress.

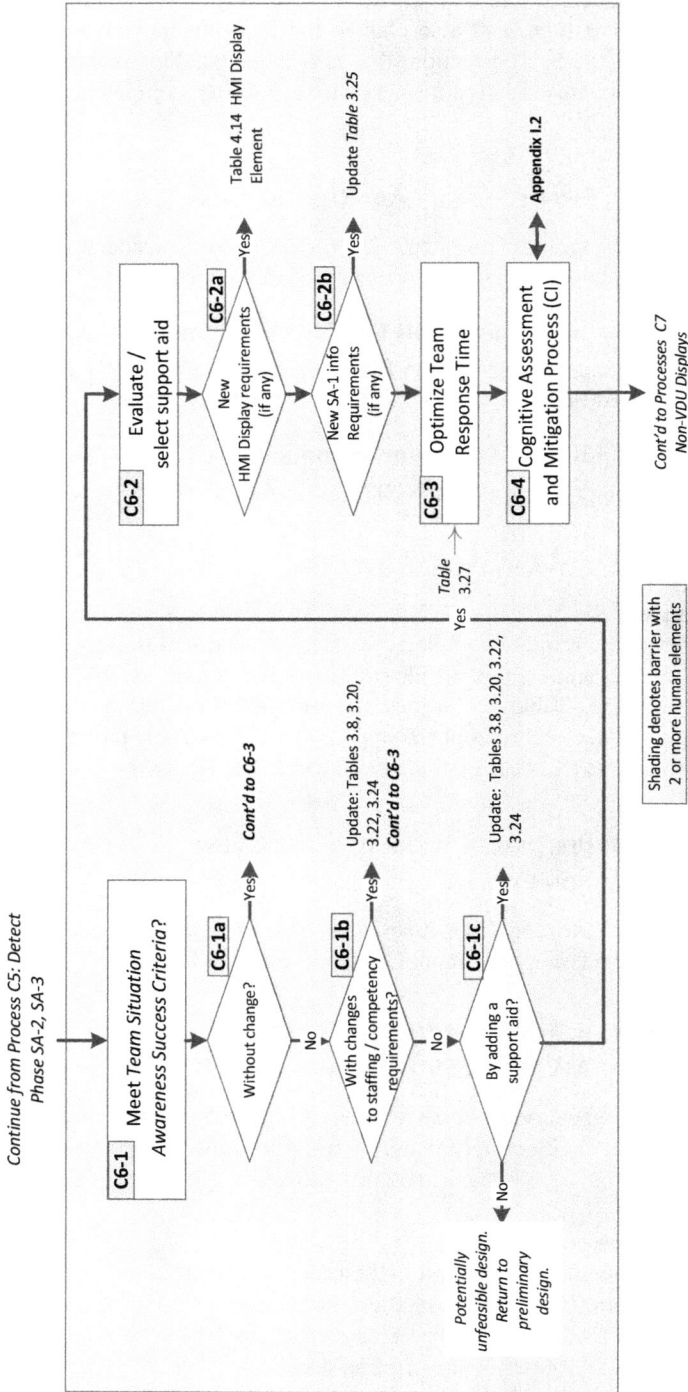

FIGURE 4.6 Detailed Design Process C-6: Team Design. Perform this Process for Every Barrier Task (multi-HE barriers).

SSA and Non-Technical Skills Success Criteria: The SSA and team functioning requirements (non-technical skills) can be reliably achieved within the specified phase safety times when the HE is under time pressure and exposed to designed-for situations and working environment conditions. The criteria are met in all modes of operations identified in Table 3.5 – *Barrier Origination Requirements.*

4.6.1 STEP C6-1, MEETS TEAM FUNCTIONING AND SSA SUCCESS CRITERIA?

See the *SSA and Non-Technical Skills Success Criteria* and input information listed in the C-6 introduction. In this step, guided by an operations expert, review and assess each SSA and NTS requirement in Tables 3.8, 3.20, 3.22, and 3.24. Consider the following positive and negative (and like) effects listed in the act phase process step C3-6.

4.6.1.1 C6-1a, Achievable Without Changes?

Confirm the stated success criteria can be reliably achieved without changes to the current design. If yes, proceed to step C6-2. If not, proceed to step C6-1b.

4.6.1.2 C6-1b, Achievable with OE or HE Changes?

1. Follow the process in C3-6b step 1. *Exception*: For the reference to Appendix I.3, refer to Table I.14 in Section I.3.7.
2. Follow the process in C3-6b step 2 with the following exceptions:
   - In Tables 3.8, 3.20, 3.22, and 3.24, add the new skill or knowledge requirement.
   - In the roster and organization chart, identify the proposed HE change.
3. If the success criteria are met by the above change(s), proceed to step C6-3. If not, proceed to step C6-1c.

4.6.1.3 C6-1c, Achievable with a PE Support Aid?

Follow the guidance in act phase step C3-6c for this step with the following exception:

- In Table 3.8, add the new skill or knowledge requirement (if any) to understand and correctly use the aid. Proceed to step C6-2.

If the success criteria cannot be met with a support aid, the barrier or task may be unfeasible. A return to the preliminary design phase may be needed.

4.6.2 STEP C6-2, SPECIFY SUPPORT AID REQUIREMENTS

Follow the guidance in act phase step C3-7 for this step. Consider the appropriate application of Meta and Compatible SA discussed in Appendix C.

4.6.2.1 C6-2a, Define Custom HMI Display Requirements (If Applicable)

If the support aid takes the form of an HMI display element, follow the process in C8-4b – VDU-Based Support Aids and Alternative Display Types.

Described in Appendix C.1.3, compatible SA should be considered in the design of displays that address the potentially unique needs of HEs who may use the same SA-1 information; however, its meaning in their assigned tasks may differ.

4.6.2.2 C6-2b, Additional SA-1 Information Required

Follow the guidance in act phase step C3-7b for this step.

4.6.3   STEP C6-3, MINIMIZE BST AND TTST BY OPTIMIZING TEAM PERFORMANCE

The step seeks additional opportunities to minimize the net time needed to execute the barrier and associated task and task phases. See Appendix J.6 for examples of options to reduce HE and team response time, a means to improve team performance.

4.6.4   STEP C6-4, PERFORM THE COGNITIVE ASSESSMENT AND MITIGATION PROCESS

This step performs the cognitive assessment and mitigation process defined in Appendix I.2 and the Evaluation Input Guidance Table I.7.

4.7   PROCESS C-7, DETECT PHASE: SA-1, NON-VDU DISPLAYS

Figure 4.7 provides an overview of the detect phase (SA-1) detailed design and engineering process specific to non-VDU-based displays. See preliminary design step B-14 for the background, references, and discussions.

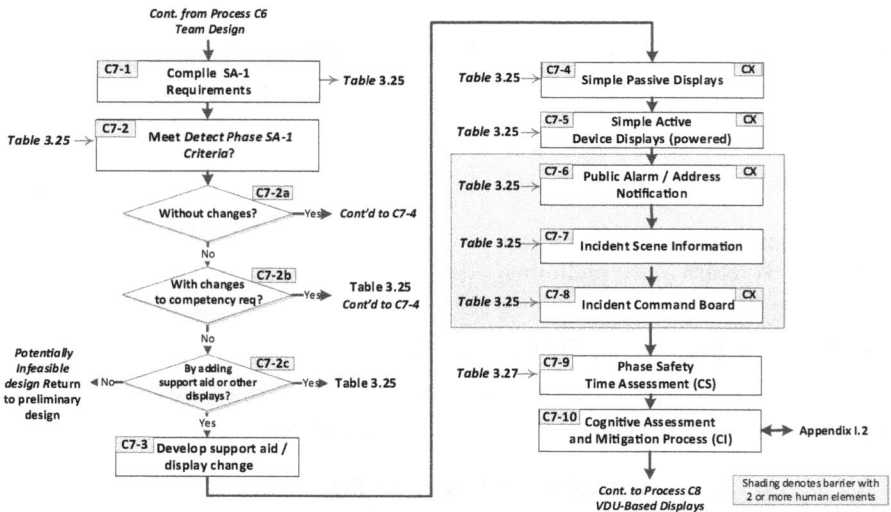

FIGURE 4.7   Detailed Design Process C-7: Detect Phase – Non-VDU Displays. Perform this Process for Every Barrier Task.

Detect Phase SA-1 Success Criteria: *The detect phase SA-1 requirements (functions and performance) can be reliably achieved within the specified phase safety times when the HE is under time pressure and exposed to designed-for situations and working environment conditions. The criteria are met in all modes of operations identified in* Table 3.5 – *Barrier Origination Requirements.*

Inputs to this process:
- See the "common inputs" listed in the process C-3 introduction.
- Table 3.8. Shared Situation Awareness Requirements.
- Table 3.12. Act Phase Response: Direct-Use PE Requirements.
- Table 3.13. Act Phase Response: Support PE Requirements.
- Table 3.16. Detect and Act Phase: In-Place PE Technical System Requirements.
- Table 3.17. Decide Phase Requirements.
- Table 3.18. Detect Phase: SA-2 Comprehension Requirements.
- Table 3.19. Detect Phase: SA-3 Projection Requirements.
- Table 3.25. Detect Phase: SA-1 Information Requirements.
- Table 4.6. Remote Monitoring of Act Phase Response and Support PE.
- Table 4.7. Design: Simple SA-1 Passive Displays (Unpowered).
- Table 4.8. Design: Simple SA-1 Active Displays (Powered).
- Table 4.9. Design: SA-1 Incident Scene Information.
- Table 4.10. Design: Incident Command Board.
- Table 4.11. Design: Received Communications.

The form of the SA-1 information varies, for example:

- Audible and/or visual alarm notification
- Visually observable information from an SA-1 device or an incident scene
- Verbally and/or visually conveyed information
- Other sensed information from an incident scene, e.g., sounds, smell, and touch.

4.7.1   STEP C7-1, COMPILE AND DEVELOP SA-1 INFORMATION REQUIREMENTS

1. Identify the new SA-1 requirements defined in the references listed in the step C-7 introduction.
2. Record this new SA-1 information in Table 3.25. (See the preliminary design step B-14 for guidance on populating the table fields.)

Note: When this step is completed, Table 3.25 *should provide a complete listing of all SA-1 information elements and requirements.*

4.7.2   STEP C7-2, MEETS DETECT PHASE SA-1 SUCCESS CRITERIA?

See the *Detect Phase SA-1 Success Criteria* and input information listed in the C-7 introduction. In this step, guided by an operations expert, review and assess each SA-1 information requirement and the detectability and direct-use interaction (perception) with each display source indicated in Table 3.25. Consider the following positive and negative (and like) effects listed in the act phase process step C3-6.

4.7.2.1   C7-2a, Achievable Without Changes?

Confirm whether the stated success criteria can be reliably achieved without changes to the current design. If yes, proceed to step C7-3. If not, proceed to step C7-2b.

4.7.2.2   C7-2b, Achievable with Changes to Competency?

1. Follow the process in C3-6b step 1. *Exception*: For the reference to Appendix I.3, refer to Table I.9 in Section I.3.2.
2. Follow the process in C3-6b step 2 with the following exception:
   - In Table 3.25, add the new skill or knowledge requirements (if any).
3. If the success criteria are met by the above change(s), proceed to step C7-4. If not, proceed to step C7-2c.

Note: The source of the SA-1 information may be an HE's visual, audible, or tactile scan of an injured person, incident scene, and so on. Access to this information may rely fully on the HE's experience (mental model) that initiates the action and them having the skills and knowledge needed to notice and accurately perceive this information.

4.7.2.3   C7-2c, Achievable with a PE Support Aid or Display Change?

Assess the limitations in the proposed display type. Example limitations may include obstructions or environmental conditions that interfere with access and attention tunneling caused by excessive stress.

If the limitation can be overcome with a support aid or a different or alternative display type, enter the selected solution (support aid, display change, or alternative display type) and function in Table 3.25.

*Note: See step C5-1c for an example of an **alternative display type**. A **display change** may require presenting displays in other locations or adding an additional display sensory form, such as by adding an audible display tone to the original visual-only display. For design considerations, see Appendix B.8 and K.5.*

If the success criteria cannot be met with a support aid or display change, the barrier or task may be unfeasible. A return to the preliminary design phase may be needed.

4.7.3   Step C7-3, Specify PE Support Aid or Display Change

If Step C7-2c recommends a support aid, follow the guidance in act phase step C3-6c with the following exception:

- In Table 3.25, add the new skill or knowledge requirement (if any) to understand and correctly use the aid. Proceed to step C7-4.

TABLE 4.7
Design: Simple SA-1 Passive Displays (Unpowered)

Element	Presented Form	SA-1 Source Examples	Info Type	SA-1 Info Function	Apply Design Process		
					SA-1 Source	Display Form	Comments
Passive device (PE)	Signage, painted markings	Signs, paint demarking a hazard boundary	Symbolic, representation, warning	**Status:** inform	See applicable design specification		For guidance, see Appendices K.2, K.5, B.5.1/2, F.3
	Windsock	Windsock	Measurement	**Status:** inform			
	Gauge indicator	Gauge	Measurement	**Status:** inform, act phase response feedback	CX		
	Mechanical position indicator	Mechanical device	Status/position	**Status:** inform, act phase response feedback	(See App. B.6)		
Other PE	Lighting systems (external support system) Insufficient lighting may impede detection and accurate perception.						

Note: For additional guidance on information types and attributes, see Appendix K.5 and Table K.7.

If step C7-2c recommends a display change, fully identify the change requirements. (For design considerations, see Appendix B.8 and K.5. See Table 3.25 for the identified sensory mode used to access the original display.)

- In Table 3.25, add the new skill or knowledge requirements (if any) to understand display changes if it is a deviation from normal practice or standards. Proceed to step C7-4.

4.7.4 Step C7-4, Simple Display (Passive, Unpowered)

This step defines the detailed design activities that apply to a simple passive, unpowered indicator. See Table 4.7 for the suggested design processes. (This step is in addition to steps C7-1 to C7-3 performed for each SA-1 item.)

The detectability of this indicator type is sensitive to area lighting, environmental conditions (e.g., reduced visibility, high noise), standardization (consistent meaning), and placement relative to access location. Minimum lighting levels may be required to ensure the necessary visibility from specified distances. Indicators should be provided in all locations where the information may be accessed.

4.7.5 Step C7-5, Simple Active SA-1 Device Indicator (Powered)

This step defines the detailed design activities that apply to a simple active, powered indicator. (This step is in addition to steps C7-1 to C7-3 performed for each SA-1 item.) See *Table 4.8* for the suggested design processes. Figure 4.8 is one example of this type of indicator.

Note: This activity is well guided by existing practice and regulatory requirements.

4.7.6 Step C7-6, Public Address and General Alarm Notification

This step defines the detailed design activities that apply to a public alarm notification-type display, which provides information throughout a defined physical area. Apply

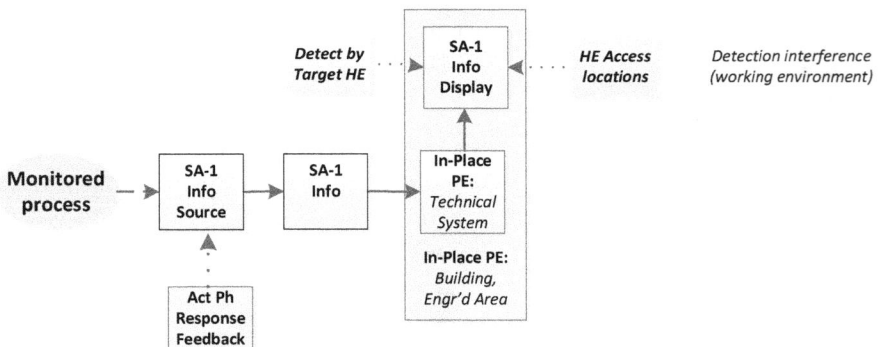

FIGURE 4.8   Example of a Simple Active Device (SA-1) Indicator.

TABLE 4.8
Design: Simple SA-1 Active Device Displays (Powered)

Element	Presented Form	SA-1 Source Examples	SA-1 Info Function	Applicable Design Process		Comments
				SA-1 Source	Display Form	
	Integral indicator	Local feedback from action response PE	– **Status:** inform, act phase response feedback			For guidance, see Appendices B.5.1/2, K.2, K.5, F.3
	Backlighting on an HE-activated switch	Feedback from AR direct-use PE Output from technical system[a]				
Simple display device (PE) **Discrete**	Audio speaker	Live or recorded voice message from technical system[a]	– **Notification:** barrier/task activation, instruct – **Status:** inform, coordinate, act phase response feedback	CX[a] (see App. B.6)	CX (see App. B.6)	For guidance, see Appendices B.5.1/2, K.2, K.3, K.5, F.3 Consider ISO 11064-5:2008, para A.6, Guidance on alarm systems
	Beacon, strobe	Output from technical system[a]	– **Notification:** barrier/task activation – **Status:** inform			
	Horn, siren					
Simple display device (PE) **Analog**	Integral full-range indicator	Analog sensing device	– **Status:** inform, act phase response feedback			For guidance, see Appendices B.5.1/2, K.2, K.5, F.3
Other PE	[a] A technical system (In-Place PE) may be required to enable/activate the *"SA-1 Source."*					

Note: For additional guidance on information types and attributes, see Appendix K.5 and Table K.7.

the design processes from C7-5 to this display type. (This step is in addition to steps C7-1 to C7-3 performed for each SA-1 item.)

This display type may combine audible information (distinct horn/siren tone, live or recorded messages) with visual information (strobe or colored beacon). The audible and visual information is coded to ensure its meaning is understood by all (standardized coding across the facility).

Note: This activity is well guided by existing practice and regulatory requirements.

4.7.7   STEP C7-7, INCIDENT SCENE INFORMATION

Apart from providing adequate lighting when required to view an incident scene, access to this SA-1 information is achieved by human and organizational elements. The HE must have the necessary mental model (training, experience, and expertise) to guide where to scan and what to scan for, and understand what the accessed information means in the context of the assigned task. Procedures should define these requirements as they apply to each barrier/task. See *Table 4.9* and the discussion below for design considerations. (This step is in addition to steps C7-1 to C7-3 performed for each SA-1 item.)

Example incident scene information:
- Scan the scene for information identified in procedures and training, such as hazard types, scale, locations, and opportunities for escalation.
- Scan and evaluate an injured person (triage) to understand the nature and extent of the injuries. For example, check their breathing, whether they are

---

TABLE 4.9
Design: SA-1 Incident Scene Information

Form	Source Examples	Info Type	SA-1 Info Function	Applicable Design Process SA-1 Source	Applicable Design Process Display Form	Comments
Real-time information sensed from an incident scene	Incident scene, personnel	Available to all senses	Status/ tracking Act ph. response FB	See Table 3.25 and "Example incident scene information" below	Raw sensory information	For guidance, see Appendices B.5.1/2, F.2, F.3, F.4, K.5

Lighting (external support system) or Support PE (flashlight) may be required to enable visual access to incident scene information.

Note: For additional guidance on information types and attributes, see Appendix K.5 and Table K.7.

conscious/unconscious, and skin color, and identify potential threat sources that may require their immediate movement from the area.

- Scan and evaluate the hazards and threats when transiting an egress/escape route, for example, check for route visibility; obstructions; and sources of danger including fire, smoke, and potential toxic gas.

The capability to reliably notice and detect the available information depends on one's knowledge of what to look for (mental model-driven), conditions that affect attention, judgment and decision-making, and limitations in the human visual system. See Appendices F.2 and F.3 for more cognitive and visual system considerations.

4.7.8 STEP C7-8, INCIDENT COMMAND BOARD

This step defines the detailed design activities that apply to the SA-1 information acquired and recorded on an Incident Command Whiteboard. See Table 4.10 and the discussion below for design considerations. (This step is in addition to steps C7-1 to C7-3 performed for each SA-1 item.)

Case Study: PG&E Natural Gas Transmission Pipeline Rupture and Fire (NTSB 2011, pp. 12–16, 97–99). A rupture in a PG&E pipeline occurred in a residential neighborhood. It and the subsequent fire were seen and reported by the public, public first responders, and PG&E personnel driving near the area. However, a 39-minute delay elapsed before the SCADA Control Room Operator issued instructions to field personnel to manually close valves that would isolate the rupture section. PG&E procedures did not designate an emergency response person (role) that had the assigned responsibility to acquire, record, and assess the incoming information, a critical step to understanding the event and determining a course of action. As a consequence, no one operator had access to all of the received and available information making it more difficult to understand what was happening and where. The Control Room Operators received both accurate and inaccurate information that were phoned in or relayed to different operators; the information was from the public, news outlets, PG&E employees driving near the rupture, and the PG&E dispatch office. The operators had full access to the real-time pipeline instrument readings and control information available at the SCADA system displays in the control room. That system did not include the capability to quickly detect a leak or rupture and identify its release rate and location. The PG&E dispatch office also received incident information though PG&E procedures did not appear to address the need for timely information updates and exchanges with the SCADA Control Room Operators. (For additional information on the accident, see Chapter 3, step B16-3.)

Emergency response may add an Incident Command Center (ICC) or emergency response center (ERC). The barrier leader monitors and controls the response from this location. In this role, the barrier leader may require extensive and very diverse information. Key information used by the leader and others may be recorded on an Incident Command Board (ICB). The ICB may be a simple whiteboard or an electronic smartboard. Within the ICC, one or more HEs may have assigned tasks

TABLE 4.10

Design: Incident Command Board

Design: Incident Command Board (Hand-Marked or Smartboard)

Element	Form	Info Type	SA-1 Info Function	SA-1 Source	Display Form	Comments
				Applicable Design Process		
Whiteboard	Text, tables, symbolic, sketch	Visual	– Barrier/task activator – Status/tracking – SA-2/3 input – Decisions – Act ph. response FB	Hand-marked Hand-marked[a] Electronic sourced	See note [b]	For guidance, see Appendices B.5.2, C.1, F.3, K.2, K.4, and K.5
Smartboard						
Room (In-Place PE)	The design and layout of the room and room lighting must support full view access by HEs requiring access to the information on the board.					
Other PE	[a] A technical system (In-Place PE) may be employed for a smartboard design or to provide live display access to a remote location, i.e., the HE(s) assigned to provide remote barrier support from a Remote Operations Center. The CX sub-process may apply to this component. [b] The display is not VDU-based and may be complex. It often includes many possible information types and forms. Text, symbol, coding and presentation should be clear, consistent, and highly readable. Those aspects and element layout and relative positioning are safety critical.					

Note: For additional guidance on information types and attributes, see Appendix K.5 and Table K.7.

to monitor for, gather, and record the required and requested information on the ICB. This may provide SA-1, SA-2, and SA-3 information to several HEs and input to several barriers. The information may be used to support barrier status tracking, planning, and decision-making, among other processes.

The ICB information should be organized and presented in a form and layout that supports quick and accurate access that directly supports how that information is used. Procedures should identify all information required for each assigned task. Others may also need access to this information, such as an HE who provides remote barrier support from a remote location.

The ICB information may take several forms, including alphanumeric text, symbols, or sketches. This makes the display complex, requiring standards and guidelines to guide how the information is presented and organized. If hand-marked, the information must be readily legible in terms of handwriting and marker type. Room lighting should enhance visual access. These considerations can have a positive or negative effect on the user, including how quickly the information can be located and accessed, and how the location and grouping of information may affect both efforts. See Appendices B.5.1/2 and K for guidance on these design topics.

Note: The effort to gather and record the information can be a complex task. This task may be assigned to a single HE or shared among several HEs. Activities may include identifying missing information, gathering information (periodically or on request) using radios and telephones, and accurately recording that information on the ICB. For cognitive reasons, the task introduces many opportunities for error. Information may be misheard, incorrectly understood, or misremembered before it is recorded. Errors in the recording process may include a slip or mistake. Furthermore, handwriting may be difficult to read or may be misread. For further insight, see Taber et al. (2012).

4.7.9 STEP C7-9, RECEIVED COMMUNICATIONS

This step defines the detailed design activities that apply to received communications that provide required SA-1 information. See Table 4.11 for design guidance and considerations that apply to device-enabled and in-person communications. (This step is in addition to steps C7-1 to C7-3 performed for each SA-1 item.)

Note: Table 3.11 *identifies communicated information such as the message content and function. Step C3-3 (act phase response process) provides design input to technical systems (In-Place PE) and selects the direct-use components (e.g., a headset). It also assesses the stated message content and communication environment to verify the viability of the suggested communication approach. See step C3-3 for additional information. Face-to-face communications are similar but include no requirements for technical systems. Refer to step C3-4 for additional information.*

4.7.10 STEP C7-10, CONFIRM PHASE SAFETY TIME

This step evaluates the detect phase SA-1 requirements and functions to confirm the phase can be reliably completed within the allocated time: $T_s + T_d$ (Table 3.27) for all modes of operation and plausible situations and conditions that can affect the phase safety time. (See Table 3.5 for the barrier modes of operation.) Follow the guidance in act phase step C3-9.

4.7.11 STEP C7-11, PERFORM THE COGNITIVE ASSESSMENT AND MITIGATION PROCESS

This step performs the cognitive assessment and mitigation process defined in Appendix I.2 and the Evaluation Input Guidance Table I.2.

TABLE 4.11

Design: Received Communications

Element	Presented Form	SA-1 Source Examples	SA-1 Info Function	Applicable Design Process	
				Direct-Use Device (e.g., Headset)	Message Form/ Content
Comms. handset or headset (radio, telephone, other)	Verbal, audio, visual	Comms sender[a]	– **Notification:** barrier/task activation, instruct – **Request** – **Status:** inform, coordinate, event feedback, feedback, support	CX (see App. B.6)	See C3-3 and Table 3.11
Face-to-face (in-person)	Verbal, audio, visual	Comms sender	– **Notification:** barrier/task activation, instruct – **Request** – **Status:** inform, coordinate, event feedback, feedback, support	—	See C3-4 and Table 3.11
Other PE					

[a] A technical system (In-Place PE) is required to enable the "*SA-1 Source*" communication.

4.8   PROCESS C-8, DETECT PHASE, VDU-BASED DISPLAYS

Figure 4.9 provides an overview of the detect phase (SA-1) detailed design and engineering process specific to VDU-based displays. (For background information, see preliminary design step B-14.)

Detect Phase SA-1 Success Criteria: The detect phase SA-1 requirements (functions and performance) can be reliably achieved within the specified phase safety times when the HE is under time pressure and exposed to designed-for situations and working environment conditions. The criteria are met in all modes of operations identified in Table 3.5 *– Barrier Origination Requirements.*

Inputs to this process: See the inputs listed for process C-7.

4.8.1   Step C8-1, Meets Detect Phase SA-1 Success Criteria?

See the *Detect Phase SA-1 Success Criteria* and input information listed in the C-8 introduction. In this step, guided by an operations expert, review and assess each SA-1 information requirement and the detectability and direct-use interaction (perception) with each display source indicated in Table 3.25. Consider the following positive and negative (and like) effects listed in the act phase process step C3-6.

4.8.1.1   C8-1a, Achievable Without Changes?

Confirm whether the stated success criteria can be reliably achieved without changes to the current design. If yes, proceed to step C8-2. If not, proceed to step C8-1b.

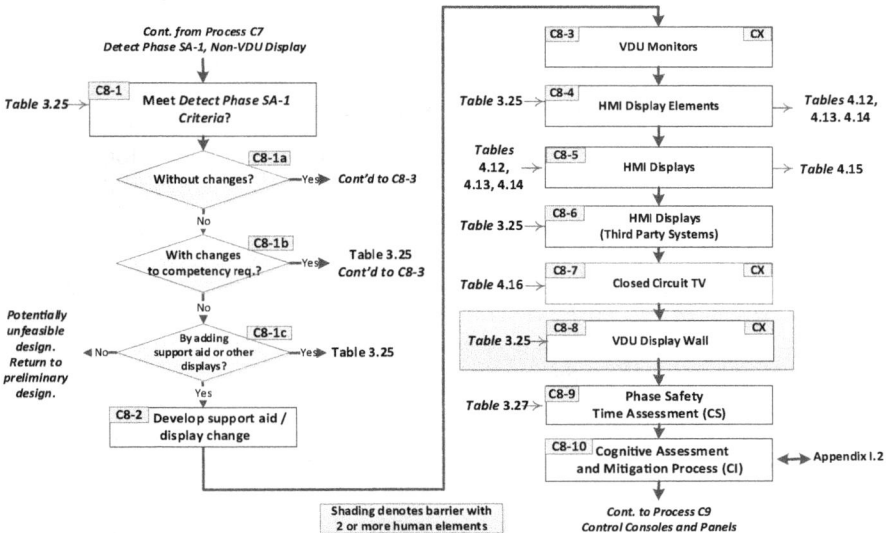

FIGURE 4.9   Detailed Design Process C-8, Detect Phase – VDU-Based Displays. Perform this Process for Every Barrier Task.

4.8.1.2   C8-1b, Achievable with Changes to Competency?

1. Follow the process in C3-6b step 1. *Exception*: For the reference to Appendix I.3, refer to Table I.9 in Section I.3.2.
2. Follow the process in C3-6b step 2 (HE or OE changes) with the following exception:
   - In Table 3.25, add the new skill or knowledge requirements (if any).
     If the success criteria are met with the above change(s), proceed to step C8-3. If not, proceed to step C8-1c.

4.8.1.3   C8-1c, Achievable by Adding a Support Aid or Alternative Display Type?

For guidance, see step C7-2c, then proceed to step C8-2.

4.8.2   Step C8-2, Specify the Support Aid or Display Change

If step C8-1c recommends a support aid, follow the guidance in act phase step C3-6c, with the following exception:

- In Table 3.25, add the new skill or knowledge requirements (if any) to understand and correctly use the aid.

If step C8-1c recommends a display change, identify the change requirements, such as adding similar displays at other locations, or adding a second display sensory mode by adding an audible display to supplement a visual display. (For additional design considerations, see Appendices B.8 and K.5.)

(If step C8-1c recommends adding an alternative HMI display element, step C8-4b addresses the design of this element type.)

4.8.3   Step C8-3, VDU Monitor Selection and Basic Design

This step examines the function, location, and physical requirements for video display units (VDU component) to confirm their type, size, capabilities, and locations. Sub-process "CX" (Appendix B.6) guides this process.

Note: See Table 3.26 *(step B-14) for information on VDU-based display types. A technical system (In-Place PE) generates the display. Other display examples may be from a CCTV system, a fire alarm panel, or an information system (email, text, or electronically stored document).*

Example considerations that may affect VDU hardware selection and component-level design:

- The display type, information/image density, resolution, layout, color accuracy, touchscreen, etc.
- The number and type of possible sources to a VDU display wall monitor.

- The display functionality and required use locations for a hand-held table (e.g., suitable for use in a wide range of possible locations and environments).
- A console-mounted touchscreen VDU provides touch targets for selecting and entering command or data.
- With console-mounted, hand-held devices and other display types, physical ergonomics principles apply. For guidance on visual access considerations, see Appendix F.3 and ISO-11064-4 (2013, Annex A). For basic workspace design considerations, see ISO-11064-4 (2013), ISO-11064-5 (2008), and Appendix K.2.

4.8.4   Step C8-4, HMI Display Element Design

This step applies to the following HMI display elements:

- Detect Phase Display Elements
- Support Aids and Alternative Display Types
- Act Phase Display Elements

For the base design process, consider NUREG (2020), ISO-11064-5 (2008), and the site/project HMI Design and Style Guide Guidelines. Refer to Appendix K.5 and Table K.7 for information types and attributes. (This step is in addition to steps C8-1 to C8-2 performed for each SA-1 item.)

Note: The documents created in this step contribute to the design package provided to the organization responsible for display element prototyping, development, obtaining end-user approval, and testing to confirm compliance with all requirements. The scope and form of this package may vary based on the implementation organization's technical and contractual requirements.

Note: The information types in Table K.7 (Appendix K.5) are normalized and employed throughout the lifecycle model.

4.8.4.1   C8-4a, Detect Phase VDU-Based Display Elements

For possible inputs to this step, see the applicable documents listed in the process C-7 introduction and the following:

- HMI Display Guidelines and Standards
- HMI Style Guide
1. This review should identify all VDU-displayed detect phase elements:
    - SA-1 element
    - SA-2 element
    - SA-3 element
    - SSA element
    a. In Table 4.12 below, record the **Element ID** and **Type, Source Table, Function, HE User, and Barrier/Task ID,** and the **Design Process** used to design this element. Applicable design processes and guidance

TABLE 4.12

HMI Display Elements: Detect Phase

Element ID	Element Type	Source Table	Element Function	HE User	Barrier/ Task ID	Comp SA Req.?	Meta SA Req.?	Selected Element Type	Element Design Docs.	Design Process
Enter unique ID for element	Enter element type: – SA-1 – SA-2 – SA-3 – SSA *One per row*	Enter source table or document (Tables 3.8, 3.18, 3.19, 3.25, 4.6)	Enter element function (from source table or reference) *See App. K.5 and Table K.7 for detail.*	Enter ID for HE user.	Enter barrier and task IDs	Enter – *Yes* – *No*	Enter – *Yes* – *No* If yes, identify function that requires recall	Enter selected display element type (from HMI design/style guide)	Enter element-specific design documents *All that apply*	Enter applicable design process(es)

may include the HMI Design Guideline, ISO-11064-5 (2008), NUREG (2020), and ANSI/ISA-101.01-2015. Also see Appendices B.8 (support aids), K.2 (physical workspace design), and K.5 (information workspace design).

  b.　Identify if the presented information is accessed and used by two or more HEs. When this occurs, proceed to step 2. Otherwise, proceed to step 3.

2.　Review information in Appendix C.1.3 on *Compatible SA*. Determine whether the information should be presented in different ways to support the needs of individual HE users. Record the selected approach in Table 4.12.

3.　Review the information in Appendix C.1.5 on *Meta SA*. Identify the SA-1 information that must be available for later, on-demand recall. Identify the requirement in Table 4.12.

4.　Evaluate and select the element type. Use an existing display template or, if required and acceptable, develop a new type. Record the **Selected Element Type** in Table 4.12. This process should consider the following:

  • Guidance provided in project documents (e.g., HMI display design guidelines, style guides, and referenced industry standards).
  • Applicable cognitive considerations included in Appendix F.
  • ISO 11064-5 (2008), for example, Section A.2.4 – Developing Formats. Example formats include mimics and diagrams, symbols, tables, text, bar chart/histograms, trend curves, graphs, data fields and forms, pie charts, and flowcharts.
  • NUREG (2020) – Human-System Interface Design Review Guidelines.
  • Applicable recommendations from the reliability assessments (Tables 3.30 and 3.31) and PIF assessment (process C-2).
  • Support aid types discussed in Appendix B.8.
  • Information attributes noted in Table K.7, Appendix K.5.

5.　Develop a display **functional specification** when needed to specify unique display functions.

6.　If needed, provide a **sketch** to show the appearance of the displayed object.

7.　Identify/specify the **performance standard(s)** that apply to this element, for example, see ISO-11064-5 (2008, Table A.2).

8.　In the Element **Design Docs**. field, enter references to the documents identified in steps 5–7.

4.8.4.2　C8-4b, VDU-Based Support Aids and Alternative Display Types

See the following documents for a required support aid or alternative display type: Tables 3.8, 3.12, 3.17–3.19, and 3.25; safety requirements specification; active human barrier design basis; and technical system specifications.

Note: For an explanation and example of an "alternative display type," see the discussion and example in process step C5-1c. For a definition, see Section 2.1 in Chapter 2.

1. The above-listed tables and the review from step C8-1c identified requirements for *support aids* or *alternative display types* to be implemented in
   an HMI display form. In *Table 4.13*, identify and record the **Element ID**
   and **Type**, **Source Table**, **Function, HE User, Barrier/Task ID**, and the
   Design Process used to design this element. Applicable design processes
   and guidance may include the HMI Design/Style Guideline, ISO-11064-5
   (2008), NUREG (2020), ANSI/ISA-101.01-2015, or EEMUA 201 (2019).
   Also see App. B.8 (support aids) and K.5 (information workspace design).
2. Evaluate and select the element type. Enter the information in the **Selected
   Display Type** field. If this is a new style type, enter *Yes* in the **New Style/
   Display Type** field. If not, enter *No*.
3. Develop/identify the documents listed in C8-4a, steps 5–7.
4. In the **Element Design Doc.** field, enter references to the docs from step 3.

4.8.4.3   C8-4c, Direct-Use Act Phase Response Controls

Possible VDU-Based Act Phase Control/Command Entry Displays: See the following documents for act phase control/command entry display elements: Tables 3.9,
3.12, 3.13, and 4.6, and other applicable documents listed in the process C-7
introduction.

1. Based on the above review, identify and confirm the act phase response
   functions initiated or implemented using a command entry field provided
   in an HMI display. Record the requirements in *Table 4.14*. Also enter the
   **Element ID, Source Table, Element Function, HE User, Barrier/Task
   ID, Associated SA-1 Feedback,** and the **Design Process** used to design
   this element. Example design processes may be owner/operator developed,
   or an industry document such as ISO-11064-5 (2008), NUREG (2020),
   ANSI/ISA-101.01-2015, or *EEMUA 201 (2019)*. Also see Appendix K.2.
2. Evaluate and select the **Selected Display Type**. Use an existing template or,
   if required and accepted, develop a new type. Record this information in
   Table 4.14.
3. Develop/identify the documents listed in C8-4a, steps 5–7.
4. In the **Element Display Docs**. field, add references to documents developed
   in step 3.

4.8.5   STEP C8-5, HMI DISPLAYS: INTEGRATED DISPLAY DESIGN

This following step allocates required display elements to the appropriate HMI displays and arranges those elements in a way that optimally supports the intended use
and user. (This step is in addition to steps C8-1 to C8-3 performed for each SA-1
item.)

*Note: Industry standard and practice guidelines and others may be applied to this
process, for example, ANSI/ISA-101.01-2015, NUREG (2020), ISO-11064-5 (2008),
or EEMUA 201 (2019). This step may be used to supplement a standard or guideline
that does not provide or adequately address the guidance provided in this step.*

TABLE 4.13

HMI Display Elements: Support Aids and Alternative Display Types

Element ID	Element Type	Source Table	Element Function	HE User	Barrier/ Task ID	Selected Display Type	New Style or Display Type?	Element Design Docs.	Design Process
Enter unique ID for element	Enter element type: – *Support Aid,* – *Alternative display type*	Enter source table or doc.	Enter element function (from source doc.)	Enter ID for HE user.	Enter barrier and task ID(s)	Enter selected display type/ form (see HMI design/style guide)	Is this a new style/display type? – *Yes* – *No*	Enter Element-specific design docs. *All that apply*	Enter applicable design process(es)

TABLE 4.14

HMI Display Element: Act Phase (Controls)

Element ID	Source Table	Element Function	HE User	Barrier/ Task ID	Selected Display Type	Associated SA-1 Feedback	Element Design Docs.	Design Process
Enter unique ID for element	Enter source table or document	Enter function – *Command request* – *Other (see source table)*	Enter ID for HE user.	Enter barrier and task IDs	Enter selected display type (from HMI design/style guide)	Enter SA-1 info feedback (if any) that confirms acceptance/ completion response to HMI display initiated command	Enter element-specific design documents *All that apply*	Enter applicable design process(es).

Integrated Display Identification, Analysis, and Design Input

1. Identify the integrated HMI displays required to support barrier tasks. For each identified display, record the following in Table 4.15: **HMI Display ID, Barrier** and **Task ID, Req'd In-Place PE** (e.g., the technical system that generates the display), and the applicable **Design Process**. Example design processes may be one developed by the owner/operator, or an industry document such as ISO-11064-5 (2008), NUREG (2020), *ANSI/ISA-101.01-2015, EEMUA 201 (2019)*. Additional employed processes or process elements may include those in Appendices K.2 and K.5.

2. **Allocate Display Elements to the Integrated Display** – Identify and allocate every display element (detect, act phase, support aids, and alternative display types) that should be available at this display. For required display elements, see Tables 4.12–4.14, and other possible sources. For elements allocated to this display, enter the **Element ID**, **Source**, **Type,** and **Function**. (Refer to the element source tables for this information.) Select and enter the selected integrated display template in the **Int. Display Template** field.

Note: The consistent application of a common set of HMI display standards and templates creates user expectations that affect their performance when interacting with the display. Those expectations are driven by the user's mental models and skill-based interactions that develop over time as they interact with similar displays and display types.

3. **Element Relationship and Display Template** – Review the elements (recorded above) to understand their functional relationship to each other. For background and selective guidance, see Appendices K.1.3 and K.2. Enter the results in the **Links To** field.

Display Layout and Design

4. **Determine element arrangement and placement** – Subject to the requirements and constraints in the selected display template, use the **Links To** information in Table 4.15 to guide the general arrangement and relative positioning of each element on the display. The functional relationship between functionally linked objects should guide their mutual placement, orientation, and positioning on the display. See Appendix K.2 and applicable industry practice publications for guidance on this design objective. (The goal is to place elements in a way that maximizes HE performance and minimizes the potential for display-induced errors.)

5. Develop a **sketch** of the display (preliminary design) that reflects the results of step 4 and the identified display standards and templates.

6. **Develop additional documents**, as required, to complete the design and information package that conveys requirements to those assigned to implement the display. For complex displays/display elements, a functional

TABLE 4.15
Integrated HMI Displays

HMI Display ID	Barrier ID	Task ID	Display Element (Info from Tables 4.12, 4.13, 4.14, other sources)				See Table 4.15 extension
			Element ID	Element Source	Type	Function	Links To
Enter unique HMI display ID	Enter barrier ID	Enter task ID	Enter display element ID from source.	Enter defining source table/doc.	Enter element type – Detect – Support Aid – Act response – Select one	Enter element function (from source doc.)	Enter ID for element that this element links to. All that apply

See Table 4.15 extension

Int. Display Template	Display Design Docs.	Design Process
Enter Selected display template	Enter List all docs created in this step	Enter applicable design process(es)

specification may be needed to describe the display presentation, functioning, and performance.

7. In the **Display Design Docs** field, include references to every design and requirements document created/modified by this process. (The documents are inputs to the technical design review and approval process and the HMI display technical package provided to the display developer/implementer.)

Note: With major projects, HMI displays are often included in a lump-sum purchase order (PO) or contract for a technical system such as a basic process control system, safety system, or packaged equipment system. The PO or contract should include pricing and options to add the plausible HMI displays, display types, and display elements. A new alternative display type may be significantly more costly than other displays, and may thus require significantly more time to define, specify, develop, and test. This should be considered in procurement planning, requisitions, and budgeting.

4.8.6 STEP C8-6, HMI DISPLAY: PACKAGED EQUIPMENT AND OTHER THIRD-PARTY SYSTEMS

This step designs and specifies barrier-specific HMI displays and elements that reside in third-party–provided systems. Systems of this type include the following:

1. Packaged equipment such as process systems, subsea monitoring and control systems, drilling control systems, and utility systems (e.g., power generation, instrument air)
2. Electrical control and monitoring systems including switchgear and motor starters
3. Specialty and instrument monitoring systems

The design process is the same as C8-4 and C8-5, though it may be subject to design constraints invoked by the third-party equipment provider. The provider may have internal display standards and standardized designs that limit design options and changes. (Opportunities to reach agreement on acceptable display changes and unit costs optimally occur in the proposal phase and before the order is placed.) The range of permitted changes may include the following:

- No change permitted.
- New display: may add a new display but may prohibit or limit design deviations from internal standards.
- Change standard display: may permit selective changes to display elements, but this is subject to limitations and constraints.
- In rare cases, the vendor may accept more significant changes like those to a color convention. (These types of changes tend to be very costly.)

(This step is in addition to steps C8-1 to C8-3 performed for each SA-1 item.)

Note: All changes to a third-party vendor's "standard" package introduce risks. Some vendors outsource the software design if they lack this internal resource. Others may seek to limit one-off designs that may contribute to potential warranty and support problems. With any change, the scope and rigor in the verification and testing processes becomes more critical. Changes may also increase project costs and delivery schedules. These are all prominent issues to consider in early planning activities.

4.8.7   STEP C8-7, CLOSED-CIRCUIT TV SYSTEM

This step performs the design activities that enable and capture the specified closed-circuit TV (CCTV) display images/views, meet specified performance requirements, and provide the required views and control functions at the specified use locations. (This step is in addition to steps C8-1 to C8-3 performed for each SA-1 item.) Perform the following step.

1. From Table 3.25, identify the specific SA-1 information required from a CCTV-captured video image. The image should identify the level of detail (camera zoom and resolution) in the viewed object, area, or scene. In *Table 4.16*, record the following information: **SA-1 ID, Barrier/Task ID,** and **HE ID** fields. Also enter the information in the CCTV camera image row: **SA-1 Info Req.,** and **Function** fields.
2. In conjunction with the CX-0.1 sub-process (Appendix B.6) and the applicable input information from step 1, progress the camera selection and design.
   - To achieve the required views, identify the required camera installation location and placement. (Different barriers may require different views from the same camera, which may be a view conflict to resolve.) Identify the required features including camera type and resolution, pan/tilt/zoom capability, night vision/infra-red/visible light only, and anti-vibration feature.
   - Identify the condition under which the HE views the information, for example, conditions that may obstruct the image.
   - Identify if the image is continuously indicated at a display or if an event automatically directs the camera based on a preset pan and zoom setting.
   - Consider additional key questions. Who requires view access to this image and from where? Is it acceptable to automatically replace an existing displayed camera or other display image at a VDU? If so, under what circumstances is this acceptable? Will the task require the HE to manually change the camera view? If so, how much time is available or needed to do so? How quickly must this occur?

Consider: Is the CCTV image used to identify and assess a gas release (release source, rate, and location) or the position of a commanded control valve? Is it used to view a fire to assess its location, type, fuel source, intensity, impingements, and potential escalation pathways? Is it used to

TABLE 4.16

CCTV Image Element and PE Requirements

SA-1 Info ID	Barrier/Task ID	HE ID	SA-1 Info Req.	Component	Design Process	Camera Requirements			
						Capabilities	**Location**	**Function**	**Design Document References**
Enter ID for this barrier or task req'd CCTV image (view)	Enter ID for barrier or task this applies to	Enter ID of HE user *One per row*	Enter SA-1 info from CCTV image	—	This step	—	Enter precise location of viewed object	**Enter** function supported by this information	Table 3.25 and other applicable source documents
			—	Enter CCTV camera ID *One per row*	CX (see App. B.6)	Enter *Examples:* – Pan, tilt, zoom with presets – Resolution – Night vision – Infra-red	Enter – Location – Elevation – View angle	Enter – Specified event directs camera to pan, tilt, zoom presents. – User manually moves camera.	Enter function and performance specifications, data sheets, elevation drawings, Cause-and-effect chart (logic functions) Others as req'd.
			—	Enter ID(s) for VDU(s) at image display location *All that apply*	C8-3	Specify minimum VDU capabilities	Enter VDU loc. (all) and HE view angle and distance	If applicable, specify controls for camera, VDU display access	See C8-3 and CX (App. B.6) processes for provided documents, e.g., VDU data sheets and location drawings. Appendix K.5

See ISO 11064-5:2008, para A.5.4, CCTV (closed-circuit TV) systems, and the presentation of pictorial images. Acquiring and presenting the required CCTV images at the required access locations likely relies on one or more technical systems.

identify the absence/presence of a person in a hazardous situation (e.g., who, where, or movement direction and speed), or their disposition (e.g., prone position with obvious injuries)? If an offshore O&G facility is temporarily abandoned because of severe weather, is the image used to monitor sea conditions, how the facility responds to severe weather and wave events, or potential collisions with wayward vessels? Is it used to monitor crane operations or personnel locations/movements during a simultaneous operations event? Does it provide a real-time view to others as a means to observe activities in other control centers and other key work locations? (Refer to Note 10a in Appendix M for an example of how this becomes safety critical information if a mitigation or emergency response barrier is active. This approach provides real-time information and may reduce the number and inopportune timing of resource-consuming two-way communications.) These examples begin to indicate the diversity and potential applications for integrating CCTV cameras (views and audio) into the barrier system.

3. CCTV Camera: Based on the information from step 2, use the CX-0.1 sub-process (Appendix B.6) to specify the CCTV camera requirements. In Table 4.16, populate the following fields: **Camera ID,** and its **Capabilities, Location,** and **Function**. Include the updated/developed design documents in the **Design Document References** field.

4. VDU display: Use the CX-0.1 sub-process and process C8-3 to define and specify the VDU requirements (where the CCTV image is displayed). For example, populate the following fields in Table 4.16: **VDU display ID** and its required **Capabilities, Location,** and **Function**. Include the updated/ developed design documents in the **Design Document References** field.

4.8.7.1   Discussion

CCTV is becoming a key component in safety critical functions, primarily because it is a flexible source of real-time, information-rich content. (See Appendix K.5, Information Workspace for background.) A barrier function may require this type of information to support a detect, decide, or act phase requirement. For those providing remote barrier support, a live video feed may help to close the experiential "being there" gap.

Remotely located HEs do not experience the background noise, tactile sensations, or chaotic and confusing visual/audio information caused by the hazard that activated the barrier. This information may contribute to differences in situation awareness and performance (positive and negative) between the HEs on the protected facility and those performing support tasks from a remote location. Providing access to live video (with audio) sound may work to reduce those differences.

Note: If an Emergency Systems Survivability Analysis (ESSA) for an offshore O&G facility determines that none of the CCTV cameras are essential, this may be a red flag indicating that active human barriers (e.g., mitigation and other emergency response barriers) are underspecified or not fully defined.

Application: An increasingly common design practice causes a CCTV camera to automatically move to a specified preset position (e.g., a pan, tilt, zoom-directed view) in response to a specified event. An example, the preset may direct the camera toward an alarming (first-out) flammable gas detector. The view is selected to encompass the more likely release points and sources. The purpose of the view is to provide the HE with information to identify the possible gas release location, size, material, and potential and do so quickly and accurately. The function is commonly underspecified (e.g., purpose not defined), and the required supporting procedures and training may be missing.

4.8.8   Step C8-8, Off-Console VDU Displays and Display Walls

This step provides design input to a VDU-based display wall that includes two or more wall-mounted VDU monitors (large screen) viewed by two or more barrier HEs.

Tables 4.15 (HMI displays) and 4.16 (CCTV displays) identify two of the several display sources that may be presented on an off-console VDU display or display wall. Review these and other documents to identify all barrier-required displays that could be presented at this location. Based on this information, record all fields in *Table 4.17.*

Note: Process C-10 (rooms) evaluates HE access (sightlines) to these displays from all plausible access locations.

TABLE 4.17
Off-Console VDU Displays

VDU ID	Display ID	Displayed Video Types	Display Source	Barrier / Task ID	HE Users (IDs)	HE Access Loc.	Permanent Display
ID for VDU 1	Enter display ID	Enter: *Examples, CCTV, HMI display, Video conference feed* All that apply	Enter display source: *CCTV, BPCS, SIS Other*	Enter barrier/ task ID	Enter HE users (IDs) *All that apply*	Enter HE VDU access location/ positions *All that apply*	Identify if display type at this VDU is permanent: Enter *– Yes* *– No*

VDU Selection/Design – See C8-3 process. VDU Display Wall Design (VDU quantity, positioning, and layout). See ISO 11064-3 (1999), Section 4.5, for design guidance.

Note: For additional guidance on information types and attributes, see Appendix K.5 and Table K.7.

4.8.9 Step C8-9, Confirm Phase Safety Time

This step evaluates the detect phase SA-1 requirements and functions to confirm whether the phase can be reliably completed within the allocated time: $T_s + T_d$ (Table 3.27) for all modes of operation and plausible situations and conditions that can affect the phase safety time. (See Table 3.5 for the barrier modes of operation.) Follow the guidance in act phase step C3-9.

4.8.10 Step C8-10, Perform the Cognitive Assessment and Mitigation Process

This step performs the cognitive assessment and mitigation process defined in Appendix I.2 and the Evaluation Input Guidance Table I.2.

4.9 PROCESS C-9, MONITORING AND CONTROL CONSOLES AND PANELS

Figure 4.10 provides an overview of the C-9 detailed design and engineering processes that apply to control consoles and panels containing barrier system elements.

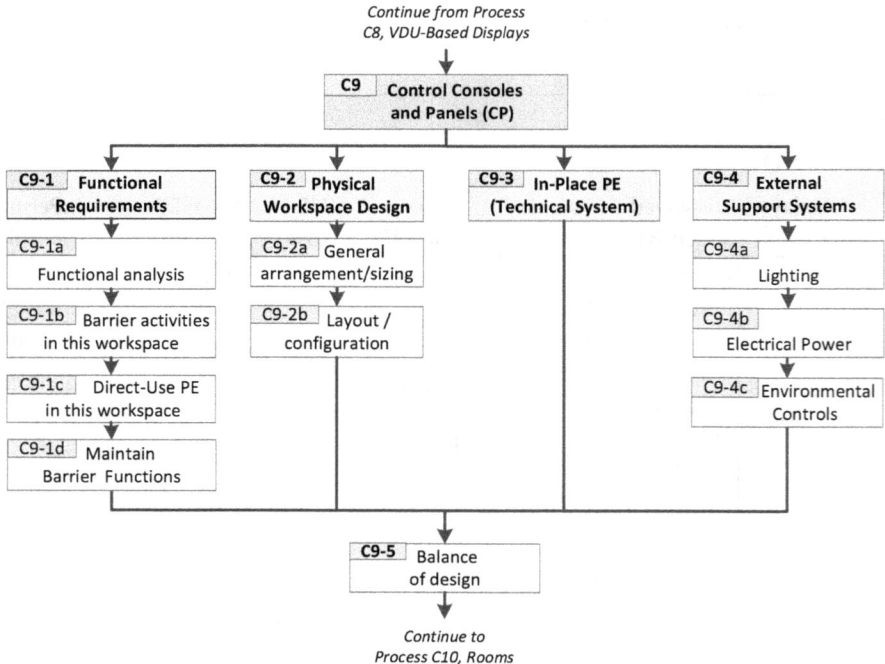

FIGURE 4.10 Detailed Design Process C-9: Control Consoles and Panels. Where Applicable, Perform this Process for Every Barrier.

Note: This process assumes that the primary control console and panel designs are governed by existing design processes such as ISO-11064-4 (2013) or EEMUA 201 (2019). The scope of process C-9 is limited to only those elements and activities that apply to active human barriers.

Design Inputs to this process:

- See the "common inputs" listed in the process C-3 introduction.
- See the "common inputs" listed in the process C-7 introduction.
- Control console and panel design documents (e.g., specifications, dimensional, layout, and general arrangement drawings).
- Appendix K.2 (workspace principles). Appendix K.1.1 addresses placement based on HE movement requirements. Appendix K.1.3 addresses placement based on functional relationships.
- Step B-2 risk assessment reports.
- Hazard assessments including fire and explosion studies.
- Table 4.16 – CCTV Image Element and PE Requirements.
- Table 4.17 – Off-Console VDU Displays.
- Applicable documents listed in:
  - Table B.11 (App. B.7.4) – Source Documents: Timing and Response Time Requirements.
  - Table B.12 (App. B.7.5.1) – Source Documents: Reliability, Availability, and Integrity Design – Physical Elements.
  - Table B.14 (App. B.7.6) – Source Documents: Survivability and Environmental Requirements.
  - Table B.15 (App. B.7.7) – Source Documents: Sizing, Layout, Location, and Placement Requirements.
- Table H.2 (App. H) – PIF Assessment: Barrier, Task, and Task Phase.
- Table H.3 (App. H) – PIF Assessment (WE): Direct-Use PE Components.

Control console or panels (CPs) covered by this process include the following:

- A monitoring and control console located in a Central Control Room. (User access is from a seated position.)
- A freestanding monitoring and control panel provided with a packaged equipment system. (User access to this field or room-installed panel is from a standing position.)
- A configurable COTS display panel, for example, a fire alarm system display panel.
- A Critical Action Panel: a custom-engineered panel that provides a hardwired display and control interface to critical systems.

Note: Processes C-3 to C-8 guide the design of stand-alone direct-use components and elements located in the engineered area. This process integrates these elements into the CP workspace.

4.9.1   STEP C9-1, FUNCTIONAL REQUIREMENTS

4.9.1.1   C9-1a, Functional Analysis

This step performs a functional analysis (FA) of the HE activities and interactions with CP-located direct-use elements and other aspects that enhance or degrade these activities. The analysis should be performed or supported by an operations specialist.

1. The process uses the guided topics and questions below. The findings of the analysis are captured and published in a functional analysis report.
   a. Identify all barrier/task activities that require HE access to the CP-located, direct-use PE. (Examples of direct-use physical elements include discrete components, VDU-based display elements, and communications equipment like radios and telephones.)
   b. Identify others who may require regular and occasional access to the CP, and situations when this may occur.
   c. Identify task activities that present the HE with potential physical or cognitive challenges that may lead to an incorrect human response. Does the task require simultaneous access to different displays and display types such as discrete components, VDU-based display elements, communication equipment, and paper-based procedures or support aids? What movements (eye, head, body) are required for simultaneous or concurrent access? Are the elements within the central vision visual view? (See Appendix F.3 for more detail.)
   d. Are CP-mounted direct-use elements (e.g., a VDU, radio, or telephone) shared with others? Do plausible situations exist wherein several persons may require use at the same time?
   e. What horizontal workspace is needed to place paper-based documents? Describe the activity. Does it require the HE to follow a stepwise procedure (paper-based) while interacting with a VDU display element or discrete component?
   f. What lighting must be present (or may interfere with) an HE interaction with direct-use PE, visual access to paper documents, or CP labels?
   g. What In-Place physical element and external support systems must be in place to maintain the required functions? For what period or timing?
   h. Do other ambient conditions exist that may interfere with the HE's access to the CP-located, direct-use PE?
2. Assess the findings in step 1 against the workspace design principles in Appendix K.2. Identify gaps that exist between these principles and the design information captured in step 1. Record each gap in Table 4.18.

   For each assessed CP, populate all fields in *Table 4.18* (excluding the **Status** field).

   Assessed FA Element – Enter the lettered topic/question (from step 1) that identified a finding/gap.

   Finding/Gap – Enter the finding, i.e., the gap (deviation) between the current design and identified workspace principle from Appendix K.2.

TABLE 4.18

Functional Analysis Recommendations

FA Element	Assessed Element	FA Seq. #	Assessed FA Item	Finding/Gap	WS Principle	Corrective Action	Status
Enter assessed element: – CP – Room – Engineered Area – Building	Enter assessed FA element	Enter finding sequence number	Enter FA assessment reference letter (from step 1 FA assessment list)	Enter gap between the current design and applied WS principle	Enter the workspace principle applied to this finding (from App. K.2)	Enter recommended corrective action	Enter – ALARP – Implement – Alternative recommendations

>   *WS Principle* – Enter the workspace principle (from App. K.2) that pro-
>   vides the design reference used to identify a gap.
>   *Corrective Action* – Enter the recommended corrective action to elimi-
>   nate or mitigate the identified gap.
>   *Status* – This field is addressed in step C9-5.
> 3. Compile and issue the functional analysis report. Include the findings from
>   step 1 and the recommendations in Table 4.18.

4.9.1.2   C9-1b, Barrier Activities in this Workspace

Based on the information developed in step C9-1a, identify and record the following
information in Table 4.19: **Console or Panel ID** and **Barrier or Task ID.**

4.9.1.3   C9-1c, Direct-Use PE in this Workspace

1. Based on the information compiled in steps C9-1a and C9-1b, and available from the applicable documents listed in the introduction, identify and
   record the HE and direct-use physical element located on or accessed at this
   CP, such as components, VDU displays, and communications equipment.
   In Table 4.19, record the information in the **Element Type** and **ID** fields.
   (Enter the ID for the specific element noted in the *Type* field.)
2. For each entered DU element type, identify its functional relationship (if any)
   with other HE(s) or direct-use elements. (See Appendix K.1.3 for guidance.)
   Record the information in the three fields under the ***Functionally "Links
   to" Element*** header. The location field identifies potential access challenges
   when movement (eye, head, body) is required to simultaneously access two
   linked elements that may reside at different physical locations or displays.

   *For example: A barrier/task action may require an HE seated at the CP
   to have an in-person communication exchange with an HE seated at the
   same CP, a different CP, or an explicitly defined nearby location.*

4.9.1.4   C9-1d, Maintain Barrier Function

1. Based on the information recorded in Table 4.19 (and other applicable documents listed in the introduction), this step identifies and records the period
   each element identified in Table 4.19 must be available to support its barrier function. Synthesizing this information, identify and record it in the
   Maintain Element Function field in Table 4.19.
2. For each element (CP located and linked to), identify the In-Place physical
   element (e.g., technical system) and external support systems (e.g., lighting,
   electrical power, and environmental controls) that must be in place to enable
   its barrier function. Record the information in the **Technical System** and
   External Support System fields.

4.9.2   Step C9-2, Physical Workspace Design and Layout

The suggested process for the following steps reviews the barrier-specific requirements and recommendations defined in step C9-1 and other requirement sources
listed in the introduction. Compare and assess these requirements against the existing

TABLE 4.19

Control Console or Panel (CP): Barrier System and Support Elements

Console or Panel ID	Barrier or Task ID	Console/Panel-Located Element (DU PE, HE)		Functionally "Links To" Element			Required for Console-Located Element		Maintain Element Function
		Element Type	Element ID	ID	Type	Location	Technical System (TS)	External Support System (ESS)	
Enter unique ID for this console or panel	Enter barrier or task ID	Enter type – HE, – VDU-based display, – Component: direct-use comms., detect/act phase PE	Enter ID for: – HE, – HMI display, – CCTV display, – Component: detect, act, or communication device	Enter element type	Enter element type/descr.	Enter other – HE, – This CP, – Off-CP component – Off-CP VDU wall	Enter TS ID and descr. *All that apply*	Enter ESS ID and descr. Lighting, electrical power, Env. Ctrls., Other *All that apply*	Enter time or period to maintain element function

Legend: CCTV – closed-circuit television; DU – direct use; Env. Ctrls – environmental controls; HMI – human-machine interface; VDU – video display unit.

CP design to confirm compliance. If a requirement or recommendation is not met, provide the requirements to those assigned CP design responsibilities for integration into the CP design. See Appendix K.2 for additional guidance.

4.9.2.1   C9-2a, General Arrangement and Sizing

Following the general guidance in step C9-2, identify the CP general arrangement and sizing requirements and recommendations not met by the current design. Provide the requirements to those assigned the CP general arrangement and sizing responsibilities for integration into that design.

4.9.2.2   C9-2b, Configuration and Layout

Following the general guidance in step C9-2, identify the CP configuration and layout requirements and recommendations not met by the current design. Provide the requirements to those assigned the CP configuration and layout responsibilities for integration into that design.

4.9.3   Step C9-3, In-Place PE: Technical Systems

Review and identify the barrier-specific requirements defined in step C9-1, Table 4.19, and other requirement sources listed in the introduction. Provide the information to those assigned the technical system (In-Place PE) design responsibilities for integration into that design. An additional requirement, namely, the **Maintain Element Function** field in Table 4.19, specifies the period the technical system must continue to provide its barrier function in the presence of the defined hazard(s).

4.9.4   Step C9-4, External Support Systems

This set of steps provides design input to those assigned ESS design responsibilities, for example, the design of lighting, power, or environmental controls. Refer to the recommendations in Table 4.18 (functional assessment recommendations from step C9-1) and Table 4.19. The **Maintain Element Function** field in Table 4.19 specifies the period the ESS must continue to support or maintain its barrier-dependent function.

In each of the following steps, existing processes address the design of the listed ESSs. The suggested process is to review the barrier-specific requirements against the current design to confirm compliance. If a requirement is not met, provide the requirements to those assigned ESS design responsibilities for integration into that design.

4.9.4.1   C9-4a, Lighting

Following the general guidance in step C9-4, identify the lighting requirements not met by the current design. Provide the requirements to those assigned lighting design responsibilities for integration into that design.

Note: The lighting design requirements for a CP in a Central Control Room may seek to prevent glare on a VDU screen and provide the lighting needed to easily read CP labels and documents placed on a CP horizontal surface. If located outdoors, the night-time lighting requirements may ensure sufficient lighting to easily read instrument tags, CP panel labels, and other non-powered displays, or equipment maintenance.

4.9.4.2   C9-4b, Electrical Power

Following the general guidance in step C9-4, identify the room electrical power (EP) requirements not met by the current design. Provide the requirements to those assigned EP design responsibilities for integration into that design.

4.9.4.3   C9-4c, Environmental Controls

Following the general guidance in step C9-4, identify the room environmental control (EC) requirements not met by the current design. Provide the requirements to those assigned EC design responsibilities for integration into that design.

4.9.5   STEP C9-5, BALANCE OF DESIGN

With the appropriate team, review the pending recommendations and corrective actions in Table 4.18 (step C9-1a, functional analysis). The path forward for each recommendation may include accepting the recommendation for implementation, accepting the current design as ALARP, or proposing a different recommendation. Forward the approved recommendations to the appropriate parties.

Note: Progressing an accepted recommendation may require a return to an earlier design process. This may warrant using the change management process to progress the design change.

4.10   PROCESS C-10, ROOMS

Figure 4.11 provides an overview of the C-10 detailed design and engineering processes that apply to rooms that contain barrier system elements. Where applicable, perform this process for each barrier.

Note: This process assumes the primary room design is still governed by existing design processes such as ISO-11064-3 (1999) or EEMUA 201 (2019). The scope of process C-10 is limited to only those elements, activities, and dependent elements and systems that apply to active human barriers.

Note: The typical control room can be a busy, work-intensive, cognitively demanding, and (at times) a stressful workspace. Under emergency conditions, a control room and other similar rooms (e.g., an Incident Command Center) should be purposely designed to support all barrier activities in these workspaces.

The common inputs to this process are as follows:

- See the "common inputs" listed in the C-9 introduction.
- Control console and panel (CP) functional analysis report including Table 4.18 (FA recommendations for CPs).

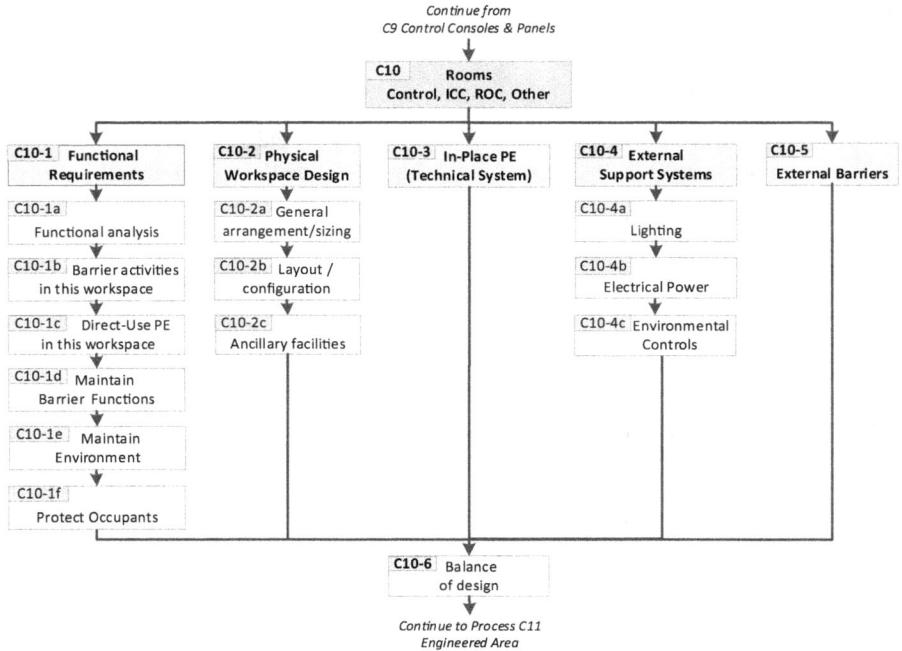

FIGURE 4.11 Detailed Design Process C-10: Rooms. Where Applicable, Perform this Process for Every Barrier.

- Table 3.15 – Detect and Act Phase: In-Place PE – Building Requirements.
- Table 4.17 – Off-Console VDU Displays.
- Table 4.18 – Functional Analysis Recommendations.
- Control console and panel functional analysis reports.
- Table 4.19 – Control Console and Panel (CP): Barrier System and Support Elements.
- Room dimensional layout and general arrangement drawings.
- Building design basis and specifications.
- Other sources listed in Table B-17 (Appendix B.7.9).

This process may apply to the following example room types:

- Control room (primary, secondary)
- Emergency response/Incident Command Center
- Medical bay
- Helicopter transfer room or stand-alone building
- Emergency response station
- Remote Operations Center (offsite location)
- External support system equipment room: lighting, electrical, and environmental controls
- In-Place PE equipment rooms (e.g., fire and gas, safety instrumented systems)

Example barrier elements in the room may include the following:

- Barrier assigned personnel (HEs)
- Control consoles and panels
- Off-console VDU-based display wall
- Other wall- and ceiling-mounted direct-use PE components
- Support PE (e.g., equipment and storage)
- Signage, instruction placards, etc.

The C-10 design process approaches a room as a workspace guided by the common workspace design principles in Appendix K.2.

Note: Stand-alone direct-use components and elements in the room are designed in processes C-3 through C-8. Process C-9 provides design input to consoles and panels. This process integrates these elements into the room workspace.

4.10.1   Step C10-1, Functional Requirements

4.10.1.1   C10-1a, Functional Analysis

This step performs a functional analysis (FA) focusing on HE activities and interactions with barrier system elements located in a room. Elements include HEs, and direct-use and Support PEs. The analysis should be performed or supported by an operations specialist, the barrier leader, or other equally knowledgeable persons.

Note: The FA examines the HE's access, interactions, and movements needed to access other HEs and the identified direct-use and Support PE. Most elements may be located on CPs, which were assessed in process C-9. Table 4.19 identifies functional linkages between a direct-use physical element located on a control console/panel to a direct-use display or act phase control element located elsewhere in the room, for example, an off-console VDU display or a different console or panel. This step completes the process using the actual (or proposed design) room dimensions and equipment placements. The assessment examines the proposed CP/PE locations; physical distances; sightlines to displays or other HEs (in-person communications); and the potential positive or negative effects of environmental conditions like lighting, distractions, and interruptions. Taken with the FA in step C9-1a, these activities provide a comprehensive FA that treats the room layout and arrangement, equipment access and placement, sightlines, and the external support systems as a fully integrated workspace.

The following are example assessments to perform within the functional analysis.

1. Assess all barrier/task activities and HE interactions with the barrier system elements located in each room. The process is guided by the topics and questions listed below. The findings and results of the analysis are captured and published in the functional analysis report.

Items a to h from step C9-1a.

i. Identify all activities that take place in this room (e.g., barrier and non-barrier activities).

j. Identify all HEs performing barrier functions in the room and if they work on a daily or occasional-only basis.

k. Identify all room-located, direct-use physical elements (stand-alone and CP-located) and Support PE.

l. Assess the layout and traffic design to confirm whether it supports or hinders the concurrent movements of personnel in the room during normal, peak, and barrier activation periods.

m. Identify if the planned room furniture adequately supports all HE activities.

n. Identify if the lighting maximizes the HE's visual access to information presented on non-VDU and VDU displays and other sources of barrier-required information.

o. Do the environmental controls maintain an environment (temperature, humidity, and air quality) suitable for the HE's activities, positively contribute to HE performance, and do not contribute to the HE's fatigue and acute stress?

p. Identify if the noise control and dampening is adequate to minimize background noise to a level that supports communications at a normal volume and limits its contribution to stress, i.e., inhibits noise from equipment fans, excessive alarms, HVAC air vents, and low-priority intercom messages.

q. Review the physical spacing and methods to limit the distraction and crosstalk from those working nearby. However, closer proximity may be needed if barrier activities require ongoing communication between HEs.

r. Identify situations of maximum occupancy. Does the room size and configuration support the maximum (design basis) number of permanent and temporary occupants, while maintaining the HE's unhindered access to barrier system elements and activities?

s. Confirm whether the room equipment placement and general arrangement minimizes the need for HE movements (e.g., eye, head, body movements) to access physical elements (e.g., equipment) and communicate with other HEs.

t. Does the room provide the ancillary facilities needed to store or hold barrier-required paper documents and Support PE?

u. Identify the maximum duration for which HEs in the room would require protection from the hazard that activated the barrier and other defined hazards.

v. If the room functions as a communication hub during barrier activation, does the configuration, design, and provisioning of the room adequately support this function?

w. See ISO-11064-3 (1999), ISO-11064-4 (2013), EEMUA 201 (2019), and like industry guidance standards for additional considerations.

2. Assess the findings from step 1 against the workspace design principles in Appendix K.2. Identify gaps that exist between these principles and the current design. Record each gap in Table 4.18. (See C9-1a for table details.)
3. Compile and issue the functional analysis report. The report includes the findings and recommendations from step 1, as recorded in Table 4.18.

4.10.1.2 C10-1b, Barrier Activities in this Workspace

From the information developed in step C10-1a (FA report and Table 4.18), identify and record the following information in Table 4.20: **Room ID**, **Barrier/Task ID**, and **Ancillary Facilities**. Ancillary facilities may include storage for Support PE and barrier/task-required paper documents such as procedures and support aids.

4.10.1.3 C10-1c, Direct-Use and Support PE in this Workspace

For each barrier/task entry in Table 4.20, identify the HE, equipment (e.g., CP or wall VDU displays) containing direct-use PE, stand-alone direct-use PE, barrier-required paper documents (types), and Support PE required for the entered barrier/task ID AND reside in the identified room. Record the following direct-use and Support PE information in Table 4.20: **Item/Descr.**, **Reference,** and **Install Type/Loc** fields. The "reference" field may be used to enter the source table for the **Item/Descr.** Entry. References to installation documents or drawings may be added to the **Install Type/ Loc** field.

4.10.1.4 C10-1d, Maintain Barrier Function

1. Based on the information recorded in Table 4.20 and available in other applicable documents listed in the introduction, this step identifies and records the period an element (identified in the **Item/Descr.** Field in Table 4.20) must be available to support its barrier function. From a synthesis of this information, identify and record the information in the **Maintain Element Function** field.
2. For each element, identify the In-Place PE (e.g., technical system) and external support system (e.g., lighting, electrical power, and environmental controls) that must be in place to enable its barrier function. Record the information in the **Technical System** and **External Support System** fields.

4.10.1.5 C10-1e, Maintain the Workspace Environment

Based on the information recorded in Table 4.20 and available in other applicable documents listed in the introduction, this step specifies the environmental control limits in the HE workspace (occupied room). Example limits include room temperature, humidity, and air quality. Add this information to the **External Support System** field in Table 4.20.

4.10.1.6 C10-1f, Protect Occupants

Table 4.20 identifies a basic requirement to maintain a specified barrier function, most of which rely on one or more HEs. Identify requirements to protect room-located personnel from potential hazards or its effects. Review Table 3.27

TABLE 4.20

Rooms: Barrier System and Support Elements

		Room-Located Barrier Element				Required for Entered Barrier Element			
Room ID	Barrier/ Task ID	Item/descr.	Reference	Install Type/Loc.	Ancillary Facilities	Technical System (TS)	External Support System (ESS)	External Protective Barrier (EB)	Maintain Element Function
Enter unique ID for this room	Enter barrier/ task ID that employs the listed barrier element	Enter barrier system element and descr. – HE – CP – VDU Wall – Wall-mount DU component – Stand-alone direct-use PE – Support PE – Paper docs., etc.	Enter reference to element source information *All that apply*	Enter install location and type: Floor mount Wall mount Ceiling mount Other	Enter required ancillary facilities for this element (if any):	Enter TS ID and descr. *All that apply See Note*	Enter ESS ID and descr. – Lighting – Electrical power – Env. Ctrls. – Other *All that apply See Note*	External protective barrier required? Enter – *YES* – *NO*	Enter time or period to maintain element function *See examples*

Note: See Table 4.19 for information for requirements identified in process C-9, control consoles and panels; CP – control console or panel; Env. Ctrls. – environmental controls; VDU – video display unit.

(safety times) and the applicable sources from Tables B.11 (App. B.7.4), B.14 (App. B.7.6), and B.20 (App. B.7.12).

1. Based on this information, determine whether an external protective barrier is required. If so, enter "YES" in the External Protective Barrier field in Table 4.20. Proceed to step 2. If not, enter "NO" and proceed to the next element or step C10-2.
2. Guided by the barrier, task, and phase safety times (Table 3.27 and other applicable information), identify the period the line-item element must continue its barrier function. Record this information in the **Maintain Element Function** field in Table 4.20.
3. Define and record information in the following fields in Table 4.21: **Protected Area, Barrier/Task ID, Protected Receptor**, and **Hazard Threat**.

Note: Table 4.21 *captures the barrier system requirements for required external protective barriers (EBs). The identified EB function may be achieved using building-provided barriers (e.g., one or more passive/active barriers) and, if required, one or more EBs external to the building. (This table is further evaluated in process C-12, buildings.)*

Note: Packaged equipment systems may provide a walk-in shelter that contains barrier-required direct-use PE. In this case, the walk-in shelter is the room addressed in this process. It may also be appropriate to address this space using process C-12 (buildings).

4.10.2   Step C10-2, Physical Workspace Design

The suggested process for the following steps reviews the barrier-specific requirements defined in step C10-1 and other requirement sources listed in the introduction. Compare and assess these requirements against the existing room design to confirm compliance. If a barrier requirement or recommendation is not met, provide the requirements to those assigned room design responsibilities for integration into that design. (For further guidance, see ISO-11064-3 (1999) and Appendix K.2.)

Note: With some industry sectors and project types, the physical dimensions and general configuration of the room may be fixed or offer few (if any) opportunities for change once it reaches a design freeze stage. This can occur early in the overall project design cycle if the room resides in an engineered, offsite fabricated building that has an early contract/purchase date (e.g., during the preliminary design phase). In this case, the challenge becomes how to maximize the design within these constraints. Preferentially, process C-10 activities occur sufficiently early to provide the best possible input before the design freeze date.

4.10.2.1   C10-2a, Sizing and General Arrangement

Following the general guidance in step C10-2, identify the room sizing and general arrangement requirements not met by the current design. Provide the requirements to those assigned room design responsibilities for integration into that design.

TABLE 4.21
External Protective Barriers

Protected Area ID/Descr. (Workspace)	Barrier/ Task ID	Protected Receptor in Protected Area	External Protective Barrier (EB)					
			Hazard Threat	ID, Descr., and Function	Type	Where implemented	Capability / Endurance Time or Period (CT) (Minutes)	CT Meets or Exceeds BSI, TTST, TPST?
Enter ID and description for protected area: Room – Occupied – Equipment Safe haven (EA) – Indoor – Outdoor Egress/escape route (EA) – Indoor – Outdoor Process Area	Enter barrier/ task ID that requires the protective function *See Note 1*	*Enter ID and description for protected receptor* – HE – Tech. Sys. – ESS – Stand-alone barrier element – Other (define) *See Note 1*	Enter the hazard/threat to receptor (each protected area) *One per row* *See Note 1*	Enter the selected barrier ID, description, and protective function *See Note 2*	Enter the EB type: – *Passive* – *Active*	Enter EB location: – Internal to building – External to building – Other	Enter the period this EB is available to provide the required protective function: – Duration of this task – Duration of barrier activation – Duration of area occupancy – Other *See Note 1*	Enter – *Yes* – *No*

CT – barrier capability/endurance time

Note 1: See Table 3.29 for requirements.

Note 2: Enter all external protective barriers required to eliminate or mitigate the hazard threat to the protected receptor (one per row).

Example – Incident Command Center: In an offshore O&G installation, HEs assigned to emergency response barriers are often located in the Central Control Room (CCR) and an Incident Command Center (ICC). Barrier activities may require in-person communications between HEs in both rooms. In addition, the barrier leader (located in the ICC) may visually monitor CCR activities as a source of information. To meet both objectives, the design may locate the ICC immediately adjacent to the CCR. A moveable window or divider could be added to separate the two rooms, an option that supports the opposing needs for face-to-face communications and to limit the barrier leader's exposure to noise and distractions generated in the CCR.

4.10.2.2   C10-2b, Layout and Configuration

Following the general guidance in step C10-2, identify the room layout and general arrangement requirements not met by the current design. Provide the requirements to those assigned room layout and configuration design responsibilities for integration into that design.

4.10.2.3   C10-2c, Ancillary Facilities

Following the general guidance in step C10-2, identify the ancillary facility requirements not met by the current design. Provide the requirements to those assigned room design responsibilities for integration into that design.

4.10.3   STEP C10-3, IN-PLACE PE: TECHNICAL SYSTEMS

Review the technical system requirements defined in Table 3.18, the functional analysis (step C10-1a), and other requirement sources listed in the introduction. Provide the information to those assigned responsibilities for the technical system (In-Place PE) design for integration into that design. An additional requirement, the **Maintain Element Function** field in Table 4.20, specifies the period the technical system must continue to provide its barrier function in the presence of the defined hazard(s).

4.10.4   STEP C10-4, EXTERNAL SUPPORT SYSTEM

This set of steps provides design input to the identified, barrier-required ESS (e.g., lighting, power, and environmental controls) that enables or maintains an HE and HE activity (e.g., environmental comfort) or a direct-use physical element function. Refer to the recommendations and requirements in Table 4.18 (functional analysis recommendations) and Table 4.20. The **Maintain Element Function** field in Table 4.20 specifies the period the ESS must continue to support or maintain its barrier-dependent function in the presence of the defined hazard(s).

In each of the following steps, an existing design process designs the listed ESS. The suggested process is to review these barrier-specific requirements against the design to confirm compliance. If a requirement is not met, provide the information to those assigned ESS design responsibilities for integration into that design.

*Note: Safety instrumented functions that reside in a safety instrumented system are typically **fail-to-safe** designs; i.e., the failure of motive power or an essential component causes the function to automatically revert to its safe state. Active human barriers (at least, many components) are **fail-to-danger**; i.e., the loss of motive power causes the loss of the supported function. Loss of electrical power does not allow an alarm beacon to light and an audible alarm tone to sound; or does not permit a remote-powered, device-enabled communication to occur. The same is true for lighting. Loss of area lighting may prevent visual access to a direct-use passive indicator under no/low light conditions. As such, the reliability of an active human barrier cannot exceed the reliability of an external support element or resource on which it acutely depends. The reliability and survivability of the barrier-dependent ESS should equal or exceed the active human barrier requirements. (See Figure D.2 in Appendix D.2.2 for a visual representation of example barrier dependencies.)*

4.10.4.1   C10-4a, Lighting

Following the general guidance in step C10-4, identify the room lighting requirements not met by the current design. Provide these requirements to those assigned EA lighting design responsibilities for integration into that design.

4.10.4.2   C10-4b, Electrical Power

Following the general guidance in step C10-4, identify the room electrical power (EP) requirements not met by the current design. Provide the requirements to those assigned EP design responsibilities for integration into that design.

4.10.4.3   C10-4c, Environmental Controls

Following the general guidance in step C10-4, identify the room environmental control (EC) requirements not met by the current design. Provide the requirements to those assigned EC design responsibilities for integration into that design.

4.10.5   STEP C10-5, EXTERNAL PROTECTIVE BARRIERS

Process C-12 (buildings) provides the evaluation, selection, and design processes that apply to the external protective barrier requirements recorded in Table 4.21.

4.10.6   STEP C10-6, BALANCE OF DESIGN

With the appropriate team, this step reviews the pending corrective actions in Table 4.18 (C10-1a) and the PIF working environment assessment (step C10-6). The path forward for each recommendation may include accepting the recommendation for implementation, accepting the current design as ALARP, or proposing a different recommendation. Forward the approved recommendations to the appropriate parties.

Note: Progressing an accepted recommendation may require a return to an earlier design process. This may warrant using the change management process to progress the design change.

4.11 PROCESS C-11, ENGINEERED AREAS

Figure 4.12 provides an overview of the C-11 design and engineering processes that apply to engineered areas. An EA is a purposely designed and provisioned safe haven or egress/escape route that may be located in a room, building, or an outdoor process area or perimeter. EAs provide and contain barrier system elements.

As applicable, perform this process for each barrier. This process assumes the primary EA design is governed by existing design processes such as NORSOK S-001 (2021), ISO 13702 (2015), and other sector-specific standards. The scope of the process C-11 is limited to only those elements and activities that apply to active human barriers.

An EA safe haven in a building may be in a room that normally serves other purposes, such as a dining area. Its design may be a combination of this process and processes C-10 (rooms) and C-12 (buildings).

The common inputs to this process are as follows:

- See the "common inputs" listed in the C-10 introduction.
- Table 3.14 – Detect and Act Phase: In-Place PE – Engineered Area Requirements.

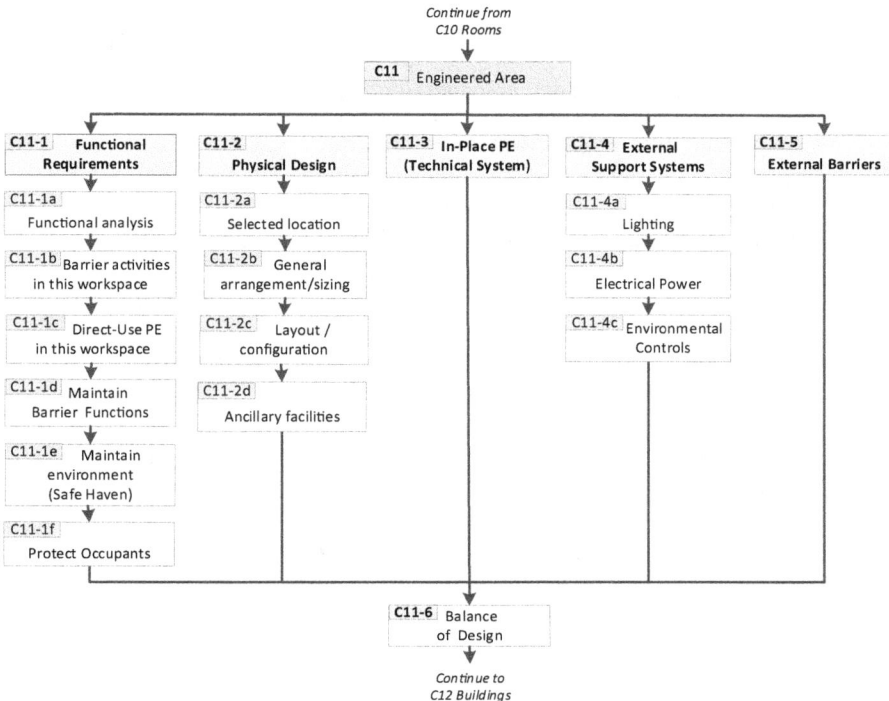

FIGURE 4.12   Detailed Design Process C-11: Engineered Areas. Note: Where Applicable, Perform this Process for Every Barrier.

- Table 4.18 – Functional Analysis Recommendations (control consoles/panels and rooms).
- Functional analysis reports (control consoles/panels and rooms).
- Table 4.20 – Rooms: Barrier System and Support Elements.
- EA dimensional layout and general arrangement drawings.
- EA design basis and specifications.
- Other sources listed in Table B-18 (Appendix B.7.10).

Note: Processes C-3 through C-8 supplement the design guidance (e.g., the CX sub-process) applied to stand-alone direct-use components and elements located in the engineered area. C-9 supplements the design of control consoles and panels. C-10 (rooms) may provide additional supplemental guidance for a safe haven located in a room. This process integrates these elements into the EA workspace design.

Case Study, Deepwater Horizon Accident: *An explosion in or near engine room #3 destroyed a section of the escape route to the aft lifeboats. Personnel assigned to these boats attempted access, found the way was blocked, and were forced to use a different route to reach the forward lifeboats. At this stage in the accident, those attempting to transit this route were likely exposed to fire, explosions, shrapnel, falling objects, and other physical dangers. Loss of emergency lighting further impeded movement and visual access to passive SA-1 indicators.*

4.11.1   STEP C11-1, FUNCTIONAL REQUIREMENTS

4.11.1.1   C11-1a, Functional Analysis

This step performs a functional analysis focusing on the HE activities and interactions that occur within an engineered area. Follow the process described in steps C9-1a and C10-1a. Record the FA findings in Table 4.18. (See step C9-1a for table details.) The analysis should be performed or supported by an operations specialist.

Key Points: When transiting an egress/escape route, personnel are physically moving through a large workspace under potentially difficult conditions. The functional analysis should consider the information in Appendices B.5.1 (look but don't see); B.5.2 (element salience/noticeability); F.1, F.2, F.3, and F.4 (automatic and conscious cognitive processes and behaviors, visual system limitations, and acute stress affects); K.1.1 (eye, head, and body movement analysis); K.1.2 (visual sightlines, optimal angles, and distances); and K.2 (physical workspace design). Transiting personnel are expected to reliably notice and perceive critical detect-phase elements placed along the route, while exposed to potentially hazardous and interfering environmental conditions. The detect phase information may be essential to movement decisions such as route and destination selection and movement speed. The hazard that activated the barrier and its potential effects and consequences (e.g., disorienting visual and audible information, shrapnel, smoke) increases the likelihood of a "look but don't see" outcome, a condition often occurs under normal conditions but becomes more prevalent under high-stress, time-constrained, and hazardous situations. The design challenge starts with selecting and locating detect-phase elements to maximize salience/noticeability. Training and developing the necessary mental models also increase detection likelihood and comprehension. (Personnel unfamiliar with the facility may lack the mental models needed to automatically scan for, notice, and understand this essential information.)

4.11.1.2   C11-1b, Barrier Activities in this Workspace

Review the information from step C11-1a and Tables 4.19 and 4.20 (control panel and room-located elements in the EA workspace). Identify and record the following information in *Table 4.22*: **EA ID** and **Type**, **Barrier/Task ID**, and **Ancillary Facilities.** Ancillary facilities are those required to hold/store the listed Support PE at the required access locations such as closets, storage areas, shelves, and cabinets. Identify the facility requirements (e.g., type, design, and locations) based on who requires access and under what conditions, and the worst-case plausible (design-case) conditions and situations when access occurs.

4.11.1.3   C11-1c, Occupants, and Direct-Use and Support PEs in this Workspace

For each engineered area and barrier/task entry in Table 4.22, identify and enter the occupants and the direct-use and support physical elements that are accessed, stored, worn, or used in this workspace (**ID/Description** Field). For each entry (as applicable), enter the element's installation type, location, and the minimum dimensions required to accommodate its storage, access, or use within the Engineered Area (**Install Type/Loc** and **Minimum Dimension** fields). The **Install Type/Loc** field may include references to the source drawings and documents that provide this information.

4.11.1.4   C11-1d, Maintain Functions

1. Based on the information recorded in Table 4.22 and the documents listed in the introduction, identify the period the direct-use or Support PE (**ID/ Description** field) must be available to support its barrier function. Record this information in the **Maintain Element Function** field.
2. For each entered element, identify the In-Place PE (e.g., technical system) and external support system (e.g., lighting or electrical power) that must be in place to enable or maintain its barrier function. Record this information in the **Technical System** and **External Support System** fields.

4.11.1.5   C11-1e, Maintain Environment

Review the HE and the direct-use and Support PEs identified in step C11-1c, the applicable information from step C11-1a, and sources listed in Table B.14 (App. B.7.6). Identify requirements to maintain the EA environmental controls to maintain element performance. For an EA located in an environmentally controlled building, the controlled limits may be temperature, humidity, and air quality. If located in an outdoor area, the controls may be limited to providing protection from weather, for example, an area rain cover or windbreak structure. Record this information in the **External Support System** field in Table 4.22.

TABLE 4.22
Engineered Area: Barrier System and Support Elements

	Engineered Area		Occupants, and Employed Direct-Use and Support PEs in EA			See Table 4.22 extension
ID	Type	Barrier/Task ID	ID/Description	Install Type/Loc.	Minimum Dimension	Ancillary Facilities
Enter EA unique ID for this EA	Enter EA type: – Egress/escape route (ER) – Safe haven	Enter barrier/task ID that employs this EA barrier element	Enter each barrier/task element and description – HE – Other personnel (no barrier role) – Stand-alone direct-use components – Control console or panel – VDU Wall – Support PE – Paper docs, etc.	Enter element installation type/location/ mount: Surface Floor mount Wall Rail/post/ structure Other	Identify minimum dimensions, e.g., bulky support or direct-use PE, other large elements transiting though EA	Enter required ancillary facilities for this element, e.g., type, location and function. All that apply

See Table 4.22 extension

	Required Support for Direct-Use and Support PE		Maintain
Technical System (TS)	External Support System	External Protective Barrier	Element Function
Enter TS ID and description Note 1	Enter ESS ID and description Lighting, electrical power, Environ. controls, Other See Note	External protective barrier required? Enter – YES – NO	Enter time or period to maintain element function See examples

Note: See Tables 4.19 and 4.20 *for CP and room requirements.*

4.11.1.6   C11-1f, Protect Occupants

Table 4.22 identifies a basic requirement to maintain a specified barrier function, most of which rely on one or more HEs. Identify requirements to protect EA-located personnel from potential hazards or its effects. Review Table 3.27 (safety times) and the applicable sources from Tables B.11 (App. B.7.4), B.14 (App. B.7.6), and B.20 (App. B.7.12).

1. Identify if an external protective barrier is required to protect the EA occupants. If so, enter "YES" in the **External Protective Barrier** field in Table 4.22. Proceed to step 2. If not, enter "NO" and proceed to step C11-2.
2. Identify the required protective period. Enter this information in the **Maintain Element Function** field in Table 4.22.
3. Enter the above information in the **Protected Area**, **Barrier/Task ID**, **Protected Receptor**, and **Hazard Threat** fields in Table 4.21 (External Protective Barriers).

Note: In areas where a toxic gas release is possible, an occupied building often includes an active barrier that detects and alarms toxic gas entry into its HVAC air inlet ducts. On alarm, the barrier function stops the HVAC fans and, if available, closes the air inlet/fire dampers. The detection and response performance of the barrier may result in some gas ingress to the building. The design may limit the internal gas concentration so it remains below the permanently disabling or fatal levels over an 8-hour exposure period. However, the value is not zero. Additional ingress sources may also exist, such as faulty exterior door seals or an ongoing need to open exterior doors to support emergency operations. The individual ingress contributions may be low. However, the collective contribution from these and other potential sources may result in an internal gas concentration that contributes to HE irritation (eyes, respiratory), distraction, and potential performance degradation. The effects may be sufficient to degrade the task/barrier reliability and performance.

4.11.2   Step C11-2, Physical Design

The suggested process for the following steps reviews the barrier-specific requirements defined in step C11-1 and other sources listed in the introduction. Compare and assess these requirements against the existing EA design to confirm compliance. If a requirement or recommendation is not met, provide the requirement to those assigned EA design responsibilities for integration into that design. See Appendix K.2 for additional guidance.

4.11.2.1   C11-2a, Selected Location

See ISO 13702 (2015), NORSOK S-001 (2021), and other sector-specific standards for guidance on where to locate an engineered area.

4.11.2.2 C11-2b, Sizing and General Arrangement

Following the general guidance in step C11-2, identify the EA sizing and general arrangement requirements not met by the current design. Provide these requirements to those assigned EA sizing and general arrangement design responsibilities for integration into that design.

4.11.2.3 C11-2c, Layout and Configuration

Following the general guidance in step C11-2, identify the EA layout and general arrangement requirements not met by the current design. (One focus area is the location and placement of detect phase elements and Support PE.) Provide these requirements to those assigned EA layout and configuration design responsibilities for integration into that design.

4.11.2.3.1 Discussion

Detect phase elements for an EA-dependent task often include passive indicators and simple active devices. (For examples, see Table 3.26 in the preliminary design step B-14.) The base design of these elements occurs in prior processes. This step further examines the element type and placement and its capability to be reliably detected by the target personnel. Detectability extends to detecting Support PE storage lockers that may be located along an egress/escape route. The design seeks to ensure all such elements are noticed and detected, an essential requirement to assure they are seen and used.

4.11.2.4 C11-2d, Ancillary Facilities

Following the general guidance in step C11-2, identify the ancillary facility requirements not met by the current design. Provide these requirements to those assigned ancillary facility design responsibility for integration into that design.

4.11.3 STEP C11-3, IN-PLACE PE: TECHNICAL SYSTEMS

Review the barrier-specific requirements from step C11-1, Table 3.16 (In-Place technical systems), and other sources listed in the introduction and Table B.21 (App. B.7.13). Provide the requirements to those assigned responsibilities for the technical system (In-Place PE) design for integration into its design. An additional requirement, namely, the **Maintain Element Function** field in Table 4.22, specifies the period that the technical system must continue to provide its barrier function in the presence of the defined hazard(s).

Identify if the application or barrier performance standards create new system-level requirements for the technical system. Example technical systems may include the following:

- Fire alarm and notification panel
- Fire and gas detection and alarm system
- Fire suppression system

- Public address/public alarm notification system
- Persons-on-board system (A system of this type may use personnel-assigned RFID cards and a computer-based database and application to track personnel. It may be used to reduce the time needed to progress and complete a muster barrier report/check-in process.)
- Device-enabled communication system
- CCTV system

4.11.4   Step C11-4, External Support System

This set of steps provides design input to the ESS (e.g., lighting, power, environmental controls) that enables or maintains HEs and HE activities (e.g., environmental comfort) or a Support PE or direct-use physical element function. Refer to the recommendations and requirements in Table 4.18 (functional assessment recommendations) and Table 4.22, and the applicable sources in Table B.19 (App. B.7.11). The **Maintain Element Function** field in Table 4.22 specifies the period the ESS must continue to support or maintain the barrier-dependent function in the presence of a defined condition or hazard.

In each of the following steps, an existing design process designs the listed ESS. The suggested process is to review these barrier-specific requirements against this design to confirm compliance. If a requirement is not met, provide the requirement to those assigned ESS design responsibility for integration into that design.

4.11.4.1   C11-4a, Lighting

Following the general guidance in step C11-4, identify the EA lighting requirements not met by the current design. Provide the requirements to those assigned EA lighting design responsibilities for integration into that design.

4.11.4.2   C11-4b, Electrical Power

Following the general guidance in step C11-4, identify the EA electrical power (EP) requirements not met by the current design. Provide the requirements to those assigned EP design responsibilities for integration into that design.

4.11.4.3   C11-4c, Environmental Controls

Following the general guidance in step C11-4, identify the environmental control (EC) requirements not met by the current design. Provide the requirements to those assigned EC design responsibilities for integration into that design.

4.11.5   Step C11-5, External Protective Barriers

External protective barrier requirements are commonly defined by regulations, industry standards, and company and regional practices. This step may contribute to the evaluation, selection, and design processes and the requirements recorded in Table 4.21. (Process C-12 identifies the external protective barriers selected to meet the active human barrier reliability and survivability requirements for physical elements and occupants within a building.)

4.11.5.1   Discussion

A petroleum refinery or chemical plant may locate a muster/rally area in an outdoor location that is normally upwind of the most likely sources of airborne hazards such as a toxic gas release. A safe haven requirement may be met by locating the EA in a blast-resistant control building that is located far from potential hazards and supported by internal and external passive and active protective functions.

An offshore O&G facility is commonly challenged by a greater range of hazards, a much smaller footprint (horizontal and vertical), and fewer options to escape danger. A safe haven is commonly located at the *cold end* of the facility, an area that maintains the greatest possible distance from the more hazardous production facilities (the hot end). If indoors, the building provides additional passive and active protective functions. (According to NORSOK S-001 (2021, cl. 22.4.4), the muster area for an oil & gas production facility *shall* be located outdoors near a lifeboat embarkation area.) A utility area (e.g., power generation systems) may be positioned between the cold and hot ends to provide an additional buffer. The cold end and utility area may be further protected by a fire wall or blast wall (passive barriers). The process area is commonly protected by additional passive and active barriers to prevent, control, and mitigate process hazards.

4.11.6   Step C11-6, Balance of Design

With the appropriate team, this step reviews the functional analysis recommendations in Table 4.18 (step C11-1a). The path forward for each recommendation may include accepting the recommendation for implementation, accepting the current design as ALARP, or proposing a different recommendation. Forward the approved recommendations to the appropriate parties for integration into the applicable design.

Note: Progressing an accepted recommendation may require a return to an earlier design process. This may warrant using the change management process to progress the design change.

4.12   PROCESS C-12, BUILDINGS

Figure 4.13 provides an overview of the C-12 detailed design and engineering processes that apply to buildings that contain barrier system elements. The building includes the spaces within the building and, as applicable, the usable space on the building roof. Where applicable, perform this process for each barrier. This process assumes the primary building design is still governed by existing design processes. The scope of process C-12 is limited to only those elements and activities that apply to active human barriers.

Note: Process C-12 is a design intersection point that highlights the need for a progressive development and fully integrated design approach that addresses every barrier system element dependency on an external support system, protective barrier, additional resource, or facility. Figure D.2 (Appendix D.2.2) provides a visual representation of possible dependencies. It offers insight into why and what barrier system element requirements should flow down to those items for integration into their design and performance standards.

The building provides a protected and controlled environment to maintain, support, and protect the active human barrier system and its dependent elements located in the building. Protective functions address hazards that may be internal or external to the building. Buildings include walk-in-type enclosures, which are common to large-scale process and power generation–packaged equipment systems.

The inputs to this process are as follows:
- Common input documents listed in the process C-11 introduction.
- Table 3.15 – Detect and Act Phase: In-Place PE Building Requirements.
- Table 4.18 – Functional Analysis Recommendations (control consoles/panels, rooms, engineered area).

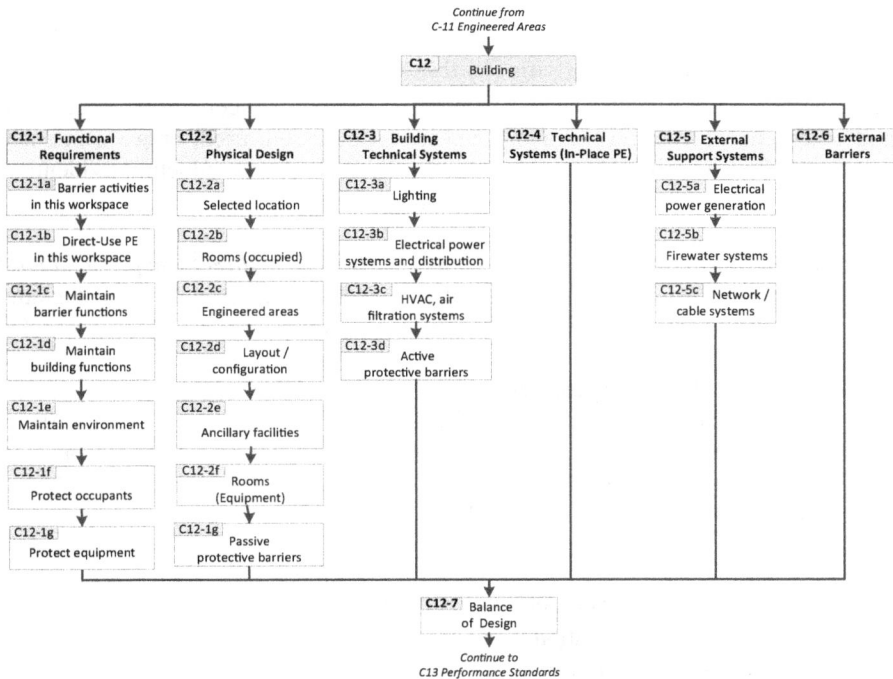

FIGURE 4.13 Detailed Design Process C-12: Buildings. Where Applicable, Perform this Process for Every Barrier.

- Functional analysis reports (control consoles/panels, rooms, engineered areas).
- Table 4.22 – Engineered Areas: Barrier System and Support Elements.
- Building and building system design documents and drawings.
- Design documents for building technical systems, ESS, and protective functions.
- Design documents for external protective barriers (active and passive).
- Other sources listed in Table B-22 (Appendix B.7.14).

Note: Processes C-3 through C-8 guide the design of stand-alone direct-use components and elements located in a building, but not in a room or engineered area. Process C-9 guides the design of control consoles and panels. C-10 may provide additional guidance for a safe haven located within a room. Process C-11 provides design guidance for an engineered area located in a building. This process integrates each of these design elements into the building design process.

4.12.1   Step C12-1, Functional Requirements

This set of steps defines the basic requirements that apply to a building that contains and supports barrier system elements and the external support systems on which it depends.

4.12.1.1   C12-1a, Barrier Activities in this Workspace

Tables 4.20 and 4.22 identify rooms and engineered areas that contain HE occupants (and others) and the direct-use and Support PEs accessed, used, or worn in each workspace. Review this and other information listed in the introduction to identify the rooms and EAs located within each building.

Note: Common with offshore O&G and other facility types, direct-use physical elements may be located on the building roof, for example, a beacon or siren. Include this area in the functional activity review, as applicable.

1. Select a building for assessment.
2. Based on Tables 4.20 and 4.22, identify and record the rooms and EAs located in the building. In Table 4.23, enter the **Building ID**, **Barrier/Task ID, Element ID/Descr.,** and **Source Reference.** Enter a descriptor for the element type.

4.12.1.2   C12-1b, Direct-Use and Support PEs in this Workspace

Review the applicable information listed in the introduction. From this information, identify the building-located direct-use and Support PEs *not* recorded in Tables 4.19

TABLE 4.23

Buildings: Barrier System and Support Elements

		Direct-Use and Support PEs Employed in Building				Required Support for Direct-Use and Support PE or Dependencies			
Bldg. ID	Barrier/Task ID	Element ID/ Description	Source Ref.	Install Type/Loc.	Ancillary Facilities	Technical System (TS)	External Support System (ESS)	External Protective Barrier	Maintain Element Function
Enter unique building ID *One per row*		Enter *occupied room* ID/description	Table 4.20	See Table 4.20 this information (process C-10)					
		Enter *EA ID/* description: *Safe Haven, Escape Route*	Table 4.22	See Table 4.22 for this information (process C-11)					
	Enter barrier/ task ID that employs this element (see right)	Enter *Element ID/* description – *Stand-alone direct-use PE* – *Support PE*	Enter reference to element info source *All that apply*	Enter installation/ mounting: floor, wall, ceiling, other *Enter ID for Install loc.:* room, EA area, etc.	Enter required ancillary facilities for this element (if any)	Enter TS ID/descr. *All that apply*	Enter ESS ID/ description – Lighting – Electrical power – Env. Controls. – Other *All that apply*	External protective barrier required? Enter – *YES* – *NO*	Enter period to maintain element function
	One per row	Enter *equipment room* ID/description	*See C12-2f*	—	—	—	*See above*	*See C12-2f*	*The most onerous case. See C12-2f*
		Enter *In-Place* TS or ESS ID/ description	*See C12-1c*	*Enter ID for equipment room, EA area, other location*	—	*See C12-1c*			*The most onerous case. See C12-1c*
		Enter *Building* TS or ESS ID/ description	*See C12-1d, C12-3*	*Enter ID for equipment room, EA area, other location*	—		*See C12-1d and C12-3*		*The most onerous case. See C12-1d*

(CPs), 4.20 (rooms), and 4.22 (engineered area). For these elements, record the following information in Table 4.23, (3rd row): **Bldg. ID, Barrier/Task ID, Element ID/ Description, Source Ref., Install Type/Loc,** and **Ancillary Facilities.** The source reference field may be used to enter the source table for the Item/Descr. References to installation documents may be added to the **Install Type/Loc.** field. Ancillary facilities may address a specific Support PE storage requirement (e.g., type, design, or location) that aligns with who requires access to this PE and the likely and worst-case design-case conditions and situations when access may be required.

4.12.1.3   C12-1c, Maintain Barrier Functions

4.12.1.3.1   Direct-Use and Support PEs
The following steps apply to the direct-use and Support PEs identified in step 12-1b.

1. For each entry from step 12-1b, enter the period this element must be available to perform its barrier function. Enter this information in the **Maintain Element Function** field in Table 4.23.
2. For each entry, identify the technical systems and external support systems (e.g., lighting, electrical power, and environmental controls) required to enable and maintain the barrier function provided by the line-item element. Record this information in the **Technical System** and **External Support System** fields.
3. In the **External Protective Barrier** field, enter "YES" if this element relies on one or more external protective barriers to protect it from hazards internal or external to the building. Otherwise, enter "NO."

4.12.1.3.2   In-Place Technical Systems (TSs) and External Support Systems (ESSs)

4. From Tables 4.19, 4.20, 4.22, and 4.23, identify the listed TS and ESS that reside in this building.
5. From step 4, enter each TS and ESS in Table 4.23. Enter the following: **Bldg. ID, Element ID/Descr., Source Ref.,** and **Install Type/Loc.** See step C12-1b for additional field entry guidance.
6. For each step 5 entry, identify the TS and ESS on which this entry depends to maintain its function. Enter the information in **the Technical System** and **External Support System** fields in Table 4.23.
7. Review the **Maintain Element Functions** entries in Tables 4.19, 4.20, 4.22, and 4.23. Identify the most onerous requirement among these entries for each technical system and ESS entry. Update the requirement in that element's **Maintain Element Functions** field in Table 4.23. (The information is additional design input to these systems.)
8. In the **External Protective Barrier** field, enter "YES" if this element relies on one or more external protective barriers to protect it from the identified hazards. Otherwise, enter "NO."

4.12.1.4   C12-1d, Maintain Building Functions

The technical systems and ESS identified in step C12-1c may rely on *other* technical systems or ESS to maintain or enable their barrier system functions. They include building TS and ESS, and other (In-Place) technical systems and external support systems that are external to the building. Identify the supporting TS and ESS. Follow C12-1c steps 5 to 8, which adds this new information to Table 4.23.

Note: Building-dedicated technical systems may themselves rely on other technical and external support systems to maintain their continued functioning and contribution to the barrier system.

Note: Refer to Figure D.2 (Appendix D.2.2). Building systems and functions commonly rely on resources that are external to the building such as electrical supply power or firewater. As such, an active human barrier may rely on a building-provided system that (itself) relies on systems and resources that are external to the building. The reliability and performance of the building, building systems, and external systems and resources should equal or exceed the active human barrier requirements.

4.12.1.5   C12-1e, Maintain Environment

A barrier-dependent, building-located room or engineered area may rely on the environmental controls and area lighting provided by building technical systems. Elements and systems that may require a controlled environment include the following:

1. HEs performing barrier activities in a building-located room or engineered area.
2. Building-dedicated technical systems that provide an external support system function on which a barrier element depends.
3. Other In-Place (building-located) technical systems and external support systems on which the building-located direct-use and Support PEs depend.

For item 1, the requirements are identified in Tables 4.19 (CPs), 4.20 (rooms), 4.22 (EA), and 4.23 (buildings) and other applicable documents listed in the introduction.

Environmental controls (ECs) are commonly required to maintain the systems mentioned in items 2 and 3 above. The EC design should be adequately addressed by existing processes. Include references to EC requirements in the SRS to ensure that the equipment is maintained to meet requirements. Provide the requirements to those assigned design responsibilities for these systems for integration into their design.

Note: A failure of an HVAC system (loss of room cooling) can cause the equipment in the room to overheat. Computer servers may contain automatic, high-temperature shutdown functions designed to prevent damage due to overheating. The shutdown results in the loss of all functions performed by that equipment. Thus, maintaining and monitoring the operational status of the HVAC and the temperature in equipment rooms is necessary.

Based on the abovementioned tables and applicable information listed in the introduction, identify and record the room and engineered area environmental control requirements, such as temperature, humidity, and air quality limits. Add or include a reference to this information in the **External Support System** field.

4.12.1.5.1   Environmental Controls as a Protective Function

HVAC and environmental control systems may also provide a building protective function. Technical systems and other equipment in the building are often not suitable for use in a hazardous environment such as in a zone 1 or 2 area. The building HVAC may include a protective function designed to maintain positive pressure in the building. This allows the continued operation of the non-classified, safety critical equipment located in pressurized areas within the building.

4.12.1.6   C12-1f, Protect Occupants

Table 4.23 identifies a basic requirement to maintain a specified barrier function, most of which rely on one or more HEs. Identify requirements to protect building-located personnel from potential hazards internal or external to the building. Review Table 3.27 (safety times) and the applicable sources from Tables B.11 (App. B.7.4), B.14 (App. B.7.6), and B.20 (App. B.7.12).

1. Identify if an external protective barrier is required. If so, enter "YES" in the **External Protective Barrier** field in Table 4.23. Proceed to step 2. If not, enter "NO" and proceed to step C12-2.
2. Identified the required protective barrier period. Enter this information in the **Maintain Element Function** field in Table 4.23.
3. Enter the above information in the **Protected Area**, **Barrier/Task ID**, **Protected Receptor**, and **Hazard Threat** fields in Table 4.21 (External Protective Barriers).

4.12.1.7   C12-1g, Protect Equipment

Table 4.23 identifies the basic requirements to maintain a specified barrier function, many of which rely on *technical systems* and *external support systems*. As such, these systems must be protected from the hazards identified in step C12-1f and potentially other hazards, as applicable.

Follow the steps from C12-1f. Record the information in Tables 4.21 and 4.23. **Exception:** Protected elements are the barrier-dependent technical systems and external support systems located in a building or on the building roof.

4.12.2 Step C12-2, Physical Design

This set of steps guides the physical design of a barrier-dependent building.

4.12.2.1 C12-2a, Selected Location

Selecting a building location that limits the threats from external hazards is an early design decision. Another design process determines and selects the HVAC air inlet (stack) locations in areas where toxic and flammable gases and other detrimental contaminants are not present. For example, see ISO 13702 (2015), NORSOK S-001 (2021), and other sector-specific standards for guidance.

4.12.2.2 C12-2b, Rooms (Occupied)

See process C-10 for guidance on the physical design of an occupied room (barrier-required) located within the building.

4.12.2.3 C12-2c, Engineered Area

See process C-11 for guidance on the physical design of an engineered area located within the building.

4.12.2.4 C12-2d, Layout and Configuration

Guided by the functional review in step C12-1a, identify and assess the suggested locations and placement of the direct-use physical elements identified in step C12-1b. Enter the information in Table 4.23. Provide these requirements to those assigned the building layout and configuration responsibilities for integration into that design.

4.12.2.5 C12-2e, Ancillary Facilities

See processes C-10 and C-11 for guidance on the physical design of ancillary facilities located within the building.

4.12.2.6 C12-2f, Equipment Rooms

Existing processes typically and adequately address the design of equipment rooms that house safety critical equipment. Examples of equipment in these rooms include building-dedicated technical systems, other technical systems (In-Place PE), electrical equipment, and uninterruptible power supply (UPS) systems.

1. Identify equipment rooms in the building that contain technical systems and external support systems identified in Tables 4.19, 4.20, 4.22, and 4.23. In Table 4.23, enter the room **ID/description**, and the barrier-dependent **Technical Systems** and **External Support Systems** located in each room.
2. Based on the information in Table 4.23, review the external protective barrier requirements for equipment in this room. If any equipment requires an external protective barrier, enter "YES" in the **External Protective Barrier** field, and proceed to step 3. Otherwise, enter "NO" and proceed to step C12-2g.

3. Based on the review in step 2, identify the most onerous **Maintain Element Function** requirement(s). Enter this information in Table 4.23.
4. In Table 4.21, enter/update the following information: **Protected Area ID/ Descr.** (In this case, the entry is the equipment room information.) From the step 3 review, enter the most onerous case information for the **Protected Receptor** (person, technical system, or ESS) and the **Hazard Threat(s)**. (If more than one threat, enter each in a separate row. The approach to mitigate each threat may differ.)

4.12.2.7   C12-2g, Passive Protective Barriers

From prior steps, Table 4.21 defines active human barrier requirements for external protective barriers (EBs). The intended EB function protects the identified receptors (personnel or equipment) from the hazard(s) recorded in Table 4.21. Receptor locations may include rooms, engineered areas, or other areas in a building or on the building roof.

This step seeks to complete Table 4.21 entries by identifying and selecting the building-provided passive barrier(s) that provide the required protective capability and function. Example passive barriers include fire-rated doors and walls and a blast-resistant building design and construction.

1. By area and receptor, identify the building-provided passive barriers that provide the required protective functions (partial or complete) identified in Table 4.21.
2. Record information for each selected barrier in Table 4.21. Record all information under the **External Protective Barrier (EB)** header (all fields).

Note: If the selected barriers do not fully achieve the required protection, steps C12-3d and C12-6 attempt to close the gap using building-provided active barriers and barriers that are external to the building.

Note: Existing standards and practices adequately address this topic. However, the unique perspective and comprehensiveness of the lifecycle model processes may provide a useful means to examine these standards and practices and identify potential gaps (or work sequences) where a safety design aspect is missed or incomplete.

4.12.3   Step C12-3, Building Technical Systems

See Tables 4.20, 4.22, and 4.23 for the technical system requirements that apply to rooms, engineered areas, and other building areas, respectively. These requirements define the period the indicated technical system must perform its defined function. For each system, review these requirements against the current technical design to confirm conformance to the active human barrier requirement. If a requirement is not met, provide the requirement to those assigned the technical system design responsibility for integration into its design. (See the notes in step C12-1d for additional background and considerations.)

Note: Figure D.2 (Appendix D.2.2) provides a visual view of the various systems and elements and the nature of the interdependencies between systems and resources.

4.12.3.1   C12-3a, Lighting

Following the information and general guidance in step C12-3, identify the lighting requirements that are not met by the current design. Provide the requirements to those assigned the lighting design responsibilities for integration into that design.

4.12.3.2   C12-3b, Electrical Power Systems

Example electrical power system (EP) systems include switchgear, motor control centers, power distribution panels, and battery-backed UPS.

Following the information and general guidance in step C12-3, identify and provide the EP requirements to those assigned EP design responsibilities for integration into their design.

Identify all active human barrier–required and barrier-dependent consumers that require UPS power. For each load, identify the situation and load period to confirm the current design has sufficient battery backup to meet this requirement in the presence of all defined loads.

Note: The facility may have a manual or automatic electrical load shed system. When the main power is limited or lost, less critical loads are disconnected to allocate the remaining power to the most critical systems. A review should confirm that the barrier systems (and the external elements and systems it relies on) are included in the list of critical systems. An ongoing review may be warranted to verify compliance. Over time, new loads may be added outside a change management process, or the load-shed logic and priorities are changed (intentionally or unintentionally). Thus, a reference to the load shed system and critical load list should be included in the safety requirements specification.

4.12.3.3   C12-3c, HVAC and Air Filtration Systems (Environmental Controls)

Following the information and general guidance in step C12-3, identify the room environmental control (EC) requirements not met by the current design. (The requirements may include HVAC functions and inlet air filtration systems, i.e., systems that may be required to reduce airborne corrosives or conductive particles.) Provide these requirements to those assigned EC design responsibilities for integration into their design.

4.12.3.4   C12-3d, Active Protective Barriers

If the required protective functions in Table 4.21 were not fully achieved in steps C12-2f and C12-2g, this step attempts to close the gap using building-provided active protective barriers.

Follow the steps from C12-2g. **Exception:** identify and select building-provided *active* barriers. Update Table 4.21 accordingly.

Example building-provided active protective barriers:

- HVAC air inlet toxic and flammable gas detection activates ESD (e.g., stops HVAC fans and closes air inlet/fire dampers).
- Building flammable gas detection activates ESD/ignition control functions (e.g., trips non-essential electrical equipment).
- Fire notification and alarm systems and act phase response actions.
- Very early smoke detection system (VESDA) and act phase response actions.
- Confirmed building fire, smoke, or high temperature activates a fire suppression system (e.g., a sprinkler or water mist system).

If a protective gap remains at the conclusion of this step, see step C12-6.

4.12.4   Step C12-4, Technical Systems (In-Place PE)

Other barrier-dependent (In-Place) technical systems and equipment may also be installed in equipment rooms. Apply the design activities in step C12-3 to these systems. Confirm the requirements (environmental controls, electrical power, external protective barriers), as they may differ from those applied to the building technical systems. Systems of this type include the following:

- General alarm/public address systems
- ESD/safety systems
- Fire and gas systems
- Telecommunication systems
- Network infrastructure

4.12.5   Step C12-5, External Support Systems (External to Building)

From the perspective of the building, a reference to an external support system requirement commonly applies to a building-provided ESS. However, a building-provided ESS commonly relies on an ESS that is external to the building (e.g., electrical power or firewater supply). In such cases, the survivability and performance of the barrier system is also reliant on an ESS that is external to the building. (For a visual representation, see Figure D.2 in Appendix D.2.2.)

4.12.5.1   C12-5a, Electrical Power Generation and Supply

Current practices typically and adequately address the design of electrical systems. A review may be needed to confirm that the electrical generation and routing of electrical power (cabling and distribution) is not affected by the hazard that activates the barrier. The reliability and survivability of the supply, building entry, and physical connection of these resources to the building systems should meet or exceed the most onerous barrier system requirements. (A failure to meet that requirement may warrant use of the change management process to address the deficiency.)

4.12.5.2   C12-5b, Firewater Systems

Current practices typically and adequately address the design of the firewater system and its supply and distribution to and within the building.

4.12.5.3   C12-5c, Network and Cable Systems

Current practices typically and adequately address the design of network and cabling systems that link technical systems to end devices and other systems external to the building. In older facilities, a review of this design may be necessary to confirm this critical infrastructure survives the hazard that activates an active human barrier and does so for the required duration. (For additional background and information, see the barrier component design step CX-0.1g in Appendix B.6.1.)

Note: This type of infrastructure may be routed underground (onshore locations) or in overhead cable trays. Overhead tray routes should avoid areas that may be affected by fire or other hazards that activate or may threaten an active human barrier. The tray system may rely on fireproofing to meet the survivability/endurance time requirement. Network communications between critical technical systems are commonly redundant, meaning they employ two (or three) separate communication channels that take different paths to limit single points of failure. However, instrumentation signal wiring and pneumatic tubing is typically simplex (not redundant). An incident that severely damages the tray containing these signal types may cause the loss (failure) of those signals and the subsequent loss of the associated barrier system element.

4.12.6   Step C12-6, External Protective Barriers (External to Building)

If the required protective functions in Table 4.21 were not fully achieved in steps C12-2f, C12-2g, and C12-3d, this step attempts to close the gap using barriers external to the building.

Follow the steps from C12-2f. **Exception:** identify and select passive (first) and active (second) protective barriers external to the building. Update Table 4.21 accordingly. (The last field in Table 4.21 is addressed in step C12-7.)

Examples of passive protective barriers include the following:
- Fire walls
- Blast walls
- Fireproofing for field-installed barrier components, essential support structures, and cable trays

Examples of external active protective barriers include the following:
- Fire and gas detection alarms that may activate:
  ○ Various levels of process, utility, and facility emergency shutdown responses

- ○ Fire suppression systems (e.g., foam, firewater deluge in process areas, and building sprinkler systems)
- ○ Ignition control actions
- • Other manual and automatically activated functions

Note: A gap that remains at the completion of this step indicates a design deficiency that may require a return to prior steps to seek a solution for the path forward. A design change may warrant using the change management process.

4.12.7   Step C12-7, Balance of Design

This step evaluates the results of the external protective barrier selection process and the results recorded in Table 4.21. (For background, see steps C12-2f, C12-2g, C12-3d, and C12-6.)

Confirm Selected Barrier Capability/Endurance Time Against Requirements

1. Based on Table 4.21, compare the **Capability Time or Endurance Period** (CT) to the applicable barrier, task, or phase safety time in Table 3.27. (The identified barrier/task/task phase elements rely on the protection provided by the external protective barrier to maintain its capability to perform the barrier function. Confirm whether the EB (individually or collectively) provides the protective function for the specified period. If yes, enter "YES" in the last column in Table 4.21. If the requirement is not met, enter "NO" and proceed to step 2.)
2. A NO response in step 1 may identify a potential discrepancy between the active human barrier requirements and the current capabilities of the selected external protective barriers (EBs). Working with those responsible for EB design (internal and external to the building), seek solutions that meet the requirements. If the gap cannot be addressed, this may identify a design deficiency that may require a return to a prior process to identify the design path forward. A change may warrant using the change management process.

Identify and Convey the Most Onerous External Protective Barrier Requirements

3. Review Table 4.21 to identify the most onerous requirements among those listed for each external protective barrier. Provide this information to those assigned responsibility for their design for integration into the applicable external protective barrier design.

Note: If the responsible person advises that the requirement is not achievable, a return to an earlier design process may be required to address the deviation. This may warrant using the change management process to progress the design change.

4.13   PROCESS C-13, PERFORMANCE STANDARDS

This process identifies and develops the performance standards (PSs) used in the life-cycle model design and verification processes. It develops the PS listed in Table 3.32, an activity started in the preliminary design step B20-3.2. Appendix D provides industry perspectives and a difference analysis. It also defines and presents the suggested PS types, content, and applicability in the lifecycle mode. Appendix G.5 presents the model verification processes the PS types proposed to support each process. See Appendices D.1 to D.3 for background information.

The inputs to this process are as follows:

- Table 3.32. Performance Standards.
- Table 4.19. Control Consoles and Panels (CP): Barrier System and Support Elements.
- Table 4.20. Rooms: Barrier System and Support Elements.
- Table 4.21. External Protective Barriers.
- Table 4.22. Engineered Areas: Barrier System and Support Elements.
- Table 4.23. Buildings: Barrier System and Support Elements.
- Applicable information and requirements listed in the source reference tables in Appendix B.7.

See Appendices D.2.2 and D.2.3 for background and the basis for the properties and subjects included in the proposed PS. Performance standards may be design-based or operations-based, discussed in Appendix D.2.1.

Note: This process does not address the performance standards mandated in regulatory or statute requirements. Such PS may tend to focus on the applied codes and standards to be employed in the design. Instead, the lifecycle model identifies PS requirements that may be missing or overlooked in current practice.

4.13.1   Background

The lifecycle model includes verification schemes that apply to the barrier system as a whole and to a barrier element or dependency. Different schemes may require different PS and PS types (design or operational). A performance standard may be required for the following:

- A selected major barrier component (e.g., a unit isolation control valve)
- A building (encloses and/or supports barrier system elements or dependency)
- Engineered areas
- A technical system (In-Place PE or building-dedicated system)
- A packaged equipment system (e.g., contains one or more barrier system elements or provides a barrier system–dependent function)
- An external support system (e.g., electrical, lighting, or environmental controls)
- An external protective barrier that may be passive or active

4.13.2　Step C13-1, Performance Standards for Physical Elements

Appendix D.3.1 and Table 4.24 present the suggested contents for a *design-based* performance standard for a physical element.

Table 4.25 presents the suggested contents for an *operation-based* performance standard for a physical element. Discussed in Appendix D.2.1, the differences from the design-based PS are in the specified requirements.

TABLE 4.24
Performance Standards for Physical Elements (Design)

Which Physical Elements	Properties to Include *(See Figure D.3 in Appendix D.3.1)*	Property Subjects to Include *(See Figure D.3 in Appendix D.3.1)*	Property/Subject Requirements *(See source tables in Appendix B.7)*
	Functionality	Functionality	Table B.10
		Response performance	Table B.11
	Integrity	Reliability	Table B.12
		Availability	Table B.12
		Integrity	Table B.12
Barrier-dependent physical elements: – Component – building/room – Engineered area – Technical system – Package equipment system – External support system – External protective barrier		Resistance to design Accident load (DAL)	Table B.14
		Resistance to environmental load (EL)	Table B.14
	Survivability (vulnerability)	Capability/ endurance time	**Component** – Table B.9, Data sheets, specifications **Engineered Area** – Table B.18, data sheets, specifications **Technical System** – Table B.21, data sheets, specifications **Building** – Table B.22, data sheets, specifications **Packaged Equipment Systems** – see "Technical System," "Building" **External support system** – Table B.19, data sheets, specifications **External protective barrier** – Table B.20, data sheets, specifications

TABLE 4.25

Performance Standard for Physical Elements (Operations)

Which Physical Elements	Properties to Include *(See Figure D.3 in Appendix D.3.1)*	Property Subjects to Include *(See Figure D.3 in Appendix D.3.1)*	Property/Subject Requirements
Barrier-dependent physical elements: – Component building/room – Engineered area – Technical system – Package equipment system – External support system – External protective barrier	Functionality	Functionality	Facility basis of design, operating philosophy, active human barrier philosophy/design basis, equipment and system data sheets and specifications For discussion and background, see App. D.2.1 and CCPS (2018, pp. 110–112)
		Response performance	
	Integrity	Reliability	
		Availability	
		Integrity	
	Survivability (vulnerability)	Resistance to design accident load (DAL)	
		Resistance to environmental load (EL)	
		Endurance time	

TABLE 4.26

Performance Standard for Human Elements (Operations)

Which Human Elements	Properties to Include *From Figure D.4 in Appendix D.3.2*	Property Subjects to Include *From Figure D.4 in Appendix D.3.2*	Property/Subject Requirements	Comments
Consider developing a performance standard for each HE Where warranted, consider separate standards that may be barrier- or task-specific (HE or HE team performance)	Functionality	Functionality	See Figure D.4 and Appendix D.3.2 for examples.	Content provided by sr. operations specialist. Technical lead having overall responsibility for barrier specification and design, collects information, and develops the PS.
		Response Performance		
	Integrity	Reliability		
		Availability		
		Integrity		
	Survivability (vulnerability)	—		
		Resistance to environmental load (EL)		
		Endurance time		

4.13.3   Step C13-2, Performance Standards for Human Elements

Appendix D.3.2 and Table 4.26 suggest the contents for an operations-based or design-based standard for barrier HEs or roles. (This PS standard type may be used in the design phase and the operate and maintain phase.)

4.13.4   Step C13-3, Performance Standards for Barrier System (Operations)

The suggested lifecycle model includes an operate and maintain phase verification process (see E-10, Chapter 7). At this stage, the barrier system is fully commissioned, operational, and expected and relied on to achieve its specified functional and safe state within the specified barrier safety time. The suggested PS to support this verification includes the same properties and subjects. However, the requirements are the specified function and performance requirements for the barrier system as a whole. See Appendix D.3.3 for guidance. (The human performance standards from step C13-2 may be included in the E-10 verification process.)

4.14   PROCESS C-14, BALANCE OF DESIGN

This process provides additional design activities in two areas not adequately addressed in other processes. The first pertains to shared systems:

- Technical systems (In-Place PE)
- External support systems
- External protective barriers

As shared systems, these elements may receive input and requirements from many disciplines. This creates a challenge in terms of compiling all requirements. Based on the review, identify the collective effect of the requirement (e.g., effects on system sizing or selection) and, where applicable, determine the most onerous requirement that may establish a primary design basis for that element, such as a worst-case survivability requirement or endurance time.

The second area pertains to:

- Additional resources
- Remote barrier support and the remote facility where this function is performed. This includes the resources and facilities that interface the remote facility equipment and systems to those located on the protected facility.
- Design review

4.14.1   Step C14-1, Other Workspace Areas

This step is a placeholder for workspace design activities for other workspaces (e.g., process and utility areas). Existing standards and practices adequately address these areas, except where noted.

4.14.2   Step C14-2, Technical Systems (In-Place PE)

Table 4.27 summarizes the design inputs and processes that may apply to an In-Place technical system that is not one of the building-dedicated systems (addressed in process C12-3). Instead, it applies to other technical systems that may be located in a building (the building provides a required protective function or ESS), in an outdoor area, or in other locations.

TABLE 4.27
Design: Technical Systems (In-Place PE)

Detailed Design Documents	Technical Systems	Dependent PE	Design Process	Input Sources
Requirements input: • Functional design • Display design • System design	**In-Place PE** • Safety instrumented systems • Fire detection/ suppression systems • Gas detection/ ESD and ignition control system • Telecomm system (e.g., CCTV, telephone, or public alarm/ public address system)	Technical system may rely on a: • Building/ equipment room • External support system • External protective barriers	**Components** C-3, C-7/8 (CX sub-process in App. B.6) **Displays** C-7, C-8 **Function** C-3 to C-8 **Control console and panels:** C-9	See Table B.21 (App. B.7.13)
Design Input: • Data sheets • Functional design drawings, diagrams, and specifications • Barrier block diagrams • Specifications • Control console and panel design • Display design				
Input to performance standard			C-13	See applicable source tables in App. B.7

Note: In this context, the term "In-Place" refers to a system that contains one or more barrier system functions but is not a building-dedicated technical system. This system may be installed in a building because it requires support from building systems (e.g., environmental controls) and external protective functions.

The suggested steps to this process are as follows:

1. Identify all barrier components and elements that reside in a specific technical system. (See the Input Sources entries in Table 4.27 for this information.)
2. Identify the barrier functional and display requirements that reside in or are enabled by the technical system.
3. Identify the most onerous performance requirements for each barrier function and element (e.g., safety times or survivability requirement).
4. Convey the information from steps 2 and 3 to those assigned primary responsibility for the technical system design for integration into its design. (Provide this information before the technical system design freeze dates.)
5. Follow up to confirm the provided information is correctly understood and integrated into the technical system design and associated documents.

4.14.3 STEP C14-3, EXTERNAL SUPPORT SYSTEMS

Like technical systems (step C14-2), external support systems must address different input requirements. This process addresses ESS requirements that apply to an engineered area, process area, and other non-building areas. *Table 4.28* summarizes the information and gives examples of the processes that apply to this design step. (For external support systems within buildings, refer to process C-12 for guidance.)

To complete this step, follow the applicable guidance and steps from step C14-2.

4.14.4 STEP C14-4, EXTERNAL PROTECTIVE BARRIERS

Process C-12 (buildings) addressed external protective barrier requirements that apply to receptors located in a building. This step addresses the requirements when receptors are located in other areas. Protective barriers of this type (passive and active) are external to a building. Table 4.21 identifies the active human barrier requirements for external protective barriers. *Table 4.29* summarizes the information and gives examples of processes that apply to this design step.

To complete this step, follow the applicable guidance and steps from step C14-2. Like technical systems (step C14-2), the review process should identify the most onerous demand/requirement placed on each external active barrier. The most onerous case may be specific to the hazard-receptor limits, location, and endurance time.

Upon completing this step, if a requirement cannot be met using the available barriers, this may identify a design deficiency that may require a return to a prior process to identify the design path forward. A change may warrant using the change management process.

4.14.5 STEP C14-5, ADDITIONAL RESOURCES

This step reviews, confirms, and updates requirements for additional resources that are part of the barrier system. (See step B20-4 for background.)

An "additional resource" may be a new capability such as an external fire response team, helicopter transport, or telemedicine resource. It may also be a new technical system or service including a satellite link, a land-based T-1 line, or an offshore fiber-optic trunk line. This step should also consider the tasks needed to activate, enable, and control this resource. (If this creates a new task, add it to Table 3.5 and progress the task design through all applicable steps, starting with those in the preliminary design process described in Chapter 3.)

The inputs to this step are as follows:
- Table 3.33. Additional Resource Requirements (see step B20-4).
- Active human barrier design basis.
- Barrier block diagrams (see step B-15).
- Safety requirements specification.
- Table 3.30. Reliability Assessment: Single Point of Failure.
- Table 3.31. Reliability Assessment: Shared Elements.
- Applicable input from processes C-1 to C-13.

TABLE 4.28
Design: External Support Systems

Detailed Design Documents	ESS	Dependent PE	Design Process	Input Sources
Requirements input: • Load demands (normal/peak, upset) • Load consumers and locations • Load shed design specification and logic Design Input: • System sizing • Specifications • Load schedule • Distribution panel schedule • Cable routing	**Electrical power systems** Switchgear Supply/distribution Battery-backup supply	This ESS may rely on: • External source power (land-based power utilities) • Cable systems, e.g., tray/cable systems • Power generation systems • Battery-backup supplies (UPS) • Buildings/rooms • Other ESS • External protective barriers	C-3, C-7, C-8, C-9 to C-12	See Table B.19 (App. B.7.11)
Input to performance standards			C-13	See applicable source tables in App. B.7
Requirements input: • Area lighting location, intensity • Lighting type (direct, indirect) Design input: see input sources Input to performance standard	**Lighting** Area Work/task Area of direct-use PE	See above electrical power systems example.		
Requirements input: • Control limits • Air quality Design input: • System sizing • Loads (each room) • Air quality requirements Input to performance standards	**Environmental controls** Temperature, humidity Air quality Building pressurization, other functions Protection from weather, direct sun	See the above electrical power systems example.		

TABLE 4.29
Design: External Protective Barriers

Detailed Design Documents	EB Type	Depends On	Design Process	Input Sources
Requirements input: • See Table 4.21 • Safety design basis	*External Protective Barrier* – **passive** • Blast wall • Fire wall • Fireproofing	This EB may rely on • Preventive maintenance • Other EBs, e.g., active protective barriers	By others	Table 4.21, Table B.20 (App. B.7.12)
Input to performance standard: • Demand scenarios • Receptor location • Capability/ endurance time				See applicable source tables in App. B.7
Requirements input: • See Table 4.21 • Safety design basis Design input: • Table 4.21 • Function and application design requirements Input to performance standard: • Demand scenarios • Receptor location • Capability/ endurance time	*External Protective Barrier* – **active** Fire detection, ESD, suppression, vessel depressurization/ isolation Gas detection, ESD, ignition control, fire suppression activation	This EB may rely on: • Technical systems • EB: passive external protective barriers • ESS: electrical systems • ESS: firewater systems • ESS: other systems and equipment		See external protective barrier-passive example.

To complete this step:

1. Review the input information for additional resource requirements.
2. In Table 3.33, enter the information for newly identified additional resources. (See step B20-4 for requirements.)
3. Update the **Resource Status** field to identify the selected provider of each resource.
4. Enter/confirm/update the method to be used to verify that each selected resource meets the requirements. Enter this information in the **Resource Verification** field. If the verification is based on a performance standard, enter the performance standard reference in the verification field. Enter

the new performance standard (if any) in Table 3.32. (See step B20-3.2 for requirements.)

5. Consider performing a risk assessment to assess the risk and consequences if an additional resource fails, degrades, or is delayed in its availability.

6. If the step 5 assessment indicates an unacceptable risk or consequence scenario, evaluate options to eliminate or mitigate the source of the failure or consequences thereof.

Note: For an applicable case study, see the preliminary design step B20-4, the DuPont La Porte, Texas, chemical plant accident.

4.14.6   STEP C14-6, REMOTE BARRIER SUPPORT (RBS)

The barrier design may include tasks (and HEs) to provide remote barrier support from a remote location. RBS may also add new facilities, equipment, and additional resources. Each should be appropriately integrated into the appropriate preliminary and detailed design steps and processes. See Appendices L.1 and L.2 for detail. See appendices L.3 and L.4 for example opportunities and areas of concern when implementing an RBS task into the barrier system design.

4.14.7   STEP C14-7, DESIGN REVIEW

This step performs the end-of-phase technical design review suggested in Appendix G.2.

4.15   PROCESS C-15, PROCEDURES

This process identifies barrier-required operating and maintenance procedures that were not identified in previous processes. It then develops a compiled procedure planning and tracking table (*Table 4.31*) that identifies the new (not yet developed) procedures, assigned provider and status.

Procedures are required barrier system elements (organizational elements). The lifecycle model identifies the task, team, and equipment usage information that includes the most challenging cognitive and physical demands placed on each HE. Much of this information benefited from operations specialist input. This new and unique information is available to guide procedure definition and development (purpose, function, and content). Figure J.1 (Appendix J.1) provides an overview of the procedure lifecycle in the prototype model. The full cycle is summarized and described in Appendix J.2. For additional guidance, see Stanton et al. (2010, Ch. 4), CCPS (2007a, Ch. 22), Edmonds (2016, Ch. 20.2), and COS (2020).

Key Point: The prototype lifecycle model demonstrates procedure traceability (content and requirements) through the entire barrier lifecycle beginning with the risk assessments and task analysis. This level of traceability is seldom possible in existing practice.

4.15.1   Step C15-1, General Guidance

The organization responsible for identifying and developing procedures should have a well-defined Procedure Development and Management Program. The lifecycle model term for this program is Procedure Development and Management System (PDMS). The PDMS should be a controlled and mature program that seeks to achieve and maintain procedure consistency, accuracy, and validity.

Note: An immature or incomplete PDMS can introduce systematic "design" errors into barrier procedures and, subsequently, into the barrier system.

A barrier-designated operating procedure informs the HE on the task activities that must occur to achieve the barrier function and safe state within the defined safety time (BST). New procedures should provide guidance that is not adequately addressed in the published standard operating procedures. Following the cognitive ergonomics approach, the procedure form and format should be consistent with existing procedures.

4.15.1.1   Who Writes and Provides Procedures?

Optimally, the owner/operator develops all procedures for every active human barrier. However, in large projects, this is often not the case. Equipment providers may be contracted to provide operating and maintenance procedures for their provided equipment. Engineering/construction contractors or others may also be contracted to provide procedures. Third-party–provided procedures tend to be more generalized and follow many different content and format standards. The barrier-required procedures should be reviewed to assess their form, format, usability, accuracy, and content. If found insufficient, the owner/operator may need to develop additional operating and maintenance procedures.

4.15.2   Step C15-2, Identify Required Procedures and Procedure Purpose

Suggested inputs to this process are as follows:

- Active human barrier philosophy and design basis.
- PDMS: procedure templates, development methodologies, approval process and authorities, modification, and update process.
- Barrier safety management plan.
- Safety requirements specification.
- Standard operating procedures (SOPs).
- Barrier organization charts and staffing plan.
- Task analysis and report (step B3-1).
- Barrier/task block diagrams (see preliminary step B-15).
- Table 3.5. Barrier Origination Requirements.
- Table 3.6. Task Origination Requirements.
- Table 3.32. Performance Standards.
- Table 4.5. Operating Procedures – Barriers/Tasks.
- Table 4.9. Access to SA-1 Incident Scene Information.
- Table 4.30. Maintenance Procedures.

TABLE 4.30
Maintenance procedures

Procedure Requirement				Procedure Detail				
Source Ref.	Proc. ID	Provider	Type/Use	Content	Format/Type	Constraint on Use	Task ID	Barrier ID
Enter source reference/ table	Enter unique ID for this proc.	Enter – O/o – Vendor – Contractor – Existing – Other	Enter type: – Preventive – Calibration – Repair/ replace	Enter required content	Enter procedure format or type (from standard templates)	Enter issues that may affect type, format, or level of detail	Enter task ID	Enter barrier ID

- Table B.5. New Operating Procedures: Direct-Use Components.
- Table B.6. New Maintenance Procedures: Direct-Use Components.
- Other task requirements tables, including those that identify barrier-employed direct-use and Support PE. (See Tables 3.2 and 4.1 for the table lists.)
- Reliability assessments (step B-19).
- Assessment of performance influencing factors (process C-2).
- Support aid requirements (processes C-3 to C-8).

4.15.2.1   C15-2a, Identify Required Operating Procedures

This step identifies any additional barrier-required *operating* procedures that were not identified in a prior process. It also migrates these to the Procedure Development Plan Table 4.31.

1. Review the input information from step C15-2 (above) to understand the barrier tasks and activities, the procedures identified in other processes (recorded in Tables 4.5 and B.5), and additional information. Also see the **Discussion** information below.
2. For each barrier/barrier task, identify new procedures required to guide and inform an HE assigned to barrier task activities. (This refers to procedures not identified in a previous process and therefore not recorded in Tables 4.5 or B.5). Populate the **Procedure Requirements** fields and (if known) the **Task ID** and **Barrier ID** fields in Table 4.5 (from step C3-3). The **Proc. ID** must be unique to that procedure. For the **Source Ref.,** enter the document/ source that was the primary basis used to identify the need for that procedure. Enter **O/o** if the procedure is provided by the facility owner/operator. If O/o is not the provider, leave the **Constraint on Use**, and (if unknown) the **Task ID**, and **Barrier ID** fields blank.

DISCUSSION

The operating procedures should adequately guide each role/HE. Identify procedures that should be written or guided by a senior operations specialist, the barrier leader, or others having a specific expertise that is unique to the procedure. See Appendix J.2 and the following for additional considerations.

HE activity–based procedure(s):

- Barrier/task activators, expectations on understanding events (SA-2) and anticipating/planning (SA-3), availability/location of other SA-1 information.
- Decision and decision-making: expectations, options, and priorities.
- Actions required to achieve a barrier/task function, safe state, and safety time.
- Complex activities. Examples: decision-making when the available information is incomplete or ambiguous, or a task includes 10 or more steps that must be performed in a proscribed sequence and at the required time.
- Activities that, if not performed correctly, may place the HE or others at risk.
- Define the task selection priorities when several barriers are concurrently active. The barriers may have the same or different priorities, or the priorities may shift due to changing conditions.

TABLE 4.31
Procedure Development Plan

Procedure ID	Procedure Type	Source Requirement	Procedure Provider	Procedure Writer	Barriers Using this Procedure	Status
Enter procedure ID *From Tables 4.5, 4.30, B.5, and B.6*	Enter procedure type *From Tables 4.5, 4.30, B5, and B.6*	Enter procedure requirement source *From Tables 4.5, 4.30, B.5, and B.6*	Enter the provider of this procedure *From Tables 4.5, 4.30, B.5, and B.6*	Enter selected procedure writer: Name/job class *Step applies to O/o provider only.*	Enter the ID for all barriers that use this procedure *All that apply*	Enter: *– Planned – Received – Defined – In development – Ready for review – Ready for approval – Approved – Update required – Rejected: revise & resubmit*

- Monitoring task: monitor an engineered area, external support system, external protective barrier, additional resource, or In-Place PE. (e.g., operational/health status or how much time the element can continue to provide its specified function or service.)
- Awareness of barrier system state: fault and alert alarms, bypass functions – off normal/normal, out-of-service systems, capacity-limited systems.
- Correct use of support aids.

Barrier team activity–based procedures:
- Leadership
- Task execution
- Communications
- Coordination
- Monitor and respond to severe stress in self and others

Support PE procedures:
- Incorrect use can lead to injury, fatality, degraded barrier performance, barrier failure (e.g., a smoke hood, self-contained breathing apparatus, or hand-held toxic or flammable gas detector.)
- An incorrectly performed remote monitoring task may lead to an injury or fatality.

Remote barrier support procedures:
- Untimely or inappropriate RBS activities or actions can degrade the barrier (team) functioning and performance. (See Appendix L.4.)

Maintenance procedures:
- If performed incorrectly, can introduce a barrier system error of failure that partially or fully disables the barrier function. (The error may remain hidden.)
- Include activities that, if not performed correctly, may expose personnel to dangers that can lead to an injury or fatality to themselves or others.
- If performed incorrectly, can result in equipment or environmental damage.

4.15.2.2   C15-2b, Identify Required Maintenance Procedures

This step identifies any additional barrier-required *maintenance* procedures that were not identified in a prior process. It also migrates these to Procedure Planning and Tracking Table 4.31.

1. Review the input information from step C15-2a, step 1 to understand the barrier tasks and activities, and the procedures identified in other processes (e.g., recorded in Table B.6).
2. For each barrier/barrier task, identify new procedures required to guide and inform those assigned to perform barrier system maintenance. (This refers to procedures not identified in a previous process and therefore not recorded in Table B.6). Populate the **Procedure Requirement** and (if known) the **Task ID** and **Barrier ID** fields in *Table 4.30*. The **Proc. ID** must be unique to each procedure. For the **Source Ref.,** enter the document/source that was

the primary basis used to identify the need for that procedure. If the provider is the facility owner/operator, enter O/o in the **Provider** field. If O/o is not the provider, leave the **Constraint on Use** field blank.

4.15.3   Step C15-3, Procedure Activity and Development Plan

This is a first step in the procedure development planning and tracking processes. It migrates selective information from Tables 4.5, 4.30, B.5, and B.6 to Table 4.31. When complete, the table identifies every new operating and maintenance required for each barrier. As such, it provides the input scope to procedure developmental process C32-2 (implementation phase). To perform this step:

1. Migrate the available information (operating and maintenance procedures) from Tables 4.5, 4.30, B.5, and B.6 to Table 4.31.
2. To the extent known at this stage, enter the known current status of each procedure by populating the **Status** field.

    Procedures may be provided by different organizations. This step is limited to gathering the available status information and entering it into this table. As such, the **Status** field entry may be limited to **Planned** or **Received**. (**Planned** may indicate a conceptual plan that assumes a third-party supplier of a barrier element will be requested to provide one or more procedures for that element, but has not as yet received a purchase order to do so.)

Note: On a large-scale capital project, procedures are often provided by many different organizations. The number of procedures can be substantial. Thus, digital tools may be needed to track and manage procedures and ensure that each is uniquely numbered/referenced.

4.16   PROCESS C-16, COMPETENCY GAP AND TRAINING NEEDS ASSESSMENT

The process summarized in Figure 4.14 performs two assessments: a competency gap assessment followed by a training needs assessment. The gap assessment evaluates the current competencies of a barrier-assigned HE against the competency requirements from the assigned tasks. An identified gap triggers a training needs assessment to determine whether and how the appropriate training can close the gap. (The results of the needs assessment guide the training design and development activities in process C32-4).

Example inputs to the two assessments:
- Barrier roster and organization chart
- Competency records for HE candidates
- Standard competencies expected (confirmed) from the HE candidate selection pool
- Barrier-specific competency requirements (Tables 3.2 and 4.1 listing the tables that may include new skill and knowledge requirements.)
- Standard training requirements for the HE candidate selection pool

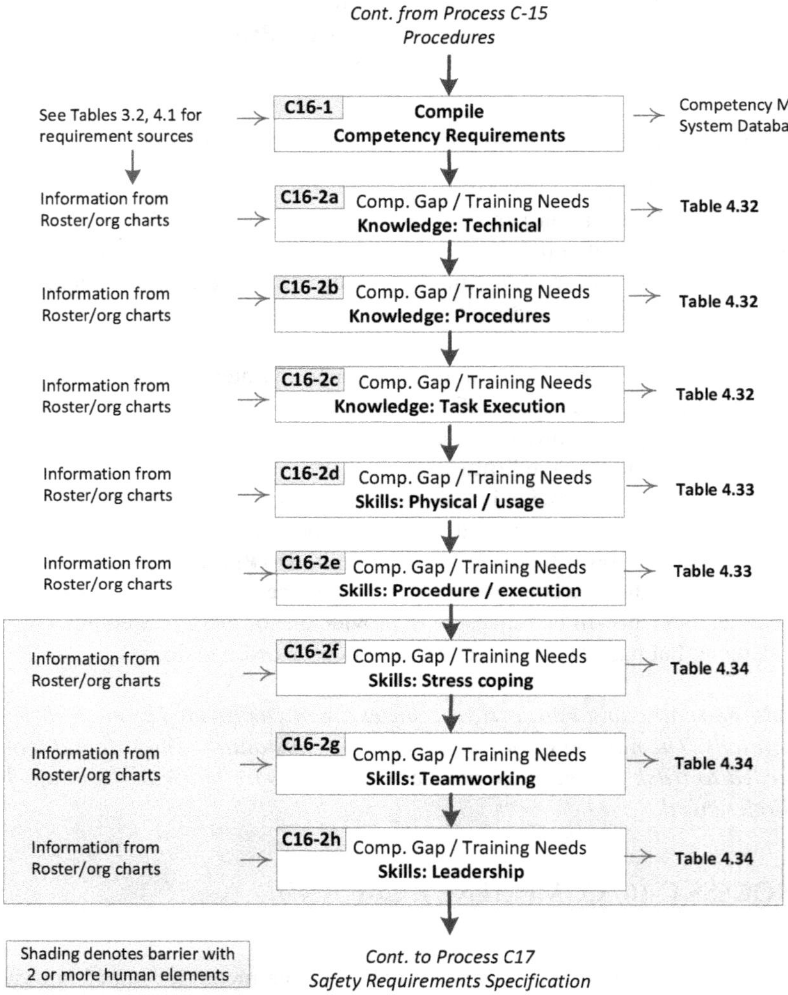

FIGURE 4.14   Detailed design Process C-16: Competency Gap and Training Needs Assessments.

- Owner/operator Competency Management System
- Owner/operator Training Development and Management System

4.16.1   BACKGROUND AND DISCUSSION

See Appendix J.3 and Figure J.1 (Appendix J.1) for background and an overview of the training activities included in the lifecycle model. Appendix J.4 provides an overview and additional information on the assumed owner/operator's competency management processes and systems included in this model. (The suggested processes and systems align with HSE 2007a). Figure F.2 (Appendix F.2) indicates the skill, knowledge, and aptitude competencies addressed or considered in the model.

Note 1: If a barrier maintenance task requires a new competency, process C-16 may be used to identify a potential competency gap and perform a training needs assessment. The assessment provides the necessary inputs to guide the training activities in process C32-4. An alternate (though less rigorous) approach applied to maintenance tasks could use EI (2020a). It employs a task analysis and risk assessment to identify the safety critical tasks (SCTs). Training tends to be developed, as required to support those SCTs.

According to Salvendy (2012, p. 491),

> training can be defined as any systematic efforts to impart knowledge, skills, attitudes, or other characteristics with the end goal being seeing improved performance. To achieve this broad goal, training must change some (or all) of the following characteristics of the trainee: knowledge, patterns of cognition, attitudes, motivation and abilities….Training efforts must have a keen eye toward the science of training and learning – it must provide opportunities to not only learn the necessary knowledge, skills, attitudes, and other characteristics (KSAOs) but also practice and apply this learning and receive feedback regarding their attempts within the training.

Stanton et al. (2010, p. 17) list attitude-specific competencies as including mutual trust, cohesion, and one's attitude toward teamwork. Encompassed within non-technical skills, these competencies are addressed in the prototype lifecycle model. They can be developed (e.g., trust and non-technical skills) through team exercises and training.

Note 2: Aptitude includes innate cognitive capabilities, some of which are not generally attainable through training and experience. HEs are assumed to have the minimum required aptitude and cognitive capabilities, which is the expected result from the candidate selection process. If this is not the case, then additional processes may be needed to identify and verify aptitude competencies.

4.16.2   STEP C16-1, COMPILE COMPETENCY REQUIREMENTS

Review the input information listed in the above step C-16 introduction for specific competency requirements. Tables 3.2 and 4.1 list example documents that provide that information. Though less likely, additional skills and knowledge requirements may also be identified in one or more of the other identified input documents. From that information, complete the following steps:

1. Gather and compile the complete list of the new skill-based competencies required for each barrier-assigned HE.

   Note: A new procedure creates a new knowledge (competency) requirement and may introduce a new skill requirement.

2. Gather and compile the complete list of the knowledge-based competencies required for each barrier-assigned HE.
3. Perform steps 1 and 2, though applied to required maintenance task competencies.
4. Enter the information from steps 1–3 into the competency management system database. (To support the activities in the lifecycle model, the CMS database should be designed to store, recall, and report this information on a per competency, per person, per barrier, and per task basis.)

4.16.2.1   Discussion

Refer to Figure F.2 (Appendix F.2). The competencies addressed in the model are those that are above and beyond those required by the standard operating procedures (SOPs) and included in the standard training regimes.

A new knowledge competency may be one of the following types. (Each may be subject to drift. See Appendix J.7.2 for detail.)

- Technical knowledge of processes, hazards, direct-use and Support PEs (usage and use hazards), potential hazard escalation pathways and rates, and others.
- Knowledge of barrier procedures.
- Knowledge of barrier task execution and activities, for example, task flow, timing, or interactions with other HEs.

A new skill competency may be one of the following types. (Each may be subject to a different skill fade period and rate. See Appendix J.7.1 for detail.)

- Skilled use of a direct-use or support physical element, or device such as an automatic (and unconscious) use of a radio or finding and calling up the desired HMI display.
- Skilled use of direct-use PE to initiate or perform an act phase action, such as correctly using and directing a fire hose or entering a command entry (number, text, or command) in an HMI display.
- Skilled performance and application of implicit coordination. (See Appendix C.3 for background.)
- Skilled *execution* performance of a complex, multi-step task or the coordinated interdependent interactions with team members.
- Non-technical skills such as leadership, teamworking, communication, maintaining situation awareness (shared and individual), or decision-making. (These skill types are needed to maintain team cohesion, coordination, and performance.)

4.16.3   Step C16-2, Competency Gap and Training Needs Assessment

The following activities are common to steps C16-2a–C16-2h. Each performs a different skill or knowledge competency gap assessment and training needs assessment.

Note: Assessing a knowledge competency may rely on computer-based testing with a follow-up assessment or demonstrated use of the required knowledge in a dynamic simulator environment. A skill-based assessment (e.g., using a device) may be an exercise where the skill is demonstrated and observed. With complex skills, a live (dynamic) demonstration may be observed by a senior operations specialist or others having the required experience and expert-level domain knowledge or skill. Both may use input from the assessed person's direct supervisor, operations management, human resources, or training department. Refer to the barrier safety management plan for additional requirements, if any. See COS (2013, Section 8.1) and HSE (2007 a/b) for additional assessment methodologies.

4.16.3.1   Competency Gap Assessment

1. For each barrier or maintenance assigned person, assess their current competencies against each new competency requirement (identified in step C16-1). Identify each competency deficiency (gap).
2. Enter each gap in the following fields in the respective Table 4.32, 4.33, or 4.34:
   - **HE ID:** The assessed HE
   - **Barrier and Task IDs**: ID for the barrier and task that required the assessed competency.
   - **Req. Source:** Enter the source of each competency requirement. (Summary Tables 3.2 or 4.1 identify the originating table that identified a competency requirement in the "New Skills or Knowledge Requirements" field in that table.)
   - **Knowledge or Skill Type:** In addition to entering the type, also enter information to clarify the specific knowledge or skill requirement. (Find this information in the identified source.)
     - Steps C16-2a–C16-2c: Enter the respective *knowledge* type (e.g., technical, procedural, or task execution).
     - Steps C16-2d–C16-2e: Enter the respective *skill* type (e.g., PE usage or task execution).
     - Steps C16-2f–C16-2h: Enter the respective *skill* type (e.g., stress coping, teamworking, or leadership).
   - **Application:** Enter the expected application of this knowledge. (See examples in the table field.)
   - **Issues to Consider:** Use this field to identify challenges that should be considered well developing the training plan. An example may include assessing if there is sufficient time to complete the training and verification process before the barrier is placed into service. Another may identify the need to address the path forward for a proposed candidate who does not have all of the competencies required by the standard operating procedures.

4.16.3.2   Training Needs Assessment

For each gap identified in the gap assessment, enter the following information in the respective Table 4.32, 4.33, or 4.34.

3. **Training Objective:** Enter the training objective specific to the skill or knowledge (competency) the trainee should be able to demonstrate upon completing training.
4. **Training Module ID:** Enter a unique training module ID or reference number.
5. **Training Provider:** Enter the proposed or planned training provider. (HE training is often provided by the owner/operator. However, training may also be commonly provided by others, e.g., a barrier equipment provider.)

Note: Closing a competency gap may be difficult or impossible if sufficient time is not available to complete the training and competency verification cycle. If this occurs, an option may be to assign a HE designation of "trainee" status that remains until the HE completes all required training, and the required competencies are verified. Perhaps, if assigned this designation, a fully competent person must actively supervise the HE's performance of barrier task activities and take over if necessary. Other alternatives may include selecting a different internal candidate, seeking a qualified external hire, or reducing the competency requirement by adding a support aid.

4.16.3.3   C16-2a, Knowledge Competency Gap/Training Needs: Technical

This step applies to a required knowledge competency specific to *technical information* (e.g., processes, equipment, fire behavior, medical triage, or hazard types and escalation pathways).

Perform the competency gap assessment described in steps C16-1 and C-16-2. Enter the findings in Table 4.32. Then, perform the training needs assessment described in step C16-2. Enter the assessment results in Table 4.32.

4.16.3.4   C16-2b, Knowledge Competency Gap/Training Needs: Procedures

This step applies to a required knowledge competency specific to *procedures* (e.g., knowledge of their existence, content, and when/how to apply). See step C16-2a for instructions to perform this step. Enter the competency gap and training needs findings in Table 4.32.

4.16.3.5   C16-2c, Knowledge Competency Gap/Training Needs: Task Execution

This step applies to a required knowledge competency specific to *barrier task execution.* (e.g., appropriate and timely team interactions.) See step C16-2a for instructions to perform this step. Enter the competency gap and training needs findings in Table 4.32.

4.16.3.6   C16-2d, Skill Competency Gap/Training Needs: PE Usage

This step applies to a required skill competency specific to the skilled use of *equipment* (e.g., appropriately direct a fire hose or navigate an HMI display system to quickly find and call up the required display). See step C16-2a for instructions to perform this step. Enter the competency gap and training needs findings in *Table 4.33*.

TABLE 4.32

HE Gap Assessment Training Plan: Knowledge (Technical, Procedure, Execution)

| HE ID | Competency Gap Assessment | | | | | Training Needs Assessment | | |
	Barrier and Task IDs	Req. Source	Knowledge Type	Application	Issues to Consider	Training Objective	Training Module ID	Proposed Training Provider
Enter HE ID	Enter ID for barrier and task requiring the entered competency	Enter ID for barrier and task that requires this competency	Enter knowledge type and detail – *Technical* – *Procedure* – *Execution*	Enter knowledge application: – *General* – *Phase activity* – *Direct-use PE use/application* – *Remote monitoring* – *Special tools* – *Maint. calibration or repair*	Enter issues to consider or address in the training plan *All that apply*	Enter training objective	Enter unique ID for this training module	Enter – O/o – Equip. provider – Other Include provider name and organization

TABLE 4.33

HE Gap Assessment Training Plan: Skills (PE Usage, Task Execution)

	Competency Gap Assessment				Training Needs Assessment		
HE ID	Barrier and Task ID	Req. Source	Skill Type	Application	Issues to Consider	Training Objective	Training Module ID	Proposed Training Provider
Enter HE ID	Enter ID for barrier and task that requires this competency	Enter Table # or doc. name/ reference that defined this req.	Enter type and detail *– PE usage* *– Task/ procedure execution*	Enter skill application: *– PE use* *– Task execution*	Enter issues to consider or address in the training plan *All that apply*	Enter training objective	Enter unique ID for this training module	Enter *– O/o* *– Equip. provider* *– Other* Include provider name and organization

TABLE 4.34
HE Gap Assessment Training Plan: Skills (Leadership, Teamwork, Stress Management)

| | Competency Gap Assessment | | | | | Training Needs Assessment | | |
HE ID	Barrier and Task IDs	Req. Source	Skill Type	Application	Issues to Consider	Training Objective	Training Module ID	Proposed Training Provider
Enter HE ID	Enter ID for barrier and task that requires this competency	Enter Table # or doc. name/ reference that defined this req.	Enter type and detail – *Stress coping/ mgmt.* – *Teamwork* – *Leadership*	Enter skill application: – Monitor/manage stress in self/others – Coordinated team execution – Communication – Learn leadership (barrier leader, other HEs)	Enter issues to consider or address in the training plan *All that apply*	Enter training objective	Enter unique ID for this training module	Enter – *O/o* – *Equip. provider* *Other* Include provider name and organization

4.16.3.7   C16-2e, Skill Competency Gap/Training Needs: Procedures and Task Execution

This step applies to a required skill competency specific to a complex *procedure* (e.g., many steps that must be promptly performed without deviation) and complex *task execution* (e.g., timely and appropriate team communications and coordination). See step C16-2a for instructions to perform this step. Enter the competency gap and training needs findings in Table 4.33.

Example: Refer to the Note in step C7-8 pertaining to the Incident Command Board. Consider the unique set of safety critical skills required to maintain awareness of what information is needed, and when and where to get that information (e.g., from people or technical systems). The information may be received in many forms (verbal, written, other) and in an uneven flow and at a potentially disruptive timing. The received information must be correctly and accurately received, remembered, and efficiently and promptly transferred to the board in the correct form such as text, sketch, and tables. Incorrect or illegible records can adversely affect the barrier result. Several skill types are needed to perform this safety critical activity.

4.16.3.8   C16-2f, Skill Competency Gap/Training Needs: Stress Coping and Management

This step applies to a required skill competency specific to *coping with and managing stress* in oneself and others. See step C16-2a for instructions to perform this step. Enter the competency gap and training needs findings in *Table 4.34.*

4.16.3.9   C16-2g, Skill Competency Gap/Training Needs: Teamworking

This step applies to a required skill competency specific to *teamworking*, which is an identified non-technical skill. See step C16-2a for instructions to perform this step. Enter the competency gap and training needs findings in Table 4.34.

4.16.3.10   C16-2h, Skill Competency Gap/Training Needs: Leadership

This step applies to a required skill competency specific to *leadership* (e.g., leadership demonstrated by the barrier leader or other barrier HEs). See step C16-2a for instructions to perform this step. Enter the competency gap and training needs findings in Table 4.34.

Note: In the O&G industry, the Offsite Installation Manager (OIM) has the senior leadership role in an offshore drilling or production facility. The OIM is commonly assigned as the emergency response barrier leader. Given the complex nature of this role, specialty training and assessment organizations are commonly required to train, assess, and verify the competency of persons assigned to this role. For an example, see OPITO (2014).

4.17   PROCESS C-17, SAFETY REQUIREMENTS SPECIFICATION

This process updates the safety requirements specification (started in preliminary design step B20-3.1) with the new information developed in this phase. For guidance on the content, development, and suggested issue dates of the SRS, see Appendix B.3.

This process should progress and seek to finalize the SRS in preparation for the final review and Issue for Construction issue.

Note: This version of the document provides important input to guide the inspection, testing, verification, and validation requirements and processes.

4.18 PROCESS C-18, PREPARE REQUISITIONS AND CONTRACTS

This process prepares the barrier-specific work scope and the technical and quality assurance requirements. It includes that information in procurement and contract documents for equipment and services that supply barrier physical and organizational elements. For an overview of this process, see Figure E.1, in Appendix E.

The following may contain barrier system elements or dependencies:
- Commercial off-the-shelf (COTS) components including individual devices or assemblies, Support PE, and others.
- Technical systems (includes VDU-based displays).
- Control consoles and panels (CPs).
- External support systems that include electrical, lighting, environmental controls, and others.
- Packaged equipment systems.
- Buildings (constructed offsite).
- Engineered areas.

The following are example documents that may be included in a procurement or contract packages. (The actual titles may differ.) Barrier requirements should be prepared and integrated into these documents. Appendix E-1 describes each document and its possible content:

1. Scope of work
2. Technical package (barrier requirements and design documents)
3. Vendor data requirements (VDR)
4. Inspection and test requirements (ITRS)

Note: Procurement can create a challenging problem to integrate the active human barrier requirements into a commercial package and do so in a way that ensures they are understood and correctly implemented and verified. Appendices A.8 and E describe a few of the challenges. A review of process C-12 (buildings) may provide further insight into the potential breadth and nature of the challenge. For example, a building may encompass many different systems, organizations, and contracts, each with its own technical, coordination, and execution challenges.

Procurement activities were introduced in step B-24 (Chapter 3, preliminary design phase). See Appendix B.6, step CX-0.3 for component-level procurement activities that include proposal content development, input, and review.

4.18.1 STEP C18-1, PREPARE INPUT TO REQUEST FOR PROPOSAL

Follow the suggested guidance in Appendix E, step E-1 to gather and prepare information to include in a request for proposal (RFP). The procurement input tables in step E-1 summarize the information (and source) developed in the prototype lifecycle model.

Based on this information, prepare the following input and content that are specific to a barrier system element:

1. Scope of work statements
2. Technical package
3. Vendor data requirements (VDR)
4. Inspection and test requirements (ITRS)

The organization assigned with the overall responsibility for the procurement or contract package solicits and interacts with many technical disciplines to compile the various inputs. The scope, VDR, and ITRS base documents may already exist. One effort is to find the most appropriate means to integrate the barrier requirements into these documents. Doing so requires awareness of when this should occur to complete that work before a potential input freeze date.

Note: With long-lead (and early need) procurement, the procurement schedule will often not align with barrier design schedule. In such cases, the barrier design schedule and activities should be adjusted to achieve the necessary alignments to the procurement schedule. Alternatively, include provisions in the purchase order or contract that support late additions at pre-negotiated pricing.

4.18.2   Step C18-2, Review the Proposal

For this step, follow the guidance in Appendix E, step E-2.

4.18.3   Step C18-3, Prepare Input to the Purchase Order or Contract

For this step, follow the guidance in Appendix E, step E-3.

4.19   PROCESS C-19, VERIFICATION #3

This process performs the verification #3 assessment. See Table 4.35 for references to the suggested verification scheme.

TABLE 4.35
Verification #3: Detailed Design Phase (PE Focus)

Verification Scheme	Example Resource Plan	Examination Form	Tangible Evidence	Performance Standards: Physical Elements	Performance Standard Basis/ Content
See Barrier Safety Management Plan	See Table G.6 (App. G.5)	See Table G.5 (App. G.5)	See Table G.5 (App. G.5)	See process C-13	See Figure D.3 in App. D.3.1

TABLE 4.36

Validation #2: Detailed Design Phase

Validation Scheme	Example Resource Plan	Examination Form	Tangible Evidence
See Barrier Safety Management Plan	See Table G.8 (App. G.6)	See Table G.7 (App. G.6)	See Table G.7 (App. G.6)

4.20   PROCESS C-20, VALIDATION (#2)

This process performs the validation #2 assessment. See Table 4.36 for references to the suggested validation scheme.

Note: The validation process described in Appendix G.6 (more common in some industries) may be less familiar or uncommon in other process industries. The validation process defined in IEC 61511-1 (2016) follows the "V" software model. As commonly applied, it tends to be an extension of the verification process. The added Functional Safety Assessment in IEC 61511-1 (2016) also tends to be an extension of the verification process. The validation scheme described in Appendix G.6 may be a significant hurdle in terms of understanding, acceptance, and successful implementation.

REFERENCES

ANSI/ISA-101.01-2015 (2015), *Human Machine Interfaces for Process Automation Systems*, American National Standards Institute/International Society of Automation

CCPS (2007a), *Human Factors Methods for Improving Performance in the Process Industries*, John Wiley & Sons, Inc.

CCPS (2018), *Bow Ties in Risk Management: A Concept Book for Process Safety*, Hoboken, NJ: John Wiley & Sons Inc., Center for Chemical Process Safety (CCPS)

COS (2013, December), *COS-3-02, Skills and Knowledge Management System Guideline*, Center for Offshore Safety, 1st Ed

COS (2020, January), *COS-3-06, Guidance for Developing and Managing Procedures*, Center for Offshore Safety, 1st Ed

Edmonds, J. (2016), *Human Factors in the Chemical and Process Industries, Making it Work in Practice*, Elsevier

EI (2020a), *Guidance on Human Factors Safety Critical Task Analysis*, London: Energy Institute, 2nd Ed, January 2020

EEMUA 201 (2019), *Control Rooms: A Guide to their Specification, Design, Commissioning and Operations*, Engineering Equipment and Materials Users Association

HSE (2007a), *Managing Competence for Safety-Related Systems, Part 1: Key Guidance*, UK Health and Safety Executive

HSE (2007b), *Managing Competence for Safety-Related Systems, Part 2: Supplementary Material*, UK Health and Safety Executive

IEC 61511-1 (2016), *Functional Safety – Safety Instrumented Systems for the Process Industry Sector – Part 1: Framework, Definitions, System, Hardware and Application Programming Requirements*, 2nd Ed, International Electrotechnical Commission

ISO 11064-3:1999 (1999), *Ergonomic Design of Control Centres – Part 3: Control Room Layout*, International Organization for Standardization, 1st Ed, 1999-12-15

ISO 11064-4:2013 (2013), *Ergonomic Design of Control Centres – Part 4: Layout and Dimensions of Workstations*, International Organization for Standardization, 2nd Ed, 2013-11-15

ISO 11064-5:2008 (2008), *Ergonomic Design of Control Centres – Part 5: Displays and Controls*, International Organization for Standardization, 1st Ed, 2008-07-01.

ISO 13702:2015 (2015), Petroleum and natural gas industries, Control and mitigation of fires and explosions on offshore production installations – Requirements and guideline, International Organization for Standardization, 2nd Ed, 2015-08

NORSOK (2021), Technical Safety, NORSOK S-001:2020+AC:2021 (en), Standards Norway

NTSB (2011), Pacific Gas and Electric Company Natural Gas Transmission Pipeline Rupture and Fire, San Bruno California, September 9, 2010, National Transportation Safety Board, Pipeline Accident Report NTSB/PAR-11/01. Washington, DC.

NUREG (2020), Human-Systems Interface Design Review Guidelines, NUREG-0700, O'Hara, J.M., Fleger, S., Rev. 3, Office of Nuclear Regulatory Research, U.S. Nuclear Regulatory Commission, Washington DC

OPITO (2014), Major Emergency Management Initial Response Training, Revision 1, OPITO Standard Code 7228, OPITO, March 13, 2014

Salvendy, G. (2012), *Handbook of Human Factors and Ergonomics*, 4th Ed, Ed. G. Salvendy and Tsinghau University, New York: John Wiley and Sons

Stanton, N.A., Salmon, P.M., Walker, G.H., Jenkins, D.P. (2010), *Human Factors in the Design and Evolution of Central Control Room Operations*, CRC Press

Taber, M.J., McCabe, J., Klein, R.M., Pelot, R.P. (2012), Development and evaluation of an offshore oil and gas emergency response focus board, *International Journal of Industrial Ergonomics*, 43, 40–51

5 Implementation
Procurement, Fabrication, Human and Organizational Elements (Model Phase C)

Figure 5.1 provides an overview of the suggested implementation process. The scope includes procurement activities (procured items that contain barrier elements), procedure and training plan development, finalizing the HE assignment to barriers, and acquiring barrier-required additional resources. In all cases, the scope of these processes is intended to be limited to only those elements that are specific to active human barriers. For inputs to this phase, see Tables 4.1–4.3. *Tables 5.1* and *5.2* summarize the assessments and documents performed and developed in this phase. Table A.2 (Appendix A.2.3) identifies the suggested participants in select activities.

TABLE 5.1

Implementation: Technical, Procurement, and Project Documents

Table	Documents	See Process/Step
—	Barrier requirements input to purchase order or contract (input to scope of work, technical package, VDR, ITRS)	C30-1, C31-1
—	Vendor data: developed under purchase order/contract	C-30, C-31
—	Design reviews and findings (procured items)	C30-2, C31-2
—	Inspection records (periodic)	C30-6, C31-6
4.5, 4.30, 4.31, B.5, B.6	Barrier procedures (developed and approved)	C32-2/3
5.7	Training modules: development, delivery, and status	C32-4
—	Barrier roster and organization chart	C-33
3.33	Additional resources: acquisition and status	C-34
—	Safety requirements specification (final update/issue)	C-36

DOI: 10.1201/9781032674476-5

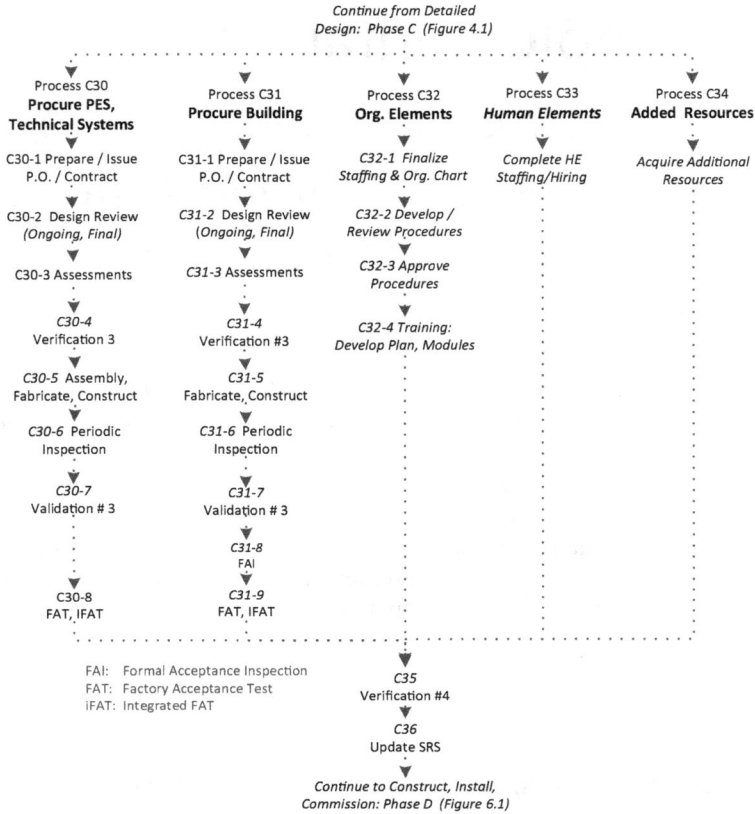

FIGURE 5.1 Implementation Phase Overview: Procure, Fabricate, Human and Organizational Elements.

TABLE 5.2
Implementation: Studies and Assessments

Table Number	Documents	See Process/Step
3.27	Phase Safety Time Assessment (see processes C-3 to C-8)	C30-3, C31-3
I.1	Cognitive Assessments (see processes C-3 to C-8)	
4.18	Functional Analysis (see processes C-9 to C-12)	
—	Verification #3	C30-4, C31-4
—	Validation #3	C30-7, C31-7
—	Verification #4	C-35
—	Factory Acceptance Inspection and Test (FAT)	C30-8, C31-8
—	Integrated FAT (IFAT)	C30-8, C31-9

5.1 PROCESS C-30, PROCUREMENT – TECHNICAL SYSTEMS AND PACKAGED EQUIPMENT SYSTEMS

This set of steps completes the procurement activities for engineered equipment, technical systems, and packaged equipment systems that contain barrier elements.

Note: A packaged equipment system skid or module may also include a stand-alone monitoring and control panel (process C-9) and walk-in room/shelter (C-10/C-12). For the latter, activities in process C-31 may also be appropriate.

5.1.1 STEP C30-1, PREPARATION OF PURCHASE ORDER, ISSUE, AND KICK-OFF MEETING

This step continues the procurement activities started in detailed design process C-18. To perform this step:

1. Follow the suggested guidance in Appendix E-3 to prepare and enter the barrier requirements into the purchase order (PO) or contract. Before package issue, review it to confirm the barrier requirements are included and appropriately integrated into the package.
2. Follow the suggested guidance in Appendix E-4 regarding attendance at the vendor kick-off meeting that occurs soon after the order is received and accepted.

Note: Project-unique situations should guide the methods used to integrate the applicable findings and recommendations (e.g., those generated in processes C-3 to C-8 and C-9 to C-12) into the PO package and design. Sufficient information may be available to complete these processes and develop the resulting requirements before the procurement cycle begins. Include this information before the PO is issued. If these processes require detailed design information from the vendor, the resulting changes may occur after the PO is issued. This can trigger a costly vendor change order. As a mitigation approach, anticipated changes should be negotiated in the bid review process with allowances and pre-negotiated pricing included in the issued PO or contract.

5.1.2 STEP C30-2, DESIGN REVIEW

A common PO provision allows the purchaser to review and comment on vendor design documents as they are being developed. The reviewer should understand and have full knowledge of the barrier-specific requirements included in the PO package. The review should confirm the vendor-provided design conforms to all requirements.

Note: As applies to all design reviews, inspections, testing, and verification and validation schemes included in the prototype lifecycle model, the scope of these activities is limited to only those elements and aspects that are part of or material to the barrier system.

Note: When the design phase is completed, a formal end-of-design review should be performed. Refer to Appendix G.2 for more information. This should occur after the assessments have been completed and recommendations implemented.

5.1.3 Step C30-3, Assessments

One or more of the following assessments apply if the procured/contracted item contains a barrier system element or a defined dependency.

If the procured item includes a detect or act phase element, perform the following assessments, as applicable:

- Conformance to the applicable phase *Success Criteria* (see assessment in processes C-3 to C-8)
- Phase safety time assessment (see assessment in processes C-3 to C-8)
- Cognitive assessment (see assessment in processes C-3 to C-8)

If the procured item is a barrier-required/dependent control console or panel, perform the applicable assessments in process C-9 (e.g., the functional analysis from step C9-1).

If the procured item is a building that includes a barrier-required element, console/panel, room, or engineered area, perform the applicable assessment in processes C-10, C-11, and C-12 (e.g., the functional analysis in step C10-1 or C11-1).

5.1.4   Step C30-4, Verification #3 – Detailed Design Deliverables

See the detailed design process C-19 and Table 4.35 for guidance on this activity.

5.1.5   Step C30-5, Assembly, Fabrication, and Construction

The PO requirements should include barrier-specific requirements that may apply to system assembly, fabrication, or construction.

5.1.6   Step C30-6, Inspections (Periodic)

During assembly, fabrication, and construction (step C30-5), the ITRS included in the PO should guide activities to perform periodic inspections to verify compliance to approved drawings and documents. See *Table 5.3* for references to the suggested plan and activities.

5.1.7   Step C30-7, Validation #3

See Table 5.4 for the suggested validation #3 scheme. The objective of this process is to detect usability issues not foreseen in the detailed design process or that have become evident in the delivered results.

Note: ISO 11064-1 (2000, para 9.8) identifies a validation requirement for the detailed design phase, but not the implementation phase. If the barrier scope is

TABLE 5.3

Periodic Inspections (Purchase Order or Contract)

Inspection Requirements	Inspection Type	Example Resource Plan	Examination Form	Tangible Evidence
See Barrier Safety Management Plan Inspection and Test Requirements	Periodic	See Table G.2 (App. G.3)	See Table G.1 (App. G.3)	See Table G.1 (App. G.3)

TABLE 5.4

Validation #3 – Procured Equipment, Systems, and Packages

Verification Scheme	Example Resource Plan	Examination Form	Tangible Evidence
See Barrier Safety Management Plan	See Table G.8 (App. G.6)	See Table G.7 (App. G.6)	See Table G.7 (App. G.6)

limited, this validation may have limited actual value. The suggested approach performs this activity at the same time as the Factory Acceptance Test (FAT) and integrated FAT.

5.1.8   STEP C30-8, FORMAL ACCEPTANCE

This set of steps performs the formal acceptance inspection and testing to verify the barrier elements meet the specified design, as well as functional and performance requirements. Acceptance testing is typically called a "Factory Acceptance Test" (FAT). The scope may include an additional test to verify an interface to an external system, which is called an "integrated FAT" (IFAT).

Table 5.5 provides the suggested plan and execution model for the FAT and IFAT. The vendor is typically responsible for scheduling, organizing, preparing for, staging, and hosting the test; recording identified deviations and corrective actions; and implementing the approved corrective actions. For the IFAT, the owner/operator or assignee may be responsible for providing the external equipment needed to perform this test.

5.2   PROCESS C-31, PROCUREMENT AND CONTRACTING – BUILDINGS

This set of steps applies to a procured building (off-site–fabricated) and the rooms and engineered areas located in the building.

TABLE 5.5
Procured Equipment and Systems – Factory Acceptance Test (FAT)

Activity	Management System	Example Resource Plan	Examination Form	Tangible Evidence	Performance Standards: Physical Elements	Performance Standards Basis/ Content
FAT Acceptance Inspection	*See Barrier Safety Management Plan*	See Table G.2 (App. G.3)	See Table G.1 (App. G.3)	See Table G.1 (App. G.3)	See detailed design process C-13	See Figure D.3 (App. D.3.1)
FAT Acceptance Testing		See Table G.4 (App. G.4)	See Table G.3 (App. G.4)	See Table G.3 (App. G.4)		
IFAT Acceptance Testing		See Table G.4 (App. G.4)	See Table G.3 (App. G.4)	See Table G.3 (App. G.4)		

5.2.1 Step C31-1, PO/Contract Preparation, Issue, and Kick-off Meeting

See step C30-1 for guidance. (The commercial document may be a PO or contract.)

5.2.2 Step C31-2, Design Review

See step C30-2 for guidance.

5.2.3 Step C31-3, Assessments

See step C30-3 for guidance.

5.2.4 Step C31-4, Design Verification #3

See the detailed design step C-19 and Table 4.35 for guidance on this activity.

5.2.5 Step C31-5, Fabrication and Construction

The PO requirements should include barrier-specific requirements that may apply to building fabrication and construction.

5.2.6 Step C31-6, Inspections (Periodic)

See step C30-6 for guidance.

5.2.7 Step C31-7, Validation #3

See step C30-7 for guidance. The difference is this scheme applies to a procured, off-site–fabricated building.

5.2.8   Step C31-8, Formal Acceptance Inspection Verification

This step performs a formal inspection to confirm the delivered building (e.g., rooms and engineered areas) complies with the approved drawings and documents. See *Table 5.6* for references to the suggested formal acceptance inspection (FAI) verification process.

5.2.9   Step C31-9, Formal Acceptance Verification (FAT, IFAT)

See step C30-8 for guidance.

The scope of a building PO or contract may include barrier-required or barrier-dependent equipment and technical systems. A FAT or IFAT may be required to verify these systems and equipment.

5.3   PROCESS C-32, ORGANIZATIONAL ELEMENTS

5.3.1   Step C32-1, Finalize HE Staffing and Hiring

This step finalizes the organization chart and roster, i.e., names the HE assigned to each barrier task/role. The organization chart should show HE backup positions; HE locations; additional resources; and the primary communication channels between HEs, to additional resources, and to other aspects as may be required by the barrier design.

Example inputs to this process:

- Barrier roster and organization chart (initially developed in preliminary design step B-25 and updated in process C-1)
- Table 3.34. Fatigue Monitoring and Management Requirements
- Barrier safety management plan
- Safety requirements specification
- HE staff and staffing policies (see Appendix J.5)

TABLE 5.6

Formal Inspection Acceptance: Fabrication/Construction

Activity	Management System	Example Resource Plan	Examination Form	Tangible Evidence	Performance Standards: Physical Elements	Performance Standards Basis/ Content
Formal acceptance inspection (FAI)	*See Barrier Safety Management Plan*	See Table G.2 (App. G.3)	See Table G.1 (App. G.3)	See Table G.1 (App. G.3)	See process C-13	See Figure D.3 (App. D.3.1)

Note: FAI may be the same as mechanical completion, though the scope is limited to barrier system elements and components only.

Note: A change to a named HE is a change in the barrier system that must be addressed. For example, replacing a named HE with a different person requires a return to the HE gap assessment process C-16. Other changes to the roster / barrier organization chart may have more significant consequences such as the deletion of an HE backup position. Changes of this scope should use the change management process to ensure the change is addressed in all applicable model processes.

5.3.2   STEP C32-2, DEVELOP AND REVIEW PROCEDURES

Table 4.31 (developed from the information in Tables 4.5, 4.30, B.5, and B.6) identifies the required new procedures, and the procedure provider and status. Steps C32-2a to C32-2c (owner/operator-provided procedures) define, schedule, and develop the procedure to a ready-for-review state. Step C32-2d (applies to all procedures) reviews and confirms the procedure readiness for approval, an action completed in C32-3. These steps should follow the guidance in the Procedure Development and Management System (PDMS) and the Barrier Safety Management Plan (BSMP).

Note: Refer to Appendix J.2 for an overview of the procedure activities and lifecycle included in the prototype lifecycle model. Figure J.1 (Appendix J.1) provides a visual representation of these activities and their interrelationship with the task analysis, training, and competency activities and the owner/operator's Procedure Development and Management Program.

Example inputs to this process:

- Procedure development schedule
- PDMS: procedure development standards, templates, and approval processes
- Appendix J.2
- Barrier safety management plan
- Standard operating procedures
- Barrier organization charts and staffing plans
- Task analysis and report (step B3-1)
- Table 3.5. Barrier Origination Requirements
- Table 3.6. Task Origination Requirements
- Table 4.5. Operating Procedures – Barriers/Tasks
- Table 4.30. Maintenance Procedures
- Table 4.31. Procedure Development Plan
- Table B.5. New Operating Procedures: Direct-Use Components
- Table B.6. New Maintenance Procedures: Direct-Use Components

5.3.2.1   C32-2a, Identify Procedure Content, Format/Type, and Constraints on Use

This step applies only to the owner/operator-assigned procedures. To prepare for this step, review the applicable information listed in the step C32-2 introduction.

Except where noted otherwise, the following steps apply to Tables 4.5, 4.30, B.5, and B.6. (These tables identify the new operating and maintenance procedures required for each barrier.)

1. Select table and procedure. If blank, enter the **Task ID** and **Barrier ID** fields indicating all barriers and barrier task that this procedure applies to.
2. Review and consider the procedure used in each barrier and barrier task. Identify the most onerous use case with consideration for all plausible use situations and environments. As applicable, enter this information into the **Constraints on Use** field. Example conditions:
   - Not enough time available to use a lengthy or overly detailed procedure.
   - Potential difficulty if required to hold/visually read the procedure while performing the procedure-identified action or activity.
3. Assess and identify the procedure content. Enter this information in the **Content** field.
   This activity should be performed or guided by a senior operations specialist with expertise in the procedure domain area or as otherwise stated in the PDMS and BSMP.
4. Select the procedure type and format. Enter this information in the **Format/ Type** field.
   The content and format of the procedure should be consistent with other procedures currently used by target users.
   Consider: A person at the low end of the experience spectrum may achieve better performance using a detailed, step-by-step procedure. Conversely, anecdotal evidence suggests this same procedure may reduce performance when used by a highly experienced person. Instead, this person may achieve better performance using an abbreviated procedure or a support aid type. (Further research is needed to confirm the validity of an adaptive procedure approach. An example system may use an electronic procedure access device that recognizes the user and provides the procedure selected (or best suited) to that user. To be viable and maintainable, the abbreviated procedure may need to be a direct subset of the detailed procedure.)
5. In Table 4.31, change the **Status** field in Table 4.31 to **Defined**.
6. Repeat the above steps for each procedure in Tables 4.5, 4.30, B.5, and B.6.

5.3.2.2   C32-2b, Identify and Schedule the Procedure Writers

This step applies only to the owner/operator-assigned procedures. This step applies to procedures where the **Status** field entry in Table 4.31 is 'ered **Defined**.' To prepare for this step, review the applicable information listed in the step C32-2 introduction.

1. For each procedure, evaluate and select the procedure writer. Enter the information in the **Procedure Writer** field in Table 4.31.
2. Based on when the procedures must achieve an approved state, determine the early and late start date to write the procedures. Schedule the procedure writer availability to align with the procedure development schedule.

Note: Achieving procedure final approval may be on the critical path to placing the barrier into service. Approved operating procedures may also be a prerequisite to progressing training development and delivery. Approval may also be required to deliver the training and guide the commissioned activities for barrier system elements. A missing procedure or using an unapproved procedure can introduce errors into a training module or the barrier system commissioning process.

5.3.2.3   C32-2c, Develop the Draft Procedure for Review

The following steps apply to procedures provided by the facility owner/operator. To prepare for this step, review the applicable information listed in the step C32-2 introduction. All entries are made into Table 4.31.

1. For each owner/operator procedure identified in Table 4.31 (and the respective procedure source table), develop the draft of the procedure for internal preliminary reviews.
2. Update the draft procedure based on internal review comments. When all draft internal reviews are complete and comments cleared, proceed to step 3.
3. Update the procedure **Status** field to **Ready for Review**.
4. Repeat the above steps for each procedure.

5.3.2.4   C32-2d, Review and Finalize Procedures (All Providers)

To prepare for this step, review the applicable information listed in the step C32-2 introduction.

1. Review the **Ready for Review** procedure draft for content, brevity, accuracy, alignment to standards, and other conformance requirements defined in the PDMS and BSMP.
2. Follow the processes and approval authorities identified in the PDMS and BSMP.
   - *Owner/operator provides procedures:* If the procedure conforms to requirements, proceed to step 3. If not, forward the comments to and repeat step C32.2c (sub-steps 2 and 3).
   - *Procedures provided by others:* If the procedure conforms to requirements, proceed to step 3. If not, determine the path forward.

   Note: A request for changes to a procedure developed by others may be limited or not permitted if that was not considered in the third-party contract or purchase order. A process should be in place to address these cases.
3. Change the procedure **Status** field in Table 4.31 to **Ready for Approval**. Proceed to step 32-3.
4. Repeat the above steps for each procedure.

5.3.3   Step C32-3, Procedure Approval

From Table 4.31, select a procedure for approval.

1. Select procedure. Confirm its **Status** field entry in Table 4.31 is **Ready for Approval**. If so, proceed to step 2. If not, select the next procedure.
2. Following the approval process and approver competency requirements in the PDMS and BSMP, confirm the procedure is ready for approval. If so, change the procedure **Status** field in Table 4.31 to **Approved**. If not, return to step 1.
3. Repeat the above steps for each procedure listed in Table 4.31.

5.3.4   Step C32-4, Training Planning and Development

Process C-16 (Tables 4.32–4.34) identifies requirements for HE training, namely, a uniquely identified training module with a specified training objective. Table B.7 (populated in the sub-process CX-0.2c, Appendix B.6) identifies operating and maintenance training requirements that are specific to direct-use components. Completing these activities reaches the readiness state needed to begin training delivery.

Note: This step applies to the training developed and delivered by the owner/operator. Additional steps may be needed to address training developed and delivered by others such as vendors and contractors. See OPITO (2014) for example training applicable to emergency response leaders. For training offshore drilling crews, see IOGP (2019c).

The activities in this series of steps (discussed in Appendix J.3.1) are as follows:

1. Review the results of the competency gap and training needs assessments recorded in Tables 4.32, 4.33, 4.34, and B.7.
2. Identify the training content.
3. Select the training delivery method.
4. Identify the required resources (e.g., training module developer, trainers, training venue).
5. Develop the training activity schedule that achieves the desired training start and/or completion dates.
6. Develop the training module and material.
7. Review and approve the completed training modules and material.

These steps and their completion are the prerequisites to beginning the training delivery indicated in process D4-1 (construct, install, and commissioning phase).

5.3.4.1   C32-4a, Review Training Needs Analysis and Other Information

The input to this process is as follows:

- HE records from the owner/operator's competency management system
- Owner/operator's published Training Development and Management System with example training modules, materials, and associated standards of practice

TABLE 5.7
Training Modules: Development, Delivery, and Status

	Training Module						Training Status		
Barrier ID	Requirement Source	Module ID	Objective	Delivery Method	Periodic Training Refresh	Module Status	Trainee ID	Status	Training Result
Enter barrier ID	Enter originating source of this training req. Enter Table 4.32, 4.33, or 4.34	Enter unique ID for this training module	Enter training module objective (from Table 4.32, 4.33, or 4.34)	Enter selected delivery method	Enter – Yes – No If Yes, enter training interval (months)	Enter module status: – Draft – QA Review – Final – Approved – Past Due (see note)	Enter ID for HE who req. this training	Enter training delivery status for this HE: – Pending – Scheduled – Complete	Enter training results: – Completed – Pass – Req. add'l training – Req. add'l; experience

Note: Reviewer/approver to meet the competency and other requirements in the PDMS and BSMP.

- Barrier safety management plan
- Competency gap assessment recorded in Tables 4.32–4.34
- Training needs assessment recorded in Tables 4.32–4.34
- Operations and maintenance training (direct-use components) recorded in Table B.7
- Approved procedures (from step C32-3)
- Other applicable documents listed in the safety requirements specification
- Desired date to start and complete training (per barrier basis)
  1. Review the inputs to understand and identify the requirements to consider and address in the remaining steps. This effort may include examining the trainees' experience, past performance, attitudes, and motivation.
  2. Populate the following fields in Table 5.7 using the information from Tables 4.32–4.34:
- **Barrier, Source and Module ID, Objective**, and **Trainee ID** – from Tables 4.32–4.34
- **Periodic Refresher Training**: if periodic retraining is recommended, enter "YES" and the interval between the refresher trainings.

Note: Retraining may be warranted for skills and knowledge subject to skill and knowledge fade and procedural use drift. (See Appendix J.7 for background and additional information.)

5.3.4.2   C32-4b, Identify Training Content

Identify the training content that must be included to achieve the training objective. Example content includes specific procedures and support aids. To this end, the CCPS (2007a, p. 79) recommends that *"The best available subject matter experts who are doers rather than supervisors should be consulted and managers should help determine content."*

5.3.4.3   C32-4c, Select the Training Delivery Method

  1. For each training module, select a delivery method appropriate to the training type and objective. Training to address a skill gap in a technical skill may differ from that needed to develop a psychometric skill. The same is true when addressing knowledge competencies. The selected training methods may also need to consider constraints identified in item 1. Examples of delivery methods are drills, simulations, exercises, and lectures.
  2. Record the delivery method in the **Delivery Method** field in Table 5.7.

From SPE (2014, p. 13),

Developing analytical and non-analytical reasoning skill has been shown to improve the quality of decision-making, as has the use of experiential training methods. In other high-risk industries and in a few other areas of the oil and gas industry, scenario-based training has been successful conducted either in simulations or as table-top exercises.

Note: Given the unique cognitive differences between skills and knowledge, the selected training method must reflect how each is learned and retained. Both are memory-based. Skills are practiced actions that become increasingly automatic with repetition in a representative environment. (A fully developed skill significantly reduces the demand on working and short-term memory.) A knowledge-based competency relies on mental models and long-term memories that are stored, developed, expanded, and updated over time. Advanced mental models, such as those required to meet a challenging SA-3 projection requirement, require extensive experience and retained knowledge. The achievement of the required (adequate) mental models and long-term memories provide the in-the-head knowledge needed to correctly perform assigned barrier tasks and do so at the required performance. This information may provide insight into why time is needed to achieve a required competency and why learning (achieving a competency) is not a linear process, from a time perspective.

Note: For a more complete discussion on competency and training, see CCPS (2022, Part 4), CCPS (2007a, Ch. 12 and 29), Flin et al. (2008, Ch. 10), Hoffman et al. (2014), Salvendy (2012, Part 3, Ch. 17), and Wickens et al. (2013, pp. 223–234).

5.3.4.4 Step C32-4d, Identify and Assign Resources

This step selects the person(s) who will develop the training modules and materials. It also identifies other resources needed to achieve readiness to begin training delivery in step D6-1. (This step may be constrained by requirements in the TDMS and BSMP.)

5.3.4.5 C32-4e, Develop Training Schedule

1. This step develops a schedule that identifies all activities that are prerequisites to achieve the desired training completion date. Completing the above steps is a prerequisite to begin training delivery as identified in process D6-1.
2. Based on the information from the competency gap assessment and training needs assessment (recorded in Tables 4.32–4.34), update Table 5.7 to show the training module requirements for each HE. Enter this information in the **Trainee ID** field. This also applies to the assigned backup roles indicated on the barrier organization chart.

5.3.4.6 C32-4f, Develop Training Module and Materials

This step develops the training modules and materials. On completion, update the **Module Status** field in Table 5.7. Example entry options are noted in the entry field.

5.3.4.7 C32-4g, Review and Approve Training (Modules and Materials)

1. This step performs reviews, verifications, and approvals as required by the TDMS and BSMP.
2. On completion of this step, update the **Module Status** field in Table 5.7. Example entry options are noted in the entry field.

Note: Allot time to address any identified corrections to the module and materials.

Note: Complex, rarely used skills are subject to skill fade, even among highly trained personnel. They include the technical and non-technical skills common to emergency response. Periodic refresher training (e.g., exercise-based) may be required to maintain these skills. For further information and guidance, see Appendix J.7.1.

5.4 PROCESS C-33, HUMAN ELEMENTS: COMPLETE BARRIER STAFFING AND HIRING

This step implements the barrier staff assignments in accordance with the approved roster, organization chart, and staffing policy, for example, shift/rotation assignments. This step includes external hires as applicable.

Note: Personnel performing this step should be fully cognizant of the methodical and systematic activities that provided its input. Any change can create a barrier deficiency, such as when an HE does not meet one or more competency requirements. A material change should be addressed using the change management process.

5.5 PROCESS C-34, COMPLETE THE ACQUISITION OF ADDITIONAL RESOURCES

This process finalizes the acquisition of the additional resources identified in Table 3.33. (See step C14-5 for background.) Deviations should be addressed using the change management process.

5.6 PROCESS C-35, VERIFICATION #4

See Table 5.8 for the suggested activities in the verification #4 scheme. This process verifies the activities attributed to fabrication, inspection, and testing of procured equipment, technical systems, packaged equipment systems, and buildings. It also verifies barrier procedure completion and approval, the planning and development of required training modules, acquisition of barrier HE staff (e.g., completion of internal transfers, hires), and the *additional resources* identified in step C14-5.

5.7 PROCESS C-36, SAFETY REQUIREMENTS SPECIFICATION (UPDATE AND ISSUE)

This process updates the SRS to include any changes or new information developed in the implementation phase. If no changes occur in phase E, this issue finalizes the barrier system in preparation for the operate and maintain phase.

Note: From this point forward, a proposed modification that changes information in the SRS (e.g., a change to barrier element, a barrier-dependent system, or a named owner/operator program) is a material change to the barrier system. The change management process should be used to process the proposed modification.

TABLE 5.8

Verification #4 – Fabricated Systems and Constructed Facilities

Verification Scheme	Example Resource Plan	Examination Form	Tangible Evidence	Performance Standards: Physical Elements	Performance Standards Basis/ Content
See Barrier Safety Management Plan	See Table G.6 (App. G.5)	See Table G.5 (App. G.5)	See Table G.5 (App. G.5)	See Process C-13	See Figure D.3 (App. D.3.1)

REFERENCES

CCPS (2007a), *Human Factors Methods for Improving Performance in the Process Industries*, John Wiley & Sons, Inc.

CCPS (2022), *Human Factors Handbook for Process Plant Operations, Improving Safety and Systems Performance*, New York: John Wiley & Sons Inc., Center for Chemical Process Safety (CCPS)

Flin, R., O'Connor P., Crichton, M. (2008), *Safety at the Sharp End: A Guide to Non-Technical Skills*, Ashgate Publishing

Hoffman, R.R., Ward, P., DiBello, L., Fiore, M., Andrews, D.H. (2014), *Accelerated Expertise, Training for High Proficiency in a Complex World*, Psychology Press

IOGP (2019c), Recommendations for enhancements to well control training, examination, and certification, London: International Association of Oil and Gas Producers, IOGP Report No 476, 2nd Ed, November 2019

ISO 11064-1:2000 (2000), *Ergonomic Design of Control Centres – Part 1: Principles for the Design of Control Centres*, 1st Ed, International Organization for Standardization, 2000-12-15

OPITO (2014), Major Emergency Management Initial Response Training, Revision 1, OPITO Standard Code 7228, OPITO, March 13, 2014

Salvendy, G. (2012), *Handbook of Human Factors and Ergonomics*, 4th Ed., Ed: G. Salvendy and Tsinghau University, New York: John Wiley and Sons

SPE (2014), The human factor; process safety and culture, SPE Technical Report, Society of Petroleum Engineers, March 2014

Wickens, C.D., Hollands, J.G, Banbury, S., Parasuraman, R. (2013), *Engineering Psychology and Human Performance*, 4th Ed, Pearson Education Inc.

6 Construction, Installation, and Commissioning (Model Phase D)

Figure 6.1 provides an overview of the suggested construction of the facility, installation of procured equipment, and pre-commissioning and commissioning activities. On completion of this phase, the barrier system and its dependent equipment, systems, facilities, and barriers are ready to start up. In all cases, the scope and activities are intended to be limited to barrier system components, elements, areas, and dependencies. For inputs to this phase, see Tables 5.1 and 5.2 for documents and information developed in the implementation phase. *Tables 6.1* and *6.2* summarize the assessments and documents developed in this phase. Table A.3 (Appendix A.2.3) identifies the suggested participants in select activities.

TABLE 6.1

Construction, Installation, and Commissioning Phase: Technical, Procurement, and Project Documents

Documents	See Process/Step
Installation records	D1-2/3, D2-2, D3-1, D4-1/2
Commissioning records	D2-7, D3-2, D4-7

TABLE 6.2

Construction, Installation, and Commissioning Phase: Assessments

Documents	See Process/Step
Inspection records (periodic)	D1-4, D2-3, D4-3
Verification #4	D1-7, D2-6, D4-6
Validation #3	D1-5, D2-4, D4-4
Formal acceptance inspection (FAI)	D1-6, D2-5, D4-5
Subsystem site acceptance test (S-SAT)	D2-8, D3-3, D4-8
Training records	D6-1
HE competency assessment	D6-2
Verification #5	D-7
Barrier site acceptance test	D-8
Validation #4 report	D-9
Pre–start-up readiness check	D-10

Continue from
Implementation: Phase C (Figure 5.1)

Process D-1	Process D-2	Process D-3	Process D-4	Process D-5	Process D-6
Facility / Site Construction	**Building (Site Constructed)**	**Install Procured Equip., Buildings**	**Construct Remote Ops. Center**	**Human Elements**	**Org. Elements**
D1-1 Engineered Areas (Site constructed)	D2-1 Construct Building		D4-1 Construct/Install (See D-1, D-2, D-3)		D6-1 Deliver Training
D1-2 Install barrier components	D2-2 Installation (See D1-2, D1-3)	D3-1 Installation	D4-2 Installation (See D1-2, D1-3)		
D1-3 Construct/Install ESS, infrastructure External Barriers	D2-3 Periodic Inspection		D4-3 Periodic Inspection		
D1-4 Periodic Inspection	D2-4 Validation # 3		D4-4 Validation # 3		
D1-5 Validation # 3	D2-5 FAI		D4-5 FAI		
D1-6 FAI	D2-6 Verification # 4		D4-6 Verification # 4		
D1-7 Verification # 4	D2-7 Commission Subsystems	D3-2 Commission Subsystems	D4-7 Commission Subsystems		
	D2-8 S-SAT	D3-3 S-SAT	D4-8 S-SAT	D-5 Verified Competent	D6-2 HE Competency Assessment

FAI: Formal Acceptance Inspection
S-SAT: Site Acceptance Test
B-SAT: Barrier Site Acceptance Test

D-7 – Verification # 5

D-8 – B-SAT   applies to barrier system

D-9 – Validation # 4

D-10 – Pre-Startup Readiness Check

Continue to Operate and
Maintain: Phase E, F, G (Figure 7.1)

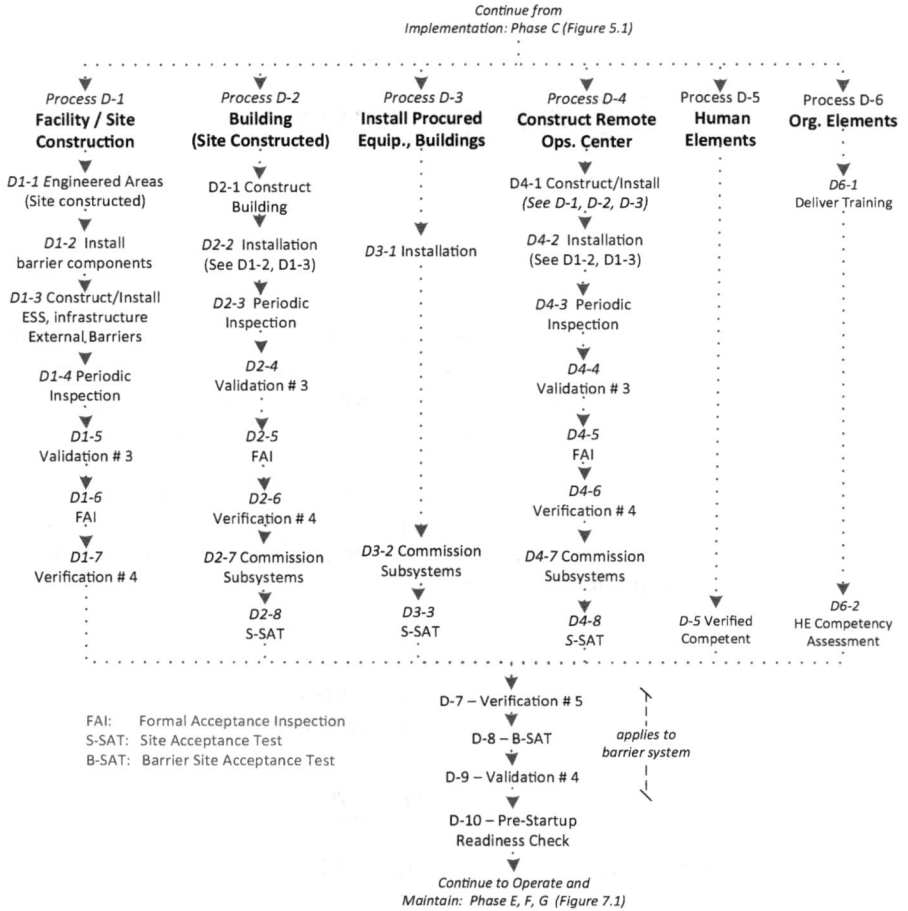

FIGURE 6.1   Construction, Installation, and Commissioning Phase Overview.

6.1   PROCESS D-1, FACILITY/SITE CONSTRUCTION

This set of steps defines the suggested facility/site installation and construction activities. The site may be land-based. If an offshore O&G facility, the location may be a primary facility module such as a process, utility, or drilling module. The scope of this process is limited to barrier system elements and identified dependencies.

Construction and installation include field-located engineered areas, direct-use displays (passive and active), and act phase components (e.g., a control valve, fire suppression system, and lifeboat). It may also include external support systems (e.g., area lighting and firewater systems), external protective barriers, and barrier critical infrastructure (e.g., power, signal, and network overhead cable trays).

6.1.1   STEP D1-1, ENGINEERED AREA (OUTDOOR AREAS)

This step constructs the engineered areas such as egress/escape routes and rally/ muster areas.

Note: Process C-11 provided barrier input to the physical design and location of engineered areas. Table 4.22 *summarizes the engineered areas such as egress/ escape routes and rally/muster areas.*

6.1.2   STEP D1-2, INSTALL BARRIER DIRECT-USE COMPONENTS (OUTDOOR AREAS)

This step installs direct-use barrier elements in engineered areas (constructed in step D1-1) and other facility/site locations.

Note: Processes C-3, C-7, C-8, and C-11 and sub-process CX (Appendix B.6) define the installation type and locations of direct-use components.

6.1.3   STEP D1-3, INSTALL AREA LIGHTING, INFRASTRUCTURE, AND EXTERNAL PROTECTIVE BARRIERS

This step installs area lighting (ESS), cable trays, and external protective barriers (EBs) in engineered areas and other facility/site areas.

Note: Processes C-3, C-7, C-8, and C-11 and sub-process CX (Appendix B.6) define the lighting requirements and installation locations.

*Note: **Infrastructure includes cable tray systems** that carry barrier-dependent power cables and control signals. Processes C-3, C-7, and C-8 and sub-process CX (Appendix B.6) specify requirements to route and physically protect cable trays and the contained barrier cables. In exposed areas, these processes may identify passive protection (e.g., fireproofing) to protect exposed trays (and the enclosed cables) from hazards (e.g., a fire) that activate the barrier.*

*Note: Process C-11 defines the requirements for **external protective barriers**. See Table 4.21 for the requirements. External protective barriers may be passive (e.g., fireproofing, a fire wall, or a blast wall) or active.*

6.1.4   STEP D1-4, PERIODIC INSPECTION

During construction, the contract requirements (e.g., the ITRS) should allow periodic inspections to verify compliance with approved drawings and documents. See *Table 6.3* for references to the suggested plan and activities.

TABLE 6.3
Periodic Inspections – Site/Facility Construction and Installation

Inspection Requirements	Inspection Type	Example Resource Plan	Examination Form	Tangible Evidence
See Barrier Safety Management Plan Inspection and Test Requirements Specification	Periodic	See Table G.2 (App. G.3)	See Table G.1 (App. G.3)	See Table G.1 (App. G.3)

TABLE 6.4
Validation #3 – Site/Facility Construction Phase

Verification Scheme	Example Resource Plan	Examination Form	Tangible Evidence
See Barrier Safety Management Plan	See Table G.8 (App. G.6)	See Table G.7 (App. G.6)	See Table G.7 (App. G.6)

6.1.5   Step D1-5, Validation #3

See Table 6.4 for references to the suggested validation #3 scheme that applies to the constructed and installed elements employed in the barrier system.

Construction practices may allow the construction contractor to make "minor" installation modifications when a design document includes a "to suit" or similar guidance statement. Without fully understanding who uses direct-use PE (e.g., visual and condition of access) and the use conditions, minor changes may be material to the barrier functioning and performance. If the validation assessment catches the deviation early in the construction process, a request to "correct" the identified deviation may be implemented within the bounds of triggering a change order.

6.1.6   Step D1-6, Formal Acceptance Inspection of Site Facility Construction

This step performs the formal acceptance inspection (FAI) to verify whether the barrier-specific installation and construction work conforms to the approved installation and construction drawings and documents. See Table 5.6 (step C31-8) for the suggested FAI scheme.

6.1.7   Step D1-7, Verification #4

Refer to detailed design process C-35 and Table 5.8 for the Verification #4 requirements.

6.2   PROCESS D-2, ONSITE-CONSTRUCTED BUILDING

This set of steps defines the suggested installation and construction activities specific to a site-constructed building that contains barrier system elements and supports barrier system activities. The status of each barrier and barrier-dependent element should be tracked through the completion of this step. Examples of barrier elements from the detailed design phase include the following:

- Stand-alone barrier system components (processes C-3 to C-8)
- Control consoles or panels (process C-9)
- Rooms (process C-10)
- Engineered areas (process C-11)
- Building-dedicated technical systems (process C-12)
- Building-provided external support systems (process C-12)
- Building-provided external protective barriers (process C-12)
- External support systems (building interface to ESS external to the building)

6.2.1   Step D2-1, Construct Building

Processes C-10, C-11, and C-12 provide guidance on the barrier critical elements and areas that reside in an onsite-constructed building or, as applicable, located on its roof. Tables 4.20–4.23 summarize example barrier and barrier-dependent elements that may be included in the building contractor's scope of supply.

Note: Many different execution models are applied to a site-constructed building. For example, the building contractor may be required to supply and install all building-dedicated technical systems. Alternatively, some or all of these systems may be free-issued (procured by others) for installation by the construction contractor.

6.2.2   Step D2-2, Install Barrier System Elements, Technical Systems, ESS, and Internal EBs

See steps D1-2 and D1-3 for guidance on this step.

Note: The difference here is that the equipment and systems are installed in a field-constructed building.

6.2.3   Step D2-3, Periodic Inspections

See step D1-4 for guidance.

Note: The difference here is that the building inspections occur at the protected facility site.

6.2.4   STEP D2-4, VALIDATION #3

See step D1-5 for guidance.

Note: The difference here is that the validation occurs in a site-constructed building.

6.2.5   STEP D2-5, FORMAL ACCEPTANCE INSPECTION (FAI) OF CONSTRUCTED BUILDING

See step D1-6 for guidance.

Note: The difference here is that the building inspections occur at site.

6.2.6   STEP D2-6, VERIFICATION #4

Refer to detailed design process C-35 and Table 5.8 for the Verification #4 requirements.

6.2.7   STEP D2-7, SYSTEM COMMISSIONING

Using existing processes, this step commissions the technical systems, external support systems, and active external protective barriers that reside within the building. The status of each barrier and barrier-dependent element should be tracked through the completion of this step.

Completion of this step brings the building systems to a state of operational readiness, a pre-condition to formal acceptance testing. Commission equipment using approved procedures. Example activities include the following:

- Pre–power-up checks
- System power-up
- Basic checks to confirm systems and equipment operate as expected
- Complete commissioning using procedures provided by the system/equipment provider
- Identify deficiencies and complete approved corrective actions if any

6.2.8   STEP D2-8, SYSTEM SITE ACCEPTANCE TESTING

This step performs the formal System Site Acceptance Inspections and Tests (S-SAT) for the barrier-dependent equipment and systems that reside in the building. The S-SAT is a critical activity in the overall verification process. It does not necessarily repeat tests performed in the FAT. Rather, it seeks to perform tests not possible at a FAT, for example, to verify system performance using the installed cable systems, source power, ambient lighting, and a live interface to other systems. (The scope of

TABLE 6.5
Buildings: System Site Acceptance Tests (S-SAT)

Activity	Management System	Example Resource Plan	Examination Form	Tangible Evidence	Performance Standards: Physical Elements	Performance Standards Basis/ Content
S-SAT Acceptance testing	*See Barrier Safety Management Plan*	See Table G.4 (App. G.4)	See Table G.3 (App. G.4)	See Table G.3 (App. G.4)	See process C-13	See Figure D.3 (App. D.3.1)

these tests should be addressed in the barrier safety management plan.) Dependent on building type and function, the subsystems include the following:

- Building-dedicated technical systems
- Building-dedicated external support systems (e.g., electrical, lighting, and environmental controls)
- Building-provided external protective barriers (active type)
- Building-located In-Place PE – technical systems

For guidance on the suggested execution plan and activities, see Table 6.5.

6.3 PROCESS D-3, INSTALL PROCURED SYSTEMS, PACKAGES, AND BUILDINGS

This set of steps applies to technical systems, packaged equipment systems, and buildings procured in detailed design processes C-30 and C-31. The timing of this step depends on the scheduling of the activities in process D-1.

6.3.1 STEP D3-1, INSTALLATION

This step completes the installation of the abovementioned equipment in accordance with the approved installation documents and drawings. Example activities include setting the equipment in place and completing electrical and utility connections. Connections include the following:

- Electrical power, signal and ground cabling and wiring
- Network and communication cables
- Process and utility connections (e.g., piping and tubing)

At the completion of this step, all physical interface connections are completed and checked. Barrier-required equipment is calibrated. A basic check confirms the equipment is ready to commence formal acceptance activities.

6.3.2 Step D3-2, Commissioning

See step D2-7 for guidance. The step applies to the equipment and buildings installed in step D3-1.

6.3.3 Step D3-3, System Site Acceptance Testing (S-SAT)

See step D2-8 and Table 6.5 for guidance.

6.4 PROCESS D-4, REMOTE BARRIER SUPPORT, REMOTE OPERATIONS CENTER

The following assumptions apply to the remote barrier support (RBS) performed from a Remote Operations Center (ROC):

- RBS tasks are designed using the prototype lifecycle model.
- The ROC and associated systems and equipment employed in the barrier function are designed using this model.
- The ROC resides in a location physically distant/separate from the protected facility.
- RBS HEs perform their assigned RBS tasks from the designated remote location (e.g., an ROC).
- The ROC contains the direct-use PE, technical systems, external support systems, and additional resources needed to enable and support the HE's assigned RBS tasks and activities. Their functionality and performance meet the requirements defined in the prototype lifecycle model.
- To the extent possible and practicable, the ROC design, layout, and technical systems should emulate the space and systems used by HEs on the protected facility.
- The direct-use PE at the ROC is the same as that used in the protected facility.
- Additional resources are required to interface the ROC to the protected facility to support all assigned HE tasks and activities in the ROC and protected location.

Refer to Appendix L for a more complete overview of the ROC regarding design, functioning, and allocation of barrier activities to HEs in the ROC.

6.4.1 Step D4-1, Construct Remote Operations Center (Site-Constructed)

See process D-2 for guidance.

6.4.2 Step D4-2, Installation of Barrier-Required/ Dependent Equipment/Systems

See steps D1-2 and D1-3 for guidance on the installation of direct-use PE and barrier-dependent systems such as external support systems.

6.4.3 STEP D4-3, PERIODIC INSPECTIONS

See step D1-4 for guidance.

6.4.4 STEP D4-4, VALIDATION #3

See step D1-5 for guidance.

6.4.5 STEP D4-5, FORMAL ACCEPTANCE INSPECTION

See step D1-6 for guidance.

6.4.6 STEP D4-6, VERIFICATION #4

Refer to Table 5.8 and detailed design step C-35 for the Verification #4 requirements.

6.4.7 STEP D4-7, COMMISSION SYSTEMS

See step D2-7 for guidance.

6.4.8 STEP D4-8, SYSTEM SITE ACCEPTANCE TEST

See step D2-8 for guidance. The scope of the S-SAT for technical systems located in a Remote Operations Center (ROC) may differ technically and functionally from the mirrored systems located in the protected facility. An ROC and RBS task introduces new HE(s), technical systems, and their interfaces to the protected facility. Additional safeguards may be added to prevent undesirable interactions. Examples of additions to the S-SAT are as follows:

- Confirm that HMI display call-up and data update times meet requirements. (ROC display and communication latency can occur if the interface and communication implementation employed to connect the two sites lacks the capacity to achieve the specified performance requirements.)
- Confirm reliable communications (e.g., radio systems, video conferencing, and other network connected systems).
- Confirm that high-bandwidth consumers (e.g., high-resolution CCTV video) do not degrade the performance of other systems that share a common communication resource.
- Confirm that an identified ROC restriction is implemented. Examples:
  - Inhibiting or preventing the capability to activate an identified function from the ROC (e.g., facility general alarm, emergency shutdown, or a fire suppression system)
  - Inhibiting or preventing the capability to override protected facility control of a CCTV camera
- Verify the performance of additional resources (e.g., fire brigade, helicopter evacuation, telemedicine).

6.5 PROCESS D-5, HUMAN ELEMENTS – VERIFY HE COMPETENCY

Process D-5 records the results of the competency assessment (performed in step D6-2) into the competency management system and the readiness status of the assessed HE or maintenance/support person. Further, the designated person identified in the barrier safety management plan (BSMP) reviews the results of the assessment and makes the final decision on HE readiness for barrier assignment and barrier activation. If the competency assessment confirms that the HE achieved all required competencies, this may be a proforma activity. In some cases, one or more competencies may not have been achieved. (Issues of this type should be addressed in the BSMP.)

6.5.1 Discussion

What if one or more required competencies were not achieved or verified? The BSMP may identify situations where an HE could be conditionally approved to perform barrier activities. In such cases, the BSMP should specify required safeguards to put in place throughout the duration of this status. For example, the person may be permitted to perform the assigned activities under the increased supervision of someone fully qualified to "take over" and provide real-time support as needed. (This may not be feasible with some barrier HE roles such as an emergency response barrier leader.) The conditional state should have a defined period during which the person is expected to gain additional experience, continue training, and achieve the required competency. If not achieved within that period, a different solution may be needed.

6.6 PROCESS D-6, ORGANIZATIONAL ELEMENTS

This process delivers the barrier-required training (HE and maintenance) and performs competency assessments to verify that barrier-assigned personnel have obtained the competencies required for their assigned tasks. See Appendices J.3 and J.4 for background information on training delivery and the lifecycle model expectations for the owner/operator's Competency Management System.

6.6.1 Step D6-1, Deliver Training

1 This step delivers the required training to the HEs (and others) identified in Table 5.7. (See implementation phase step C32-4 for background.)
2 As the training progresses, update the HE training progress in the **Training Status** field in Table 5.7. Example entry options are noted in the entry field.
3 Using the selected means for confirming the training results, enter the results in the **Training Result** field in Table 5.7. Example entry options are noted in the entry field.

Note: This step applies to training delivered by the owner/operator. An equivalent process is needed to address the training developed and delivered by other organizations such as vendors and contractors. For training examples for an emergency response barrier leader, see OPITO (2014). For training examples for offshore drill-ing personnel, see IOGP (2019c).

Note: Delivering training and verifying the achievement of the required competen-cies may be on a critical path to participate in the formal site acceptance inspection and testing and achieve barrier operational readiness. Training activities should be scheduled to align with these dates.

6.6.2   Step D6-2, HE Competency Assessment

This step performs the competency assessments that verify whether barrier-assigned personnel achieve and demonstrate all required competencies. Record the results of this assessment in the competency management system (CMS) database. (Step D-5 uses the results of this process.)

Note: Once identified in detailed design step C16-1, the barrier-specific competency requirements should be gathered and entered in the CMS database. The com-petency requirement provides the basis (measurable evidence) to support the competency (skill and knowledge) verification process. The barrier safety management plan and CMS should define the plan and methodology used to perform this step.

Note. For additional information and guidance, see CCPS (2022, Part 4), COS (2013 section 8), HSE (2003, 2007 a/b), and Stanton et al. (2010, Ch. 2).

6.7   PROCESS D-7, BARRIER SYSTEM – VERIFICATION #5

The verification scheme #5 verifies barrier system readiness before it is placed into service. See Table 6.6 for guidance on this verification scheme.

6.8   PROCESS D-8, BARRIER SYSTEM SITE ACCEPTANCE TEST

The Barrier System Site Acceptance Test (B-SAT) is a full demonstration of the barrier system, i.e., of PE, HE, and OE. See Table 6.7 for references that define the activities and basis of the B-SAT.

TABLE 6.6
Verification #5 – Operational Readiness

Verification Scheme	Example Resource Plan	Examination Form	Tangible Evidence	Performance Standards: Physical Elements	Performance Standards Basis/ Content
See Barrier Safety Management Plan	See Table G.6 (App. G.5)	See Table G.5 (App. G.5)	See Table G.5 (App. G.5)	See detailed design process C-13	See Figure D.3 (App. D.3.1)

TABLE 6.7
Barrier System Site Acceptance Test (B-SAT)

Activity	Management System	Example Resource Plan	Examination Form	Tangible Evidence	Performance Standards: Physical, Human Elements	Performance Standards Basis/Content
B-SAT Acceptance testing	See *Barrier Safety Management Plan*	See Table G.4 (App. G.4)	See Table G.3 (App. G.4)	See Table G.3 (App. G.4)	See detailed design process C-13	See App. D.3.3 and Figure D.4 (App. D.3.2)

6.8.1 DISCUSSION

A fully automated safety instrumented function (designed to the IEC 61511 standards) is typically tested many times before being placed into service. Regular equipment monitoring and testing occurs once it is put in service. This is commonly not true for the more complex active human barrier; a barrier type may have many interdependencies and fail-to-danger elements and relies on one or more humans to achieve its safety function. In the Deepwater Horizon accident, the more frequently tested, fully automated SIFs failed. The scale, nature, and consequences of these failures were profound. (See Appendix M for more information.) The degraded and less rigorously tested emergency response barriers and prompt response of the nearby Damon Bankston were the primary contributors that saved the lives of 115 souls.

Note: The B-SAT may identify PE, HE, or OE deficiencies that may prevent the barrier from achieving its intended function or performance. If HE competency contributes to that result, additional time, repetition, and training may be needed to meet requirements. Operations leadership and personnel should remain aware of this risk. For the period that this condition remains, resolving barrier deficiencies should be prioritized over a short-term production objective.

6.9 PROCESS D-9, BARRIER SYSTEM – VALIDATION #4

Validation scheme #4 seeks to validate the now fully operational barrier system. See Table 6.8 for references that provide the suggested guidance on validation scheme #4.

This scheme seeks to assess adequacy and usability issues in the task design; workspace design; layout and provisioning; design, placement, and accessibility of direct-use PE and support PE; and organizational elements like procedures, training, and defined competencies. The scheme can also be used to examine other areas including trust, morale, and motivation. The suggested timing is concurrent with the B-SAT (process D-8). This "live" demonstration may reveal issues that were not otherwise apparent or considered.

TABLE 6.8
Validation #4 – Barrier System

Validation Requirements	Validation Type	Example Resource Plan	Examination Form	Tangible Evidence
See Barrier Safety Management Plan	Validation #4	See Table G.8 (App. G.6)	See Table G.7 (App. G.6)	See Table G.7 (App. G.6)

6.10   PROCESS D-10, PRE-START READINESS CHECK

The local regulatory regime and industry practice may require a start-up readiness check. This suite of checks includes those identified in the steps described above.

REFERENCES

CCPS (2022), *Human Factors Handbook for Process Plant Operations, Improving Safety and Systems Performance*, New York: John Wiley & Sons Inc., Center for Chemical Process Safety (CCPS)

COS (2013, December), *COS-3-02, Skills and Knowledge Management System Guideline*, 1st Ed, Center for Offshore Safety

HSE (2003), *Competence Assessment for the Hazardous Industries, RR086*, London: Safety and Health Executive

HSE (2007a), *Managing Competence for Safety-Related Systems, Part 1: Key Guidance*, UK Health and Safety Executive

HSE (2007b), *Managing Competence for Safety-Related Systems, Part 2: Supplementary Material*, UK Health and Safety Executive

IOGP (2019c), Recommendations for enhancements to well control training, examination, and certification, London: International Association of Oil and Gas Producers, IOGP Report No 476, 2nd Ed, November 2019

OPITO (2014), Major Emergency Management Initial Response Training, Revision 1, OPITO Standard Code 7228, OPITO, March 13, 2014

Stanton, N.A., Salmon, P.M., Walker, G.H., Jenkins, D.P. (2010), *Human Factors in the Design and Evolution of Central Control Room Operations*, CRC Press

7 Operate and Maintain, Modify, and Decommission (Model Phases E, F, and G)

This chapter addresses the remaining phases in the barrier lifecycle including operate and maintain (phase E), modify (phase F), and decommissioning (phase G). Figure 7.1 provides an overview of the suggested processes. The process activities should be specified in the barrier safety management plan (BSMP), SRS, and the owner/operator's published policies and programs. In all cases, the scope of these

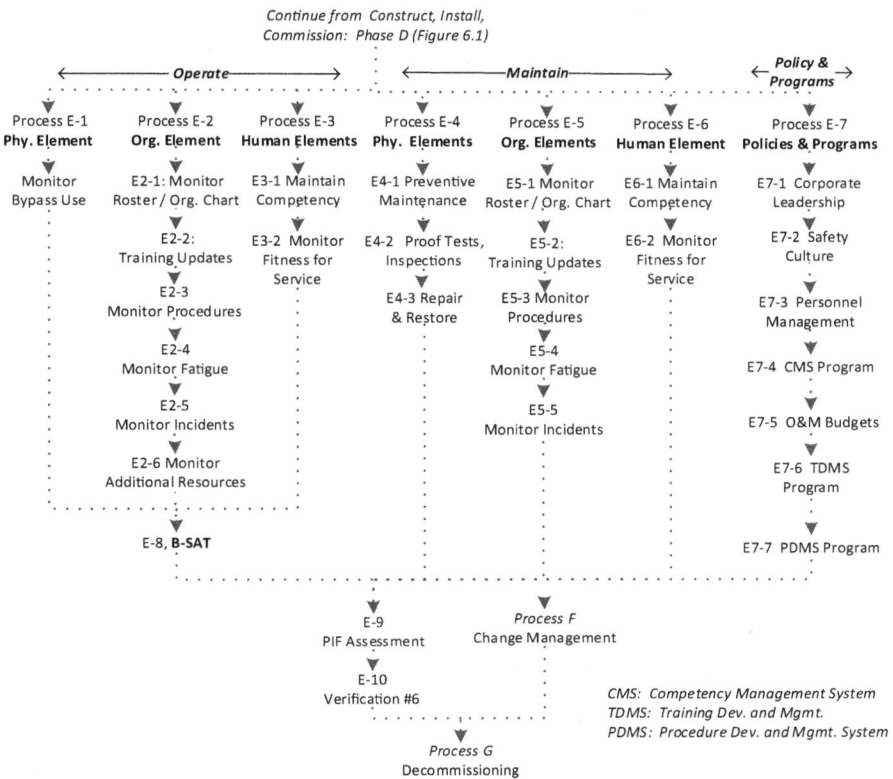

FIGURE 7.1 Operate and Maintain, Modify, and Decommissioning Phase Overviews.

DOI: 10.1201/9781032674476-7

TABLE 7.1
Operate and Maintain Phase: Technical and Project Documents

Documents	See Process/Step
Barrier bypass use status reports	E-1
Maintenance records and reports	E4-1, E4-2, E4-3
Barrier roster and organization chart	E2-1, E5-1
Training records	E2-2, E5-2
Updated procedures (if any)	E2-3, E5-3
Change management documents (design, installation, construction)	F
Decommissioning records	G

TABLE 7.2
Operate and Maintain Phase: Assessments

Table Number	Documents	See Process/Step
—	Bypass use assessment	E-1
—	Procedure use assessment	E2-3, E5-3
—	Fatigue assessments	E2-4, E5-4
—	Incident assessments	E2-5, E5-5
—	Additional resource assessment	E2-6
—	Training updates	E2-2, E5-2
—	Competency verification report	E3-1, E6-1
—	Fitness for service assessment	E3-2, E6-2
—	Barrier System Site Acceptance Test (B-SAT) report	E-8
H.2	PIF Assessment – Barrier, Task, Workspace PE Assessment	E-9
—	Verification #6	E-10

processes is intended to be limited to only those elements that are specific to active human barriers (or safety critical tasks, as applicable). Tables 7.1 and 7.2 summarize the documents and assessments developed in these phases. Table A.4 (Appendix A.2.3) identifies the suggested participants in select activities.

7.1   PROCESS E-1, OPERATIONS – PHYSICAL ELEMENTS, MONITOR BYPASS USE

This step, guided by IEC 61511-1 (2016, cl. 16), suggests a process to monitor bypass use. The activity seeks to identify misuse and the use conditions required in response to an unforeseen process situation or barrier design limitation. All these conditions require corrective action.

To support this activity, the SRS should specify the necessary bypass design features, for example, automatically recording bypass activation and return to normal service, the event date and time, and who performed these actions (based on user login ID). The requirements and suggested timing of this process should be addressed in the BSMP and SRS.

Case Study: Deepwater Horizon Accident: The facility-wide general alarm that initiated the muster barrier was designed to automatically activate on detection of a confirmed flammable gas alarm. On the day of the accident, this feature was placed into a bypass mode, a not uncommon occurrence. In this mode, the operator must actively monitor gas alarms and manually activate the general alarm when gas detection is confirmed. The actual alarm activation occurred late after the blowout started and several catastrophic explosions occurred. With the alarm absence, post-event testimony highlighted the crew's uncertainty in how to respond to the rapidly evolving and confusing events. A different, active barrier was also in a bypass mode. The barrier was never activated. Refer to Appendix M, Notes 4 and 11 for further detail.

7.2 PROCESS E-2, OPERATIONS – ORGANIZATIONAL ELEMENTS

7.2.1 Step E2-1, Monitor Barrier Roster and Organization Chart

This step suggests an event-driven process that responds to every proposed change to the barrier roster or organization chart.

To change a named person on a barrier roster, consider using the applicable design steps to appropriately address that change. For example, start with the HE gap assessment and training plan (detailed design process C-16), and then use the applicable follow-on processes.

A change to the organization chart (e.g., change in backup positions, communication channels) may be a material change to the barrier design. This type of change should be addressed using the change management process. See process F for guidance.

7.2.2 Step E2-2, Training Update

This step suggests performing the training updates (refresher training) identified in Table 5.7. (See detailed design step C32-4a). High-consequence but infrequently activated barriers may require periodic training to prevent or mitigate the undesirable effects of skill fade and drift. (See Appendix J.7 for information on both topics.) The CMS and TDMS identify additional periodic training requirements and recommended or required training intervals.

Note: The training update may also require a follow-on competency assessment. This requirement should be defined in the BSMP, CMS, and/or TDMS.

7.2.3   Step E2-3, Monitor Procedures

Over time, O&M practices may drift in terms of the correct use of published procedures, information typically provided in formal training. The procedure deviations may occur when O&M personnel:

- Identify and use a workaround to perform a task or activity that was not adequately addressed in the published procedure.
- Identify and use a "shortcut" process that takes less time or effort to achieve an end. This shortcut, if effective, could be adopted in a revised procedure. However, it could also be an unsafe deviation that puts the barrier function and performance at risk.

The suggested timing may be periodic and/or event-based. A drift in procedural deviations may have positive or negative results. All deviations should be identified and assessed. It may be appropriate to use the change management process for findings that could necessitate a change to a barrier procedure. (For more information on drift, such as its forms and mechanisms, see Appendix J.7.)

Note: The prototype lifecycle model seeks to achieve a key objective, namely, to reduce the gap between work-as-imagined and work-as-done. This gap is a red flag that may indicate a potential design error or drift in operational practice, both known contributors to incidents and accidents. A contributor to the Deepwater Horizon accident, personnel deviated from a documented safety procedure. As had become normal practice, on a kick the crew lined up a divert system (receives fluids from the kicking well) to a low-pressure mud gas separator instead of to a divert overboard pipeline. Procedures discussed the safety purpose of the divert overboard lines but did not adequately define the conditions on when they should be used. Further, the implied conditions in the procedures were not easily discerned. Had a periodic procedure review been conducted that engaged and received input from the DWH crew, this issue may have been identified and corrected. (For additional background and details, see Appendix M, Note 2.)

7.2.4   Step E2-4, Monitor Fatigue

Table 3.34 (step B-26, preliminary design) specifies a maximum fatigue limit and identifies conditions that can lead to an exceedance thereof. This step suggests periodic and event-based processes to prevent or mitigate exceeding the fatigue limit.

The suggested process is as follows:

- Current state (periodic review) – Monitor and verify whether the sleep opportunities for barrier-assigned personnel are sufficient to remain below the maximum fatigue limit specified in Table 3.34. Perform this review based on the intervals defined in the BSMP or SRS. A detected exceedance may require a corrective action.
- Proactive (event-driven review) – Review every proposed change that may limit sleep opportunities to a level that would cause a fatigue limit

exceedance or chronic fatigue condition. The response to a projected exceedance is to seek alternate solutions. If the exceedance is short-term, establish limits for the maximum permitted exceedance duration. During this exceedance period, consider additional monitoring to ensure affected personnel do not experience fatigue to a level that materially degrades their performance.

Conditions that can lead to excess fatigue (controlled by the owner/operator) include situations where:

- Rotation patterns do not support or allow for adequate sleep in a 24-hour period (i.e., ignores the effects of the known disturbance to human circadian cycles).
- Insufficient staffing leads to chronic over-reliance on excessive overtime.
- Poorly planned events (e.g., plant turnarounds or start-up).

Note: For further information, see step B-26, CCPS (2022, Ch. 15), and Edmunds (2016, Ch. 22).

7.2.5   Step E2-5, Monitor Incidents

This step suggests an additional activity to monitor, evaluate, and learn from barrier-attributed and barrier-affected incidents. This process should use the existing owner/operator processes but may require adaptations to address the prototype lifecycle model and its cognitive-focused approach.

Suggested Scope and Framing: Incidents that (1) can degrade or unexpectedly enhance barrier activation, functioning, or performance, (2) cause an unexpected or problematic outcome that results from barrier activation, or (3) can degrade a barrier system dependency (e.g., an external support system or protective barriers). The findings may present opportunities or identify deficiencies.

Suggested Timing: Periodic and/or event-based (as noted in the BSMP)

1. Gather and review incident information (pertaining to the stated scope).
2. Evaluate the information
   a. Did the required barrier activate, and did so within the specified safety times?
   b. Did the activated barrier conform to requirements in the safety requirements specification and the applicable performance standards? If not, what requirements were not met?
   c. What actions were performed correctly?
   d. What actions were taken that deviated from requirements but also reduced the duration or the undesired consequences of the incident?
   e. Did the effects of the incident/hazard event degrade or disable one or more elements in the barrier system or a dependency (external support system or barrier)? If so, describe the mechanism(s) that caused or contributed to degradation or failure.
3. Identify deficiencies in the barrier system or a dependent external support system or barrier, if any.

4. Identify, evaluate, and recommend corrective actions to address the identified deficiencies.
5. Address the recommendations using the change management process (see process F for details).

Case Study: HOSL Buncefield Petrol Tank Overfill and Fire (COMAH 2011, pp.13–14, 19, 32–33). Padlocks provided on the tank-mounted, independent level alarm switches were commonly removed and not replaced. This was not perceived to be an actionable incident (e.g., a weak signal). Had that occurred, the resulting investigation might have identified its safety critical purpose. The level switch included a test lever used to test the switch function and alarm. The padlock held the lever in its operational position. Movement from that position disabled the switch-activated level alarm and a shutdown function that stopped inflow to the tank. This was the case on the day of the accident; the control room supervisors and maintenance personnel remained unaware that the uncontrolled movement of the test level had disabled the switch-generated alarm and the associated shutdown function. The condition was not monitored or alarmed in the control room.

Information from the automatic tank gauging system (ATG) was critical to safe process operations. Information from this system was essential to the supervisors' ability to monitor tank levels and, when required, take the necessary manual action to prevent a tank overfill event. However, the repeated ATG failure impeded the supervisor's ability to actively monitor tank levels and take timely actions to prevent a tank overfill event. It was not clear if the repeated failures were perceived to be an acute safety threat, or if those failures triggered an internal process to assess if additional compensating safeguards were needed. (For background information on the accident, see Chapter 3, step B-14. Also see the maintenance repair and restore process E4-3 for related information.)

7.2.6   STEP E2-6, MONITOR ADDITIONAL RESOURCES

This step suggests a process that monitors the *additional resources* recorded in Table 3.33. (See step C14-5.) The monitoring function may be periodic, event-based, or both.

Suggested Scope: Review every event or change that may materially affect an additional resource recorded in Table 3.33.

Suggested Timing: Periodic and/or event-based (as noted in the BSMP)

1. Identify a potential change or deviation attributed to the resource. Examples of changes are as follows:
   a. The resource does not meet barrier requirements (e.g., in terms of function or performance).
   b. A proposed change to the resource provider, contract, or resource location.
   c. A potential change in external conditions that may degrade the resource (e.g., increased exposure to cyber incidents, the provider's exposure to supply chain problems and issues).

    d. New or expanding demands regarding a resource shared with other use applications and users (e.g., a shared satellite communication link that connects an ROC to the protected facility).
2. Ascertain recommended corrective actions to address the identified deficiencies (if any).
3. Address the recommendations using the change management process (see process F for details).

7.3 PROCESS E-3 OPERATIONS – HUMAN ELEMENTS

7.3.1 STEP E3-1, MAINTAIN COMPETENCY

Based on the requirements in the BSMP or SRS, perform periodic competency assessments to confirm that barrier-assigned personnel have obtained the defined and required competencies. (A periodic assessment may be warranted for a competency more susceptible to skill fade or drift. For additional background, see Appendix J.7.)

For additional information and guidance, see CCPS (2022, Part 4), Edmunds (2016, Chapter 20.2), HSE (2003, 2007 a/b), COS (2013, Section 8), and Stanton et al. (2010, Ch. 2).

7.3.2 STEP E3-2, MONITOR FITNESS FOR SERVICE

This step is recommended but not addressed in the prototype lifecycle model. Periodic monitoring should occur to verify the health (physical and mental) of barrier-assigned personnel, as health conditions can materially degrade their capabilities and performance. An owner/operator program that provides this function should be referenced in the BSMP.

7.4 PROCESS E-4, MAINTENANCE – PHYSICAL ELEMENTS

The following activities should be defined and included in the BSMP and SRS. The processes are limited to barrier system elements and barrier system dependencies, for example, external support systems and external protective barriers.

7.4.1 STEP E4-1, PREVENTIVE MAINTENANCE

This step suggests an assurance process to monitor the completion of the barrier-required preventative measure at the prescribed intervals.

7.4.2 STEP E4-2, PERIODIC INSPECTION AND PROOF TESTING

This step suggests an assurance process to monitor the completion of the barrier-required inspections and proof testing at the prescribed intervals.

    Inspection procedures should identify the inspection scope, objectives, methods, and basis for verification. Examples of findings include verifying that an explosion-proof enclosure that protects a barrier element during barrier activation is unable to perform that function if the cover is missing or not installed correctly. Furthermore, a field instrument may fail or become unreliable or inaccurate if the required heat tracing or purge is missing or not operating to the required performance. Finally, degraded fireproofing may directly or indirectly reduce the achievable barrier endurance time.

The barrier may contain active elements that can experience failures that are not self-revealing and employ a fail-to-danger design. In this case, it cannot perform its intended function if the device, control system, or electrical power to the device fails. Examples include lamps, beacons, sirens, and speakers. If the SRS defines a Safety Integrity Level (SIL) target to an element or subsystem, it should also define the required test type (online, offline), test scope, and maximum permitted intervals between tests. Performing those activities is the means to verify compliance with the SIL requirement.

Note: For further information, see para 16.3 in IEC 61511-1 (2016).

7.4.3 Step E4-3, Repair and Restore

This step suggests the monitoring process for repair and store activities (barrier system PE and dependencies). Minimum performance targets should be included in the SRS, such as the minimum average mean time to repair and availability of spare parts. Identified deviations should be assessed, and corrective actions identified and implemented to align activities to ensure the specified requirements are attained. Instances of repetitive failures should initiate high-priority actions that may include interim solutions and permanent corrective actions. Such failures (frequent occurrences or a failure is not quickly corrected) may require use of the change management process to find interim means to mitigate the failure effects, e.g., add a new monitoring (task change or addition), modify a procedure, or develop a new interim use procedure. The change management process should address all associated changes such as training or changes to roles.

Case Study: HOSL Buncefield Petrol Tank Overfill and Fire (COMAH 2011, pp. 19). The repeated failure of the tank gauging system did not appear to elevate the maintenance priority of this safety critical system and allocate the resources needed to promptly confirm the failure root cause and implement a permanent corrective solution. (With continued occurrence, the supervisors found workaround actions that provided a temporary fix.) In a different incident, a regularly used tank remained in service over a nine-month period, while its independent tank switch alarm and shutdown function remained inoperable. The COMAH report indicated this situation should have been addressed using a change management process. (For background information on the accident, see Chapter 3, step B-14. For related information, see the operations incident monitoring process E2-5.)

7.5 PROCESS E-5, MAINTENANCE – ORGANIZATIONAL ELEMENTS

7.5.1 Step E5-1, Monitor Roster or Organization Chart Change

See step E2-1 for guidance.

7.5.2 Step E5-2, Training Update

See step E2-2 for guidance.

7.5.3 Step E5-3, Monitor Procedures

See step E2-3 for guidance.

7.5.4 Step E5-4, Monitor Fatigue

See step E2-4 for guidance.

7.5.5 Step E5-5, Monitor Incidents

See step E2-5 for guidance.

7.6 PROCESS E-6, MAINTENANCE – HUMAN ELEMENTS

7.6.1 Step E6-1, Maintain Competency

See step E3-1 for guidance.

7.6.2 Step E6-2, Monitor Fitness for Service

See step E3-2 for guidance.

7.7 PROCESS E-7, PROGRAMS AND POLICIES

To be comprehensive, the prototype lifecycle model includes direct and indirect interfaces to the identified owner/operator programs and policies. Figure J.1 (Appendix J.1) indicates their interfaces to lifecycle model processes, step, or activity. An owner/operator program or policy can positively or negatively affect the functioning and performance of a barrier designed and managed by the prototype lifecycle model. Further, a change in the program or policy can also have a positive or negative effect. Process E-1 provides insight into how each contributes to the barrier design and lifecycle over the life of the barrier.

Figure C.1 (Appendix C introduction) provides some insight into where and how owner/operator leadership, safety culture, policies, and programs can affect human performance regarding barriers and safety critical tasks. Figure F.2 (Appendix F.2) provides another representation indicating the effects of the results of these policies and programs (e.g., trust, time pressure, fatigue, staffing policy) on human performance at the cognitive and physical activity levels.

Verification #6 includes verifying the operational state of the programs and policies in process E-7 and whether other activities in phase E are performed to the stated intervals and schedules.

Note: An undetected deficiency in a program or policy may lead to a latent error in the barrier design or systems that operate and maintain the barrier. An error in a program-developed procedure or training module template may contribute to human error and injury. The prototype lifecycle model assumes the owner/operator programs and policies are mature, effective, published, and professionally managed. Corporate leadership words and actions can be misinterpreted, and as such, the safety culture may be incorrectly understood. A myopic or keyhole view may create a false or incomplete perspective of an environment that may be rife with conditions that contribute to undesirable human behavior and a degrading process safety state. The processes in E-7 are not intended to suggest how to design and manage these programs, policies, and practices. Instead, they are included to highlight the effect of blunt-end elements on barrier-assigned personnel and their contribution to barrier reliability and performance.

7.7.1   STEP E7-1, CORPORATE LEADERSHIP

No role has a greater collective effect on barrier performance than that created by corporate leadership decisions, words, and actions. This step suggests the importance of maintaining a closed-loop awareness of the consequences, results, and environment that the words and actions of corporate leadership have on barrier-assigned personnel, and the resulting effects on barrier reliability and performance. Corporate leadership affects all steps in process E-7.

***Case Study: HOSL Buncefield Fuel Tank Overfill, Fire and Explosion** (COMAH 2011, p. 28). From COMAH (2011, p. 28),*

> Total Head Office in Watford had considerable influence over system of work of the HOSL site and was supposed to provide the necessary engineering support and other expertise. In reality that support was lacking. Both the Operations Manager and the Terminal Co-ordinator had too much to do. The latter was given insufficient direction on how to prioritise and had insufficient expertise and resources to cope with the duties placed upon him. In particular, he was given little help in implementing the safety management system.
>
> *(For background information on the accident see Chapter 3, step B-14)*

7.7.2   STEP E7-2, SAFETY CULTURE

Safety culture is the environment created by corporate leadership, policies and programs, safety management systems, employee involvement, feedback and feed-forward processes, and actual practice. Individual and group interactions in this environment contribute to perceptions, beliefs, intentions, and motivations that collectively contribute to safety culture. Ideally, a robust safety culture is achieved and self-sustaining. Unfortunately, given default (unconscious) human behaviors, changing motivations, and innate cognitive attributes discussed in Appendix F, organizational culture is inherently entropic. (The *chronic unease* behavior achieved in a high reliability organization is at odds with the innate *cognitive ease* behavior and motivations discussed in Appendix F.2.5. As a counterweight, persistent, intentional efforts and reinforcement are needed to achieve and maintain the desired safety culture. As such, all corporate culture is inherently dynamic and constantly changing, and requires continuous and focused effort to maintain).

The PIF assessment in process E-8 is a means to reveal aspects of this environment, such as threat stress, morale, attitude, motivation, and workload. Surveys could also be employed to further reveal individual beliefs, attitudes, job-based realities, and perceptions that may positively or negatively affect barrier performance.

The "system" needs to constantly push against ubiquitous and persistent production bias. This is a natural outcome (i.e., think mental models) when others notice a risk-taker's promotion and when management overlooks a procedure violation that achieved a key production goal or provided a ready solution to an immediate problem. Attitude, cultural norms, a perceived threat of job loss, loss of prestige, or promotional opportunities can affect the complex decision-making required by emergency response barriers. These are intended and unintended by-products of the climate created by an organization's safety culture.

Note: For additional background and guidance, see Edmonds (2016, Ch. 18) and Stanton et al. (2010, Ch. 15).

Case Study: Deepwater Horizon Accident*: On the day of the accident, several BP executives were onboard to celebrate and give an award to the Deepwater Horizon crew for outstanding safety performance (no lost time accidents, behavior-based safety). Process safety was not a basis for the award. Before the accident occurred, the results of an outside survey indicated that 50% of the responders did not feel empowered to initiate a "stop work" for fear of reprisal. In the lead-up to the accident, the number of process safety incidents and shortcuts was increasing, many of which are discussed in this book. Feedback on personnel expressions of concern was perceived as unwelcome (Konrad, 2011, pp. 187–189). Readers might also have experienced extended periods of cost cutting, excessive overtime (fatigue), deferred maintenance, layoffs and staff shortages, and a declining safety culture. These are examples of conditions that often lead to a noticeable increase in incidents and a perception that "the wheels are about to come off."*

7.7.3 Step E7-3, Personnel Management

This step suggests the importance of understanding the effects of changes in personnel management practices, e.g., changes to staffing, shift assignments, basis for promotion, and reward and recognition programs. A sudden loss of a barrier-assigned HE (promotion, reassignment, or leaving the company) creates a barrier deficiency. The deficiency remains until the position is filled and the assigned personnel acquires and demonstrates the necessary competencies. Understaffing (temporary or chronic) can lead to excessive fatigue that may be acute and/or chronic. These conditions can degrade the cognitive capability and performance of barrier-assigned personnel.

Note: For additional background and guidance, see Edmonds (2016, Ch. 20.1 – staffing and workload), Stanton et al. (2010, Chs. 6 and 9 – staffing and workload, shift patterns), and CCPS (2022, Ch. 15 – fatigue).

7.7.4 Step E7-4, Competency Management System

The prototype lifecycle model assumes the existence and reliable continuation of a mature and published competency management system (CMS). The included processes are linked (have interfaces) to this program and therefore affected by its efficacy, maturity, and reliable application.

Note: For additional background and guidance, see the references in the note in Step E3-1.

7.7.5 Step E7-5, Operating and Maintenance Budgets

This step suggests the importance of understanding the effects of incremental and step changes to operating and maintenance budgets and practice. Sudden budget cuts or chronically underfunded operating and maintenance budgets can lead to short-term (and potentially long-term) reductions in staffing, maintenance, and the various barrier-specific monitoring processes included in this lifecycle phase. Readers might have experience with the cyclical and often disruptive upheaval that occurs when an organization responds to a major step change in the external business, and regulatory or legal environment. These responses seldom enhance policies and programs in the short term. More often, they have the opposite effect. These effects can (and often do) degrade barrier reliability and performance. Barrier-assigned personnel become increasingly distracted (attention capture) when concerns over future employment increase.

7.7.6 Step E7-6, Training Development and Management System

The prototype lifecycle model assumes the existence and reliable continuation of a mature, published, and managed Training Development and Management System. The included processes are linked (have interfaces) to this program and therefore are affected by the efficacy, maturity, and reliable application thereof.

Note: For additional background and information, see Appendix J.3.

7.7.7 Step E7-7, Procedure Development and Management System

The prototype lifecycle model assumes the existence and reliable continuation of a mature, published, and managed Procedure Development and Management System. The included processes are linked (have interfaces) to this program and therefore are affected by the efficacy, maturity, and reliable application thereof.

Note: For additional background and information, see Appendix J.2.

7.8 PROCESS E-8, B-SAT

Refer to Table 6.7 in process D-8 (Chapter 6) for requirements. At this stage, personnel should have gained more experience performing the assigned barrier activities.

The objective of the test is to verify that all barrier system tasks are correctly performed, and the specified functions and safe states are achieved within the specified safety times. Re-performing the B-SAT at this stage should reflect the normal performance expected, i.e., the baseline. This test should be repeated periodically at an interval defined in the BSMP or SRS.

Note: HEs gain additional experience if the barrier activates. If this does not occur and no other means are provided to maintain skills, the test may indicate a degradation from the prior test. This outcome can occur with skill fade or drift. These conditions indicate the need for additional training and competency assent.

7.9 PROCESS E-9, PIF ASSESSMENT

This step suggests performing the PIF assessment process described in Appendix H.2. The assessment requirements and timing (e.g., periodic or event-triggered) should be defined in the BSMP.

The purpose of this activity is to identify operational and organizational changes that may be missed in other monitoring processes and may materially affect the barrier system functioning and performance in positive or negative ways. The assessment could identify a finding that requires a corrective action. A finding that indicates a material deviation from the original assessment (indicating signs of drift, skill fade, or an unmanaged barrier system modification) may warrant using the change management process for further evaluation.

7.10 PROCESS E-10, VERIFICATION #6

See Table 7.3 for the suggested activities in verification #6 scheme. This scheme seeks to verify whether the activities in lifecycle phase E are performed to the defined schedules and intervals. It also confirms if the findings of these activities are addressed according to published processes and procedures.

TABLE 7.3
Verification #6: Operate and Maintain Phase

Verification Scheme	Example Resource Plan	Examination Form	Tangible Evidence	Performance Standards: Physical Elements	Performance Standards Basis/Content
See Barrier Safety Management Plan	See Table G.8 (App. G.5)	See Table G.7 (App. G.5)	See Table G.7 (App. G.5)	See Process C-13	See App. D.3.3 and Figure D.3 in App. D.3.2

7.11   PHASE F, MODIFICATIONS (CHANGE MANAGEMENT)

Owner/operators have existing change management processes in place. The BSMP should reference this process to ensure that it is integrated into the barrier lifecycle design. All references to change management assume the existing process will be used.

A change that may warrant the use of the formal change management process are those that add, delete, or modify any information (by inclusion or reference) in the barrier safety management plan or the safety requirements specification.

Case Study: PG&E Natural Gas Transmission Pipeline Rupture and Fire (NTSB 2011, pp. 54–57, 90–91). On the day of the accident, work was underway to replace a battery-backed power system. While performing that work, an incident caused an uncontrolled pressure increase in a pipeline section that had an undetected weld defect. The increase was sufficient to rupture the weld. The incident resulted from inadequate job planning, risk assessments, preparation, and execution. Examples:

- *Did not identify every planned work step and the potential hazards associated with each step. (A technician removed an electrical circuit to a circuit breaker that was not mentioned in the plan. Unbeknownst to the technician, the action removed electrical power to a critical pipeline instrumentation and control cabinet.)*
- *Technicians did not notify the control room before disconnecting power to the abovementioned breaker, nor provide immediate notification when an unexpected response occurred (the loss of power to a pipeline control cabinet).*
- *Did not adequately address contingencies and emergency plans to address possible unforeseen events. (Technicians made unplanned control changes in their attempt to recover from the prior, unexpected pressure upset.)*
- *One of the Control Room Operators was assigned to be a single point of contact to communicate and coordinate with the technicians. However, different operators picked up and responded to their calls defeating the safety purpose of the single point contact.*
- *None of the Control Room Operators appeared to increase their vigilance in monitoring pipelines that could be affected by the work.*

    For additional background information on the accident, see Chapter 3, step B16-3.

7.12   PHASE G, DECOMMISSIONING

The suggested barrier lifecycle model includes the decommissioning of a barrier that is no longer required. For guidance on this process, see IEC 61511-1 (2016, clause 18).

REFERENCES

CCPS (2022), *Human Factors Handbook for Process Plant Operations, Improving Safety and Systems Performance*, New York: John Wiley & Sons Inc., Center for Chemical Process Safety (CCPS)

COMAH (2011), "Buncefield: Why Did it Happen?", COMAH Competent Authority. Downloaded August 23, 2023 from https://www.hse.gov.uk/comah/buncefield/index. htm. Follow web link: *'Buncefield: Why Did It Happen? (PDF)'*

COS (2013, December), *COS-3-02, Skills and Knowledge Management System Guideline*, 1st Ed, Center for Offshore Safety

Edmonds, J., (2016), *Human Factors in the Chemical and Process Industries, Making it Work in Practice*, Elsevier

HSE (2003), *Competence Assessment for the Hazardous Industries, RR086*, London: Safety and Health Executive

HSE (2007a), *Managing Competence for Safety-Related Systems, Part 1: Key Guidance*, UK Health and Safety Executive

HSE (2007b), *Managing Competence for Safety-Related Systems, Part 2: Supplementary Material*, UK Health and Safety Executive

IEC 61511-1 (2016), *Functional Safety – Safety Instrumented Systems for the Process Industry Sector – Part 1: Framework, Definitions, System, Hardware and Application Programming Requirements*, 2nd Ed, International Electrotechnical Commission

Konrad, J., Shroder, T. (2011), *Fire on the Horizon, the Untold Story of the Gulf Oil Disaster*, HarperCollins Publishers

NTSB (2011), Pacific Gas and Electric Company Natural Gas Transmission Pipeline Rupture and Fire, San Bruno California, September 9, 2010, National Transportation Safety Board, Pipeline Accident Report NTSB/PAR-11/01. Washington, DC

Stanton, N.A., Salmon, P.M., Walker, G.H., Jenkins, D.P. (2010), *Human Factors in the Design and Evolution of Central Control Room Operations*, CRC Press

8 The Case for Developing a New Standard for Active Human Barriers

8.1 INTRODUCTION

One of the stated purposes of this book, discussed in the introduction, is to present the case for commencing work on a new global consensus standard dedicated to active human barriers. This is one of the two suggested paths forward to redress acknowledged deficiencies in the design and lifecycle management of this type of barrier. Unlike other active barriers (e.g., those governed by the IEC 61511-1 standard), active human barriers are not supported by a standard specific to this barrier type, nor are they adequately supported by a generalized standard that adequately addresses cognitive ergonomics. Section 8.2 examines some of the profound deficiencies in current standards and practice and the unacceptable pace in recognizing and addressing those issues. Section 8.3 presents the case for developing a new consensus standard for active human barriers. It also presents a novel approach for framing, guiding, and creating that standard. The new standard should address each of the issues raised in Section 8.2.

8.2 DEFICIENCIES IN CURRENT KNOWLEDGE, STANDARDS, AND PRACTICE

8.2.1 OVERVIEW – CURRENT ISSUES AND POTENTIAL CONSEQUENCES

The Deepwater Horizon accident investigation reports and the industry's response to the accident cited human factors as primary causal contributors to the accident. (Unfortunately, the recommendations from these reports and their subsequent implementation continue to demonstrate their ineffectiveness in addressing these issues in barrier system design.) Section 1.4 provides example industry findings, concerns, and deficiencies in current practice and standards. Those areas and the following may help to understand the historical and current lack of timely progress toward improving the reliability and performance of human-dependent barriers and safety critical tasks.

- *Missing Consensus Standard for Human-Dependent Barriers:* A global consensus standard that is dedicated to active human barriers, lifecycle-based, and underpinned by cognitive science and cognitive ergonomics does not currently exist.

DOI: 10.1201/9781032674476-8

- *Diverging Definitions, Constructs, and Practice:* See Section 8.2.2.
- *Exclusionary and Deficient Definitions and Guidance:* See Section 8.2.3.
- *Recognized (and Unrecognized) Deficiencies in Industry Standards:* See Section 8.2.4.
- *Progressing Incremental Change When a Profound Change is Needed:* See Section 8.2.5.
- *Seeking Solutions from Other Industries...It is Not a Panacea:* See Section 8.2.6.
- *Lack of Focus on Cognitive Ergonomics, the Elephant in the Room:* See Section 8.2.7.
- *A Major Source of Systemic Risk in the U.S. O&G Industry:* See Section 8.2.8.
- *Continued Failures in Human-Dependent Barriers Remain Likely:* See Section 8.2.9.

8.2.2   DIVERGING DEFINITIONS, CONSTRUCTS, AND PRACTICE

Globally, the lack of progress in reaching consensus on the most basic aspects of active human barrier design is contributing to the diverging paths being taken by industry organizations, country regulatory regimes, and owner/operators. (Recognizing the problem, Sklet (2006) presented suggested terms and definitions to move the discussion forward.)

Consider the recent book from the Center for Chemical Process Safety (CCPS) and Energy Institute dedicated to bowtie application and diagrams (CCPS, 2018). This was one of the many industry responses to the Deepwater Horizon accident. The book presents guidance and recommendations for developing bowtie diagrams and integrating human and organizational elements into those diagrams. Tellingly, the authors included a set of barrier definitions, criteria, framing, and explanatory information. Most were similar to the industry standards though some were different. The implications are as follows:

- By providing the information, it appears the authors foresaw the need to establish this foundational basis as a necessary precursor to addressing the book's primary objective. That objective was to "establish a set of practical advice on how to conduct bow tie analysis and develop useful bow tie diagrams for risk management" (CCPS 2018, p. 1).
- Unique to the book, it provides yet another (different) set of base barrier definitions, information, and guidance for active human and other barrier types. Depending on one's perspective, this may be an example of a process of divergence. Being authored by two different organizations, some may view it to be an effort toward convergence.

The CCPS book is one of many post–Deepwater Horizon accident publications that demonstrates the progression (and consequences) of the diverging views and guidance. Without consensus, where and how does this end? How does it affect:

- Industry organization efforts to develop a consensus standard for active human barriers?

- Multination organization's desire to standardize barrier design and management systems across its global facilities?
- Industry and owner/operator efforts to develop internal standards and best practice guidance documents?
- Regulatory efforts to update or develop new regulations?
- The challenge (and training cost) for engineering/design companies and technical system providers expected to comply with these varied requirements?

Consider the effect if this latest set of diverging changes (or those from other sources) are adopted into a country-specific regulation, regional guidance standard, or an owner/operator's internal standard. Each divergence creates a different path with varying levels of inertia and resistance to change. This ongoing process increases the challenge to those promoting and seeking to develop a common global consensus standard for active human barrier design.

Note: Before the publication of the IEC 61511-1 standard, and its precursor ANSI/ ISA S84.01, similar diverging views and practices were common. Had that continued unabated, regulatory agencies, engineering contractors, product suppliers, insurers, and others would now be facing a more fractured and diverse range of regulatory and safety system implementation challenges. In addition, less progress would have been made toward improving the reliability and performance of the barriers guided by this document.

8.2.3 EXCLUSIONARY AND DEFICIENT DEFINITIONS AND GUIDANCE

Guidance standards may present a one-size-fits-all approach that ignores common situations and applications where one or more defined property requirements cannot be met. *Independence* is an example. The preventive and mitigation barriers may be chosen to address a specific hazard scenario, provide a definable (quantitative or qualitative) level of risk reduction, or support a defense-in-depth philosophy. Compliance may require that no two or more barriers in this set be permitted to share a common element (HE, component, or system), or mutually rely on the same external support system or protective barrier. The objective is to have no common points of failure or degradation. According to CCPS (2015, p. 35),

> Independence is achieved when the performance of one IPL is not affected by the IE or by the failure of any other IPL to operate. When the performance of one component is dependent upon the successful operation of another, the devices are not independent. Independence is an important concept, although absolute independence is generally not achievable.... However, IPLs should be sufficiently independent such that the degree of independence is not statistically significant.

> *(IE = Initiating Event, IPL = Independent Protection Layer.)*

Understanding the "degree of independence" concept is problematic if an explicit criterion is not stated and clear guidance is not provided on how to address such cases.

Unfortunately, independence is often not possible, as is common to emergency response and other mitigation barrier types, and some preventive barriers.

The ubiquitous emergency response barrier is commonly required though seldom has an alternative option that is independent. In O&G, well-control preventive and mitigation barriers (same hazard) often fail to meet the independence requirement because they share the same HE, technical system, or electrical power system. Major accidents like Deepwater Horizon (DWH) continue to demonstrate a facility reliance on barriers that do not meet an overly rigid independence requirement. Discussed in Appendix M, most if not every active (technology only and independent) barrier in the accident causal chain failed. It is therefore important to recognize that the emergency response barriers, though poorly executed, were the last line of defense that saved the lives of 115 souls. To those persons and their families, friends, and co-workers, this outcome simply cannot be stressed enough. *Accidents continue to prove that ER barriers often become the final arbiter of who survives an accident.* (See Appendix M for additional information on the nature and consequence of the active and active human barrier successes and failures.)

To be an effective guidance standard or practice, it should recognize and clearly address conditions and situations where independence or other property requirements cannot be met (in total or in part), and alternatives are not available. The document should provide clear acceptance/performance criteria, and the guidance needed to achieve the maximum possible shared element integrity, reliability, and survivability designs. Resilience methodologies or approaches may offer additional options. Guidance may also be needed to establish the appropriate application and limits in using an ALARP approach. Additional guidance may include the following:

- Promote processes that accurately and fully identify barriers that have shared elements (e.g., preliminary design process B19-2) and the elements they share (HEs, components, systems, and external support or protective systems).
- Ensure barriers that do not meet a specified independence requirement are explicitly identified as such in widely accessed documents and displays, for example, a bowtie diagram or a dynamic (live) bowtie status display.
- If the capacity or endurance time of the shared element is a primary constraint on barrier safety time (BST), include that information in widely used documents and accessed displays.
- For the shared element (a safety critical element):
  - Identify the most onerous function and performance requirement among all of the dependent barriers and apply that requirement to the shared element.
  - Define the process that identifies the most onerous requirements, and state where that requirement is captured (e.g., in a specification or data sheet) and applied.
  - Identify the design and management solutions required to maximize the element reliability, capacity/endurance time, and survivability. Ensure the element has the design capability and capacity to address the most demanding, plausible scenario (e.g., a demand load based on

the simultaneous activation of likely concurrent barriers and other demand loads).

- Assess all plausible conditions and events that may cause element degradation or failure and identify the specified safeguards that address each condition and event.
- Provide a dedicated performance standard for the element. See Appendix D for guidance.
- Identify if a shared element should automatically be classified as a safety critical element (SCE). Examples may include utility systems (e.g., electrical or pneumatic power), technical systems (e.g., a radio or general alarm system), or reliance on a common person. See EI (2020b), for example, guidance.

- Enhance the operational/health status monitoring requirements for shared elements. (Monitoring becomes a time-sensitive, safety critical task when the affected barriers are activated.) As applicable, purposely designed displays should present the current state, directional status movements, and remaining availability time. Consider symbolic display types that provide near-instant comprehension while placing a minimal cognitive demand load on the monitoring personnel.
- Capture the above information in the safety requirements specification or similar document. Doing so is essential to a lifecycle approach and change management.

8.2.4 Recognized (and Unrecognized) Deficiencies in Industry Standards

An article by Aas et al. (2010, pp. 286–291) states,

- The Petroleum Safety Authority (PSA) Norway has emphasized the importance of understanding the interaction between human, technology, and organization in their Human-Technology-Organization (HTO) approach, which is more balanced than HCD.
- We suggest two radical changes to the current regulation of CC design. The first is to make ISO 11064 more goal based with only one normative part. The second is to create a new HTO standard which embraces the entire lifecycle and has a stronger focus on organizational factors, which appears to be one of the main challenges in CC design today.

The ISO 11064-1:2000 introduction describes the standard as follows:

This part of ISO 11064 includes requirements and recommendations for a design project of a control centre in terms of philosophy and process, physical design and concluding design evaluation, and it can be applied to both the elements of a control room project, such as workstations and overall displays, as well as to the overall planning and design of entire projects. Other parts of ISO 11064 deal with the more detailed requirements associated with specific elements of a control centre.

A control center project (capital project) may encompass a control building, control and ancillary rooms, control consoles and HMI displays, and the workspace lighting and environmental control design. Activities often include identifying and assessing the activities performed in each workspace including personnel movements and communications (in-person and device-enabled). This information guides the workspace sizing, configuration and layout, equipment placement, etc.

8.2.4.1   Human-Centered Design – Background, Capabilities, and Limitations

Principle number 1 in ISO 11064-1:2000 (cl 4.2) refers to its "Application of human-centered design approach." Human-centered design (HCD) is a methodology and approach (with many variants) used by a diverse range of industries. Its origin dates to a Stanford University program started in 1958.

The article by Aas et al. (2010, p. 286) mentions ISO 13407, a standard later replaced by ISO 9241. ISO 9241-220:2019(e) (cl. 3.11), defines HCD as an "approach to system design and development that aims to make interactive systems more usable by focusing on the use of the system; applying human factors, ergonomics and usability knowledge and techniques." It includes the terms "ergonomics" and "human factors" but provides little evidence of its intent to actively address cognitive ergonomics. If that were the intent, the standard provides no guidance beyond the basic HCD processes.

Many HCD principles have been widely used for decades. Safety system designers commonly engage and get input from end-users (operators) when developing requirements. Prototypes for new HMI display objects are often guided by input from end-users, who then participate in preliminary and final display product testing with validation opportunities. End-users also commonly participate in the factory and site acceptance tests performed to verify that a new or modified technical system conforms to requirements. These tests typically focus on verification activities, though they do provide an environment for validation activities.

An important observation is that *the past application of HCD principles did not prevent the Deepwater Horizon accident. Nor has its use prevented latent design errors (cognitive-attributed) of the type discussed throughout this book.* HCD standards are missing many (if not most) of the cognitive-focused processes, tools, and guidance provided in this book. To meaningfully address cognitive ergonomics, a viable HCD standard should include or address the following (or similar) in its provided guidance and processes:

a. Identify the underpinning cognitive science and elements addressed in its processes and tools. (See Appendices C and F for examples.)
b. Using an accepted task analysis process, define all tasks required to achieve the barrier function, and then identify task assignee or role. (This is included in some standards.)
c. Decompose each task to its constituent task phases (detect: SA-1, SA-2, SA-3, decide, and act). Define the task phase requirements that establish the basis for further design. (See Chapters 2 and 3 for examples.)

d. For each task phase, identify, select, and specify the human, physical, and organizational elements needed to achieve the task phase, task, and barrier requirements. (See Chapters 2 and 3 for examples.)

e. Employ a cognitive science–driven assessment process that reveals cognitive-attributed design errors and suggests corrective solutions matched to the identified error type. (See Appendix I for an example process and supporting tools.)

HCD variants (e.g., ISO 9241-220:2019 and ILO/IEA 2021) do not include guidance, processes, or tools equivalent to those noted above. For this reason, it seems unlikely their use will significantly reduce the cognitive-attributed type design errors that remain common contributors to incidents and major accidents.

8.2.4.2 Recommendation for a New HTO Standard

Aas et al. (2010, p. 291) presented the case that "An HTO standard like we suggest should define HTO and related concepts and state the top-level goals to be achieved." On its face, this author supports the statement. One of the intents is that the proposed HTO standard should not repeat the more detailed guidance provided in existing standards such as ISO 11064. However, that approach would seem to miss a core issue. Current standards and practices lack the capability to reliably identify and prevent latent design error types that are cognitive-attributed. The process industry has not been successful in defining the new methods and tools needed to realize that capability. To address that issue, a new HTO standard should consider and perhaps introduce processes and activities like those suggested in Chapters 2–7.

Note: The prototype lifecycle model provides ideas, concepts, and methodologies to consider in a new HTO standard. It appears to address all key points raised in the abovementioned article. It encompasses the HTO frame and presents a full life-cycle model. It comprehensively addresses organizational elements using processes that are typically missing in existing barrier design and management standards. Though developed specifically to address active human barriers, most processes in the model are task-based. As such, they should be equally applicable to all safety critical tasks. To a limited extent, it addresses selected maintenance activities, a topic mentioned in the paper but not discussed in detail.

8.2.5 Progressing Incremental Change When a Profound Change Is Needed

Discussed in Chapter 1, the referenced process industry reports, articles, and call-to-action white papers identify and acknowledge major deficiencies in current practice. (Their issue dates range from 2012 to 2023.) Most lie in the fields of cognitive and organizational ergonomics. Johnsen et al. (2017) discussed concerns with recent changes to the Norwegian standard (NORSOK S-002). Changes made to public-facing documents did not markedly address cognitive ergonomics. Incrementalism, a common approach in engineering, makes incremental changes to existing standards and guidance documents based on lessons learned from projects and new learnings.

This may not be the right approach when addressing the less familiar fields of cognitive ergonomics and cognitive system engineering. Surprisingly, little of the information in Appendices F and I is widely known or effectively applied. Given its profound potential in safety and design, it may be appropriate to view these fields in ways that are similar to a new and potentially disruptive technology. Perhaps that provides the necessary perspective that drives the incentive, motivation, and push for aggressive paths that can more rapidly apply this "new technology."

8.2.6 Seeking Solutions from Other Industries...It Is Not a Panacea

When major accidents occur, the process industries commonly look to other industries for solutions, such as those adopted from the aviation and nuclear industries. In response to the Deepwater Horizon accident, efforts are underway to adopt and deploy Crew Resource Management (CRM, aviation industry) into select O&G training programs. CRM addresses non-technical skills, decision-making, situation awareness, and other areas that have a known positive effect on the aviation cockpit crews and others. Considering the O&G starting point, positive effects should and likely will be realized. However, from the perspective of the complete set of design and management processes presented in Chapters 2–7, integrating CRM into a training program addresses one of the many deficient elements in barrier system designs and management systems. It would be unfortunate if that initiative draws attention from the larger need to develop a more comprehensive and integrated design methodology applied to human-dependent barrier system design and management.

When looking for solutions from other industries, what types of deliberations should occur? Consider the CRM adoption as an example. CRM has been proven effective in the aviation industry. It is important to consider the culture and environment of that industry. Compared to O&G and other process industries, the aviation industry is more highly regulated, has better defined competency requirements, and requires more frequent competency verifications using advanced, costly, and high-fidelity dynamic simulators. Further, aviation places a significantly greater emphasis, with increased regulatory requirements, on procedure use, development, validation, management, monitoring, and enforcement. In contrast, the process industries place less reliance and less stringent controls on procedures and the associated development and management systems. It may be important to examine and determine the extent to which the success of CRM in aviation is tied to and a product of this more controlled and regulated culture and environment. The same analysis should be applied when considering other solutions being considered for adoption. So, how might those outcomes change when the solution (any solution) is adopted by a significantly different, and less regulated, controlled, and standardized industry? It may be reasonable to envisage that the results may well differ in the degree to which benefits may be realized in practice.

In closing thoughts, the process industries may benefit from further introspection on their approach to difficult problems (barrier assessment, design and lifecycle management methodologies) and topics (cognitive science and cognitive ergonomics). The Deepwater Horizon and other major accidents appear to demonstrate that historical methods and practice are unable to reliably prevent the cognitive type (designed-in) errors that remain persistent incident and accident contributors. Continued reliance on solutions from other industries should not be the default or the dominant approach. Given past history and the differences in the process industries discussed earlier, it may be time to consider and pursue organically developed solutions that are better suited to addressing its unique challenges. (The prototype lifecycle model, developed by a process industry practitioner, presents one example of that pursuit.)

8.2.7 LACK OF FOCUS ON COGNITIVE ERGONOMICS, THE ELEPHANT IN THE ROOM

It may seem surprising the extent to which existing standards of practice are significantly to profoundly deficient in their application of cognitive ergonomics to safety system design and management. An employed standard may use the term *ergonomics*, but then fail to clarify if it includes cognitive ergonomics or is limited to physical ergonomics and a few aspects specific to human performance (perhaps a subset of non-technical skills). To gain insight into where, how, and at what level cognitive science should be addressed and applied in safety critical design, refer to the cognitive assessment and mitigation process in Appendix I. Appendices C and F provide the models and cognitive science that fully underpins that process. Unfortunately, the perceived need to design and assess at this level remains sparse.

Refer to the crude but useful cognitive ergonomics maturation scale described in Section 9.1. The *sparse* reference in industry documents may indicate a maturity level in the initial stages 1 and 2 (of 6). In stage 1, the organization has little or perhaps no knowledge of the cognitive science information included in Appendix F. In stage 2, the organization may have awareness and knowledge of that information, but has not made significant inroads into the gathering, assessing, and packaging that information as a first step toward making it available for use in a cognitive ergonomics–focused application. Without a significant new push from a forward-looking industry or owner/operator champion or the issue of new cognitive-focused standards, progress in this area may continue to stall or move at an anemic pace. At that pace, marginal changes may achieve a few modest gains over the next five to ten years. It also assures cognitive-type (designed in) errors in barrier systems and safety critical tasks will continue to be causal contributors to incidents and accidents.

8.2.8 A MAIN SOURCE OF SYSTEMIC RISK REMAINS IN THE U.S. O&G INDUSTRY

NASEM (2023) reported on the current state of the O&G industry operating in the U.S. Gulf of Mexico. It responds to a request to assess "whether systemic risk management by industry and regulators designed to avoid another offshore disaster was improving over time" (p. xi). The findings indicate improvements that reduced risks in many areas. However, this was not the case with contingent barriers.

> There is little industry or regulatory guidance available for contingent barriers, which the committee views as a main source of industry systemic risk…. Contingent barriers are those that require active human interaction or intervention to prevent the accidental release of hydrocarbons.

(NASEM, 2023, p. 124)

Even today, neither regulation nor industry standards sufficiently reflect modern terminology, thinking, and practice about barriers and use of the bowtie model.

(NASEM, 2023, p. 124)

In response to the Deepwater Horizon accident, regulatory changes were made to U.S. 30 CFR 250, entitled "Oil and Gas and Sulfur Operations in the Outer Continental Shelf – Revisions to Safety and Environmental Management Systems" (SEMS). The new regulation addressed stop work authority, ultimate work authority, and employee participation. It also mandated that organizations implement the performance-based SEMS programs defined in the standard, API RP 75 (2004, 3rd edition). The standard mirrors an earlier U.S. process safety management (PSM) regulation that was promulgated in 1992, namely, 29 CFR 1910.119. It defined PSM elements and processes required for managing hazards attributed to highly hazardous chemicals. From its opening statement, the regulation "contains requirements for preventing or minimizing the consequences of catastrophic releases of toxic, reactive, flammable, or explosive chemicals. These releases may result in toxic, fire, or explosion hazards."

The U.S. Chemical Safety Board (CSB) investigation of the DWH accident concluded the following: "To successfully minimize undesirable consequences, therefore, industry must shift from correcting individual 'errors' identified post-incident to a systematic approach for managing human factors" (CSB 2016, p. 242). In response, API published a revised version of the RP 75 standard (2019, 4th edition) adding a few new human factor requirements. This version was not universally well received. From NASEM (2023, p. 118), "Both the current (Third) and new RP 75 editions (Fourth) still refer to human factors only generally and do not mention contingent barrier management. Hence, BSEE regulations and audits are in the same situation." Regulatory agencies are limited in their ability to audit in areas that are missing or poorly defined in the regulatory code.

It may be worth reflecting on the reported deficiencies in active and active human barriers that were causal contributors to the accident. CSB (2016) and USCG (2010) identified deficiencies in procedures, training, decision-making, and clarification of roles, all areas addressed in the SEMS program. Therefore, it may appear that SEMS is the appropriate solution to prevent these deficiencies. However, a review of the past may present a somewhat different view. In the 1980s, a series of catastrophic accidents triggered a regulatory and industry response. Example accidents include the 1984 toxic chemical release from a Union Carbide chemical plant in Bhopal, India (2,259 or more fatalities, 558,125 injuries), and the 1989 explosion in a Phillips 66 chemical complex in Texas (23 fatalities, 314 injuries). Directed toward these industry sectors, U.S. regulatory agencies promulgated the PSM regulation discussed earlier, namely, 29 CFR 1910.119. In this same period, a U.S.-based professional practice and standards organization began work on a new design and lifecycle standard for technology-based active barriers referred to by the term *safety instrumented functions*. In 1996, the standard was formally adopted and published as ANSI/ISA S84.01. A few years later, a near-identical version was adopted and published as IEC 61511-1. This and later versions are now globally applied and widely recognized as generally accepted good engineering practice.

It is important to recognize that SEMS and the IEC standard have different purposes. Process safety management programs (e.g., SEMS) are proven vehicles for ensuring safety critical systems are operated and maintained in accordance with their approved and published requirements. SEMS promotes/requires incident investigations and that corrective actions/changes be managed using a published change management process. In contrast, the IEC 61511 standard is a proven vehicle for ensuring a governed safety system is designed and managed using a performance-based lifecycle model. (A standard of this type is the primary means to prevent the introduction of errors into the barrier system design. SEMS does not have this capability.) The information in Appendix M (Deepwater Horizon accident) identified *design errors* (physical and organizational) that caused or contributed to barrier failures that were causal contributors to the accident. Unfortunately, the accident investigators did not identify the *absence* of an IEC 61511-equivalent standard for active human barriers as a possible root cause and explanation for the widely ranging failures. Consequently, the issue was not addressed in the subsequent recommendations discussed in Section 8.2.7. Integrating additional human factor guidance into API RP 75 will not prevent the introduction of errors into the initial barrier design or address the other critical functions and activities provided by a dedicated design and lifecycle standard. For these reasons, contingent barriers seem poised to remain a main source of systemic risk in the U.S. offshore O&G industry in the future.

8.2.9 Continued Failures in Human-Dependent Barriers Remain Likely

Investigation reports from major accidents continue to identify the profound consequences of the human-dependent barrier and safety critical task failures in complex sociotechnical systems. Historically, the reports commonly found human error to be the causal contributor. That finding may appear to identify the root cause of the failure. However, in most cases the true root cause is somewhat different. The failure may have been the product of an undesired or unmet human action. However, that action was the consequence of a design error (often cognitive-attributed) that existed within the barrier system design, or the systems used to manage the barrier over its lifecycle.

Discussed in the preceding sections and throughout this book, a common barrier system design error is a designed-in cognitive/capability mismatch. The mismatch results when the design places a cognitive demand (expectation) on the task assignee (HE) and that person does not have the ability or capability to reliably meet that demand at the required time. As often occurs, that effort may be further impeded by a foreseeable task/workload conflict, poor display design, or a degraded work environment. Current standards and practice do not have the tools and processes needed to reliably reveal and prevent these design error types. As a consequence, hidden errors in existing and new human-dependent barriers will remain persistent contributors to incidents and major accidents.

Consider the information in Appendix M. It provides a timeline and details for the active and active human barrier failures that were causal contributors to the Deepwater Horizon accident.

- Using current standards and practice, what is the likelihood that one or more of these same errors and failures (or similar error types, similar or different applications) could currently exist and occur in facilities the reader works or visits?
- If the response is failure-seems-probably or failure-seems-likely:
  - What is the basis for that response?
  - Was that basis guided by: A deficient or misguided regulation, standard, or practice? An insufficient expertise in cognitive ergonomics or little to no knowledge of the information in Appendix F?
- If the response is we-are-fine...no problem here:
  - What is the basis for that response?
  - Was that basis guided by: Past experience...nothing has happened yet? Having adequate knowledge of the information in Appendix F and cognitive ergonomics? The recent adoption of a new standard or changes to an existing practice? An improved focus and controls on organizational elements that may include procedures, training, competency assessments, or a competency management program?
- What are the perceptions when the above questions are applied to other organizations (e.g., peers, other countries, or regulatory regimes)?
- Consider the maintenance on a human-dependent barrier or a high-consequence active barrier type, for example, a SIL 3, high-integrity pressure protection system (HIPPS). Do site practices consider the potential for / and the consequence of a cognitive-type error in the performance of a safety critical task? Are processes in place that could reliably detect the error and its consequence in a timely period?

8.3   A SUGGESTED PATH FORWARD – DEVELOP A NEW GLOBAL CONSENSUS STANDARD

8.3.1   JUSTIFICATION FOR DEVELOPING THE STANDARD

Sections 8.2 and 1.4 (Chapter 1) provide evidence and reasons to take a more aggressive approach to address the acknowledged deficiencies in existing designs, standards, and practice. This section promotes the proposition that a new global consensus standard, one dedicated to active human barriers, provides the required means to prevent or mitigate many (if not most) of the identified deficiencies. The suggested approach is to consider the content and experience gained from a widely or globally used standard for a different active barrier type, i.e., IEC 61511-1 (2016). Section 8.3.2 identifies questions that may arise when first attempting to define the scope, framing, and content of the new standard. Section 8.3.3 outlines the pros and cons of using the standard as a potential template. Section 8.3.4 discusses the novel and insightful proof-of-concept approach taken to further demonstrate the viability of the suggested approach.

8.3.2   Defining the Standard Scope, Boundaries, and Content

Example questions and challenges presented to those attempting to develop a new global consensus standard for active human barrier may include the following:

1. Scope and Objectives
   * What are the core objectives for this new standard?
   * What is the proposed scope?
   * Who is the audience?
   * What problem is it attempting to solve or address?
   * Will this present a lifecycle model approach?
2. Terms, definitions, and constructs
   * Which terms and definitions?
   * Which constructs and core models?
3. Alignment with existing regulations and practices
   * Define the philosophy and approach if attempting to accommodate the various regulatory requirements worldwide.
   * Define the philosophy and approach if attempting to accommodate the various global practices and conventions used worldwide.
   * Define the process for selecting or rejecting existing or new methods and practices to include in the standard. What criteria should be used to guide this process?
4. Topic, issues, and boundaries
   * Should the standard be prescriptive, performance-based, or guidance only?
   * Define the active human barrier types included in the standard, for example, preventive, mitigation, and emergency response barriers. Should it also address human-dependent safeguards and safety critical tasks?
   * What lifecycle phases should be included? Duplicate those in IEC 61511-1 (2016)?
   * Should it address all barrier elements (e.g., physical, organizational, and human)? If so, to what level of detail?
   * Should it address barrier dependencies? If so, to what level of detail? Using what approach?
   * Should it address the barrier system relationship and linkage to owner/operator policies and programs?
   * Should it address performance influencing factors, performance standards, or verification, validation, and other assessments?

8.3.3   Consider IEC 61511 as a Conceptual Model

This book presents the case for using IEC 61511-1 (2016) to guide the conceptual framing and development of a new standard dedicated to the design and lifecycle of active human barriers. This section examines the pros and cons of this approach.

Pros

- The standard applies to active barrier types.
- It integrates and aligns with the global, industry-agnostic, functional safety standard IEC 61508 (series).
- The standard is proven and mature. (The standard originated from the 1996 publication of ISA 84.01 adopted by ANSI in 1997).
- Globally used and widely accepted. Used in O&G, chemicals, petroleum, petrochemicals, and other sectors, with applications in production, manufacturing, storage, distribution, shipping, and other areas.
- Widely accepted as good engineering practice. (The degree of compliance varies according to local regulatory and statutory requirements, joint venture partners, insurer requirements, company practice, etc.)
- Includes and defines two documents that are essential to designing, operating, and maintaining a barrier through the full lifecycle, namely, the functional safety management plan and safety requirements specification.
- It reduces the likelihood of incidents and accidents.
- If fully implemented and correctly operated and maintained, it reduces the net operating cost and the number and scale of incidents and accidents.
- It addresses competency requirements for activities in different lifecycle phases.
- It defines quality requirements (e.g., design reviews, inspection, testing, verification, and validation).
- It is amenable to digitization (e.g., design development, requirements management, verification tools, and maintenance support).
- Normative and informative guidance addresses different areas and topics.

Cons

- The primary guidance applies to barriers that reside fully within technical systems and equipment. Guidance on organizational and human elements is limited and uneven.
- Cognitive ergonomics is an acute weakness (all lifecycle phases).
- It describes barrier bypass functions that are commonly provided to support online testing and maintenance. Given the information in this book, the guidance provided for design, application, and use is significantly incomplete. (Once placed into the bypass mode, the barrier reverts to an active human barrier. The mode is often misused and employed in situations that were not considered in the original design.)
- Complex to learn and gain experience. (This should be expected from any new standard that introduces a lifecycle approach. Many years and projects are often needed to gain experience with the activities in all lifecycle phases.)
- It requires new technical expertise (e.g., certified functional safety specialists).
- It often increases the capital cost of the barrier system.
- A failure to operate and maintain the barrier system as designed can increase the type and scale of an incident or accident.
- For large barrier installations, it creates a demand for new barrier design and management database tools.

Note: Engineering and operating companies developed many of the initial database tools for internal use. Commercial products eventually followed.

- Tends to increase the technical and execution complexity of projects that design, develop, and implement barriers governed by this standard. The required execution often creates out-of-sequence activities that affect work performed by other technical disciplines and organizations. (The same will likely occur with some activities in the prototype lifecycle model.)
- Stated validation requirements are activities that others may view to be verification activities. For example, a site acceptance test is identified as a validation activity.
- It takes a different approach to performance standards. The safety requirement specification may serve the function of a performance standard in a verification process.

8.3.4   GUIDANCE PROVIDED BY THE PROTOTYPE LIFECYCLE MODEL

The prototype lifecycle model, described in Chapters 2–7, adopted a novel approach to demonstrate the viability of using IEC 61511-1 (2016) as a guide to inform the development of a new active human barrier standard.

1. From its inception, the lifecycle model was purposely guided by several core concepts from this standard.
   - It adopts the same lifecycle phasing (phases and framing). Table 8.1 shows the mapping of the lifecycle model to the IEC standard.
   - It adopts the two essential documents needed to support a lifecycle approach, namely, the functional safety management plan and safety requirements specification. (For details, see Appendix B.2 and B.3, respectively. Table B.1 shows the lifecycle model mapping to the SRS requirements from cl. 10.3.2.)
   - It adopts and adapts other elements from the standard, such as verification and selected assurance activities.

   Note: The lifecycle model also delineates where and why specific deviations from the IEC standard may be appropriate. Validation, discussed in Appendix G.1.2, is one example.

2. The lifecycle model is comprehensive and highly detailed. Its processes and tools are guided by recognized industry experts. Further, the entire model is underpinned by cognitive science and cognitive ergonomics. As such, the model may prove to be an invaluable resource to those charged with developing the new standard.
   - It identifies new technical, quality, and execution issues and topics to consider.
   - It analyzes the variations between industry guidance and regulatory requirements and proposes paths forward aligned to a cognitive ergonomics–based

approach and methodologies. For examples, see Appendices D.2, G.1, and H.4, which address performance standards, verification and validation, and performance influencing factors, respectively.

- It identifies execution challenges that may be unique to a cognitive ergonomics approach and presents possible solution options. For examples, see Appendices A.8, D.4, E, G.1, and L and the topical notes included in many of the steps and processes presented in Chapters 3–7.
- It demonstrates the potential validity of adopting two key documents from the standard, namely, the barrier safety management plan (described in Appendix B.2) and safety requirements specification (described in Appendix B.3).
- Used as a reference guide, the model and the extensive information in the appendices provide a unique perspective and material to examine the standard for potential safety critical errors and omissions. (For further discussion on this topic, see Chapter 9, Section 9.11, New Ability to Detect Errors and Omissions.)

TABLE 8.1

Mapping the Lifecycle Model to IEC 61511-1 (2016)

| Figure 3.3, Ref. Box Number | **Safety Lifecycle Phase or Activity** | | |
	Title	Req. Clause	Prototype Lifecycle Model
1	Hazard and risk assessment	8	Phase B (Ch. 3)
2	Allocation of safety functions to protection layers	9	Phase B (Ch. 3)
3	SIS safety requirements specification	10	Phases B, C, D (Ch. 3, 4 and 5), App. B.3
4	SIS design and engineering	11, 12, 13	Phase D (Ch. 4, 5)
5	SIS installation, commissioning, and validation	14, 15	Phase D (Ch. 6)
6	SIS operation and maintenance	16	Phase E (Ch. 7)
7	SIS modification	17	Phase F (Ch. 7)
8	Decommissioning	18	Phase G (Ch. 7)
9	SIS verification	7, 12.5	Appendix G
10	SIS FSA	5	Appendix G *Replaced by modified V&V process*
11	Safety lifecycle structure and planning	6.2	Phase A (Ch. 3), Appendices A and B.2 (Barrier Safety Management Plan)

SIS – safety instrumented system; FSA – functional safety assessment; V&V – verification and validation.

REFERENCES

Aas, A.L., Johnsen, S.O., Skramstad, T. (2010), Experiences with human factors in Norwegian petroleum control centre design and suggestions to handle an increasingly complex future. In R. Briš, C. Guedes Sooares, & S. Martorell (Eds.) *Reliability, Risk, and Safety, Theory and Applications* (pp. 285–292), 1st Ed, CRC Press.

API RP 75 (2004), *Safety and Environmental Management System for Offshore Operations and Assets*, 3rd Ed, American Petroleum Institute

API RP 75 (2019), *Safety and Environmental Management System for Offshore Operations and Assets*, 4th Ed, American Petroleum Institute

CCPS (2015), *Guidelines for Initiating Events and Independent Protection Layers in Layer of Protection Analysis*, New York: John Wiley & Sons Inc., Center for Chemical Process Safety (CCPS)

CCPS (2018), *Bow Ties in Risk Management, A Concept Book for Process Safety*, Hoboken, NJ: John Wiley & Sons Inc., Center for Chemical Process Safety (CCPS)

CSB (2016), Drilling rig explosion and fire at the Macondo well, Investigation report volume 3, Report no 2010-10-1-OS, U.S. Chemical Safety and Hazardous Investigation Board, Washington, DC.

EI (2020b), *Guidelines for Management of Safety Critical Elements (SCE)*, Energy Institute London, 3rd Ed, January 2020

IEC 61511-1 (2016), *Functional Safety – Safety Instrumented Systems for the Process Industry Sector – Part 1: Framework, Definitions, System, Hardware and Application Programming Requirements*, 2nd Ed, International Electrotechnical Commission

ILO/IEA (2021), *Principles and Guidelines for Human Factors/Ergonomics (HFE) Design and Management of Work Systems*, Geneva, Switzerland: International Labor Organization

ISO 11064-1:2000 (2000), *Ergonomic Design of Control Centres – Part 1: Principles for the Design of Control Centres*, 1st Ed, International Organization for Standardization, 2000–12–15

ISO 9241-220:2019(e) (2019), *Ergonomics of Human-System Interaction – Part 220: Processes for Enabling, Executing, and Assessing Human Centred Design within Organizations*, 1st Ed, International Organization for Standardization

Johnsen, S.O., Kilskar, S.S., Fossum, K.R. (2017), Missing focus on human factors – organizational and cognitive ergonomics – in the safety management for the petroleum industry, *Journal of Risk and Reliability*, 231(4), 400–410

NASEM (2023), *Advancing Understanding of Offshore Oil and Gas Systematic Risk in the U.S. Gulf of Mexico: Current State and Safety Reforms Since the Macondo Well Deepwater Horizon Blowout*, National Academies of Science, Engineering and Medicine, Prepublication Copy (Downloaded from nap.nationalacademies.org website on April 7, 2023)

Sklet, S. (2006), Safety barriers: definition, classification and performance, *Journal of Loss Prevention in the Process Industries*, 19, 494–506

USCG (2010), Report of Investigation into Circumstances Surrounding the Explosion, Fire, Sinking and Loss of Eleven Crew Members Aboard the MODU Deepwater Horizon in the Gulf of Mexico April 20–22, 2010, Volume 1, United States Coast Guard

9 Concluding Observations, Findings, and Remarks

This chapter is organized as follows:

9.1 COGNITIVE ERGONOMICS: DEVELOPMENTAL STAGES AND CHALLENGES

The introduction to this book describes the evolution and maturity of active human barrier design, which significantly lags other active barrier types. The following are suggested stages in this cycle. Achieving progress in cognitive ergonomics occurs as an individual, organization, and industry sector moves through each stage, each of which presents its own unique challenges.

1. Organizations (e.g., regulatory, owner/operators, engineering, and service companies) and project teams have limited or no awareness and knowledge of cognitive ergonomics and cognitive science. The missing consensus on definitions contributes to confusion. Are there material differences between

the term's human factors, human performance, human factor engineering, and human and organizational performance?

2. Information on cognitive science is scattered in sources that many do not read, such as books, journal articles, and reports (Shorrock & Williams, 2017, pp. 128–130). To apply this information, it must be discovered and assessed to evaluate its readiness and application to active human barriers and safety critical tasks.

The NUREG (2016, p. 3) elaborates,

> this information is scattered throughout the broad fields of cognitive psychology, behavioral psychology, neuropsychology, human factors, and human performance; even review articles in the literature typically only focus on one narrow aspect of human performance (e.g., attention, situation awareness (SA)) without systematically documenting the available information for the full range of human performance. Moreover, the majority of literature focuses the description of results and conclusions on how humans can successfully perform given tasks without explicitly delineating the conditions under which humans would fail the tasks (i.e., it often needs analysis and inference to identify the information about human failures from literature).

3. Once identified and validated, the information from step 2 must then be compiled and developed into a more usable form. The process of doing so may identify information from the research community that is not useful, practical, adequately progressed beyond the research phase, or easily migrated to a more usable form (Shorrock & Williams, 2017, pp. 130–133).

Note: Appendix F is an example product from this phase. The information in Table F.1 required many years to discover, compile, and (through trial and error) determine the appropriate presentation form and format.

4. To meaningfully assess and define barrier task requirements, each task must first be decomposed into its constituent detect, decide, and act phases. From that output, the detect phase should be further decomposed into its sub-phases (detect, comprehend, and project/anticipate). This level of decomposition provides the necessary starting point for applying processes that can reliably identify and prevent or mitigate cognitive (and physical) mismatch design error types that occur in the task phase and sub-phase.

Note: Globally, defining an active barrier as tasks remains less common (e.g., limited to a few organizations and regions), though is slowly increasing over time. The further decomposition to the task phase level is not common.

5. Processes must then be developed that integrate and apply the information from steps 3 and 4 into a fully integrated barrier design and lifecycle model. Different processes and adaptations may be needed if applied to existing barriers.

Note: The lifecycle processes presented in Chapters 2–7 are examples.

6. A global consensus standard is needed to consolidate, advance, and normalize active human barrier design and lifecycle processes.

Note: When this book was published, the standard did not exist. It is unclear whether initiatives may be underway to begin development. Until that occurs, wide global disparities will persist regarding the terms, basic frames, concepts, and guidance. Existing standards and practice will continue to lack the capability to reliably prevent latent errors (cognitive type) in barrier system design. Seen in guidance documents developed in response to the Deepwater Horizon accident, poorly formed definitions (cognitive ergonomics relevant) are being introduced into these documents. Because existing standards are not full lifecycle models, an ongoing and typically undetected degradation in barrier function, performance, or integrity seems likely as the barrier progresses through its lifecycle phases. The emergence of regional standards may create new challenges for multinational corporations and organizations seeking consistency across their areas of global practice and facilities.

The process industries are at varying stages, many of which may still be in stage 1 or 2. An extended period in one step may delay progress in achieving progress in other steps. It remains unclear whether any industry organization (e.g., in the O&G sector) has reached a critical nexus that accelerates this progression. If the above stages and framing are indicative, completing this progression will take many years if not decades. *A primary objective of this book is to present possible solution options and paths to accelerate this process and shorten the timeline.*

9.2   WHERE TO START?

Having reviewed the book, the reader may be interested in exploring how to proceed. The following are a few example ideas to consider:

1. A company or organization may have a goal or a broadly focused objective to prevent or markedly reduce the potential for the next major accident. For those who recognize latent design errors (cognitive type) to be primary contributors, the company or organization may see value in evaluating the prototype lifecycle model for its potential efficacy, validity, and achievable results. Alternatively, the focus of the evaluation may be a comparison to current practice to identify where those practices may be deficient.

Note: The details, examples, and explanations readily support such efforts.

2. A further interest may progress to a cursory cost-and-benefit analysis. Section 9.3 provides suggestions to consider, support, and guide that effort. The greater specificity and detail in the lifecycle model (processes and step descriptions, data, and examples) provides sufficient information to

develop a conceptual-level cost estimate (person-hours) to apply the model or model elements. It may also provide sufficient information to identify potential opportunities to improve barrier reliability, performance, and integrity.

3. Further interest may warrant a demonstration project that attempts to confirm the benefits, validate the model processes and methods, develop improved cost estimates, and use the learnings to improve future projects. The project scope may be limited to lifecycle phases A and B (Chapter 3), or selected steps within those phases. It may also be limited to a single barrier or an especially complex barrier (or safety critical task) that warrants use of a deeper, more structured, and systematic methodology. Regardless of the project type and scope, it should be planned and progressed in a manner that provides measurable results.

Note: SPE (2015) may provide additional insights on how to plan and progress a demonstration project. The presented project performed a human reliability analysis and other processes to assess safety critical operations and maintenance tasks.

4. Prepare the project plan:
   a. Define project objectives, scope, and deliverables.
   b. Define the project success criteria and measurable outcomes used to determine whether the project objectives were achieved.
   c. Review the prototype lifecycle model. Select the model processes and activities to be included in the project.
   d. Develop the schedule.
   e. Identify the required staff and resources, and the required competencies (see Appendix A).
   f. Develop the cost estimate (e.g., estimated people-hours to perform identified activities).
5. Perform the project according to the project plan.
6. Evaluate the results:
   a. Assess the results against the defined success criteria.
   b. Identify the lessons learned and suggestions to improve a potential future project.
   c. Gather and develop metrics to estimate the next project.

Note: Based on experience from the initial rollout of the IEC 61511-1 standard, early project compliance was selective and limited. Over time, owner/operators increased their compliance with the various standard requirements. The added activities increased project effort and cost, a reflection of the new work. Eventually, the cost increases capped out and began to decline as personnel, contractors, and others gained experience implementing the standard. Cost benefits were also realized by the use of more efficient execution methodologies and new digitization tools for design automation and information management. The trajectory for implementing the lifecycle model processes may be similar. It may be possible to truncate this timeline through the early development of digitalization (e.g., database) tools.

9.3 ASSESSING POTENTIAL COST AND BENEFITS

9.3.1 POTENTIAL COSTS

An initial effort to estimate the cost for implementing the lifecycle model, or some part of it, may be the product of a person (one having the requisite experience or knowledge) who develops a conceptual-level estimate for the resources and level of effort needed to complete the identified activities. (The dominant or the total cost may be attributed to people.)

Should a more accurate estimate be needed, that may be achieved through a demonstration project that provides actual cost data and the potential or realized benefits. Section 9.2 identifies example demonstration projects. This book provides considerable detail and input to those attempting to evaluate and estimate the project cost (e.g., the resources, effort, and durations). Deploying the cognitive ergonomics–focused approach relies on the participation of those having expertise in cognitive ergonomics, operations, and, where applicable, maintenance. Appendix A suggests the appropriate participation in the various lifecycle activities.

9.3.2 POTENTIAL IMPROVEMENTS TO BARRIER RELIABILITY AND PERFORMANCE

The prototype lifecycle model offers the following potential opportunities and avenues for improving active human barrier and safety critical task reliability and performance.

1. **The model responds to the industry recommendations and recognized deficiencies in the call-to-action white papers published in response to the Deepwater Horizon accident.** (1) It does so by adopting many of the proposed recommendations, while rejecting those deemed to be regressive or counterproductive. (2) Recognized by Johnsen et al. (2017), the responding industry changes tended to be incremental and therefore unable to adequately achieve the desired results. The model goes well with those recommendations by presenting a comprehensive solution (the lifecycle model) what is fully grounded and underpinned by cognitive ergonomics and cognitive science.
2. **Reduce Latent Errors Known to be Contributors to Incidents and Accidents.** The potential benefit of deploying the prototype lifecycle model lies in its many processes that are purposely designed to reliably identify, prevent, and mitigate the error types in design and lifecycle management processes that cause the undesired human actions and behaviors that have been primary contributors to major accidents.
3. **Enhance Human Reliability and Performance.** The performance and reliability of humans assigned to perform barrier and other safety critical tasks should measurably improve if designed and managed using the lifecycle model. This can be confirmed by examining before and after information. For a task, capture performance information based on its original design. Redesign the task using the lifecycle model processes. Using the same assessment criteria, assess and compare the revised task performance to the original design. For an active human barrier, perform the barrier

system acceptance test (B-SAT) (see Appendix G.3) on the original barrier. Redesign the barrier using the lifecycle model processes. After being placed into operation for a nominal period, repeat the B-SAT and compare the results to the original design.

4. **Develop a More Advanced Barrier Status Checklist.** As the prototype lifecycle model is well structured, it provides a rich basis for developing a more comprehensive checklist to assess existing active human barriers (e.g., design errors, design deviations, or operational state).

5. **Provides a Basis to Examine Existing Standards and Practice.** The comprehensive and advanced information in the book provides a rich source of information to examine and assess an existing standard or practice for potential weaknesses.

Note: In response to a request from a standards committee, this author was asked to comment on the potential systemic sources of errors in a published standard from a cognitive ergonomics perspective. Given the foundation and materials in this book, the errors were readily apparent and easily identified.

6. **Provides a Foundational Basis and Material to Guide the Development of a New Global Consensus Standard.** Chapter 8 presents the case on why a new global standard for active human barriers is needed, and provides suggestions on its potential scope, framing, content, and development. The latter is addressed in two ways. Section 8.3 proposes that an existing, globally accepted standard for a different active barrier type (IEC 61511) be used as a potential model and template. To demonstrate the validity of that approach, the lifecycle model adopted its lifecycle phasing, several processes, and the two key documents required to enable a lifecycle approach. The model also provides a rich source of technical, quality, and project execution information that can be used to inform the scope, topics, and the technical and project execution issues to consider in the new standard. (The same information should be of similar value to an organization that has plans to develop or update their internal standards for active human barriers.)

9.4   ACHIEVABLE HUMAN RELIABILITY AND PERFORMANCE

History provides us with information on human reliability and performance in various barrier environments and situations given existing standards and practice. What we do not know (yet) is the maximum and sustainable level of human reliability and performance that may be achieved if tasks and organization elements are developed and managed by using a more efficacious and cognitive-focused set of processes and methods. It seems plausible and reasonable to expect improvements in both areas.

Note: See Fisher (2022, Ch. 2) for an expanded conversation on the quantitative limits on human performance (e.g., the baseline achievable error rates for skill-based, rule-based, and knowledge-based activities).

A review of the barrier failures in the Deepwater Horizon accident (discussed in Appendix M) may provide additional insights. Using a "What If" approach, consider if and how use of the prototype model might have reduced the likelihood of the various barrier failures. The following questions are examples to consider. (All pertain to operate and maintain phase processes unless noted.)

- Would the "Monitor Procedures" in process E2-3 have reduced the likelihood of the procedure deviation that failed to divert the kicking well to the overboard lines?
- Would the PIF assessment process in the detailed phase process C-2 (societal element PIF) identify the procedure deviations as plausible or likely?
- Would the apparent competency deficiencies that delayed the control room response to the flammable gas alarms have occurred? Would the competency review process E3-1 have identified and responded to that gap?
- Would the incident reporting and investigations in processes E2-5 and E5-5 have changed any of the outcomes given the incidents that occurred in the months leading up to the accident?
- Would the more comprehensive approach to barrier operations and management have contributed to awareness and tracking of the potential damage to the upper annular and its potential effect on barrier functions? Would it have resulted in repairing the failed control POD in the blowout preventer, consistent with regulatory requirements?
- Would the model's increased focus on the barrier leader have changed when the barrier leader first arrived at the control room where inadequately trained personnel delayed activation of the blind shear ram, the emergency disconnects system, and the required responses to flammable gas alarms?

Note: This type of assessment carries potential risks. The accident information may be incomplete or incorrect. Hindsight can contribute to an inaccurate or incomplete understanding. Both points should give pause in how this information is used, as it does not provide definitive proof and could be misleading. With that caveat, the assessment may still provide useful insight if seeking to determine whether a particular process in the lifecycle model should be considered and included in a new consensus standard.

9.5   APPLICATION TO SAFETY CRITICAL TASKS

Indicated in Table 2.1 (Section 2.2), active human barriers and safety critical tasks share similarities and a few key differences. Many if not most of the task-oriented activities in the lifecycle model may also be applied to safety critical tasks. The use of this model, when contrasted to existing practice (e.g., EI 2020a), should yield improved results. Its more extensive task decomposition process helps to expose task activities where cognitive mismatch errors are more likely to reside. Model processes also create a more complete set of task requirements and design documents that contribute to improved design traceability and the ability to better manage and maintain the task over its full lifecycle. Other model processes monitor for undesirable deviations that can occur over time, for example, drift and skill fade.

Case Study (Hypothetical), a Safety Critical Maintenance Task Performed on a High-Consequence Safety System. Consider the design and management of a safety critical maintenance task for a high-consequence (e.g., SIL 3) safety instrumented function (SIF). For reasons discussed in Chapter 8, the original design was likely underspecified (e.g., display design or procedure and competency requirements). That increases the likelihood that one or more errors exist in the original design. Over time, drift and skill fade (discussed in Appendix J.7) can also introduce degradation processes. Perhaps these events contribute to an increased number of false activations. If not corrected, confidence in the barrier may degrade causing an increased use of its bypass/inhibit feature. An unsupported increase in the intervals between proof tests may occur. These events decrease the likelihood the SIF will correctly function when requested. Collectively, they contribute to a drift-to-danger progression that, unfortunately, is not uncommon. When this occurs, the true state of the SIF is not actually known, a fact that often remains hidden. The process industries have seen a rapid rise (percentagewise) in the number of the installed high-consequence (SIL 3) systems. Some and perhaps many are operated by companies that have limited technical resources and little or no knowledge of cognitive ergonomics. An error in a safety critical maintenance task can cause or contribute to an incident and potentially to the next major accident. (In addition to the 11 fatalities, 17 injuries, and long-term environmental damage to the region's sea life and coastal wetlands, the costs and penalties incurred by BP as a consequence of the Deepwater Horizon accident exceeded sixty billion U.S. dollars.)

9.6   CLOSING THE GAP: WORK-AS-IMAGINED VERSUS WORK-AS-DONE

A stated goal in the prototype lifecycle model is to minimize the work-as-imagined (WAI)/work-as-done (WAD) gap. A key approach for doing so is the inclusion of end-users and the appropriate experts in key design and lifecycle activities. (Refer to the tables in Appendix A.2.3 for suggested participants). Operations specialists and personnel having assigned roles in a barrier system (e.g., the barrier leader) are recommended participants. For example, their input is keenly important in the task analysis and the task definition and requirements activities that frame and guide the remaining lifecycle activities.

Many of the model monitoring processes in the operate and maintain phase can identify WAI-WAD gaps (e.g., the procedure monitoring processes E2-3 and E5-3). An existing procedure may be out of date or have a known deficiency that forces personnel to adopt ad hoc workarounds. Perhaps a post–start-up change occurred that requires a modification to an existing procedure. The modification was not identified because the original change did not use the change management process. Unsafe procedure deviations may also begin with an unsafe shortcut that eventually becomes normalized over time.

9.7   WHY IS EXECUTION SUCH A CHALLENGE?

Implementing the lifecycle model will present many of the same challenges common to other human factors–type work performed in a capital project environment. This book provides helpful execution guidance for many lifecycle activities. Work of this type requires varying levels of experience and expertise in different areas. Understanding the technical and operational knowledge requirements may be more straightforward. Modern projects increasingly require broad and deep project execution knowledge referring to the how, what, when, and by whom. Appendix A.8 discusses example challenges in these areas. Some of the greater challenges (common to all technical disciplines) occur when the project schedule forces non-optimal execution and out-of-sequence work. This often occurs when a major equipment item/package is purchased very early in the project cycle, and the package contains barrier elements. Situations of this type often require early definition and development of those element requirements and their integration into the procurement package at a time that aligns with the procurement schedule. Analogous situations occur when an early project freeze date forces a critical sizing decision (e.g., firming up the primary dimensions and layout of a procured building), or the placement of major equipment and other primary facility layout decisions (e.g., information that affects the location of egress/escape routes and safe havens.) These situations are common in a large capital project for a floating, offshore O&G facility. The timing and consequences of these events may limit the opportunities to make significant workspace changes (sizing, configuration, or layout) to the preliminary design phase. Attempting the same change in a later project phase may be prohibitively costly or limit the change to non-optimal solutions. In such cases, preliminary design phase efforts and decisions may need to rely on data from a prior project (hoped to be indicative) and judgement-based decisions informed by incomplete information. Proactive planning, well-developed tools and methods, and experience guided judgement are among the essential tools and methods needed to progress the barrier design and execution process in these environments.

9.8   REFLECTING ON COGNITIVE SCIENCE AND APPENDIX F

Appendix F provides examples of the cognitive science and cognitive ergonomics that underpin the cognitive ergonomics–focused prototype lifecycle model. The information was sourced from many books, peer-reviewed journals, and other technical documents. The effort to discover, comprehend, crosscheck, and compile this information occurred over a 10-year period. The information in Table F.1 provides a uniquely compelling, contrasting, and more complete picture of human cognitive system attributes, capabilities, limitations, and behaviors. The information is essential to understanding and appreciating the innate differences between two fundamentally different cognitive processes (automatic and conscious). Once absorbed and internalized, readers may discover they have a new lens and a keener insight for understanding the mechanisms and attributes that create human perceptions and comprehension, drive decisions, and affect behaviors (rational and non-rational) and actions. For further insight, see Figure F.2 (Contributors to Human Performance and Behaviors, Appendix F.2). Appendix B.5.1 explores some of the drivers, mechanisms, and reasons why a human can direct their eyes towards an object and then fail to "see" it. Appendix B.5.3 reviews some of the ways and situations where time (clock time) becomes a problem when humans interact with sociotechnical systems.

Figure F.1 (Appendix F.1) provides insight into the non-deterministic and asynchronous interactions that occur in autonomic, automatic, and conscious cognitive processes, each operating at different response *rates*. Autonomic and automatic processes can modify, delay, or inhibit a conscious process. A conscious process can also halt, interfere, and influence an automatic process (e.g., a skilled-based action or behavior). This information sheds light on why and how situationally unique conditions contribute to a very wide range of possible outcomes. *The outcome depends on which cognitive process becomes the final arbiter that guides and controls perception, comprehension, decisions, and actions.*

Common sources of task variability (discussed throughout the book) result from the inherent and innate capabilities, limitations, and functioning of human memory, both short- and long-term. Current practice seldom recognizes the extent to which human memory contributes to barrier and safety critical task success and failures. Section F.2.2 and step B-10 (Chapter 3) provide glimpses into these limitations. Some of the non-rational biases discussed in F.2.5 can also be attributed (in full or in part) to the attributes and normal functioning of short- and long-term memory. Section F.5 provides additional insight into long-term memory functions and mental models.

Collectively, this foundational information should be used to guide the design and management of human-dependent, safety critical activities performed in time-pressured and high-consequence environments. It provides the essential underpinnings in the lifecycle model processes, methods, and tools that are first principles based, meaning they are derived from and guided by cognitive science and cognitive psychology.

9.9 WORKSPACE DESIGN

Existing industry standards and guidance documents (e.g., the ISO 11064 series) provide workspace design processes and tools. Appendix K presents cognitive-centric and cognitive ergonomics additions that extend their capabilities. Workspace design can be more challenging for emergency response barriers. Appendix F.3 provides information to better understand the capabilities and limitations in the human visual system. This information guide the science-based, workspace design processes (e.g., display selection, layout, and sightlines). Similarly, Appendix B.5.1 reviews the reasons why a human may look at an object but fail to notice/see it. These are important considerations when selecting and placing display objects along an egress/escape route. Both topics are integrated into the physical workspace guidance in Appendix K.2. Appendices K.3, K.4, and K.5 (communications, human, and information workspace, respectively) provide unique perspectives and tools to consider when designing these very different workspace types.

9.10 INSIGHTS ON APPENDIX M AND THE DEEPWATER HORIZON ACCIDENT

From NSIA (2021, p. 11),

> The purpose of clarifying sequence of events and circumstances is to be able to understand and describe what happened and describe what and who were involved in such a way that this affected the events that led to the accident and the pertaining damage and loss.

Drawing from an extensive range of reputable sources, Figure M.1 (Appendix M) and the included notes attempt to compile and integrate information on the active and active human barriers that were causal contributors to the Deepwater Horizon accident. The expanded information contributes to a more complete story of the events and how each barrier contributed (positively or negatively) to the accident, timing, escalation pathways, and outcomes. A compilation at this level is essential if the goal is to explore and identify possible new findings, conclusions, and recommendations missed in the individual accident investigations and reports. The prototype lifecycle model may provide additional insights and ideas that may contribute to the further evolution of accident investigatory methods, tools, and efficacy. ***Much more can be learned from this accident!***

9.11 NEW ABILITY TO DETECT ERRORS AND OMISSIONS

Readers may gain a powerful new or extended capability that develops from absorbing and fully internalizing the information in this book. That capability provides a new lens (a distinct perspective) to examine and assess new and existing barriers, regulations, and industry standards and guidance documents. By simple inspection, this new lens may allow one to readily identify missing content, incorrect use of terms, and potential errors in the presented material or guidance. Knowledge-wise, it also provides the means to identify and understand how and why a design mismatch error (cognitive attributed) can cause or contribute to task and barrier degradation and failure. That same knowledge also provides the means to prevent those errors and enhance human, task, and barrier performance.

9.12 TERMS AND LANGUAGE

Readers may notice differences in some of the language and adopted terms. The global industry has not agreed on a common set of terms and definitions for active human barriers. A concerted effort was made to select, define, and use terms consistently throughout the book. It is hoped the added effort contributes to a more consistent, precise, and insightful understanding of the material, processes, and issues presented and discussed in this book. A few new terms provide shorthand descriptors for common object types addressed in many lifecycle processes. Examples include *In-Place PE*, *direct-use PE*, *Support PE*, and *engineered area*. (Common terms such as human-system interface (HSI) were considered, but given their varied definitions did not seem to provide the desired clarity and preciseness. Readers may decide otherwise.)

The terms *skill* and *knowledge* have a specific and science-based meaning in the book. Lifecycle model processes identify and specify task-specific skill and knowledge requirements. A specified skill competency is achieved when the required skilled action or behavior (physical or cognitive) is repeatedly practiced (learned) to where it can be repeatably and accurately performed automatically, and therefore places little to no demand on working memory/short-term memory resources. A knowledge competency also resides in the head but is fundamentally different. A specified knowledge competency (content, depth, breadth, and accuracy) is achieved when that material is adequately and correctly understood, retained in long-term memory and mental models, and can be reliably recalled on demand. Discussed in Appendix F.5, the content of this in-the-head knowledge reflects how an individual understood the information when it was encoded into memory. Verifying a knowledge competency should use methods that can reliably and accurately assess the recalled memory to verify it meets the competency requirement.

The precision and clarity in the model-employed terms (often missing in regulatory and industry practice publications) are integral to the cognitive ergonomic processes, methodologies, and descriptions adopted in the lifecycle mode. Consider how these terms (e.g., as apply to a "skill" or "knowledge" requirement) may provide better guidance to those charged with selecting training delivery and competency verification methods that are best suited to each competency type and requirement. Further, the selected terms contribute to a lifecycle model that seeks to be fully traceable, verifiable, and cognitive science guided. Readers may want to examine how these (and like) terms are defined and used in their current practice and practice standards. Consider how each informs, and guides, or potentially limits, misleads, or degrades understanding, execution, and the product of those practices.

9.13   MODEL TABLES: DATA, GUIDANCE, AND REFERENCE

Page wise, a sizable portion of the book is allocated to tables. This is one of the attributes that sets this book apart from others. Most capture the information defined and developed in lifecycle processes. Others provide guidance or baseline information used in those processes. A third type, the cross-reference tables in Appendix B-7, aids the reader seeking the model process or activity that develops or modifies specific information or information types. A fourth type, the procurement support tables in Appendix E-13, is an aid that provides suggestions for integrating barrier requirements into requisitions, purchase orders, and contracts.

Data tables in the lifecycle model capture requirements, design and constraint data, and other information generated by and used in one or more model processes. These processes identify where and how this information is used in each phase of the barrier lifecycle. This approach, and the greater detail and information specificity in each table, demonstrates the more advanced level of design traceability achieved in the model. The tables also provide the reader with the means to compare the information developed and captured in the lifecycle model and in existing practice. Model table information that is not developed or captured in an existing practice may indicate a potential deficiency in the practice that places the barrier at risk.

Traceability is one of the key attributes that helps to establish the validity of the lifecycle model. Any barrier element that cannot be traced to a definitive basis requirement reveals a potential error entry point into the design and management of a barrier or task. The model data tables and cross-reference tables provide the means to validate the model from that perspective. (Few industry practice and guidance standards can make that claim.)

As a final thought, the data tables provide an invaluable starting point for progressing a digital solution. Most tables include key fields to support a relational database format. The database (DB) becomes a common information source that is available for use by all disciplines and organizations. The information is in forms and terms that most can readily understand. The DB form may also contribute to a digital twin solution, and new computer-aided design, verification, data management, and distribution tools.

Note: The data-intensive standard IEC 61511 drove the need for database solutions used to support and manage the active barriers guided by this standard. Emergency response barriers are often more similar than different within a given owner/operator organization. This appears to make this barrier an ideal candidate for migrating the barrier data (or perhaps the entire model) to a computer-aided design and management solution. If it proves to be viable and practicable, that solution may markedly reduce engineering cost, improve the capability to meet extremely tight project schedules, and provide efficient and efficacious tools for tracking and managing the barrier in the operate and maintain phase.

9.14 NEW PROCESSES AND ADAPTATIONS TO EXISTING PRACTICE

Chapters 2–7 present the complete prototype lifecycle model and its core constructs. The topic-based appendices (page wise nearly half the book) provide foundational information, analysis, tools, additional processes, and guidance that support many lifecycle model assessments and quality processes. Several review and analyze differences between commonly used industry standards and guidance documents. Those analyses guide the approaches and solutions selected or adapted in the lifecycle model. For examples, see Appendices D (performance standards), G (verification and validation), H (performance influencing factors), and J (human and organizational elements). In some cases, a conventional process was modified or expanded to better support the more rigorous cognitive ergonomics approach. For examples, see Appendices D, G, H, and K (workspace design).

9.15 COGNITIVE ASSESSMENT AND MITIGATION PROCESS

The cognitive assessment and mitigation process in Appendix I may be a first-of-kind. No similar approach was identified in literature searches performed when preparing to write this book. The idea for the tool was inspired by NUREG (2016). (Its stated purpose was to provide a sound technical basis to model human performance and behavior. Doing so should improve the accuracy and reduce the variability in the human reliability contribution input to risk assessments.) Material and ideas from the NUREG document was repurposed, reconfigured, and additional information was added to create the processes and tools in Appendix I. The book often refers to it as a *first principles–guided process*, referring to a process that is innately underpinned by the science of human cognition and behavior (information of the type included in Appendix F).

The process and tools in Appendix I are presented as "white paper" prototypes. *The cognitive assessment tables (I.3) are intended to be example starting points only.* Before use, experts from the organization (e.g., the owner/operator) should review and update the table information in Appendix I.3 to reflect the latest available information and the organization's accepted practice and knowledge base. Each table presents a range of options to mitigate an identified cognitive mismatch finding. The organization may choose to add proven options, identify the order of preference, or remove options that should not be considered.

The results from using this process may differ from historical approaches, some of which may be deemed a best practice. The value of the best practice approach has been proven over time. It is the product of *past experience* and the available knowledge base at that time. A hurdle may result if a rigid view of a best practice prevents the consideration, uptake, and application of new methods driven by new knowledge. (The introduction of new knowledge differs from introducing a new technology. The latter commonly creates new and often unforeseen problems, an outcome that tends to be less so with new knowledge.) Best practices, for example, those included in an organization's internal HMI display standard or style-guide, should be re-examined using the cognitive-type knowledge and processes included in this book. Doing so may reveal hidden deficiencies (cognitive attributed) and where, why, and how improvements can be made that more directly and intuitively support a user assigned to perform barrier tasks under time pressure and difficult environments. As an example of the latter, see the "Alternative Display Type" in process C5-1c. Because the *lifecycle model* performs design activities at a fundamentally new and deeper level, it should and likely will reveal issues in historical designs and practice.

9.16 APPROACHING PROCEDURES FROM A DISPLAY PERSPECTIVE

Note: The ideas discussed in this section are not included in the lifecycle model. They are included here for future consideration and discussion.

A procedure or a support aid is a critical element in the barrier system. It also is a unique "display" type in the barrier system (a view that is not commonly held). Adopting that view may motivate some to explore if and how lifecycle model methodologies (e.g., the display related methods in Chapters 3 and 4) could be applied to their design and assessment. Doing so may provide a new means to identify and prevent latent design and cognitive mismatch errors that lie within them.

Consider the use of a hardcopy procedure while performing a safety critical task. Every use is situational. The nature of the task affects how (e.g., when, and how often) the user must switch attention between the task being performed and viewing/comprehending the information in the procedure. A need for frequent task switches may identify problematic scenarios of the type discussed in Appendix F.2.3. A physical complexity, the user may attempt to hold the procedure while executing the task or find a way to position the procedure that maintains hands-free visual access (e.g., maintains the necessary viewing distance and angle). That access may be degraded by the environment such as insufficient lighting or inclement weather. The text size may be insufficient for that viewing angle and distance. Unique to each user, the user may have a complex relationship with procedures in general or one in particular. Example issues may include one's trust in the procedure or a belief the work can be correctly performed without it. At a deeper level, situations, motivations, expediencies, and perceived goals may also affect use. These simple examples may indicate the need for cognitive ergonomic focused tools that can supplement existing procedure development and assessment methodologies. Those solutions may include the workspace design principles in Appendix K and the assessment process presented in Appendix I (cognitive assessment). Other lifecycle model tools and processes may also apply.

9.17 MODEL IS NOT OPTIMIZED

No effort was made to optimize the prototype lifecycle model. It has varying degrees of overlap between processes and assessment tools. When developing a prototype model, the goal and intent is to present the widest range of ideas and methods to demonstrate what can be achieved. As is always the case, practitioners will identify, optimize, and modify these methods in ways that are the most suitable (best fit) to their projects, organization, and internal practices and capabilities.

REFERENCES

Fisher, R. (2022), *Understanding Mental Models*, Kindle Direct Publishing

EI (2020a), *Guidance on Human Factors Safety Critical Task Analysis*, Energy Institute, London, 2nd Ed, January 2020

IEC 61511-1 (2016), *Functional Safety – Safety Instrumented Systems for the Process Industry Sector – Part 1: Framework, Definitions, System, Hardware and Application Programming Requirements*, 2nd Ed, International Electrotechnical Commission

Johnsen, S.O., Kilskar, S.S., Fossum, K.R. (2017), Missing focus on human factors – organizational and cognitive ergonomics – in the safety management for the petroleum industry, *Journal of Risk and Reliability*, 231(4), 400–410

NSIA (2021), Framework and analysis process for systematic safety investigations, 3rd Ed., Norwegian Safety Investigation Authority

NUREG (2016), *Cognitive Basis for Human Reliability Analysis, NUREG-2114*, Whaley, A.M., Xing, J., Boring, R.L., Hendrickson, S.M.L., Joe, J.C., LeBlanc, K.L., Morrow, S.L., Office of Nuclear Regulatory Research, U.S. Nuclear Regulatory Commission, Washington DC

Shorrock, S., Williams, C. (2017), *Human Factors & Ergonomics in Practice, Improving System performance and Human Well-Being in the Real World*, CRC Press

SPE (2015), Safety critical tasks: Identification of human error to control risks from major accidents, Petrie, SPE-175449-MX, Society of Petroleum Engineers, Presented 2015 @ SPE Offshore Europe Conference & Exhibit, Aberdeen Scotland

Appendix A
Project Execution and Planning

This appendix is organized as follows:

- A.1 Lifecycle Phases – Project Stage Gate Planning
- A.2 Assessments and Quality Processes
- A.3 Procurement
- A.4 Verification, Validation, Inspection, and Testing
- A.5 Procedures and Training
- A.6 Remote Barrier Support and Remote Operations Center
- A.7 Cognitive Assessment and Mitigation Process
- A.8 Example Execution Challenges

Success in planning for and performing the activities that implement the suggested lifecycle model begins with developing a viable execution plan. The model introduces new processes and methods and suggests changes to the timing of several proposed activities. Each presents different challenges to consider and address in the execution plan and planning stages. This appendix provides information and suggestions on selected topic areas. The purpose of this guidance is to provide a means to contribute to a larger design goal, such as including the appropriate personnel in key activities to reduce the work-as-imagined and work-as-done gap.

Section A.8 summarizes some of the challenges when implementing the prototype lifecycle model. These will be familiar to those working in large or complex capital projects or were early implementers of new industry practice standards.

A.1 LIFECYCLE PHASES – PROJECT STAGE GATE PLANNING

The typical stage-gated process for capital projects requires planning, budgeting, and other information included in a project plan that guides the next project phase. Common plan elements are as follows:

- Identify the input information and guidance needed to progress and guide the work.
- Define the work scope, including work product deliverables and supporting activities. (Both are addressed in the model.)
- Identify the human resources needed to perform the work. Include an organization chart that shows the respective roles and interfaces.
- Develop a time and cost estimate (budgets) for each deliverable and activity.
- Identify requisitions or contracts to acquire resources such as a contracted facilitator or scribe, cognitive ergonomics specialist, or an independent assessor to perform the verification process.

- Develop a preliminary schedule that shows activity start dates and durations, estimated issue deliverable dates, and requisition and contract dates. The schedule must align to activities in the primary facility schedule. (Section A.8.3 discusses schedule challenges that create non-optimal, out-of-sequence work scenarios. Issues of this type should be considered and reflected in the preliminary schedule and execution plan.)

A.2   ASSESSMENTS AND QUALITY PROCESSES

This section provides example execution guidance specific to assessments and quality processes included in the prototype lifecycle model.

A.2.1   SCOPE AND APPLICATION

The following tables summarize the included assessment and quality processes:

- Table 3.4. Preliminary Design: Studies and Assessments
- Table 4.3. Detailed Design: Studies and Assessments
- Table 5.2. Implementation: Studies and Assessments
- Table 6.2. Construct, Install, and Commission Phase: Assessments
- Table 7.2. Operate and Maintain Phase: Assessments

A.2.2   USE OF A FACILITATOR AND SCRIBE

Some assessments may require a facilitator and scribe, such as the task analysis. Using a facilitator and scribe may be warranted if the activity:

- Includes a diverse group with many participants in a live setting.
- Is constrained by a limited participation window for key participants (e.g., operations specialists) and should be completed within that window.
- Generates a significant amount of information that is captured and immediately displayed to the group to confirm accuracy (scribe).

The facilitator guides a structured process that engages the team to identify and address areas of interest, manages time, and maintains team focus. The scribe accurately captures key information. The facilitator is guided by a scope of work and terms of reference, documents that define the scope, objectives, methodology, information to be captured, and the form and content of the assessment report. If the facilitator and scribe roles are outsourced, procurement activities should be proactively planned and budgeted.

A.2.3   PARTICIPANTS

Tables A.1–A.4 identify possible assessment and monitoring process participants for distinct stages in the barrier lifecycle. (The tables assume a process safety engineer has the oversight or coordination role that maintains alignment with requirements in the barrier safety management plan and safety requirements specification.)

Note: A cognitive system engineer (CSE) has knowledge and expertise on the material in Appendices C and F. The CSE may have different job titles such as "human factors engineer" or "human performance engineer."

TABLE A.1
Example Assessment Participants – Preliminary Design (Phase B)

Analysis or Assessment Description	Design Process	Facilitator/Scribe	Ops Specialist	Barrier HE (Lead)	Barrier HE (Other)	Technical Disc.	Facility Engineer	Process Safety	Cognitive Sys. Engr.	Human Factor Engr.	Training Dept.	Human Resources	O&M Leadership	Safety Dept.
Task analysis	B3-1	R	A	A				R	A/C	C				
Verification #1	B3-3, App. G.5					C		R						
Task phase analysis	B-5 to B-14		A	A	C	C	C	R	A/C	A/C				
Non-technical skills	B-13		A	C				R	C	C				
Functional allocation	B-15		A	A		A	C	R	C	C				
Dependency assessments	B-17, B-18		CA			A		R						
Reliability assessments	B-19		A	A			C	C	R					
Verification #2	B-21, App. G.5		A				C	R	I	I				
Validation #1	B-22, App. G.6		A				A	R	I	I				
Staffing and roles	B-25	R	C	C									C	C
Fatigue management	B-26	R	C	C				C					C	C
Design review	B-27, App. G.2	TBD												

(Row label at left margin: B – Preliminary Design)

Note: R – primary responsibility to organize, facilitate, document, and issue study/assessment results and manage changes; A – attend/participate; C – consult; I – inform.

TABLE A.2

Example Assessment Participants – Detailed Design and Implementation (Phase C)

	Analysis or Assessment Description	Design Process	Facilitator/Scribe	Ops Specialist	Barrier HE (Lead)	Barrier HE (Other)	Technical Disc.	Facility Engineer	Process Safety	Cognitive Sys. Engr.	Human Factor Engr.	Training Dept.	Human Resources	O&M Leadership	Safety Dept.
C – Detailed Design	PIF assessment	C-2, App. H.2	R	A	A		C/A	A	R	C	A				
	Phase safety time	C-3 to C-8		A	A		C	C	R	C	C				
	Cognitive assessment	C-3 to C-8		A	A		C	C	A	R	I				
	PIF – component WE factors	C-3, C-7, C-8	TBD	A	C/A		C/A	C	R	I	A				
	HE gap assessment	C-16		A	A	A			R	C					
	Functional analysis	C9-1, C10-1, C11-1		A	A	A	C	C	R	C	C				
	Verification #3	C-19, App. G.5		C			C	C	R	I	I				
	Validation #3	C-20, G.7		A	C				R	C	C				
	Design review	C14-7, App. G.2							TBD						
C – Implementation	Design review	C30-2, C31-2, App. G.2							TBD						
	Verification #3	C30-4, C31-4, App. G.5		C			C	C	R	I	I				
	Validation #3	C30-7, C31-7, App. G.6		A	C				R	C	C				
	Formal acceptance inspection (FAI)	C30-8, C31-8/9, App. G.3			C			R	I	I	I				
	Factory Acceptance Test (FAT)	C30-8, C31-9, App. G.4			C/A		R	R	I	I	I				
	Integrated FAT (IFAT)	C30-8, C31-9, App. G.4			C		R	R	I	I	I				
	Verification #4	C-35, App. G.5		A	A	C	C	C	R	I	I				

Note: R – primary responsibility to organize, facilitate, document, and issue study/assessment results and manage changes; A – attend/participate; C – consult; I – inform.

TABLE A.3
Example Assessment Participants – Construct, Install, and Commission (Phase D)

D– Construct, Install, Inspect, and Test

Analysis or Assessment Description	Design Process	Ops Specialist	Barrier HE (Lead)	Barrier HE (Other)	Technical Disc.	Facility Engineer	Process Safety	Operations Dept.	Maintenance Dept.	Cognitive Sys. Engr.	Training Dept.	Human Resources	O&M Leadership	Safety Dept.
Periodic inspections	D1-4, D2-3, D4-3, App. G.3		C			R	I			I				
Verification #4	D1-7, D2-6, D4-6, App. G.5		C		C	C	R	C	C					
Validation #3	D1-5, D2-4, D4-4, App. G.6		R	A		R	R			C				
Formal acceptance inspection (FAI)	D1-6, D2-5, D4-5, App. G.3		C			R	I	C/A	C/A	I				
System site acceptance test (S-SAT)	D2-8, D3-3, D4-8, App. G.4		C/I		A	A	I	R		I				
HE competency assessment	D6-2		C/I	C/I			I	I		I	R			
Verification #5	D-7, App. G.5		C	C	C	C	C	R	C	C	C		C	
Barrier site acceptance test (B-SAT)	D-8, App. G.4	A	A	A			I	R	I	A	C			
Validation #4	D-9, App. G.6	A	C	C			I	R		C	I			
Pre–start-up readiness check	D-10		A	C	A	A	A	R	A	C	C		C	

Note: R – primary responsibility to organize, facilitate, document, and issue study/assessment results and manage changes; A – attend/participate; C – consult; I – inform.

TABLE A.4

Example Assessment Participants – Operate and Maintain (Phases E, F, and G)

Phase	Assessment or Assurance Description	Design Process	Ops Specialist	Barrier HE (Lead)	Barrier HE (Other)	Technical Disc.	Facility Engineer	Process Safety	Operations Dept.	Maintenance Dept.	Cognitive Sys. Engr	Human Factors Engr.	Training Dept.	Human Resources	O&M Leadership	Safety Dept.
E – Operate & Maintain Phase	Monitor bypass use	E-1		C	C			I	R	C						C
	Roster/Org chart change	E2-1, E5-1		C/I	I			I	R	R				C	I	
	Training update	E2-2, E5.2		I	I			I	I	I			R		I	
	Monitor procedures	E2-3, E5.3		C	C			I	R	R					I	
	Monitor fatigue	E2-4, E5-4		C	C			I	R	R					I	
	Monitor incidents	E2-5, E5-5	A/C	A/C	A/C			I	A/C	A/C	A/C	A/C			I	R
	Monitor add'l resources	E2-6		I	I	C/I	C/I	I	R	I					I	
	Maintain competency	E3-1, E6-1		I				I	C/I	C/I	C/I		R		I	
	Fitness for service	E3-2, E6-1		I	I			I	I	A				R	I	
	Monitor preventative maintenance	E4-1						I	R	R					I	
	Monitor inspections and proof tests	E4-2						I	R	R					I	
	Monitor repair & restore	E4-3						I	C	C					I	
	B-SAT	E-8 App. G.6	A	R	A	C	C	A	R		C/I	C/I			I	
	PIF assessment	E-9 App. H.2	A	A	C	C	C	R	C	C	C/A	A/C			I	
	Verification #6	E-10 App. G.5	C/I	C	C	C	C	R	C	C	C	C	C		I	
F	Change management	F								TBD	TBD					
G	Decommissioning	G								TBD	TBD					

Note: R – primary responsibility to organize, facilitate, document, and issue study/assessment results and manage changes; A – attend/participate; C – consult; I – inform.

A.3   PROCUREMENT

See Appendix E for suggested execution models and guidance that apply to the procurement of equipment, systems, and buildings that contain barrier system elements.

A.4   VERIFICATION, VALIDATION, INSPECTION, AND TESTING

See Appendix G for execution guidance that applies to the various verification, validation, inspection, and testing activities included in the prototype lifecycle model.

A.5   PROCEDURES AND TRAINING

See Appendices J.2.2 and J.3.2 for execution guidance that may apply to procedures and training development and delivery activities.

A.6   REMOTE BARRIER SUPPORT AND REMOTE OPERATIONS CENTER

See Appendix L for execution guidance that may apply to Remote Barrier Support and a Remote Operations Center activity.

A.7   COGNITIVE ASSESSMENT AND MITIGATION PROCESS

See Appendix I.4 for execution guidance that applies to the cognitive assessment and mitigation process.

A.8   EXAMPLE EXECUTION CHALLENGES

A.8.1   INCREASED RELIANCE ON OPERATIONS EXPERTISE AND PERSONNEL

Unsurprisingly, progressing a user-centered design methodology requires greater input from the end-user (e.g., the operations and maintenance personnel assigned to perform, support, and maintain barrier system functions and elements). These personnel are commonly in high demand. Commitment and follow-through from the owner/operator (or others) to provide them is needed. Adaptations of the prototype lifecycle model may be needed to address constraints regarding when personnel can be made available (which may conflict with the proposed timing and duration of the activity.)

Key Point: Operations specialists with the necessary knowledge, experience, and expertise should participate in the task analysis (step B3-1) and task requirements definition steps B-4 to B-16 (preliminary design phase). These steps create the foundation on which all remaining processes are built. As a requirement, an operations specialist should participate in the task analysis, the starting point for active human barrier design. If not performed in the preliminary design phase, the last remaining opportunity to do so occurs in the detailed design process C-1, which updates the entire preliminary design. A possible but less desirable option is to have the operations specialist review the product of these processes to include his/her input into a design guided or defined by less qualified personnel. A concern with this approach is a tendency to comment only on what is written and not on what may be missing.

A.8.2 New Expertise Requirements (Cognitive Ergonomics)

Some activities in the prototype lifecycle model (e.g., the cognitive assessment and mitigation process in Appendix I) require expertise in cognitive ergonomics. This skillset should encompass the information in Appendices C, F, and I. In the near term, this competency may be in short supply globally since it is not one that a traditional human factors engineer typically has.

Note: The requirement for personnel with new expertise is not an unfamiliar issue. A similar challenge occurred with the issue of the IEC 61511-1 standard, which included requirements for the new discipline/expertise (a functional safety expert). While personnel with this expertise were in short supply for many years, with the increased demand the gap slowly closed. The same may be true for those with expertise in cognitive ergonomics. Early demands may rely on outsourcing to acquire this expertise.

Project managers are often hesitant about adding a new resource type that is unfamiliar. Owner/operators may need to press a project management organization to acquire the necessary expertise to achieve the intended outcome.

A.8.3 Out-of-Sequence Work

Increasingly common, capital projects have compressed project schedules. Compressed schedules often create challenging situations that require a technical discipline to perform work at a non-optimal time and use non-optimal methods to generate input information required by others. This can occur with early procurement and other early-stage key project activities, for example, the base facility layout and major equipment placement. Procurement may begin in the preliminary design phase for items that have a long delivery schedule, or the project needs design information from the vendor (e.g., equipment dimensions) to progress a larger project activity, such as the initial base facility layout and location plans. These situations create out-of-sequence work scenarios. When they occur, barrier-related work may need to start early. That effort may place a high reliance on the best available information at the time, experience, and judgment. Input information to others may include barrier system requirements that should be integrated into the facility layout, technical and packaged equipment systems, external support systems, and external protective barriers. Execution success requires early planning and (possibly) early staffing. That effort includes tracking and maintaining awareness of what others need and the date that information is needed. It may also require knowledge to understand which input information has the greatest importance, and to maintain awareness of key design freeze dates that can prevent a future change (or correction) to a critical barrier design aspect.

A.8.4 Interacting with Other Disciplines and Organizations

As discussed elsewhere in this book, barrier system elements often reside in many equipment items, systems, buildings, and facility areas. Thus, other disciplines and organizations may have primary responsibility for their design and oversight. Progressing the barrier effort requires interacting and coordinating with these varied persons and organizations to gain information, provide them with the barrier requirements (those aspects that affecting their scope of supply), and the provided requirements are understood, captured, and appropriately implemented. Coordination is needed to support the various assessments and verification and validation processes. The interactions should be timely, productive, and mutually efficient (time and effort). This work may start by identifying the responsible parties and acquiring basic information such as key schedule milestones and the potential challenges with each interface. The effort may then move to confirming the processes, agreements, and potential constraints for inserting barrier requirements into the appropriate technical, project, and commercial documents. It also includes agreeing on who participates in key activities like design reviews and assessments. Preferably, these activities are codified in the execution plan of the larger project. If not, progress may increasingly become more dependent on developing the necessary project (person-to-person) relationships that enables that work.

Appendix B
Technical Design, Resources, and Topics

This appendix is organized as follows:

- B.1 Introduction
- B.2 Barrier Safety Management Plan (BSMP)
- B.3 Safety Requirements Specification (SRS)
- B.4 Remote Barrier Support Design Basis
- B.5 Technical Design Topics
- B.6 CX Sub-Process – PE Component-Level Design
- B.7 Source Inputs to Design, Project Documents, Procurement, and Quality
- B.8 Support Aids (Job Aids)

B.1 INTRODUCTION

This appendix includes various technical topics and information, design processes (components), and useful resource tables.

B.2 BARRIER SAFETY MANAGEMENT PLAN (BSMP)

This section identifies content that could be included in the proposed barrier safety management plan (BSMP). Many of the model lifecycle processes may provide additional guidance on the suggested content and purpose of this document.

Note: Guided by IEC 61511-1:2016 (the global standard for a different active barrier type), the prototype lifecycle model includes the BSMP and safety requirements specification. Both are essential for managing the barrier system over its full lifecycle from conceptual design up to and including decommissioning.

1. **Purpose of This Document**
   - To describe the document's purpose and objectives.
   - To describe the facility where the listed barriers reside.
2. **Scope**
   - To identify the barriers included in this plan.
   - To identify the lifecycle phases included in this plan.
   - To identify the organizations and parties included in this plan.

3. References

The following documents, programs, and policies integral to this plan serve as references:

- Safety requirements specification (identified barriers)
- Owner/operator safety management system
- Owner/operator management programs (e.g., staffing and staff management, fitness for service, competency, training, and procedure management)
- Information repositories that maintain and manage the barrier system records
- Change management system

4. Safety Management System

Provide an overview of the facility's safety management system and describe its application to the development, management, and performance of this plan.

5. Technical Safety Planning

Identify the activities required to update, maintain, and successfully implement the planning and plan for each lifecycle phase. Planning for the next lifecycle phase may include the following:

- Updating this document, as required, to define and guide the required and expected phase activities, timing, and responsibilities.
- Identifying and defining the required phase activities, expected outputs, and desired results.
- Identifying the resources and organizations required to implement and support the identified plan activities and their assigned responsibilities.

6. Management of Technical Safety

6.1 Function of the Safety Requirements Specification

Include a statement on the purpose and function of the safety requirements specification (SRS) as it applies to this plan.

- a. The SRS is the single document that defines and specifies the entirety of each barrier system.
- b. The SRS and referenced documents should all be maintained and managed as a version-controlled document.
- c. The SRS, essential to barrier lifecycle management, provides the definitive reference to determine whether:
  - i. A finding from a quality process is a deviation that requires corrective action to return the barrier element to its original state and condition.
  - ii. A proposed modification that changes any information in the SRS, a condition that warrants using the change management process.

6.2 Management System Activities and Requirements

Identify and delineate the activities, responsibilities, programs, and systems employed to achieve, manage, and maintain each barrier system.

- a. Barrier System Requirements: Identify the documents and sources that comprehensively and explicitly define each barrier system.

b. Responsible Organization, Department, and Roles
    i. Include an organization chart that identifies every organiza-
        tion, department, and role that has a specified responsibility
        for the barrier system or barrier-dependent equipment, systems
        and workspaces.
    ii. Identify the responsibilities and activities assigned to each
        named organization, department, and role.
c. Verification and Validation: Define the activities and responsibili-
    ties that verify and validate the specified barrier system confor-
    mance and usability requirements. Activities may include formal
    design reviews, inspections, functional and performance testing,
    and verification and validation assessments.
    i. Identify each required activity that will be the basis for verify-
        ing and validating the barrier system. This includes the O&M
        phase, and monitoring activities identified in Chapter 7 and
        indicated in Figure 7.1.
    ii. Define the objective and purpose of each activity.
d. Competency Management: Define the competency requirements
    and management system.
    i.   Provide an overall statement of the objectives and processes of
         the competency management system.
    ii. For every identified role in b above, define each required
         competency.
    iii. Identify how each competency is verified, for example, through
         an observed drill or computer-aided test.
    iv. Identify when competency verification should occur, for exam-
         ple, prior to commencing a defined activity that requires a
         specified competency.
    v.   Identify the actions and procedures that apply when a required
         competency is not met.
e. Procedure Management
    i. Identify the names of programs and standards used to develop,
        verify, monitor, and maintain procedures.
    ii. Define the system used to monitor, update, and manage proce-
        dures in each phase of the barrier lifecycle.
f. Training Development and Management
    i.   Identify the named programs and defined practices used to
         develop, verify, and deliver training.
    ii. Define the system and processes used to track, maintain, and
         manage the training and training modules employed to achieve
         and maintain the identified barrier competencies.
    iii. Define requirements for periodic retraining to address skill
         fade and drift.
g. Maintenance and Repair
    i. Identify the source of the barrier system maintenance require-
        ments and management.

    ii. Identify the minimum expected practice for maintenance activities that may take a barrier element out of service during these activities.

B.3 SAFETY REQUIREMENTS SPECIFICATION (SRS)

Included in this section:

* B.3.1 Suggested SRS Contents
* B.3.2 Alignment to IEC 61511-1 (2016, cl 10.3.2)
* B.3.3 Suggested SRS Issue Timing

This section suggests content to include in the proposed safety requirements specification (SRS). The prototype lifecycle model includes additional content suggestions and guidance.

Note: Guided by IEC 61511-1:2016, the prototype lifecycle model includes this document and the barrier safety management plan. Both are essential for managing the barrier system over its full lifecycle from conceptual design up to and including decommissioning.

The SRS is the singular, comprehensive document that specifies every aspect of a barrier system and does so directly or by reference. In addition to function, performance, and system requirements, it includes the quality and operate/maintain phase requirements. The SRS is the essential resource when considering a potential change to any element in the barrier system (PE, OE, or HE) or in an external support system, external protective barrier, or workspace on which it depends. A change to any content or requirements included or referenced in this document can potentially degrade or contribute to barrier failure. Therefore, all such changes should be addressed using the change management process.

Note: The amount of information for an active human barrier may be considerably greater than that for other active barrier types. The capture and management of this information in a secured database form may reduce the time needed to develop, maintain, distribute, and manage the SRS and is contents.

B.3.1 SUGGESTED SRS CONTENT

1. Purpose of This Document
* To describe the document's purpose and objectives.
* To describe the facility where the listed barriers reside.

2. Scope
To identify the barriers included in this specification.

3. References
General applicable requirements:
* Regulatory and statutory requirements
* Industry standards
* Company standards and practices

Documents that provide or define barrier-specific requirements include the following:

- Risk studies (preliminary design step B-2)
- Task analysis and report (preliminary step B3-1)
- Specifications
- Functional requirements documents (e.g., cause-and-effect charts, logic diagrams, and control narratives)
- Data sheets
- Equipment lists
- Drawings (e.g., general arrangement, layout, location, installation, interconnection, termination)
- Assessments and reports
- Deficiency and corrective action reports with final dispositions

Note: For a complete listing of source documents, see Summary Tables 3.2–3.4, 4.1–4.3, 5.1, 5.2, 6.1, 6.2, 7.1, and 7.2.

4. Common Requirements

This section defines the common standard requirements that apply to all barriers included in this document.

1. Basic/common design requirements:
   a. Barrier and task activators
   b. Direct-use PE
   c. Support PE
   d. Bypass, override, and inhibit functions
   e. Fault detection and notification
   f. Fault tolerance and voting design
   g. Failure response (e.g., fail-to-safe and fail-to-danger designs)
   h. System and component-level reliability, availability, and integrity requirements and implementation methods
   i. Requirements and design to support online testing and maintenance
2. Component selection, design, and base requirements (see Table B.9 in Appendix B.7.2 for examples)
3. Reliability, availability, and integrity requirements (see Tables B.12 and B.13 in Appendix B.7.5 for examples)
4. Survivability requirements (see Table B.14 in Appendix B.7.6 for examples)
5. Common display requirements
6. Workspace design requirements
7. General performance standards
8. Organizational elements (design basis)
   a. Common procedures: assessment, quality, operate, maintain, change management (version-controlled documents)
   b. Staff policies and plans (e.g., rotation, overtime)
   c. Maximum fatigue limits
9. Human element requirements (design basis)
   a. Fitness for service
   b. Competency compliance

10. Accepted methods to realize and quantify achievement of a specified Safety Integrity Level or similar reliability requirement

5. Barrier Systems (separate section for each barrier)
5.1 Barrier System

This section defines the overall barrier system function and performance requirements. Suggested system information includes the following:

1. Barrier origination (Table 3.5)
2. Requirements and information noted in Section B.3.2 (this appendix)
3. Barrier organization chart and roster (roles)
4. Additional requirements for the barrier leader (e.g., leadership and monitoring requirements)
5. Display requirements and design
6. Every barrier system document and drawing, for example:
   a. Applicable risk assessment information and recommendations
   b. Task analysis and report (step B3-1)
   c. Specifications
   d. Data sheets
   e. Equipment lists
   f. Drawings
   g. Certifications
   h. Assessments and reports
   i. Deficiency and corrective action reports with final dispositions
7. Performance standards
8. Barrier-specific organizational elements (design basis)
   a. Organizational chart and roster (indicate primary and backup roles, normal locations, etc.)
   b. Competency requirements (each HE/task) and status
   c. Training records (e.g., modules, dates)
   d. Procedures (e.g., those required for the operate and maintain phase, change management, version-controlled documents; see Tables 4.5, 4.30, B.5, and B.6)
9. Performance standards (see Tables 3.7 and 3.32) and other performance measures (e.g., reliability, availability, integrity, and survivability)
10. Environmental extremes at each plausible HE location (see direct-use PE locations)
11. Operating and maintenance procedures (refer to unique procedure identifiers for these version-controlled documents)
12. Barrier dependencies: Identify systems, endurance time, monitoring requirements
    • External support systems (see Table 3.28)
    • External protective barriers (see Table 3.29)
    • Additional resources (see Table 3.33)
    • Building including rooms and walk-in shelters (see Tables 4.20 and 4.23)
    • Engineered area (see Table 4.22)

DETECT (SA-1)

Monitor SA-1 Information
- Barrier / task activator
- All SA-1 info to support SA-2, 3, SSA, decisions, act response, monitor feedback, etc.

Source and Access Locations for all specified SA-1 Information

Physical Elements
- Non-VDU displays, VDU-based displays, communication equipment, etc.
- Engineered area
- In-Place PE: Alarm system, VDU-based display systems, etc.
- Monitor support PE

Organizational Elements
- New knowledge & skills requirements
- Procedures, training, etc.

Dependencies
- External support systems.

Phase Target Response Time
Performance Standards

DETECT (SA-2, SA-3)

SA2-Comprehension
Specify comprehension requirements

SA3-Project Future State
Specify projection requirements.

Organizational Elements
- See Detect (SA-1)
- Non-technical skills

Phase Target Response Time
Performance Standards

DECIDE

Decisions
Define all required decisions

Organizational Elements
- See Detect (SA-1)
- Non-technical skills

Phase Target Response Time
Performance Standards

ACT

Action Response(s)
Define all required actions, including outbound communications

Physical Elements
- Direct use: fire hose, radio, etc.
- Support: smoke hood, self-contained breathing apparatus, etc.
- In-place: technical system, public alarm system, etc.
- Engineered area

Organizational Elements
- See Detect (SA-1)
- New knowledge & skills: PE usage

Dependencies
- External barriers, e.g., blast wall
- External support systems, e.g., backup power system or firewater system

Phase Target Response Time
Performance Standards

TASK GOAL: Enter the task goal
Task ID: Enter a unique task ID
Task Assignee: Person assigned to this task, responsible for task success
Task Activator: what triggers the activation of this task
Target Task Response Time (TRT): Target time to complete task

---

BARRIER INFORMATION

Barrier ID: Enter barrier ID
Barrier Function: Enter barrier function and safe state
Barrier Activator: Enter barrier activator
Barrier Response Time: Enter BRT

Team Coordination & Communication
Unique Exchange ID: Inbound Task ID(s):   Outbound Task ID:
Exchange Message Content: enter msg   Conveyance, e.g., 1-way, 2-way, broadcast to all
Exchange Form: e.g., verbal, video   Exchange Media: e.g., radio, DCS display, face-to-face
Dependence on External Support System: e.g., emergency power
Organizational Elements: Procedures, training, drills, non-technical skills, etc.   Performance Standards

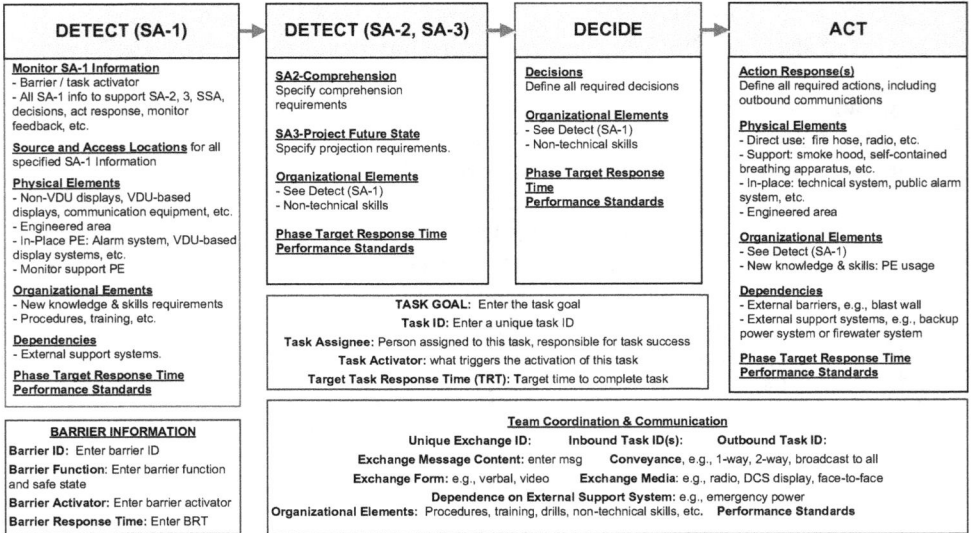

FIGURE B.1   Overview of Barrier and Barrier Task Requirements.

- Cable tray and support systems (see Appendix B, sub-process CX-01g)

13. Interfaces to other systems (e.g., In-Place PE, dependent systems)

5.2 Barrier Task (separate section for each barrier task)

This section defines the individual barrier task function and performance requirements. Suggested system information includes the following:

1. Task origination requirements (see Table 3.6).
2. Task block diagrams (see preliminary design process B-15).
3. Task and team requirements (see Tables 3.2 and 4.1 for a summary of requirements).
4. Direct-use PEs, for example:
   - Displays (see Tables 3.25, 4.6, 4.7, 4.8, 4.10–4.17).
   - Support aids (see Tables 3.8, 3.12, 3.17–3.19, 3.25, 4.13).
   - Act phase response PE (see Tables 3.9, 3.12).
5. Listing of all direct-use PE locations (detect and act phase PE).
6. Listing of Support PE and use locations (see Tables 3.9, 3.13).
7. List all plausible HE locations (see above mentioned tables for direct-use PE use locations).
8. Figure B.1 summarizes much of the content to include for each barrier task.

B.3.2 ALIGNMENT TO IEC 61511-1 (2016), CL. 10.3.2

IEC 61511-1 (2016, clause 10.3.2) identified 29 requirements to include in the SRS. Table B.1 indicates where each is addressed in the prototype lifecycle model.

TABLE B.1

Mapping to SRS Requirements in IEC 61511-1 (2016, clause 10.3.2)

Safety Requirement Specification

Requirements from IEC 61511-1 (2016, cl. 10.3.2) *(Paraphrased unless indicated as a direct quote)*	Source Information from Lifecycle Model (processes, steps)	
	Preliminary Design Phase B	**Detailed Design Phase C**
1. Description of the SIF (active human barrier) that achieves the safety function, e.g., a logic narrative, or cause-and-effect chart	Tables 3.5, 3.6 (B2-5, B3-2), barrier block diagram (B-15), cause-and-effect charts (B20.1). See SRS Section 5 (B20-3.1)	Updated preliminary design (C-1). From processes C-3, C-7, C-8 (invokes sub-process CX-0.1c in App. B.6). Updated SRS Section 5 (C-17).
2. A list of the input and output devices connected to each SIF (active human barrier function), e.g., field device tag list	Tables 3.9, 3.11, 3.25 (B-6, B-7, B-14), barrier system block diagram (B-15), cause-and-effect charts (B20.1), process and instrument diagrams (P&IDs)	Updated preliminary design (C-1). Updated SRS Section 5 (C-17).
3. Identify sources of common cause failure	Tables 3.30, 3.31 (B-19, reliability studies)	Updated preliminary design (C-1). Updated SRS Section 5 (C-17).
4. Define the safe state that achieves the safety function	Table 3.5 (B2-x)	Updated preliminary design (C-1). Updated SRS Section 5 (C-17).
5. Define "…individually safe process states which, when occurring concurrently, create a separate hazard (e.g., overload of emergency storage, multiple relief to flare system)"	Tables 3.5 (B2-x) and 3.30, 3.31 (B-19, reliability studies)	Updated preliminary design (C-1). Updated SRS Section 5 (C-17).
6. Define "…the assumed source of demand and demand rate…" on the barrier system	Risk assessment reports (B2-x)	Updated preliminary design (C-1). Updated SRS Section 5 (C-17).
7. "Requirements related to proof test intervals" 8. "…requirements relating to proof test implementation"	Tables 3.30, 3.31 (B-19, reliability studies). See SRS Section 5 (B20-3.1)	From processes C-3, C-7, C-8 (Each invokes sub-process CX-0.1h in App. B.6). Updated SRS Section 5 (C-17).

(Continued)

TABLE B.1

(Continued)

Safety Requirement Specification

Requirements from IEC 61511-1 (2016, cl. 10.3.2) *(Paraphrased unless indicated as a direct quote)*	Source Information from Lifecycle Model (processes, steps)	
	Preliminary Design Phase B	**Detailed Design Phase C**
9. Response time requirements to achieve the required safe state within the process safety time	Tables 3.5, 3.27 (B2-x, B-16)	Updated preliminary design (C-1). Updated SRS Section 5 (C-17).
10. "…the required SIL and mode of operation (demand/continuous) for each SIF"	Table 3.5 (B2-x). See SRS Section 5 (B20-3.1)	Updated preliminary design (C-1). Updated SRS Section 5 (C-17).
11. "…a description of SIS process measurements, range, accuracy, and their trip points"	Instrument data sheets. See SRS Section 5 (B20-3.1)	Updated preliminary design (C-1). Updated SRS Section 5 (C-17).
12. "…a description of SIS process output actions and the criteria for successful operation, e.g., the leakage rate for valves…"	Tables 3.5 (B2-x) and 3.7 (B-4, base performance std.). See SRS Section 5 (B20-3.1)	Updated preliminary design (C-1). Tables 3.7, 3.32 (performance standards.) From process C-3 (invokes sub-process CX-0.1c/h/j in App. B.6). Updated SRS Section 5 (C-17).
13. "…the functional relationships between inputs and outputs including logic, mathematical functions and any required permissives…"	Barrier block diagram (B-15) cause-and-effect charts, logic diagrams, control narratives (B20-1). See SRS Section 5 (B20-3.1)	Updated preliminary design (C-1) Voting design from processes C-3, C-7, C-8 (invokes sub-process CX-0.1c/h in App. B.6). Updated SRS Section 5 (C-17).
14. "…requirements for manual shutdown for each SIF…"	Tables 3.7 (B-4, base performance std.), and 3.30, 3.31 (B-19, reliability studies). Recorded in SRS Section 5 (B20-3.1)	Updated preliminary design (C-1). Updated SRS Section 5 (C-17).
15. "…requirements relating to energize or de-energize to trip…"	P&IDs, cause-and-effect charts (B20-1). See SRS Section 5 (B20-3.1)	Updated preliminary design (C-1). From processes C-3, C-7, C-8 (invokes sub-process step CX-0.1c/h in App. B.6). Updated SRS Section 5 (C-17).

(Continued)

TABLE B.1

(Continued)

Safety Requirement Specification

Requirements from IEC 61511-1 (2016, cl. 10.3.2) *(Paraphrased unless indicated as a direct quote)*	Source Information from Lifecycle Model (processes, steps)	
	Preliminary Design Phase B	**Detailed Design Phase C**
16. "…requirements for resetting each SIF after a shutdown (e.g., requirements for manual, semi-automatic, or automatic final element resets after trips)."	Cause-and-effect charts, logic diagrams, control narratives (B20-1). See SRS Section 5 (B20-3.1)	Updated preliminary design (C-1). From processes C-3, C-7, C-8 (invokes sub-process step CX-0.1c/h in App. B.6). Updated SRS Section 5 (C-17).
17. "…maximum allowable spurious trip rate…"	Tables 3.7 (B-4, base performance std.), and 3.32 (B20-3.2, performance standards). See SRS Section 5 (B20-3.1)	Updated preliminary design (C-1). Tables 3.32 (C-13). Updated SRS Section 5 (C-17).
18. "…failure modes for each SIF, and the desired response of the SIS, e.g., alarm, automatic shutdown…"	P&IDs, cause-and-effect charts (B20-1). See SRS Section 5 (B20-3.1)	Updated preliminary design (C-1). Updated SRS Section 5 (C-17).
19. Identify requirements for procedures for starting up and restarting the barrier system, for example, components, technical systems.	–	Tables 4.5, 4.30, 4.31, B.5, B.6. From process C-15 (procedures). Updated SRS Section 5 (C-17).
20. All interfaces between the SIS and any other system (including the BPCS and operators)	Barrier Function block diagram (B-15). See SRS Section 5 (B20-3.1)	Updated preliminary design (C-1). Updated SRS Section 5 (C-17).
21. "…a description of the modes of operation of the plant and requirements…" of the barrier when operating in each mode	Table 3.5 (step B2-x). See SRS Section 5 (B20-3.1)	Updated preliminary design (C-1). Tables 4.5, 4.30, B.5, B.6 (C-15). Updated SRS Section 5 (C-17).
22. "…the application program safety requirements specification…"	Not specifically addressed in the lifecycle model processes, though may be appropriate for atypical and complex software applications	

(Continued)

TABLE B.1

(Continued)

Safety Requirement Specification

Requirements from IEC 61511-1 (2016, cl. 10.3.2) *(Paraphrased unless indicated as a direct quote)*	Source Information from Lifecycle Model (processes, steps)	
	Preliminary Design Phase B	**Detailed Design Phase C**
23. "…requirements for bypass including written procedures to be applied during the bypassed state which describe how the bypasses will be administratively controlled and then subsequently cleared…"	See SRS Section 5 (B20-3.1)	Updated preliminary design (C-1). Tables 4.5, 4.30, 4.31, B.5, B.6. Procedures (C-15). Updated SRS Section 5 (C-17).
24. "…the specification of any action necessary to achieve or maintain a safe state of the process in the event of fault(s) being detected in the SIS, taking into account all relative human factors…"	See SRS Section 5 (B20-3.1)	Updated preliminary design (C-1). Tables 4.5, 4.30, 4.31, B.5, B.6. Procedures (C-15). Updated SRS Section 5 (C-17).
25. "…the mean repair time which is feasible for the SIS…"	Tables 3.7 (B-4, base performance std.) and 3.32 (B20-3.2, performance standards). See SRS Section 5 (B20-3.1)	Updated preliminary design (C-1). Performance standards (C-13). Updated SRS Section 5 (C-17).
26. "…identification of dangerous combinations of output states of the SIS that need to be avoided…"	Tables 3.7 (step B-2) and 3.30, 3.31 (B-19, reliability studies). See SRS Section 5 (B20-3.1)	Updated preliminary design (C-1). Updated SRS Section 5 (C-17).
27. "…identification of extreme environmental conditions that are likely to be encountered by the SIS during shipping, storage, installation, and operations."	Basic environmental design data, instrument data sheets, specifications. Recorded in SRS Section 5 (B20-3.1)	Updated preliminary design (C-1), vendor data (C-18, procurement), C-3, C-7, C-8 (see CX-0.1h, App. B.6). Updated SRS Section 5 (C-17).

(Continued)

TABLE B.1

(Continued)

Safety Requirement Specification

Requirements from IEC 61511-1 (2016, cl. 10.3.2) *(Paraphrased unless indicated as a direct quote)*	Source Information from Lifecycle Model (processes, steps)	
	Preliminary Design Phase B	**Detailed Design Phase C**
28. "…identification of normal and abnormal process operating modes for both the plant as a whole (e.g., plant start-up) and individual plant operating procedures (e.g., equipment maintenance, sensor calibration or repair). Additional SIFs may be required to support theses operating modes…"	Tables 3.5 (step B2-x), 3.30, 3.31 (B-19, reliability studies), and 3.29 (B-18, external barriers). See SRS Section 5 (B20-3.1)	Updated preliminary design (C-1), Tables 4.5, 4.30, 4.31, B.5, B.6. Procedures, C-15. Table 4.21 (external barriers, C14-4). Updated SRS Section 5 (C-17).
29. "…definition of the requirements of any SIF necessary to survive a major accident event, e.g., time required for a valve to remain operational in the event of a fire."	Instrument data sheets and specifications, Tables 3.7 (B-4, base performance std.), 3.32 (B20-3.2, performance standards), and 3.30, 3.31 (B-19, reliability studies). See SRS Section 5 (B20-3.1)	Updated preliminary design (C-1). Performance standards (C-13). C-3, C-7, C-8 (see CX-0.1i, App. B.6). Tables 4.19, 4.20, 4.21, 4.22, 4.23 (C-9 to C-12). Updated SRS Section 5 (C-17).

Note: SIS is analogous to barrier system, SIF is analogous to active human barrier.

B.3.3 SUGGESTED SRS ISSUE TIMING

The SRS is a "living document" that undergoes periodic updates at different stages in the barrier system lifecycle. The prototype lifecycle model proposes the document be updated and issued at the following times or when the indicated events occur.

1. Nearing completion of the preliminary design phase (Figure 3.3, step B20-3.1)
2. Nearing completion of the detailed design phase (Figure 4.1, process C-17)
3. Nearing completion of the implementation phase (Figure 5.1, process C-36)

4. A change that modifies the requirements (e.g., change management process)
5. A change to a version-controlled document that is part of the barrier system or its dependencies. (A change to a version-controlled document may be a change to the barrier system, and thus warrants using the change management system.)

B.4   REMOTE BARRIER SUPPORT DESIGN BASIS

The barrier system may include one or more remote barrier support (RBS) tasks. The tasks are performed from a designated remote location. The term used to identify this location is *Remote Operations Center* (ROC). The RBS functions are performed by one or more HEs located at the ROC. To do so requires:

- An ROC facility, which includes its base infrastructure (e.g., electrical, lighting, environmental controls).
- PEs in the ROC used by an HE to perform the RBS-assigned tasks.
- Interfaces between the ROC and protected facility that enables the above-mentioned PEs at the specified performance.

This is a multi-disciplined, multi-organizational document.

1. **Purpose**
   This document provides the technical and functional design basis for the facilities and equipment needed to enable and support the performance of remote barrier support (RBS) tasks from a Remote Operations Center (ROC).
2. **Scope**
   The scope in this document is limited to only those requirements necessary to enable and support RBS tasks performed from the ROC. This scope includes the following:
   - The ROC and the equipment and systems needed to enable and maintain the specified capabilities. (Barrier-assigned personnel perform RBS tasks at this location.)
   - Technical systems and equipment located in the ROC that allow a barrier-assigned HE to perform RBS tasks from this facility.
   - The additional resources and real-time interfaces that connect the above systems and equipment to the source systems and equipment located on the protected facility.
3. **Applicable Regulations, Industry Codes and Standards, Other Documents**
   Examples of documents include the following:
   - Regulatory requirements and statutes
   - Industry codes and standards
   - Company standards and guidance documents
   - Barrier system documents that identify the detect and act response phase, direct-use requirements, and elements located at the remote location
   - Performance standards (applicable standards in Tables 3.7 and 3.32)

4. Design Philosophy and General Requirements

Describe the philosophy that defines the degree and manner to which the ROC facilities should duplicate or mimic those located on the protected facility. For example, this may apply to the physical layout and provisioning of the ROC and its included monitoring and control consoles, VDU-based display wall, and communications equipment.

Define the following:

- Basic design requirements for the ROC (e.g., facility location, accessibility, sizing, and layout). If required by the barrier task, include ancillary facility requirements (if any) such as a separate video conferencing room.
- ROC facility performance requirements (e.g., availability and survivability).
- Technical system requirements (e.g., availability, reliability, integrity, survivability).
- Additional resource requirements (see step B20-4 for background).
- Performance requirements that apply to the ROC, ROC-located technical systems and equipment, and the associated interfaces to the protected facility.

5. Organizational Requirements

Define the technical disciplines, departments, and organizations responsible for:

- Managing and maintaining the ROC facility and its dependent external support systems and external protective barriers (if any).
- Managing and maintaining the technical systems and equipment located in the ROC.
- Managing and maintaining the additional resources and technical system interfaces.
- Providing and managing the ROC-located HEs charged with performing RBS tasks.

6. Identify the Functional Capabilities From the ROC

Enabled by the ROC, identify the functional capabilities needed to support the RBS tasks identified in this document. For example:

- Communication access (TX/RX, receive only) to telecommunication systems (e.g., radio systems, telephone).
- View access (full, selective) to protected facility monitoring and control HMI displays.
- Control access (full, selected, none) to enter and execute real-time control actions at the abovementioned displays.
- View access (all, selected, none) to the protected facility's CCTV cameras and recorded video.
- Control access (full, selective, none) to CCTV cameras (e.g., pan, tilt, and zoom controls).
- Live audio/video conferencing from the ROC or conference room to the protected facility control room, Incident Command Center, a conference room, etc. Identify others who may also need the capability to link to this system from other locations.

- Other communication systems (e.g., email, text, or video conferencing).
- View access to other VDU-based displays available on the protected facility.
- View and control access (define) to radar displays and controls, and similar systems.

7. Identify the Required Technical Systems

Identify the required technical systems and equipment that provide the functional capabilities defined in Section 6. For example:

- Facility alarm monitoring and display systems that provide the specified access (display and control) to the basic process control system, safety shutdown system, fire and gas systems, emergency safety systems, and other like systems.
- Displays that enable access to select and control the CCTV video feed and cameras (as permitted in Section 6)
- Facilities to access communication systems (e.g., radio, telephone, radar, video conferencing, satellite, weather)

8. Technical System Interfaces and Additional Resources

Given the information in Sections 6 and 7, identify each interface between the technical systems located in the ROC and in the protected facility. Identify the performance standards and requirements that apply to each interface.

9. Appendix

9.1 Appendix A – Remote Barrier Support Modes of Operations and Required System Capabilities

Table B.2 captures the information used in the above sections. (Sections 6 and 7 identify the more onerous requirements from this table.)

RBS Task ID and Description: Identify every RBS task – ID and description. (See Table 3.6 for task information.)

RBS Modes of Operation: Enter each mode of operation possible for this function as performed from the ROC. (This information is only relevant for RBS tasks that may be affected by the current operating mode of the protected facility, the ROC, or activities performed within the ROC.)

HE Capability: Enter every HE functional capability needed to perform the entered task in each mode of operation. Section 6 provides examples of this information type.

Required Equipment and Systems (Each Capability): Enter every equipment item (direct-use) and technical system (In-Place PE) required to enable each capability.

Capability Constraint: Enter constraints (if any) that apply to each capability. (See Section 6 for examples.)

Required Infrastructure and Additional Resources: Enter all required infrastructure and additional resources needed to realize, enable, and maintain identified capabilities.

TABLE B.2

Design Basis: Remote Barrier Support – Systems, Modes, and Infrastructure

Barrier ID	RBS Task ID and Description	RBS Modes of Operation	HE Capability (Detect and Act Phase)	Required Equipment and Systems (Each Capability)	Capability Constraint	Required Infrastructure and Additional Resources
Enter ID for barrier with RBS function	Enter the RBS task ID and description	Enter all modes of operations that may affect this RBS task.	Enter every HE functional capability needed to perform this RBS task in the indicated mode of operation.	Enter the equipment and systems needed to enable/support each HE capability	Enter constraints that apply to this capability *Enter all that apply*	Identify all infrastructure and additional resources required to enable/maintain the RBS functions in the ROC *Enter all that apply*

TABLE B.3

Design Basis: Remote Barrier Support – Network Interface Requirements

Technical System (Located at the ROC)	Technical System (Protected Facility)	Worst-Case Bandwidth Demand Scenario	Hazard Scenarios
Enter the ID and description for each technical or communication system that requires a real-time interface to the source system located on the protected facility	Enter the ID and description for the source system	Enter the most plausible, worst-case bandwidth demand required by this interface	Identify the hazards/ environmental conditions (ROC and protected facility) that may place this interface at risk *All that apply*

9.2 Appendix B – Telecommunication Requirements

Table B.3 summarizes the potential requirements and considerations that apply to the network and communication interfaces between systems located at the ROC and protected facility. (The reliability and performance of the RBS task function and the barrier system depend on this interface.)

B.5   TECHNICAL DESIGN TOPICS

B.5.1   What Causes Humans to "Look but Not See"?

Many conditions contribute to the common question, "Why do we fail to see what may be directly in front of us?" This section reviews a few of the more common issues. Each is addressed to varying degrees in the prototype lifecycle model. The following topics are discussed in this section:

1. Inherent limitations in the human visual system
2. Attentional issues (too focused on a particular object or object type)
3. Top-down scanning (based on mental models)
4. Bottom-up scanning (attention grabbed by a "shiny object")
5. Effects of excessive fatigue or stress, health issues, or problems at home
6. Obstructed by environmental conditions

Issue 1 – Visual System Limitations

Refer to the information in Appendix F.3, beginning with Figure F.4. What the human eye can "see" depends on the visual angle from the eye's line of sight. SA-1 information located in visual zones B and C is less likely to be detected, a limitation inherent to the human visual system. (Appendix K.1.2 provides design guidance to consider when locating barrier objects in these zones.)

Issue 2 – Attentional Issues

A person can miss display information located in any visual zone, including zone A. (See Figure F.4.) This can occur if one's attention is highly focused on a particular object. As discussed in Appendix F.2.2, working memory (short-term memory) limits the information that can be processed in a conscious cognitive process. If performing a task to "find" a particular object in a workspace, automatic processes can block visual information not consistent with the sought object. A shift away from the focused attention view is needed to notice other objects located in the workspace, some of which may also be material to the active task. (Also see "Access to Sensory Data in Table F.1, Appendix F.1.)

*Note: Another applicable term is **Inattentional Blindness**, which describes a type of change blindness. For a more complete discussion, see Wickens et al. (2013, pp. 55–56). The likelihood of change blindness lessens with increased salience.*

Issue 3 – Top-Down Scanning

Visual scanning is commonly driven by mental models that identify which information is important and where it may reside. One may fail to "see" (detect/notice) information in a clear visual view if it is not perceived as important or expected to be in the viewed location, as can occur if not included in the activated mental model.

Issue 4 – Bottom-Up Scanning

One's attention can be captured by a "shiny object"; i.e., the object is highly salient relative to adjacent objects. The diverted attention capture may or may not be beneficial. It is beneficial if the information is situationally important. Conversely, the attentional shift may be hazardous if it diverts attention at a critical moment or if critical information is missed.

Issue 5 – Excess Fatigue and Stress, Health Issues, Problems at Home

High fatigue and stress, degraded health, or problems at home can cause one to be distracted and less attentive to vital information.

Issue 6 – Access Obstructed by Environmental Conditions

Environmental conditions, which can be transient/situational, can impede visual access to information. The following situations are a few examples:

* Insufficient lighting can hinder access to non-active displays like signs, paint markings, and labels. With this display type, the display and area lighting are inseparable from a design perspective.
* Too much lighting can hinder access to some electronic displays.
* Smoke, steam, ice, and other environmental conditions may interfere with the visual access to a display.
* Environmental eye irritants can degrade and interfere with human eyesight.

B.5.2 ENHANCING DISPLAY SALIENCE

The term "salience" refers to the noticeability and detectability of an object. As such, increasing the salience of an object increases the likelihood that it will be noticed and detected. Displays (audible, visual, verbal, tactile) that activate a barrier or task should employ a design approach that enhances display salience. The same is true for other essential barrier information, although care is needed to balance display salience relative to other important. Table B.4 provides design options to improve display salience.

Because of the challenges associated with change blindness, seeing a light turn on or off increases its salience. However, this is only true if one views the transition.

Note: The SEEV model is a useful tool to understand what drives visual scanning and detectability (Wickens et al. 2013, pp. 50–53, 2023, pp. 49–53). In this model, a measure of detectability = sS − efEF + exEX + vV. The weighting factors are normally set at 1. Detectability is a function of **salience**, **effort** *(degree of eye, head, or body movement required to the object),* **expectancy** *(does the object reside in an expected location or area of known high activity?), and* **value** *(the usefulness or importance of the information). On effort (the only negative value), even a simple eye movement invokes a small degree of resistance to that movement. That effort may be countered by the other three factors.*

For example: An HMI display that includes a "lamp" function should always indicate the lamp's presence in both the on and off state. Both contribute to a mental model that retains knowledge of the lamp's existence and location on the display, and its appearance and behavior in both states.

B.5.3 TIME AS A PRIMARY ADVERSARY

Important objectives in barrier design are to:

1. Achieve the barrier/task function
2. Achieve the barrier/function within the specified safety time
3. Achieve the required reliability and performance (e.g., minimize human error)

For many reasons, time can negatively affect each of these elements; thus, time is a primary adversary to address in the barrier design and other lifecycle phases. Further, humans are not reliable clocks. This places the time sensitive aspects of barrier performance at risk if not addressed in the design. The barrier safety time (BST) limits the time available to perform and complete barrier functions and achieve the specified safe state. An attempt to "make up time" by increasing one's pace consumes working memory, which may slow or degrade other cognitive activities. As such, it is a potentially problematic trade-off. Also, a perception that one is "running out of time" may lead to inappropriate shortcuts that place the task or barrier at risk.

The challenge is to achieve a barrier system design that creates slack, i.e., a time contingency budget. This time may be needed to respond to unforeseen events and conditions or recover from a mistake (an error).

TABLE B.4

Design Options to Improve Display Salience

More Salient	Less Salient	Reference
Movement toward you	Static (not moving)	Endsley & Jones (2012, p. 36)
Blinking/flashing light	Static light display	Endsley & Jones (2012, p. 36), Wickens et al. (2013, p. 51)
The color red	Other colors	Endsley & Jones (2012, p. 36)
Larger text or display object size	Smaller text or display object size	Endsley & Jones (2012, p. 36), Wickens et al. (2013, p. 51)
Clear text/text font is easy to read	Text less easy to read	–
Object in foveal view and not surrounded by clutter	A busy display, e.g., many elements that are adjacently located with little spatial or functional differentiation	–
Brief, clear, verbal instruction using standardized terms and phrasing	A muddled, wordy, or potentially ambiguous message instruction	–
Conveyed information or instruction is a combination of visual and audible or verbal	Conveyed information or instruction is only audible or only visual	–
More salient than other objects (make the object brighter, larger, or clearer than all adjacent objects)	The object salience is the same for all adjacent objects	
High display contrast (text, object) to background	Low, poor, or reduced display contrast to background	Wickens et al. (2013, p. 51)
High light intensity relative to background/ambient lighting	Lower light intensity	Wickens et al. (2013, p. 52)
A rapid change (attention perspective)	A slow change over time	Wickens et al. (2013, pp. 52–53)
Adjacent (proximity) to concurrently or sequentially used (next step) information	Concurrently or sequentially used information not adjacently placed (proximity)	–

The following provides additional information to consider on this topic:

- In preliminary design step B-10, see the applicable notes, examples, and discussion.
- Appendix B.8 suggests example support aids that reduce the cognitive resources and effort (time) needed to perform time-sensitive tasks within the available time.
- Appendix F includes topics that address the effects and possible human responses to time-pressured situations. Example responses may include attention tunneling (ignore information), task switch error (failure to switch to higher priority task), and others.
- Figure F.2 (Appendix F.2) identifies factors that can positively and negatively affect human performance (such as capacities, capabilities, response performance, and behaviors) in ways that directly and indirectly effect the time taken to progress barrier task activities.
- Appendix F.5 identifies temporal (time) and event order behaviors, capabilities, and limitations that are inherent to long-term memories and mental models.
- Table J.3 (Appendix J.6) provides suggestions to improve HE performance in ways that reduce the time needed (clock time) to perform various task activities.

The task definition and design should seek a least-time (often a least-effort) approach that achieves the barrier function or task goal. This approach may increase the use of automated tools, employ support aids that reduce the time needed for SA-2/3 and decision activities, and maximize and exploit the appropriate use of skill-based competencies such as skill-based team interactions, coordination, and communications.

Additional opportunities of the abovementioned types are addressed throughout this book. Barrier designers must always remain aware that time is an unrelenting and often the primary adversary to active human barrier reliability and performance. Design processes should aggressively seek to maximum the use of the available time and, where possible, create slack in the available time.

Essential Point: Barrier, barrier task, and task phase safety times have uncertainties. Things often go wrong or do not go as planned (the known unknowns). For these reasons, barrier safety time is a critical resource that must thoughtfully addressed in the design process and actively managed in the operational phase (i.e., when the barrier is activated).

B.6   CX SUB-PROCESS – PE COMPONENT-LEVEL DESIGN

The CX sub-process, summarized in Figures B.2–B.4, applies to the design of direct-use (detect and act phase) and Support PE. This sub-process is performed in the following detailed design phase processes:

- Process C-3 (act phase design) as noted in Table 4.4.
- Process C-7 (non-VDU displays and received communications) as noted in Tables 4.7, 4.8, 4.10 and 4.11.

*Cont. from detailed design
processes C3, C7, C8
(Component design)*

CX-0.1	Physical component selection and design	See Figure B.3 for detail
CX-0.2	Organizational element requirements	See Figure B.4 for detail
CX-0.3	Develop requisition input/documents	Appendix E (E.1, E.2)

Return to source process

Note: CX-0.1 includes the PIF assessment from App. H.3

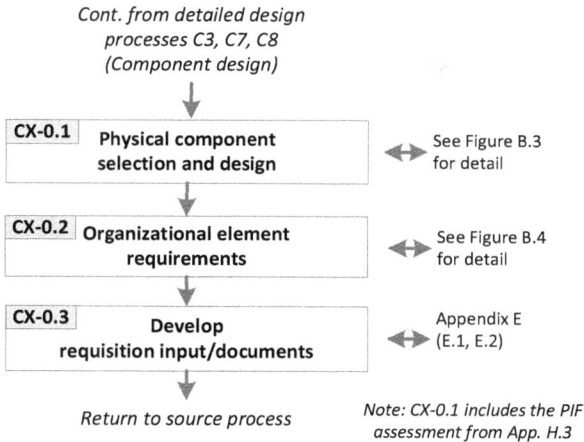

FIGURE B.2 CX Process Overview (Direct-Use Component Design).

- Process C8-3 (VDU monitors).
- Process C8-7 (CCTV cameras) as noted in Table 4.16.

The components addressed in this design process are commercial-off-the-shelf (COTS) devices or elements. They may include intelligent devices with limited configuration options. This process does not apply to engineered displays such as an HMI display.

Note: The CX sub-process does not replace existing practice and design activities that select, design, and implement these components. Instead, it provides design input to those with primary design responsibility or confirms that the design complies with the barrier requirements. As such, it is a check process.

B.6.1   Sub-Process CX-0.1 – PE Component Selection and Design

Figure B.3 summarizes the design activities performed in step CX-0.1. The suggested approach is to provide the information from this process to those with primary responsibility for their specification, design, and procurement. The timing of this handover should align with the responsible party's scheduled activities, allowing time to integrate these requirements into the component selection and design process. If this timing is not possible, the effort may need to shift to one that checks the output from the responsible party's work to confirm that the CX-0.1 requirements are correctly integrated into the component selection, specification, design, and procurement documents.

B.6.1.1   Step CX-0.1a – Component Selection

The components addressed in the CX process are the direct-use or Support PE (worn or carried). Tables 3.2 and 4.1 provide a list of tables that identify all instances

From detailed design processes
C-3, C-7, C-8
Act and Detect Phase Design

| CX-0.1a | Physical element component selection |

↓

| CX-0.1b | Location and placement |

↓

| CX-0.1c | Functional design |

↓

| CX-0.1d | Response time design |

↓

| CX-0.1e | PIF Assessment: Working Environment (See Note) |

| CX-0.1f | External support systems |

↓

| CX-0.1g | Interconnects and cable routing |

↓

| CX-0.1h | Reliability, availability, and Integrity design |

↓

| CX-0.1i | Environmental and survivability design |

↓

| CX-0.1j | Component Performance Standard |

↓

To CX-0.2 Organizational Element Design

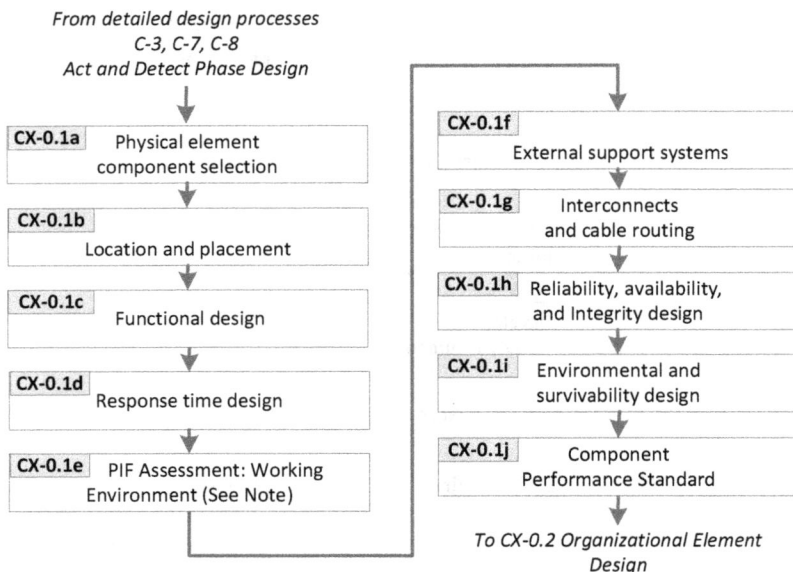

FIGURE B.3 CX Sub-process: Step CX-0.1 Activities. See Appendix H.3 for PIF Assessment Details.

of direct-use PE. Table 3.13 identifies the Support PE. Examples of direct-use PE include the following:

- Non–VDU-based act phase activation devices (e.g., pushbuttons, switches, radio or telephone handsets, fire hoses)
- Non-VDU displays (see Table 3.26 for the display types included in this category)
- CCTV cameras
- VDU monitors
- Incident command display board (e.g., a smartboard)

Table B.9 (Appendix B.7.2) lists documents that may define the requirements applicable to the assessed components. Review the applicable documents to better understand the requirements.

Note: If the component depends on an In-Place PE (e.g., technical system), external support system, or external protective barrier, the component requirements may also apply to the dependent equipment, systems, and barriers.

If the detect phase component is a barrier or task activator, then review and consider the information in Appendices B.5.1, B.5.2, K.1, and K.2.

If the component is a Support PE that requires local or remote monitoring, then review and consider the additional information in Table 4.6 (remote monitoring developed in the detailed design phase step C3-8b.).

Additional considerations that may affect component selection include the following. The component:

- Meets or exceeds all applicable requirements in the SRS and Table 3.7 (base barrier performance standards)
- Is suitable for use as a barrier or task activator (e.g., salience; see Appendix B.5.2)
- May be considered if included in an Accepted/Approved Vendor List
- Meets the specified functional requirements
- Meets the specified barrier/task/phase safety time requirements (see the safety time requirements in Table 3.27). Other similar requirements may be included in the SRS or base performance standards in Table 3.7
- Meets or can be implemented to meet the specified reliability and availability requirements
- Meets or can be implemented to meet the specified integrity requirement (the barrier or a barrier component may be assigned a Safety Integrity Level target; the component and its implementation must meet this requirement.)
- Is or can be implemented in a way that makes it suitable for operation and use in the installed environment
- Meets or can be implemented to meet the specified survivability requirements
- Is compatible with the recommendations from the following assessment processes:
  - Single Point of Failure Assessment (see step B-19, Table 3.30)
  - Shared element reliability assessment (see step B-19, Table 3.31)
  - Assessment of performance influencing factors (see process C-2)
  - Safety time assessment (see processes C-3 to C-8)

Note: Component selection may be on the project critical path (e.g., a control valve with an exotic material requirement has a long delivery schedule). In such cases, the procurement cycle must start much earlier than that for other equipment. This timing may be out of sequence with the preferred timing for the CX process (i.e., the timing of the C-3 process). Its timing, when possible and practicable, should be adjusted so the information is available for timely inclusion in the requisition technical package. Also see Appendix A.8.3 and A.8.4 for additional guidance.

B.6.1.2 Step CX-0.1b – Location and Placement

Table B.15 (Appendix B.7.7) lists the documents that define the location and placement requirements for direct-use PE. This step seeks to confirm that the components are correctly located and placed or positioned to meet barrier requirements. For guidance, see the applicable workspace design guidance in Appendix K.

B.6.1.3 Step CX-0.1c – Functional Design

This step seeks to confirm that the selected component and its proposed implementation meet the barrier functional requirements. For example, achieving the barrier safe state may rely on a component (e.g., control valve) that stops process flow within the maximum permitted leakage rate (e.g., tight shutoff). The required function may be linked to its location and placement (checked in step CX-0.1b).

Table B.10 (Appendix B.7.3) lists documents that may define functional requirements that apply to the assessed component. These requirements may need to be extended to dependent elements, for example, if the function relies on a technical system (an identified In-Place PE).

B.6.1.4   Step CX-0.1d – Response Time Design

This step seeks to confirm the response time achieved by the selected component and the proposed implementation can achieve the barrier, task, and phase safety times in Table 3.27 and other applicable documents in Table B.11 (Appendix B.7.4).

With active components, the safety/response time for a detect or act phase element is a function of the device performance, application (e.g., the nature or attributes of the monitored or controlled process), implementation (e.g., the interface and implementation in an In-Place technical system), and installation.

If the component is a CCTV camera, the response time may include the time needed to control a camera (manually or automatically control its pan, tilt, or zoom) to achieve the specified view. An assessment of an act phase response element should consider the time needed for the component to attain the required safe state (e.g., the movement time a control valve takes to reach and achieve the required position and safe state).

B.6.1.5   Step CX-0.1e – PIF Assessment (Working Environment)

This step performs a working environment PIF assessment for the selected component. See Appendix H.3 for the suggested assessment process. Table H.1 defines the PIF working environment criteria to consider in the assessment. The results are recorded in Table H.3. Table A.2 (Appendix A.2.3) suggests the assessment participants. The H.3 process addresses recommendations and the integration of the approved recommendations into the previous and remaining CX-0.1 steps.

B.6.1.6   Step CX-0.1f – External Support Systems

The step seeks to confirm that the barrier-dependent external support systems' (ESS) requirements in Table 3.28 (External Support System Requirements) are addressed in the design. It should also identify ESS requirements that may be missing in Table 3.28 and similar design documents. If this occurs, update the requirements accordingly in the appropriate documents. Table B.19 (Appendix B.7.11) lists other documents that identify other external support system requirements.

ESS design information includes the lighting required to perceive a passive detect phase display accurately and quickly, such as the lighting type, location, and intensity. The ESS may include the electrical power source required by the component or its dependent In-Place PE (e.g., technical system).

Note: Figure D.2 (Appendix D.2.2) provides a visual representation to show the dependencies and why a component requirement should be applied to other systems and equipment on which its reliable and continued functioning depends. Examples of performance requirements are endurance time and resistance to environmental and design accident loads.

The following are examples of the ESS types on which the selected barrier component may depend:

- Electrical power sources and distribution systems (e.g., sourced from a UPS or emergency power bus)
- Instrument air supply and distribution system
- Hydraulic supply and distribution system
- Nitrogen supply and distribution system
- HVAC (e.g., control ambient temperature and humidity)
- Environmental air filtration system

B.6.1.7 Step CX-0.1g – Interconnections, Cables, and Routings

This step reviews the information from step CX-0.2b (component placement) and the cable connections to In-Place PE (e.g., a technical system) or an external support system (e.g., electrical power, instrument air, or instrument ground system). It seeks to identify if a required or identified connection (signal, network, power, tubing, or mechanical linkages) can be damaged or destroyed by the hazard event that activated the barrier. It may also seek to confirm that the cable system meets the applicable performance standard requirements specified in Table 3.7 (base performance standard), the SRS, and applicable documents listed in Table B.14 in Appendix B.7.6. Options to address an identified deficiency in a cable system may include redundant and diverse cable routings to ensure a single event cannot damage both cables.

Many active human barrier components require electrical power and may be activated by a single (non-redundant) control signal, such as a lamp, audio speaker, or horn. Available design options for simplex connections, though limited, include adding fireproofing to cable trays, selecting a route with the least exposure to identified hazards, and adding additional devices supplied through a different cable route. The endurance time for tray fireproofing should meet or exceed the barrier activation duration.

B.6.1.8 Step CX-0.1h – Reliability, Availability, and Integrity Design

This step confirms that the component design and proposed implementation meet the specified performance requirements for reliability, availability, and integrity. Table B.12 (Appendix B.7.5.1) lists the documents that specify these requirements. (The requirements may limit which components can be selected.) The same requirements may also need to be applied to the component dependencies (e.g., In-Place PE, ESS, external protective barriers, and the power and signal cabling).

If a Safety Integrity Level is assigned to the barrier or component (e.g., defined in the SRS), then this step confirms if the requirement can be met. Include all assumptions used to verify the requirements, such as type of proof testing and interval. Record this information in the SRS.

B.6.1.9 Step CX-0.1i – Environmental and Survivability Design

This step confirms whether the selected device and its proposed implementation and installation location can achieve the specified environmental and survivability requirements. Table B.14 (Appendix B.7.6) lists the documents that define these requirements. These requirements may also apply to other elements on which the component depends, as discussed in prior steps.

B.6.1.10   Step CX-0.1j – Component Performance Standard

This step determines whether the assessed component warrants the development of a dedicated performance standard to use in a verification process. If so, see Appendix D.3.1 for guidance.

B.6.2   Sub-Process CX-0.2 – OE Component Requirements

This series of activities define organizational element requirements that may apply to the selected component.

B.6.2.1   Step CX-0.2a – New Operating Procedures (Component-Level)

This step determines whether a new component-level operating procedure is required to guide the operational use of the barrier system component. The procedure may be required if this is a first use, or a one-off (materially different) application or installation in the facility. Different procedures may be needed for different operating modes. A new process may also be warranted if the operational use of the component introduces a new skill or knowledge requirement. These or other skill or knowledge requirements are identified in Tables 3.12 (act phase response PE), 3.13 (Support PE), and 3.25 (detect phase SA-1 display).

If a new component-level operation procedure is required, enter the procedure, and populate all associated fields in Table B.5. The procedure identifier (**Proc. ID**) must be unique. For the **Source Ref.** field, enter the document/source that was the primary basis used to identify the need for the procedure. If the provider is the facility owner/operator, enter O/o in the **Provider** field. If O/o is not the provider, leave the **Constraint on Use**, **Task ID**, and **Barrier ID** fields blank. (See Appendix J.2 for background.)

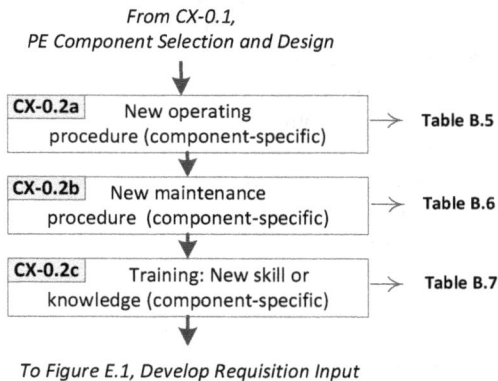

FIGURE B.4   CX Sub-process: Step CX.02 Organizational Element Design.

TABLE B.5

Operating Procedures: Direct-Use Components

| | **Procedure Requirement** | | | | **Procedure Detail** | | | |
Source Ref.	Proc. ID	Provider	Type/Use	Content	Format/Type	Constraint on Use	Task ID	Barrier ID
Enter source reference or table	Enter unique ID for this proc.	Enter – O/o – *Vendor* – *Contractor* – *Existing* – *Other*	Enter Type: – *Start-up* – *Normal* – *Abnormal* – *Shutdown*	Enter required content	Enter procedure format or type (from standard templates)	Enter issues that may affect type, format, or level of detail	Enter task ID. *All that apply*	Enter barrier ID

Note: A direct-use, detect or act phase component includes a wide range of possible devices and equipment. Examples may include a fire hose, lifeboat (act phase response PE), a self-contained breathing apparatus (Support PE), or a radio, beacon, or CCTV camera.

Note: With third-party systems, operating and maintenance procedures may be provided in whole or in part by the equipment provider. The procedures, often generic in nature, may be inadequate based on how the component is operated or maintained. In such cases, additional procedures may be required.

Note: New skill or knowledge requirements are reviewed and assessed in process C-16, competency gap, and the training needs assessment.

B.6.2.2   Step CX-0.2b – New Maintenance Procedures (Component-Level)

This step determines whether a new component-level maintenance procedure is required to guide the preventive, normal, or specialized maintenance/repair and recovery of the selected component. Reasons to require a new procedure may be similar to those noted in step CX-02a, an unfamiliar software configuration method, calibration routine, or online testing approach. Maintenance activities may require a new special tool or introduce the potential for a new maintenance/handling hazard. The applicable processes in EI (2020a) may be an appropriate alternate approach to assessing and identifying those tasks that may be safety critical. Safety critical tasks may warrant new procedures.

If a new component-level maintenance procedure is required, enter the new procedure, and populate all associated fields in Table B.6. The procedure identifier (**Proc. ID**) must be unique. For the **Source Ref.** field, enter the document/source that was the primary basis used to identify the need for that procedure. If the provider is the facility owner/operator, enter O/o in the **Provider** field. If not, leave the **Constraint on Use**, **Task ID**, and **Barrier ID** fields blank. (See Appendix J.2 for background.)

B.6.2.3   Step CX-0.2c – Training Requirements (Component-Level)

The first-time use, or first-of-kind installation or application of a component in the facility may warrant the addition of a new operating or maintenance procedure in steps CX-0.2a and CX-02b. Those procedures may introduce new component-specific skill and knowledge requirements. If so, add that information to Table B.7. *Each new competency creates a new training requirement.* Enter each new training requirement and populate all associated fields in Table B.7. See process C32-4 (implementation phase) for guidance on how to populate the **Training Detail** fields. Use an entry of *O/o* in the **Provider** field to identify training provided by the owner/operator. (See Appendix J.3 for background.)

Training provided by component equipment providers and others is often generalized. Different or additional training may be needed if the component implementation is unique and thus not adequately addressed in the generalized training.

TABLE B.6

Maintenance Procedures: Direct-Use Components

Procedure Requirement				Procedure Detail				
Source Ref.	Proc. ID	Provider	Type/Use	Content	Format/Type	Constraint on Use	Task ID	Barrier ID
Enter source reference or table	Enter unique ID for this proc.	Enter – O/o – *Vendor* – *Contractor* – *Existing* – *Other*	Enter Type: – *Preventive* – *Calibration* – *Repair/ replace*	Enter required content	Enter procedure format or type (from standard templates)	Enter issues that may affect type, format, or level of detail	Enter task ID. *All that apply*	Enter barrier ID

TABLE B.7
Training Requirements: Direct-Use Components (Operate and Maintain)

Barrier ID	Task ID	PE Component or Element ID	Training Source		Objective and Delivery Method	Training Result	Periodic Retraining Req.?	New Skills and Knowledge
			Training Source	Module ID				
Enter barrier ID	Enter task ID	Enter PE ID and description	Enter – O/o – 3rd-party vendor – Contractor	Enter training module ID/ description	Enter training objective and delivery method	For training other than "new," identify how training result was verified	Enter – Yes – No If Yes, enter training interval (months)	Enter new skill or knowledge to use this procedure

Training Source spans "Training Source" and "Module ID" columns; Objective and Delivery Method, Training Result, Periodic Retraining Req.?, and New Skills and Knowledge fall under Training Detail.

B.6.3   Sub-Process CX-0.3 – Requisition Development

This step performs the procurement activities, which include those identified in Appendix step E-1 (e.g., develop and provide input to a Request for Proposal – RFP) and, if applicable, step E-2 (e.g., review the RFP response). When required to insert barrier requirements into an RFP package, the source tables in Appendix E-13 identify the barrier information created by the lifecycle model (and the model processes and steps that created that information) and suggests where to insert the information into the RFP package. Discussed in Appendix E, the following are example document types included in an RFP package developed within a capital project environment. (In practice, RFP packages may vary, e.g., different document types, names, and organization.)

1. Scope of Work statements
2. Technical package
3. Vendor data requirements (VDR)
4. Inspection and test requirements (ITRS)

B.7   SOURCE INPUTS TO DESIGN, PROJECT DOCUMENTS, PROCUREMENT, AND QUALITY

This section provides the following tables that summarize the potential source of input information applicable to design/project activities:

- Table B.8. Source Documents: Common Information Requirements
- Table B.9. Source Documents: Component Selection, Design, and Implementation Requirements
- Table B.10. Source Documents: Functionality Requirements
- Table B.11. Source Documents: Timing and Response Time Requirements
- Table B.12. Source Documents: Reliability, Availability, and Integrity Requirements (Physical Elements)
- Table B.13. Source Documents: Reliability, Availability, and Integrity Requirements (Human Elements)
- Table B.14. Source Documents: Survivability and Environmental Requirements
- Table B.15. Source Documents: Sizing, Location, Layout, and Placement Requirements
- Table B.16. Source Documents: HMI Displays (VDU-Based) Requirements
- Table B.17. Source Documents: Room Requirements
- Table B.18. Source Documents: Engineered Area Requirements
- Table B.19. Source Documents: External Support Systems Requirements
- Table B.20. Source Documents: External Protective Barriers Requirements
- Table B.21. Source Documents: Technical Systems Requirements
- Table B.22. Source Documents: Building Requirements

B.7.1 Common Information Requirements

Table B.8 (located in Section B.7.15) summarizes the common source project and technical documents that may provide input to the design and implementation of the barrier system.

B.7.2 Component Selection, Design, and Implementation Requirements

Table B.9 (located in Section B.7.15) summarizes the source documents that may provide input to the component selection, design, and implementation requirements of the barrier system.

B.7.3 Functionality Requirements

Table B.10 (located in Section B.7.15) summarizes the source documents that may provide or define the functionality requirements of the barrier system.

B.7.4 Timing and Response Time Requirements

Table B.11 (located in Section B.7.15) summarizes the source documents that may provide the requirements for the barrier, task, task phase, and active-level timing, and response time.

B.7.5 Reliability, Availability, and Integrity Requirements

Tables B.12 (physical elements) and B.13 (human elements) (both located in Section B.7.15) summarize the source documents that may provide the reliability, availability, and integrity requirements for the physical and human elements of a barrier system and their dependencies (e.g., external support systems and external protective barriers).

B.7.5.1 Physical Elements

Table B.12 (located in Section B.7.15) summarizes the physical elements.

B.7.5.2 Human Elements

Table B.13 (located in Section B.7.15) summarizes the human elements.

B.7.6 Survivability and Environmental Requirements

Table B.14 (located in Section B.7.15) summarizes the source documents that may provide the environmental and survivability requirements that apply to the physical and human elements of the barrier system and their dependencies (e.g., external support systems and external protective barriers).

B.7.7 Sizing, Layout, Location, and Placement Requirements

Table B.15 (located in Section B.7.15) summarizes the source documents that may provide information on sizing, location, layout, and placement requirements.

B.7.8 HMI DISPLAY (VDU-BASED) REQUIREMENTS

Table B.16 (located in Section B.7.15) summarizes the source documents that may provide guidance and requirements for HMI display requirements, design, and other activities.

B.7.9 ROOM REQUIREMENTS

Table B.17 (located in Section B.7.15) summarizes the source documents that may provide design guidance and requirements for a room occupied by one or more HEs performing assigned barrier activities.

B.7.10 ENGINEERED AREA REQUIREMENTS

Table B.18 (located in Section B.7.15) summarizes the source documents that may provide design guidance and requirements for engineered areas (e.g., egress/escape route or safe haven) occupied by one or more HEs (or others) performing assigned barrier activities.

B.7.11 EXTERNAL SUPPORT SYSTEM REQUIREMENTS

Table B.19 (located in Section B.7.15) summarizes the source documents that may provide design guidance and requirements for the external support systems (e.g., lighting, electrical power, environmental controls, HVAC) on which one or more barrier elements depend.

B.7.12 EXTERNAL PROTECTIVE BARRIER REQUIREMENTS

Table B.20 (located in Section B.7.15) summarizes the source documents that may provide design guidance and requirements for the external protective systems (e.g., passive, active) on which one or more barrier elements depend.

B.7.13 TECHNICAL SYSTEM REQUIREMENTS

Table B.21 (located in Section B.7.15) summarizes the source documents that may provide the requirements for the technical systems identified as *In-Place PE* or a dedicated building technical system. (See the preliminary design step B-6 for examples of technical systems.)

B.7.14 BUILDING REQUIREMENTS

Table B.22 (located in Section B.7.15) summarizes the source documents that may provide design guidance and the requirements for the external protective systems (e.g., passive, active) on which one or more barrier elements depend.

B.7.15   REFERENCE TABLES

TABLE B.8
Source Documents: Common Requirements

Source Document	Developed/Used in Lifecycle Phase	Applicable Information
Facility basis of design	All	Facility/project purpose, location, capacity, throughput, primary processes, selected key technology, regulatory, etc.
Basic environmental design data (BEDD)	All	Ambient environmental conditions (e.g., weather) that may affect personnel and equipment
Approved Vendor List (AVL)	All	Identifies permitted components, equipment, service providers, etc.
Facility/site plot plan	All	Base plot layout, location of major equipment, buildings, roads, etc.
Operating and maintenance philosophies	All	High-level guidance on the owner/operators preferred/expected basis for O&M activities
Discipline design philosophies (each discipline)	A, B, C, E	Design philosophies from other disciplines to consider in area of design overlap (e.g., facilities, emergency shutdown and blowdown, fire protection, packaged equipment systems, safety, and automation systems)
design basis (each discipline)	B, C, D, E	Basis-level design guidance for facilities, equipment, systems, packaged equipment systems, utility systems, buildings, egress/escape routes, fire protection systems, etc.
Alarm philosophy	B, C, D, E	Guides alarm priority selection as it may apply to facility alarm systems, control room monitoring and alarm systems, etc.
Instrument Index	All	Provides device tag names, make/model, etc.
Safety studies	B, C, D, E	Fire and blast studies, noise studies, etc.
Safety equipment list	B, C, D, E	Listing of tagged safety equipment

(Continued)

TABLE B.8

(Continued)

Source Document	Developed/Used in Lifecycle Phase	Applicable Information
Facility and equipment general arrangements	B, C, D, E	Facility drawings showing major equipment, buildings, roads, egress/escape routes, etc. Input to the design of engineered area, equipment placement/location, etc.
Process flow diagrams	A, B, C, E	Input to technical system design, component selection, etc.
P&IDs – process and instrument diagrams	All	Identifies equipment, systems, connections, process data, etc.
Electrical area classification drawings	B, C, D, E	Input to component and equipment selection and placement, etc.
Fire zone drawings and schematics	C, D, E	Input to component and equipment selection/placement, engineered area design, etc.
Risk assessments	All	Basis for barrier
Task analysis	B, C, D, E	Basis for task and phase design
Barrier safety management plan	All	Provides plan information that applies to design phase quality, verification, and validation processes, O&M phase verification, competency management requirements, roles, and responsibilities
Safety requirements specifications	All	Provides the complete barrier system requirements as information becomes available
Active human barrier design basis	B, C, D	Provides design-basis–level guidance to barrier system design, i.e., required regulatory and industry standards, owner/operator standards, base-level design, and performance requirements and guidance
Cause-and-effect charts, logic diagrams	B, C, D, E	Provides the logical relationship between detect phase input and act phase response outputs

TABLE B.9

Source Documents: Component Selection, Design, and Implementation

Source Document	Step/ Process	Document Title	Applicable Information	Input to
Table B.8	Common Information Requirements			
Table 3.5	B2-x	Barrier Origination Requirements	Component function in barrier system, activator, achievement of safe state, etc.	Component Selection Component Design Component Implementation Org. Element – Procedures – Training – Competency – Maintenance requirements
Table 3.6	B3-1	Task Origination Requirements	Component function in barrier system, activator, achievement of safe state, etc.	
Table 3.7	B-4	Barrier Performance Standards: Base Requirements	Base performance requirement to consider in selection, placement, implementation	
Table 3.9	B-6	Act Phase Response Physical Element Requirements	Identifies direct-use and Support PE	
Table 3.11	B-7	Communication Requirements (Outbound)	Identifies communication equipment	
Table 3.12	B-8	Act Phase: Direct-Use PE Requirements	Identifies direct-use PE components (SA-1 info source, act phase response PE), Support PE, and functions	
Table 3.13	B-8	Act Phase: Support PE Requirements		
Table 3.14	B-9	Detect and Act Phase In-Place PE – Engineered Area (EA) Requirements	Direct-use PE, Support PE, and HEs in engineered areas, EA endurance time, maximum occupants of area, etc.	
Table 3.15	B-9	Act Phase In-Place PE – Building Requirements	Direct-use PE, Support PE, and HEs in building, endurance time, maximum occupants of area, etc.	
Table 3.16	B-9	Act Phase: In-Place PE: Technical System Requirements	Direct-use PE reliance on identified technical system (TS), TS function and endurance time, SA-1 status monitoring, etc.	

(Continued)

TABLE B.9

(Continued)

Source Document	Step/ Process	Document Title	Applicable Information	Input to
Table 3.18	B-11	Detect Phase: SA-2 Requirements	Identifies SA-1 inputs to support requirement, a possible PE component	
Table 3.19	B-12	Detect Phase: SA-3 Requirements	Identifies SA-1 inputs to support requirement, a possible PE component	
Table 3.25	B-14	Detect Phase—SA-1 Information Requirements	Identifies all SA-1 input sources, required access locations, presented form, required In-Place PE, etc.	
Block Diagram	B-15	Barrier/Task Function Block Diagram	Identifies primary barrier components/equipment and interconnections	
Table 3.27	B-16	Barrier, Task, and Phase Safety Time Requirements	Input to component selection and implementation	
Table 3.28	B-17	External Support System Requirements	Identifies PE dependencies, endurance times, etc.	
Table 3.29	B-18	External Protective Barrier Requirements	Identifies PE dependencies, endurance times, etc.	
Table 3.30	B-19	Reliability Assessment: Single Point of Failure	Review recommendations for component/design recommendations	
Table 3.31	B-19	Reliability Assessment: Shared Elements		
Table 4.6	C3-8	Remote Monitoring of Act Phase Response and Support PE	Identifies SA-1 info source and monitoring function	
Table 4.16	C8-7	CCTV Image Element and PE Requirements	Identifies camera and placement requirements	
Table 4.17	C8-8	Off-Console VDU Displays	Identifies information/ displays presented at each VDU monitor	

TABLE B.10

Source Documents: Functionality Requirements

Source Document	Step/Process	Document Title	Applicable Information	Input to
Table B.8	Common Information Requirements			**Physical Element**
Table B.9	Component Selection, Design, and Implementation Requirements			
AHB design basis	B-1, C-1	Active human barrier design basis	Provides guidance on typical functionalities (e.g., fault detection, voting logic)	– Functional specification – Cause-and-effect charts – Logic diagrams – Control narratives
RBS design basis	B-1, C-1	Remote barrier support design basis	May define functionality required at remote location, specific to system interfaces or technical additional resources, and other functions required to enable or restrict access to functions	
SRS	Step B20-3.1, C-17, C-36	Safety requirements specifications	Provides guidance on typical functionalities (e.g., fault detection, voting logic) and other functionalities	**Human Element** – Assigned barrier/task actions and activities – Required skills and knowledge to support task activities (e.g., mental models, long-term memory)
Table 3.5	B2-x	Barrier Origination Requirements	Identifies relationship between barrier activator and attainment of barrier safe state/function	
Table 3.6	B3-1	Task Origination Requirements	Defines task activities, the basis for defining task phase functions and achieving the task goal or safe state	
Table 3.7	B-4	Barrier Performance Standards: Base Requirements	Review for possible requirements	**Org. Element** – Procedures – Training
Table 3.8	B-5	Shared Situation Awareness Requirements	Identifies SSA function	
Table 3.11	B-7	Communication Requirements (Outbound)	Identifies act phase response actions	

(Continued)

Table 3.12	B-8	Act Phase: Direct-Use PE Requirements	Identifies outbound communication "function"
Table 3.13	B-8	Act Phase: Support PE Requirements	Identifies act phase response function
Table 3.14	B-9	Detect and Act Phase In-Place PE – Engineered Area (EA) Requirements	Identifies EA function, sizing, PE/HE in area
Table 3.15	B-9	Act Phase In-Place PE – Building Requirements	Identifies building function, sizing, PE/HE in area
Table 3.16	B-9	Act Phase: In-Place PE – Technical System Requirements	Identifies system function, use conditions, endurance time, etc.
Table 3.17	B-10	Decide Phase Requirements	Identifies decide phase functions
Table 3.18	B-11	Detect Phase: SA-2 Requirements	Identifies SA-2 comprehension functions
Table 3.19	B-12	Detect Phase: SA-3 Requirements	Identifies SA-3 projection functions
Table 3.20	B-13	NTS Requirements: Teamworking Requirements	Identifies NTS teamworking functions
Table 3.22	B-13	NTS Requirements: Leadership	Identifies NTS leadership functions
Table 3.24	B-13	NTS Requirements: Monitor and Management Acute Stress	Identifies NTS coping/stress mgmt. function
Table 3.25	B-14	Detect Phase – SA-1 Information Requirements	Identifies function of SA-1 information
Barrier/task block diagram	B-15	Barrier or task function block diagram (e.g., see Figure 3.6)	Identifies functions and interfaces
Table 3.27	B-16	Barrier, Task, and Phase Safety Time Requirements	Defines safety times applied to functions

TABLE B.10
(Continued)

Source Document	Step/Process	Document Title	Applicable Information	Input to
Table 3.29	B-18	External Protective Barrier Requirements	Defines EB function in barrier system, SA-1 monitoring requirements for active barriers	
Table 3.30	B-19	Reliability Assessment: Single Point of Failure	Review for new functional requirements	
Table 3.31	B-19	Reliability Assessment: Shared Elements		
Table 3.33	B20-4	Additional Resources	May define new functions attributed to an additional resource (e.g., monitoring function, new task)	
Table 4.6	C3-8	Remote Monitoring of Act Phase Response and Support PE	Defines remote monitoring function and functionality	
Table 4.16	C8-7	CCTV Image Element and PE Requirements	Identifies camera image and capability placement requirements	
Table 4.17	C8-8	Off-Console VDU Displays	Input to component selection	
Table 4.18	C-9, C-10, C-11	Functional Analysis Recommendations	Review for requirements	
Table 5.7	C32-4	Training Modules: Development, Delivery, and Status	Identifies training objectives and trainees	

Table B.2	App. B.4	Design Basis: Remote Barrier Support – Systems and Infrastructure	Identifies HE functions and functions of Support PE, ESS, and additional resource requirements
Table B.3	App. B.4	Design Basis: Remote Barrier Support – Network Interface	Design input to technical requirements, functionality, survivability, etc.
Table B.4	App. B.5	Design Options to Improve Display Salience	Input to display selection, design, location, and placement
Table B.5	C-3, C-7, C-8 (CX-0.3, App. B-6)	New Operating Procedures: Direct-Use Components	Identifies new procedures and procedure types, driven by new skills or knowledge requirements
Table B.6		New Maintenance Procedures: Direct-Use Components	
Table C.1	App. C-1	Team Situation Awareness – Design Objectives and Applications	Input guidance to PE display design, OE training, procedure
Table H.2	C-2, E-11 (see App. H.2)	PIF Assessment: Barrier, Task, Task Phase	Review for applicable recommendations
Table H.3	C-3, C-7, C-8 (see App. H.3)	PIF Assessment (WE): Direct-Use PE Components	
Table I.1	C-3 to C-8 (see App. I.2)	Cognitive Assessment Evaluation and Recommendations	

(Continued)

TABLE B.10
(Continued)

Source Document	Step/Process	Document Title	Applicable Information	Input to
Table K.1	C-3, C-7, C-8 (CX-0.1b sub-process, App. B.6)	Functional Analysis: Sensory Access to Object While Transiting through a Workspace	Input to guide device selection, location, placement, and relative positioning within a workspace	
Table K.2	C9-1, 10-1,	Functional Analysis: Object Access from Seated Location at Control Console		
Table K.3	C11-1 (see App. K.1, K.2)	Functional Relationship Between PE Objects	Applicable to direct-use (detect and act phase) non-VDU– and VDU-based displays	
Table K.4		Application of Workspace Principles: Physical Elements		
Table K.7	B-14, C3-8, C-7, C-8 (see App. K.5)	Barrier Information Types and Attributes	Input to guide design of information capture, retention, display form and design, etc.	

TABLE B.11

Source Documents: Timing and Response Time Requirements

Source Document	Step/Process	Document Title	Applicable Information	Input to
Table B.8	Common Information Requirements			**Task Design**
Table B.9	Component Selection, Design, and Implementation Requirements			
RBS design basis	B-1, C-2	Remote barrier support design basis	May provide guidance on RBS functions, performance of additional resources, etc.	– Task design – Functional allocation (PE, HE) – HE number/locations
SRS	Step B20-3.1, C-17, C-36	Safety requirements specifications	Provides all barrier system response requirements	**Physical Elements**
Table 3.5	B2-x	Barrier Origination Requirements	Process safety time from risk assessment	– Component selection – Functional design
Table 3.7	B-4	Barrier Performance Standards: Base Requirements	Review for possible requirements	– HMI display design – Communications design
Table 3.8	B-5	Shared Situation Awareness Requirements	Identifies functions that may contribute to RT, SA-1 timing, etc.	– Layout and location drawings
Table 3.9	B-6	Act Phase Response Physical Element Requirements	Target act phase safety time	**Human Elements**
Table 3.11	B-7	Communication Requirements (Outbound)	Communication duration	– Skills and knowledge requirements
Table 3.12		Act Phase: Direct-Use PE Requirements		*Note:*
Table 3.13	B-8	Act Phase: Support PE Requirements	PE use duration	*Requirements may be defined by the most onerous of different barrier system requirements*

(Continued)

TABLE B.11
(Continued)

Source Document	Step/Process	Document Title	Applicable Information	Input to
Table 3.14	B-9	Detect and Act Phase In-Place PE – Engineered Area (EA) Requirements	PE endurance time	
Table 3.15	B-9	Act Phase In-Place PE – Building Requirements		
Table 3.16	B-9	Act Phase: In-Place PE – Technical System Requirements	Technical system endurance time	
Table 3.27	B-16	Barrier, Task, and Phase Safety Time Requirements	Defines target safety times for barrier, task, task phase, and task activities	
Table 3.28	B-17	External Support System Requirements	ESS demand period	
Table 3.29	B-18	External Protective Barrier Requirements	EB protection period/endurance time	
Table 3.30	B-19	Reliability Assessment: Single Point of Failure	May define response requirements	
Table 3.31	B-19	Reliability Assessment: Shared Elements	May define response requirements	
Table 3.33	B20-4	Additional Resources	Response times	
PIF assessment	C-2	PIF assessment report (task-based)	Recommendations may add new requirements	

Table 4.6	C3-8	Remote Monitoring of Act Phase Response and Support PE	RT for RM function
Table 4.17	C8-8	Off-Console VDU displays	Refer to CX-0.1d component design for response performance
–	C-3 to C-8	Phase safety time assessment report	Refer for RT possible input
Table 4.18		Functional Analysis Recommendations	Review for applicable recommendation
Table 4.19	C-9, C-10, C-11	Control Console or Panel (CP): Barrier Systems and Support Elements	Maintain element function period or duration
Table 4.20		Rooms: Barrier Systems and Support Elements	
Table 4.21		External Protective Barriers	Protective period
Table 4.22	C-11	Engineered Area: Barrier Systems and Support Elements	Maintain element function period or duration
Table 4.23	C-12	Buildings: Barrier Systems and Support Elements	
Table B.2	App. B.4	Design Basis: Remote Barrier Support – Systems and Infrastructure	Review for requirements
Table B.3	App. B.4	Design Basis: Remote Barrier Support – Network Interface Requirements	

(Continued)

TABLE B.11
(Continued)

Source Document	Step/Process	Document Title	Applicable Information	Input to
Table H.2	C-2, E-9 (see App. H.2)	PIF Assessment: Barrier, Task, Task Phase	Review for applicable recommendations	
Table H.3	C-3, C-7, C-8 (see App. H.3)	PIF Assessment (WE): Direct-Use PE Components		
Table I.1	C-3 to C-8 (see App. I.2)	Cognitive Assessment Evaluation and Recommendations		
Table K.1	C-3, C-7, C-8 (CX-0.1b sub-process, App. B.6)	Functional Analysis: Sensory Access to Object While Transiting through a Workspace		
Table K.2	C9-1, 10-1,	Functional Analysis: Object Access from Seated Location at Control Console		
Table K.3	C11-1 (see App. K.1, K.2)	Functional Relationship Between PE Objects	Review for background/input	
Table K.4		Application of Workspace Principles: Physical Elements		
Table K.5	B-7, C3-3 (see App. K.3)	Selecting Communication Methods		
Table K.7	B-14, C3-8, C-7, C-8 (see App. K.5)	Barrier Information Types and Attributes		

Note: RT – response time.

TABLE B.12

Source Documents: Reliability, Availability, and Integrity – Physical Elements

Source Document	Step/Process	Document Title	Applicable Information	Input to
Table B.8	Common Information Requirements			**Performance Standards**
Table B.9	Component Selection, Design, and Implementation Requirements			**Specifications**
Table B.10	Functionality Requirements		PE locations	**Reliability Design**
Table B.11	Timing and Response Time Requirements			
Table B.14	Survivability and Environmental Requirements		Environmental/accident loads	– Component selection
AHB Design Basis	B-1, C-1	Active human barrier design basis		– Error detection/fault-tolerant design
RBS design basis	B-1, C-2	Remote barrier support design basis		
SRS	Step B20-3.1, C-17, C-36	Safety requirements specifications		**Availability Design**
Table 3.7	B-4	Barrier Performance Standards: Base Requirements		– Redundancy
Table 3.30	B-19	Reliability Assessment: Single Point of Failure		– Minimize downtime – Component selection – Fault tolerance design – Design to support online maintenance
Table 3.31	B-19	Reliability Assessment: Shared Elements	Review for requirements and applicable recommendations	– Resistance to accidents (DAL) and environ. (EL) loads (see Figures D.3 and D.4, App. D.3)
Table 3.33	B20-4	Additional Resources		
—	C3, C7, C8	Components designed/implemented to meet performance requirements (see CX-0.1h, Appendix B-6)		
Table 4.18	C9-1, C10-1, C11-1	Functional analysis recommendations for control consoles and panels, rooms, and engineered areas		**Integrity Design**
Table 4.19	C9-1	Control Consoles and Panels: Barrier Systems and Support Elements (Maintain dependent TS, ESS)		– Safety Integrity Level (SIL) verification – Verification requirements and processes

(Continued)

TABLE B.12
(Continued)

Source Document	Step/Process	Document Title	Applicable Information	Input to
Table 4.20	C10-1	Rooms: Barrier Systems and Support Elements (Maintain dependent TS, ESS, EB)		
Table 4.22	C11-1	Engineered Area: Barrier Systems and Support Elements (Maintain dependent TS, ESS)		
Table 4.23	C11-1	Buildings: Barrier Systems and Support Elements (Maintain dependent TS, ESS, EB)		
Table G.1		Inspection requirements (see App. G.3)		
Table G.3		Testing requirements (see App. G.4)		
Table G.5	various	Verification requirements (see App. G.5)		
Table G.7		Validation requirements (see App. G.6)		
Table B.2	App. B.4	Design Basis: Remote Barrier Support – Systems and Infrastructure		*Note:*
Table B.3	App. B.4	Design Basis: Remote Barrier Support – Network Interface Requirements		*Requirements may be defined by the most onerous of different barrier system requirements*
Table H.2	C-2, E-9 (see App. H.2)	PIF Assessment: Barrier, Task, Task Phase		

Table H.3	C-3, C-7, C-8 (see App. H.2)	PIF Assessment (WE): Direct-Use PE Components
Table I.1	C-3 to C-8 (see App. I.2)	Cognitive Assessment Evaluation and Recommendations
Table K.1	C-3, C-7, C-8 (invokes CX-0.1b sub-process, App. B.6)	Functional Analysis: Sensory Access to Object While Transiting through a Workspace
Table K.2		Functional Analysis: Object Access from Seated Location at Control Console
Table K.3	C9-1, 10-1, C11-1 (see App. K.1, K.2)	Functional Relationship Between PE Objects
Table K.4		Application of Workspace Principles: Physical Elements

TABLE B.13

Source Documents: Reliability, Availability, and Integrity Requirements (Human Elements)

Source Document	Step/Process	Document Title	Applicable Information	Input to
				Performance Standards (Human Elements)
Table B.8	Common Information Requirements			
Table B.10	Functionality Requirements		HE locations	
Table B.11	Timing and Response Time Requirements			
				PE/OE Design
Table B.14	Survivability and Environmental Requirements		Environmental/accident loads	– Eliminate source of cognitive-attributable latent design errors
Barrier Org. Chart	C32-1	Barrier organization chart	Identifies HE roles, backup positions	– Enhance tolerance to HE errors
Health program	–	Organization wellness programs to maintain personnel health and fitness for service	Fitness for service information	**Integrity/Resilience**
Staffing policies	–	Staffing policies and procedures	Ensure adequate sleep schedules	– HE resistance to environmental loads, task demand (see Figure D.4, App. D.3)
Competency database	C32-1	Defined competency database (see Tables 3.2 and 4.1 for source of new skills and knowledge requirements)	Review database for requirements, verification status	**PE Usability**
Verified competency	D-5	Competency verification reports from competency management database	Review competency verification reports	– Physical ergonomics – Design validation
–	B-25, C-1, C32-1, D-5, E2-1, E5-1	Roster and organization chart	Review for HE and backup status	**HE Competency**
–	D-5, E3-1, E6-1	Competency assessment verifications	Review competency confirmation	– Defined competency requirements – Periodic training updates, competency assessment

AHB design basis	B-1/C-1	Active human barrier design basis		
RBS design basis	B-1/C-1	Remote barrier support design basis		
SRS	Step B20-3.1, C-17	Safety requirements specifications	Review for requirements	
Table 3.7	B-4	Barrier Performance Standards: Base Requirements		**HE Fitness for Service**
Table 3.30	B-19	Reliability Assessment: Single Point of Failure		– Fatigue Management
Table 3.31	B-19	Reliability Assessment: Shared Elements		– Monitor health
Table 3.34	B-26	Fatigue Monitoring and Management Requirements	Input on fatigue target	**HE Availability**
Table 3.33	B20-4	Additional Resources		– Role backup
Table I.1	C-3 to C-8 (see App. I.2)	Cognitive Assessment Evaluation and Recommendations	Review for applicable recommendations	– Protect HE personnel (e.g., Support PE, external protective barriers)
Table H.2	C-2, E-9 (see App. H.2)	PIF Assessment: Barrier, Task, Task Phase		– Maintain environment
Table H.3	C-3, C-7, C-8 (see App. H.3)	PIF Assessment (WE): Direct-Use PE Components		**Minimize/Mitigate PIFs/PIF Effects**
Table K.1	C-3, C-7, C-8	Functional Analysis: Sensory Access to Object While Transiting through a Workspace	Review for reliable access to direct-use PE, unobstructed	*Note:* *Requirements may be defined by the most onerous of different barrier system requirements*
Table K.2	(invokes CX-0.1b sub-process in App. B.6)	Functional Analysis: Object Access from Seated Location at Control Console		
Table K.3	C9-1, 10-1, C11-1	Functional Relationship Between PE Objects	movements, appropriate proximal locations for related elements, etc.	
Table K.4	(see App. K.1, K.2)	Application of Workspace Principles: Physical Elements		

TABLE B.14

Source Documents: Survivability and Environmental Requirements

Source Document	Step/Process	Document Title	Applicable Information	Input to
Table B.8		Common Information Requirements		**Environmental Design**
Table B.9		Component Selection, Design, and Implementation Requirements		
Table B.11		Timing and Response Time Requirements	Demand period or endurance time	– Personnel – Rooms – Buildings
Table B.12		Reliability, Availability, and Integrity Requirements (Physical Elements)	PE/HE locations, accident/environmental loads	– Engineered areas – Technical systems
B.13		Reliability, Availability, and Integrity Requirements (Human Elements)		– Other PE
Barrier Org. chart	B-25, C-1, C32-1, D-5	Barrier organization chart	Identifies HE roles with assigned backups	
AHB design basis	B-1/C-1	Active human barrier design basis		**Survivability Design**
RBS design basis	B-1/C-1	Remote barrier support design basis	Review for requirements	– Personnel – Components
SRS	Step B20-3.1, C-17, C-36	Safety requirements specifications		– Technical systems – Buildings
Table 3.5	B2-x	Barrier Origination Requirements	Identifies hazard that activated the barrier	– Engineered areas – External support systems
Table 3.7	B-4	Barrier Performance Standards: Base Requirements	Defines base survivability requirements	– External protective barriers – Other PE
Table 3.9	B-6	Act Phase Response Physical Element Requirements	Identifies direct-use and Support PE use locations	*Note:*
Table 3.11	B-7	Communication Requirements (Outbound)	Identifies comms: in-person, direct-use comms, PE and use locations, durations	*Requirements may be defined by the most onerous of different barrier system requirements*

Table 3.12	Act Phase: Direct-Use PE Requirements	Identifies direct-use PE use, use locations and hazards, PE usage hazards
Table 3.13	Act Phase: Support PE Requirements	Identifies Support PE function and use locations
Table 3.14	Detect and Act Phase In-Place PE – Engineered Area Requirements	Identifies HEs in this area and activities
Table 3.15	Act Phase In-Place PE – Building Requirements	Identifies HEs in this area and activities
Table 3.16	Act Phase: In-Place PE: Technical System Requirements	Identifies required In-Place PE that must survive hazards and environmental loads
Table 3.25	Detect Phase – SA-1 Information Requirements	Identifies SA-1 information access locations
Table 3.28	External Support System Requirements	Identifies barrier-dependent external support systems and demand period
Table 3.29	External Protective Barrier Requirements	Identifies protected HE/PE, protective function, and duration
Table 3.30	Reliability Assessment: Single Point of Failure	Review for potential survivability requirements/design from design accident/ environmental loads
Table 3.31	Reliability Assessment: Shared Elements	
Table 3.33	Additional Resources	
Table 4.6	Remote Monitoring of Act Phase Response and Support PE	Defines Support PE hazard to user (the basis for the RM function)

(Continued)

TABLE B.14
(Continued)

Source Document	Step/Process	Document Title	Applicable Information	Input to
Table 4.19	C9-1	Control Console or Panel (CP): Barrier Systems and Support Elements	Maintains element function period or duration (e.g., HE, PE, technical system)	
Table 4.20	C10-1	Rooms: Barrier Systems and Support Elements		
Table 4.22	C11-1	Engineered Area: Barrier Systems and Support Elements		
Table 4.23	C12-1	Buildings: Barrier Systems and Support Elements		
Table 4.21	C-9, C-10, C-11, C-12	External Protective Barriers	Identifies hazard threat and protective function, endurance time, etc.	
—	C-3, C-7, C-9 (invokes CX.01 sub-process in App. B.6), C13	Review generated information on component design and implementation	Review for requirements	
Table B.2	App. B.4	Design Basis: Remote Barrier Support – Systems and Infrastructure	Review for requirements	
Table B.3	App. B.4	Design Basis: Remote Barrier Support – Network Interface Requirements	Identifies hazard scenarios	
Table H.2	C-2, E-9 (see App. H.2)	PIF Assessment: Barrier, Task, Task Phase	Review for applicable recommendations	
Table H.3	C-3, C-7, C-8 (see App. H.3)	PIF Assessment (WE): Direct-Use PE Components–		

TABLE B.15

Source Documents: Sizing, Layout, Location, and Placement requirements

Source Document	Step/Process	Document Title	Applicable Information	Input to
Table B.8	Common Information Requirements			
Table B.9	Component Selection, Design, and Implementation Requirements			**Sizing, Layout**
Table B.10	Functionality Requirements			
–	–	Facility/building layout and general arrangement drawings	Input to design activity	– Sized for maximum occupants, optimal HE/PE placement
		Noise Studies		– Supports ancillary facilities
AHB design basis	B-1/C-1	Active human barrier design basis	Review for possible requirements	– Optimal location relative to hazard and environmental loads
RBS design basis	B-1/C-1	Remote barrier support design basis		
SRS	Step B20-3.1, C-17, C-36	Safety requirements specifications	Review for requirements	**HE/PE Location, Placement**
Table 3.7	B-4	Barrier Performance Standards: Base Requirements	Review for possible requirements	– Placed for optimal HE visual access/sightlines
Table 3.9	B-6	Act Phase Response Physical Element Requirements	Provides location information for direct-use	– Placed to support optimal movements, least distractions and noise, etc.
Table 3.11	B-7	Communication Requirements (Outbound)	PE	
Table 3.13	B-8	Act Phase: Support PE Requirements	Identifies Support PE that may affect sizing	
Table 3.14	B-9	Detect and Act Phase In-Place PE – Engineered Area (EA) Requirements	Identifies EA function, sizing, PE/HE in area	*Note:*
Table 3.15	B-9	Act Phase In-Place PE – Building Requirements	Identifies building function, sizing, PE/HE in area	*Requirements may be defined by the most onerous of different barrier system requirements*

(Continued)

TABLE B.15
(Continued)

Source Document	Step/Process	Document Title	Applicable Information	Input to
Table 3.25	B-14	Detect Phase – SA-1 Information Requirements	Identifies access locations for direct-use PE	
Table 3.28	B-17	External Support System Requirements	Identifies dependencies, input to sizing, etc.	
Table 3.29	B-18	External Protective Barrier Requirements	Identifies protected element and the nature of that protection	
Table 3.30	B-19	Reliability Assessment: Single Point of Failure	Review for requirements	
Table 3.31	B-19	Reliability Assessment: Shared Elements		
Table 4.6	C3-8	Remote Monitoring of Act Phase Response and Support PE	Defines remote monitoring location	
Table 4.7	C7-4	Design Simple SA-1 Passive Display	Sub-process CX-0.1b specifies the location and placement of display (direct-use) component	
Table 4.8	C7-5	Design Simple SA-1 Active Display Device		
Table 4.10	C7-8	Design: Incident Command Board		
Table 4.11	C8-4	Video Display Unit: Component Selection and Design	See detailed design process C-10 for guidance.	
Table 4.16	C8-7	CCTV Image Element and PE Requirements		
Table 4.17	C8-8	Off-Console VDU Displays		
Table 4.18	C9-1, C10-1, C11-1	Functional Analysis Recommendations for control consoles and panels, rooms, engineered areas, and buildings	Review for requirements	

Table 4.19	C9-1	Control Console or Panel (CP): Barrier Systems and Support Elements	Provides layout/location information, dimensional guidance, ancillary facilities, etc.
Table 4.20	C10-1	Rooms: Barrier Systems and Support Elements	
Table 4.22	C11-1	Engineered Area: Barrier Systems and Support Elements	
Table 4.23	C12-1	Buildings: Barrier Systems and Support Elements	
Table B.2	App. B.4	Design Basis: Remote Barrier Support – Systems, Modes, and Infrastructure	
Table B.3	App. B.4	Design Basis: Remote Barrier Support – Network Interface Requirements	Design input to interface locations
Table B.4	App. B.5	Design Options to Improve Display Salience	May affect location or placement of displays
Table H.2	C-2, E-9 (see App. H.2)	PIF Assessment: Barrier, Task, Task Phase	Review for possible recommendations
Table H.3	C-3, C-7, C-8 (see App. H.3)	PIFAssessment(WE):Direct-UsePEComponents–	
Table K.1	C-3, C-7, C-8, (invokes CX-0.1b sub-process in App. B.6)	Functional Analysis: Sensory Access to Object While Transiting through a Workspace	Input to guide device selection, location, placement, and relative positioning within a workspace
Table K.2		Functional Analysis: Object Access from Seated Location at Control Console	
Table K.3	C9-1, 10-1, C11-1	Functional Relationship Between PE Objects	
Table K.4	(see App. K.1, K.2)	Application of Workspace Principles: Physical Elements	

TABLE B.16

Source Documents: VDU-Based HMI Displays

Source Document	Step/Process	Document Title	Applicable Information	Input to
Table B.8	Common Information Requirements			**Physical Element**
Table B.10	Functionality Requirements			– Functional specification
Table B.11	Timing and Response Time Requirements			– Cause-and-effect charts
HMI display design	B-1, C-1	HMI Display Guidelines and Style Guides	HMI display guidelines, style guides	– Logic diagrams
AHB design basis	B-1, C-1	Active human barrier design basis	Review for potential guidance	– Control narratives
RBS design basis	B-1, C-1	Remote barrier support design basis	Review for potential guidance	**Human Element**
SRS	B20-3.1, C-17, C-36	Safety requirements specifications	Review for requirements	– Assigned barrier/task actions and activities
Table 3.5	B2-x	Barrier Origination Requirements`	Identifies barrier activator	– Required skills and knowledge to support task activities (e.g., mental models, long-term memory)
Table 3.6	B3-1	Task Origination Requirements	Identifies task activator	
Table 3.7	B-4	Barrier Performance Standards: Base Requirements	Review for requirements	
Table 3.8	B-5	Shared Situation Awareness Requirements	Identifies potential HMI Display supported function	**Org. Element**
Table 3.12	B-8	Act Phase: Direct-Use PE Requirements	Identifies potential function at HMI display	– Procedures
Table 3.17	B-10	Decide Phase Requirements	Identifies decision phase functions	– Training
Table 3.18	B-11	Detect Phase: SA-2 Requirements	Identifies potential function	
Table 3.19	B-12	Detect Phase: SA-3 Requirements	supported by the HMI display	
Table 3.25	B-14	Detect Phase – SA-1 Information Requirements	Identifies SA-1 information function	

Table 3.27	B-16	Barrier, Task, and Phase Safety Time Requirements	Defines safety and response times applied to detect and act phase functions
Table 3.30	B-19	Reliability Assessment: Single Point of Failure	Review for HMI display requirements
Table 3.31	B-19	Reliability Assessment: Shared Elements	
Table 4.6	C3-8	Remote Monitoring of Act Phase Response and Support PE	Defines remote monitoring function and functionality
Table 4.12	C8-4	HMI Display Elements: Detect Phase	
Table 4.13	C8-4	HMI Display Elements: Support Aids and Alternative Display Types	
Table 4.14	C8-4	HMI Display Elements: Act Phase (Controls)	Display design requirements
Table 4.15	C8-5	Integrated HMI Display	
—	C8-4	HMI Display sketch	
		HMI Display functional specification	
Table I.1	C-3 to C-8 (see App. I.2)	Cognitive Assessment Evaluation and Recommendations	Review for applicable recommendations

TABLE B.17

Source Documents: Room Design Requirements

Source Document	Step/Process	Document Title	Applicable Information	Input to
Table B.8	Common Information Requirements			**Room Design**
Table B.9	Component Selection, Design, and Implementation Requirements			– Sizing
Table B.14	Survivability and Environmental Requirements			– Layout
Table B.15	Sizing, Layout, Location, and Placement Requirements			– Equipment placement
Table B.21	Technical Systems (In-Place PE) Requirements			– Lighting
AHB design basis	B-1, C-1	Active human barrier design basis	Review for requirements	
RBS design basis	B-1, C-1	Remote barrier support design basis	Review for applicable requirements for Remote Operations Center	
SRS	B20-3.1, C-17, C-36	Safety requirements specifications	Review for requirements	**HE Design Inputs**
Table 3.5	B2-x	Barrier Origination Requirements	Identifies hazard that required the barrier	– Access to direct-use PE
Table 3.6	B3-1	Task Origination Requirements	Identifies HE task assignments, potential room activities	– Movements – Sightlines
Table 3.7	B-4	Barrier Performance Standards: Base Requirements	Review for possible requirements	– Personal comfort and protection
Table 3.9	B-6	Act Phase Response Physical Element Requirements	Identifies direct-use and Support PE in this workspace	
Table 3.11	B-7	Communication Requirements (Outbound)	Identifies HE communications in/from this workspace	**Dependencies**
Table 3.12	B-8	Act Phase: Direct-Use PE Requirements	Identifies direct-use PE that may be in this workspace	– HVAC/environmental control
Table 3.13	B-8	Act Phase: Support PE Requirements	Identifies Support PE that may be worn or carried in this workspace	– Lighting systems – Electrical power systems – External protective barriers

Table 3.14	B-9	Detect and Act Phase In-Place PE – Engineered Area (EA) Requirements	If the room is an engineered area, table provides design information applicable to the room	
Table 3.15	B-9	Act Phase In-Place PE – Building Requirements	Building may provide info specific to the room, endurance time, etc.	
Table 3.25	B-14	Detect Phase – SA-1 Information Requirements	Identifies SA-1 information accessed in this workspace	
Table 3.27	B-16	Barrier, Task, and Phase Safety Time Requirements	Defines response and safety times that may affect movements, equipment placement, etc.	
Table 3.28	B-17	External Support System Requirements	Identifies room ESS functions and requirements	
Table 3.30	B-19	Reliability Assessment: Single Point of Failure	Review for requirements	*Note:* *Requirements may be defined by the most onerous of different barrier system requirements*
Table 3.31	B-19	Reliability Assessment: Shared Elements		
Table 4.18	C10-1	Functional Analysis Recommendations	Requirements for room functions, design, layout, HE location, PE placement, sightlines, etc.	
Table 4.20	C10-1	Rooms: Barrier Systems and Support Elements		
Table 4.21	C10-1	External Protective Barriers	Identifies required protective functions	
Table B.2	App. B-4	Design Basis: Remote Barrier Support – Systems, Modes, and Infrastructure	Potential input to Remote Operations Center	
Table H.2	C-2, E-9 (see App. H.2)	PIF Assessment: Barrier, Task, Task Phase	Review for applicable recommendations	
Table H.3	C-3, C-7, C-8 (see App. H.3)	PIF Assessment (WE): Direct-Use PE Components–		

TABLE B.18

Source Documents: Engineered Area Design Requirements

Source Document	Step/Process	Document Title	Applicable Information	Input to
Table B.8		Common Information Requirements		**EA Design** – Sizing – Layout – Provisioning – Lighting – Environmental controls – Optimal sightlines to displays – Route markings (paint, signs, etc.) **PE in this workspace** – Stand-alone PE – Support PE – Other
Table B.9		Component Selection, Design, and Implementation Requirements		
Table B.14		Survivability and Environmental Requirements		
Table B.15		Sizing, Layout, Location, and Placement Requirements		
Table B.19		External Support System Requirements		
Table B.20		External Protective Barrier Requirements		
AHB design basis	B-1, C-1	Active human barrier design basis	Review for requirements	
SRS	B20-3.1, C-17, C-36	Safety requirements specifications		
Table 3.5	B2-x	Barrier Origination Requirements	Identifies hazard that required the barrier	
Table 3.6	B3-1	Task Origination Requirements	Identifies HE task assignments, potential EA activities	
Table 3.7	B-4	Barrier Performance Standards: Base Requirements	Review for requirements	
Table 3.9	B-6	Act Phase Response Physical Element Requirements	Identifies direct-use and Support PE in this location	
Table 3.11	B-7	Communication Requirements (Outbound)	Identifies HE communications to/from this workspace	
Table 3.12	B-8	Act Phase: Direct-Use PE Requirements	Identifies direct-use PE components (detect and act phase), Support PE, and functions	
Table 3.13	B-8	Act Phase: Support PE Requirements		
Table 3.14	B-9	Detect and Act Phase In-Place PE – Engineered Area Requirements	Identifies EA function, sizing, PE/HE in area	

Table				Human Element
Table 3.16	B-9	Act Phase: In-Place PE: Technical System Requirements	Identifies technical system function, use conditions, endurance time, etc.	
Table 3.25	B-14	Detect Phase – SA-1 Information Requirements	Identifies SA-1 information accessed in this workspace	– HEs and others in this workspace
Table 3.27	B-16	Barrier, Task, and Phase Safety Time Requirements	Defines response and safety times applied to activities in this workspace	– Physical movements in workspace
Table 3.29	B-18	External Protective Barrier Requirements	Identifies protected HE/PE, protective function, etc.	– Barrier activities in this workspace
Table 3.30	B-19	Reliability Assessment: Single Point of Failure		**Protect/Maintain functions**
Table 3.31	B-19	Reliability Assessment: Shared Elements	Review for requirements	– Maintain PE environment
Table 4.18		Functional Analysis Recommendations	Identifies PE/HE/ancillaries in this workspace; dependent TS, ESS, and EB; and endurance time	– Maintain PE environment
Table 4.22	C11-1	Engineered Area: Barrier Systems and Support Elements	PE/HE locations, access requirements, etc.	– Protect HEs
Table 4.21		External Protective Barriers	Identifies required protective functions	(See Table 4.22)
Table B.4	App. B.5	Design Options to Improve Display Salience	Input to display selection, design, location, and placement	Note: *Requirements may be defined by the most onerous of different barrier system requirements*
Table H.2	C-2, E-9 (see App. H.2)	PIF Assessment: Barrier, Task, Task Phase		
Table H.3	C-3, C-7, C-8 (see App. H.3)	PIF Assessment (WE): Direct-Use PE Components	Review for applicable recommendations	

TABLE B.19

Source documents: External Support System Requirements

Source Document	Step/Process	Document Title	Applicable Information	Input to
Table B.8	Common Information Requirements			Performance Standards
Table B.9	Component Selection, Design, and Implementation Requirements			Procurement Specifications
Table B.12	Reliability, Availability, and Integrity (Physical Elements)			Load lists
Table B.14	Survivability and Environmental Requirements			Capacity Sizing
Table B.21	Technical System Requirements			Technical and Functional Specifications
AHB design basis	B-1, C-1	Active human barrier design basis		Lighting Location / Layout
RBS design basis	B-1, C-1	Remote barrier support design basis		Drawings and Specifications
SRS	B20-3.1, C-17, C-36	Safety requirements specifications	Review for requirements	– Location – Type – Function
Table 3.7	B-4	Barrier Performance Standards: Base Requirements		
Table 3.28	B-17	External Support System Requirements	Identifies ESS requirements for listed PEs	HVAC, Environmental Controls
Table 3.29	B-18	External Protective Barrier Requirements	Review for active EBs requirements, e.g., electrical power	– Temperature / humidity limits, setpoint – Air quality – As active protective barriers
Table 3.30	B-19	Reliability Assessment: Single Point of Failure		
Table 3.31	B-19	Reliability Assessment: Shared Elements	Review for ESS requirements	*Note:* *Requirements may be defined by the most onerous of different barrier system requirements*

—	Component ESS requirements (see CX-0.1f sub-process in App. B.6)	ESS requirements for barrier system components
Table 4.19	Rooms: Barrier Systems and Support Elements	Identifies ESS requirements for CPs
Table 4.20	External Protective Barriers (EB)	Identifies ESS requirements for rooms
Table 4.22	Engineered Area: Barrier Systems and Support Elements	ESS support to EBs (active type)
Table 4.23	Buildings: Barrier Systems and Support Elements	Identifies ESS requirements for EA-located HE/PE
Table B.2	Design Basis: Remote Barrier Support, Network Interface Requirements	Identifies ESS required to support RBS functions, PE equipment
Table B.3	Design Basis: Remote Barrier Support – Network Interface Requirements	Design input to external network systems

TABLE B.20

Source Documents: External Protective Barrier Requirements

Source Document	Step/Process	Document Title	Applicable Information	Input to
Table B.8	Common Information Requirements			**Performance Standards**
Table B.9	Component Selection, Design, and Implementation Requirements			**Procurement Specifications**
Table B.12	Reliability, Availability, and Integrity (Physical Elements)			**Lighting Location/ Layout**
Table B.14	Survivability and Environmental Requirements			**Drawings**
Table B.21	Technical System Requirements			**Load lists**
AHB Design Basis	B-1, C-1	Active Human Barrier Design Basis		**Capacity Sizing**
SRS	B20-3.1, C-17, C-36	Safety Requirements Specifications	Review for requirements	**Technical and Functional Specifications**
Table 3.7	B-4	Barrier Performance Standards: Base Requirements		**Lighting Design** Type
Table 3.29	B-18	External Protective Barrier Requirements	Review for active EBs, i.e., electrical power requirement	Location Function
Table 3.30	B-19	Reliability Assessment: Single Point of Failure		**HVAC/Environmental**
Table 3.31	B-19	Reliability Assessment: Shared Elements	Review for EB requirements	**Controls** Temperature/ humidity limits and setpoint
—	C3, C7, C8	Component EB requirements (see CX-0.1i sub-process in App. B.6	EB requirements for barrier system components	Air filtration Active barrier element
Table 4.20	C10-1	Rooms: Barrier Systems and Support Elements	Identifies EB requirements and protected elements	*Note:* *Requirements may be defined*
Table 4.21	C-10, C-11, C-12	External Protective Barriers	Defines EB requirements, identifies hazard threats, endurance times, etc.	*by the most onerous of* *different barrier system*
Table 4.22	C-11, C-12	Engineered Area: Barrier Systems and Support Elements	Identifies EB requirements and protected elements	*requirements*
Table 4.23	C-12	Buildings: Barrier Systems and Support Elements		

TABLE B.21

Source Documents: Technical Systems (In-Place PE) Requirements

Source Document	Step/Process	Document Title	Applicable Information	Input to
Table B.8	Common Information Requirements			
Table B.9	Component Selection, Design, and Implementation Requirements			
Table B.11	Timing and Response Time Requirements			
Table B.12	Reliability, Availability, and Integrity (Physical Elements)			
Table B.16	HMI Displays (VDU-Based) Requirements			
AHB design basis	B-1, C-1	Active human barrier design basis		**Performance Standards**
RBS design basis	B-1, C-1	Remote barrier support design basis		**Procurement Specifications**
SRS	B20-3.1, C-17, C-36	Safety requirements specifications		**Technical and Functional Specifications**
Table 3.7	B-4	Barrier Performance Standards: Base Requirements	Review for requirements	**Technical Design and Layout Drawings**
Table 3.9	B-6	Act Phase Response Physical Element Requirements		*Note:*
Table 3.12	B-8	Act Phase: Direct-Use PE Requirements		*Requirements may be defined by the most onerous of different barrier system requirements*
Table 3.16	B-9	Act Phase: In-Place PE: Technical System Requirements	Direct-use PE reliance on identified technical system (TS), TS function and endurance time, SA-1 status monitoring, etc.	
Table 3.30	B-19	Reliability Assessment: Single Point of Failure	Review for ESS requirements	
Table 3.31	B-19	Reliability Assessment: Shared Elements		

(Continued)

TABLE B.21
(Continued)

Source Document	Step/Process	Document Title	Applicable Information	Input to
Table 4.8	C7-5	Design: Simple SA-1 Active Display Device	Processes that define TS inputs/outputs, technical design guidance, etc.	
Table 4.10	C7-8	Design: Incident Command Board	Review TS requirements if using a SMART board	
Table 4.11	C8-4	Video Display Unit: Component Selection and Design	Review requirements for presented different video sources to VDU	
Table 4.15	C8-5	Integrated HMI Displays	Review TS requirements for generating an HMI display	
Table 4.16	C8-6	CCTV Image Element and PE Requirements	Review requirements for CCTV image display presentation	
Table 4.18	C9-1	Functional Analysis Recommendations (Control consoles and panels)	Review for requirements applicable to technical system	
Table 4.19	C9-1	Control Console or Panel (CP): Barrier Systems and Support Elements	Input to console/panel sizing, configuration, design, layout, etc.	
Table 4.20	C10-1	Rooms: Barrier Systems and Support Elements		
Table 4.22	C11-1	Engineered Area: Barrier Systems and Support Elements	TS required to support barrier function in these workspaces	
Table 4.23	C12-1	Buildings: Barrier Systems and Support Elements		
Table B.3	App. B-4	Design Basis: Remote Barrier Supporting Telecommunication System Requirements	Identifies technical systems located in Remote Operations Center	

TABLE B.22

Source Documents: Building Requirements

Source Document	Step/Process	Document Title	Applicable Information	Input to
Table B.8		Common Information Requirements		
Table B.9		Component Selection, Design, and Implementation Requirements		
Table B.14		Survivability and Environmental Requirements		**Rooms**
Table B.15		Sizing, Layout, Location, and Placement Requirements		Design, layout, equipment placement, provisioning
Table B.17		Room Requirements		**Engineered Areas**
Table B.18		Engineered Area Requirements		Design, layout, equipment placement, provisioning
Table B.19		External Support Systems Requirements		**Building-dedicated Technical Systems**
Table B.20		External Protective Barrier Requirements		– HVAC, environmental controls
Table B.21		Technical Systems (In-Place PE) Requirements		– Lighting systems
AHB design basis	B-1/C-1	Active human barrier design basis	Review for requirements applicable to buildings	– Electrical power systems – Active protective barriers
RBS design basis	B-1, C-1	Remote barrier support design basis	Review for applicable guidance on the Remote Operations Center	**Protective Functions**
SRS	B20-3.1, C-17, C-36	Safety requirements specifications	Review for requirements	– Personnel
Table 3.7	B-4	Barrier Performance Standards: Base Requirements	Review for possible requirements	– Equipment
Table 3.30	B-19	Reliability Assessment: Single Point of Failure	Review for requirements	
Table 3.31	B-19	Shared Element Reliability Assessment		

(Continued)

TABLE B.22
(Continued)

Source Document	Step/Process	Document Title	Applicable Information	Input to	Dependencies
Table 4.18	C-9, C-10, C-11	Functional Analysis Recommendations	Requirements for building functions, design/layout, equipment placement, etc. (includes workspace requirements: control consoles and panels, rooms, engineered areas)		
Table 4.23	C12-1	Buildings: Barrier System and Support Requirements			**Dependencies** – External support systems – External protective barriers
Table 4.21	C12-1	External Protective Barriers	Identifies the EBs (internal and external) required in the building to protect personnel and equipment		
Table B.2	App. B.4	Design Basis: Remote Barrier Support – Systems, Modes, and Infrastructure	Potential input to Remote Operations Center		*Note:* *Requirements may be defined by the most onerous of different barrier system requirements*
Table H.2	C-2, E-11 (see App. H.2)	PIF Assessment: Barrier, Task, Task Phase	Recommendations on OE design and programs, task design (functions)		

B.8 SUPPORT AIDS (JOB AIDS)

Detailed design processes C-3 through C-8 may identify the requirements for a support aid needed to achieve the barrier function. This section provides guidance on support aids. For a full discussion, see CCPS (2022, Part 2) and Salvendy (2012, pp. 518–520). The information in the support aid should coordinate and complement that in the barrier/task procedure.

Salvendy (2012, p. 518) explains that a support (job) aid type may be informational, procedural, decision-making, or coaching. Informational aids may also be selected/designed to support or enhance comprehension (SA-2), projection (SA-3), or task execution. A procedural aid may provide step-by-step instructions to guide a complex or seldom performed task or task sequence. A checklist may be used to track and checkoff step completion for a lengthy or complex procedure. The decision-making aid provides a method (flowchart or heuristic) to support the decision-making process. Salvendy (2012, Table 7, p. 519) provides the following conditions under which a support aid may be needed:

- The performance of a task is infrequent, and the information is not expected to be remembered.
- The task is complex or has several steps.
- The costs of errors are high.
- Task performance is dependent on knowing a large amount of information.
- Performance depends on dynamic information or procedures.
- Performance can be improved through self-correction.

The CCPS (2022, p. 46) elaborates that "Job aids provide people with the necessary information and knowledge to perform tasks. They help people perform tasks in a 'rule-based' mode of human performance rather than having to rely on their general knowledge." The appropriate job aid (support aid) should improve cognitive outcomes by supporting/supplementing challenging cognitive processes and by minimizing/eliminating an inappropriate reliance on working/short-term memory or an unachievable mental model/long-term memory (e.g., the available candidates lack the domain expertise required to meet a challenging SA-3 requirement.).

The CCPS (2022, Ch. 6) discusses job aids including their types, application and considerations for use. Figure 6.1 in this reference provides an aid selection tool that considers task criticality, complexity, frequency, and the time available to complete the task.

A support aid for a time-critical, emergency response task (a time-of-the essence event) should be brief and easy to read. An appropriate aid may be a *grab card* in the form of a laminated card or sheet of paper. An emergency response task that is complex (e.g., has many steps or requires a high level of time-sensitive coordination between several people) may warrant the use of a checklist or step-by-step instruction. Developing the form may become more challenging if time available for use is short (CCPS 2022, p. 54).

Support aids can be accessed in various ways, including paper, an HMI display, or other electronic access/delivery methods. Considerations may necessarily include the ability to access the aid from different locations. The aid content and form must be appropriate to its use in the supported task activity. CCPS (2022, Sections 6.2 and 6.3) identifies additional aid types and provides considerations and guidance for use. CCPS (2022, Section 6.4) provides a brief discussion of electronic aids and when this approach may be appropriate.

Support aids specific to time should be considered. Time is safety critical and often a challenge to track. All active human barrier activities tend to be time-critical. Aids that enhance the use and awareness of time should be considered. From a barrier reliability and performance perspective, humans under stress often become unreliable timekeepers. Common support aids include countdown timers, count-up timers, time-based alerts and reminders, and other time-based displays as needed to support a time-sensitive decision or action. Depending on the application, the time-based aid may include an audible alarm or other easily detected alert. (This feature should only be added if an action is expected in response to the alert.)

When selecting the support aid (e.g., the aid presentation type and form), consider the *Compatible SA* guidance in Appendix C.1.3.

Appendix C
Teams and Teamworking

This appendix is organized as follows:

- C.1 Team Situation Awareness
- C.2 Non-Technical Skills
- C.3 Implicit and Explicit Coordination
- C.4 Trust

A barrier that requires two or more HEs creates a teamworking environment. The barrier/task functioning and performance depend on the nature, efficacy, and efficiency of the interactions and interdependencies within the team. Barriers of this type introduce two additional areas of design, namely, ***Team Situation Awareness*** and ***Non-Technical Skills***.

From Hollnagel (2003, Ch. 31, p. 754),

> The result of the team design effort is a team structure that specifies both the structure and the strategy of the team, including who owns resources, who takes actions, who uses information, who coordinates with whom, the tasks about which they coordinate, who communicates with whom, who is responsible for what, and who shall provide backup to whom.

The design of the barrier task and team develops, integrates, and implements human, physical, and organizational elements in a way that reliably achieves the barrier function at the specified performance level. Figure C.1 shows the relationships between individual competencies and attributes, barrier-level design, task-level design and execution, and conditions that impact team performance. Figure F.2 (Appendix F.2) identifies key internal and external factors that affect the cognitive and physical performance of each team member. As such, they can also affect team interactions, coordination, and collective performance.

C.1 TEAM SITUATION AWARENESS

C.1.1 INTRODUCTION

Situation awareness (SA, individual and team) is commonly identified as a required non-technical skill. As elaborated below, team SA requirements can affect HE, PE, and OE design and development. There are many models of team SA, which have both similarities and differences. The team SA models adopted for this book include M. Endsley's shared SA construct (Endsley and Jones, 2001, Ch. 1) and the compatible, transactive, and Meta SA constructs from the distributed situational awareness model proposed by Salmon et al. (2009).

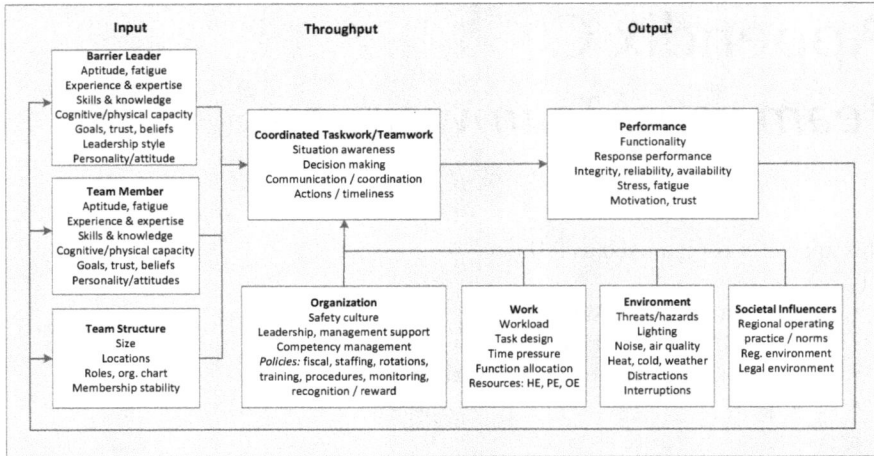

FIGURE C.1   Barrier Team Performance Model.

Source: Adapted and extensively modified from Flin et al. (2008, Figure 5.1).

1. **Shared SA:** See the preliminary design step B-5 and C.1.2 for the topic introduction and detail.
2. **Compatible SA** – Different from shared SA, compatible SA

> is based on the notion that no two individuals working within a collaborative system will hold exactly the same perspective on a situation. Compatible SA therefore suggests that, due to factors such as individual roles, goals, tasks, experience, training and schema, each member of a collaborative system has a unique level of SA that is required to satisfy their particular goals.
>
> *(Salmon et al. 2009, p. 190)*

> *Note: As applied in the model, compatible SA is extended to address cases where the same information (different users) should be presented in the form that is best suited to how each user uses the information. This may apply if the form of the information presentation causes misinterpretation or places an unnecessary demand load on working memory/short-term memory. The presentation form may be right for one person (performing Task A) but wrong for another (performing Task B). See step C-8, HMI display design, for the application of compatible SA. See Endsley and Jones (2012, p. 216) for additional background and support for this approach.*

3. **Transactive SA,** as applied in the model, refers to a human element transaction (the requesting agent) with an external source of information (the external sourcing agent) that leads to the acquisition of information needed to support an SA requirement (Salmon et al. 2009, pp. 67–69, 192–193, 220). As examples, the external agent may be a person, procedure, support aid, email, HMI display, or information available in the environment.

Information from the latter may be visible, audible, or tactile feedback from a fire or an injured person. Within teams (HE-HE transactions), Salmon et al. (2009, p. 193) inform:

> Team members may exchange information with one another (through requests, orders and situation reports); the exchange of information…leads to transactions in the SA being passed around; for example, the request for information gives clues to what the other agent is working on. The act of reporting on the status of various elements tells the recipient what the sender is aware of.

4. **Meta SA** is an "awareness of what other agents in the system know" (Salmon et al. 2009, p. 220).

Note: Why is Meta SA important? Chiappe et al. (2012) informs us, "perception research reveals that people rely on the structure of the world and the action in it to limit internal storage and processing." This awareness should work to enhance task performance because one is learning to work with the barrier system as it has been designed. The task design (e.g., procedures, training, and physical element design) determines where information is stored, when it should be accessed, and by whom. It may expect the HE to access and acquire information at a fixed point (step or time) and hold it in working/short-term memory until needed. Alternatively, the design may expect and allow for on-demand access when needed, for example, just-in-time (JIT). (JIT supports optimal execution performance, though can become a liability when things go wrong.)

The above discussed design options invoke questions and considerations. Does the information change frequently? Does the external system (information source/holder) automatically update this information? Does task success require the use of the most recent information? What if the external source becomes unavailable? Is the information difficult to remember as can be the case if the information is not relatable/memorable or a string of random characters? Is it reasonable to expect one to hold the information in working/short-term period given different situations, for example, during peak workloads, extended periods of time (many seconds or minutes), or the occurrence of persistent or poorly timed interruptions and distractions? What happens if the information source is an HE (the request interrupts the provider's work in progress)? Table K.7 (Appendix K.5) may offer additional questions and considerations.

Regardless of the approach taken, people who are elements in the barrier system will learn where information in the system resides, when it is available or updated, and how to get it when needed. With drills and repetition, each member of the barrier team develops their unique Meta SA, information stored in one's long-term memories. (With repetition, this complex interplay of memory and behaviors progresses and develops into automatic skilled actions.) That memory typically improves individual and team performance. Those same memories may become a liability if the

experiences that formed those memories did not include the abovementioned disturbance conditions and situations and their knock-on effects.

C.1.2 Shared SA

With shared SA (SSA), no two HEs share a complete and mutually represented
picture of the event in progress. A requirement to do so is unrealistic if it places
an unsustainable cognitive and workload demand on one or more team members.
Instead, step B-5 (preliminary design) defines the minimum SSA requirements each
HE needs to maintain the level of team cohesiveness and coordinated response that
achieves the barrier function within the specified safety time.

To this end, Hollnagel (2003, Ch. 31, p. 759) notes that "the shared mental model
explains from a cognitive point of view what is required to design a robust and flexible team capable of handling unexpected disturbances in dynamic situations."

A common picture develops when the team receives the first information from the
barrier leader. SSA contributes to mutual understanding and mutually compatible
team actions. Shared mental models develop as an outcome of repeated team drills
and exercises that create and contribute to a shared experience and environment.
Teams that remain together over longer periods develop shared mental models, as
they experience how each member and the team as a whole performs in different
situations (Sneddon et al. 2006). Cross-training and drills can help to develop shared
views among team members as they gain experience performing their assigned roles,
interact with each other, and observe and experience how and when each performs
their assigned tasks. These activities also contribute to learning who holds the needed
information, who is likely to request that information, and the timing of the request.

For an offshore O&G facility, owner/operators commonly seek to reduce the
number of persons residing (24x7) on the facility to reduce the risk exposure and
cost. The varied nature of many emergency response barriers requires a wide range
of team capabilities. To accommodate both objectives, the barrier team should be
designed to have the fewest number of individuals needed to reliably achieve the
barrier function and specified performance. Different tasks often require different
skills, knowledge, and expertise. As such, the typical barrier team is heterogeneous,
an important fact to address in team design and role assignments. For an offshore
O&G facility, the barrier leader (typically the Offshore Installation Manager) guides
and maintains team focus, coordinated actions, and priorities. Key roles should have
assigned competent backups that are ready to assume the role on a moment's notice.

Team situation awareness (team SA) encompasses an enabling design (PE, OE,
and HE) that transforms a heterogeneous group of individuals into a functioning
and performing team. The barrier tasks, task assignments, procedures, communication protocols, training, and competency requirements affect how the team interacts,
coordinates to achieve the barrier function, and responds to change. A share team
understanding, and picture is needed to maintain the capability to provide a timely
and appropriate collective response to certain event types and requests. For example,
a common shared team awareness is needed to collectively (lockstep) response if a
barrier leader activates a new barrier (e.g., a search and recovery barrier), or changes
the status of an active barrier or the relative priority between two barriers.

C.1.2.1 Remote Barrier Support

For a remote barrier support function performed from a remote location, achieving and maintaining the required level of SSA by the RBS HE is challenging. Endsley (2021, pp. 7–8) states that a *shared environment* is one of the Team SA devices that contribute to SSA. Being physically remote from the protected facility, the RBS HE does not have access to the rich sources of sensed ambient information (through visual, audio, smell, or tactile senses) available to the HEs at the protected facility before and during barrier activation. Depending on the assigned RBS task, access to this information may be important to the performance of that task. Options to access this information may include having live access to radio communications and live CCTV camera views of the facility, Central Control Room, Incident Command Center, and muster areas. Not having this information may increase the RBS HE's reliance and use of communications (radio, telephone), an action that increases the workload of HEs located at the protected facility. An inability to access these rich sources of information may degrade the performance of the RBS task. (For background information on the richness of different information sources, see Table K.5 in Appendix K.3.)

Other methods to close the *shared environment* gap may include the following:

- Implementing ROC physical facilities that duplicate or closely mirrors those at the protected facility, for example, the room layout and provisioning, control consoles, and access to displays available on the protected facility. (This approach is especially important if the barrier staffing/rotation practice swap roles of those rotated between the ROC and the protected facility.)
- If the ROC is in a different time zone, consider visual cues to indicate the time of day at the protected facility.
- Monitor the weather at the location of the protected facility.

C.1.3 Compatible SA

An important aspect of compatible SA, two HEs assigned to different tasks may use the same information. However, the meaning and application of that information to each HE may change based on their assigned tasks.

The following is a suggested process that applies when two or more HEs use the same SA-1 information accessed from an HMI display:

1. Identify SA-1 information used by two or more HEs assigned to the barrier.
2. Examine the function of the information as it applies to each task. For example, the information may be provided to support a feedback monitoring function, an SA-2 or SA-3 requirement, or a decision or act phase response. (Table 3.25 provides this information.)
3. When the use of the SA-1 information differs, evaluate whether the proposed display approach provides the appropriate presentation, given how the information is used and how it contributes to the desired interpretation. Will the proposed display adequately support a stated comprehension

requirement or contribute to cognitive bias (Appendix F.2.5) or keyhole effect (Appendix F.2.6)? If the answer to any of these questions is *YES*, different display presentations may be needed.

4. When it becomes necessary to propose a different display presentation (step 3), confirm that the recommendation does not violate display standards (e.g., a one-of-a-kind display may create new and different problems).

C.1.4 TRANSACTIVE SA

Transactive SA applies to an HE interaction with an information source. The source may be an HE, a passive or active display, or sensory information available from the environment.

A product of transactive SA, an exchange between two HEs provides clues to the receiver about what others may be doing. The exchange should require less time and effort to convey information if the level of shared SA is high (Endsley 1995, p. 39). The exchanged or conveyed information should be limited to only that needed to perform an assigned task or meet a shared SA requirement.

Select the communication protocols, terms, and syntax that best ensure the information is correctly understood while minimizing the exchange duration (Gasaway 2013, Ch. 7). A two-way exchange can improve communication accuracy but consumes the attention resources of the engaged parties for the exchange duration. This action may divert attention from potentially higher priority actions. Using pre-determined and mutually understood terminology and code words can reduce the exchange duration without reducing the exchange effectiveness.

See Appendices K.3–K.5 for guidance on the design of the communication workspace.

C.1.5 META SA

The definition for Meta SA employs the term "agent." Here, the agent refers to the HE, PE (e.g., passive or active display), OE (e.g., procedures), or ambient environment source (e.g., an incident scene) that holds and provides the required information. With experience, the HE learns where this information resides and how and when it can be accessed and made available. An important consideration is that "SA may sometimes involve simply knowing where in the environment to find a particular piece of information, rather than remembering what the piece of information is" (Durso 1999, pp. 1–2). Meta SA also refers to knowing the information that others may need and when they need it.

Stress, excessive workload, frequent interruptions, and other task conditions common to active human barriers may reduce the amount of information the HE can reliably retain in the working memory and its storage duration. These conditions increase the likelihood that information in the working memory (i.e., short-term memory) is forgotten or incorrectly remembered.

Chiappe et al. (2012) presents a situated SA model that posits, "Individual operators off-load computation and information storage to the environment as much as possible to limit what they have to do internally" (p. 630). Many activities in the

TABLE C.1
Team Situation Awareness – Design Objectives and Applications

Shared SA	Compatible SA	Transactive SA	Meta SA
To function and perform, teams need a minimum shared understanding, e.g., SSA. (See C.1.2 for background.)	Two or more HEs may need the same SA-1 information to support different tasks. (See C.1.3 for background)	Through transactions with other HEs, displays, and the ambient environment, the HE acquires the needed information. Each transaction should be necessary and limited to the time needed to complete the transaction. (See C.1.4 for background.)	Identify information that each HE needs, when it may be needed, when it may be available, and if the information can be stored or retained in a repository easily accessed in the future without undue burden on the assessor or others. (See C.1.5 for background.)
Design Objective: Define the minimum required shared understanding each team member must have. Develop the mental models needed to develop this shared understanding. Identify the SA-1 information needed to achieve this end.	**Design Objective:** As appropriate, present information in the form that best reflects how the information applies to each HE-assigned task.	**Design Objective:** Identify and understand the nature/form of each HE transaction. "Design" the transaction to minimize the physical and cognitive effort and duration. The selected transactive approach should be the option best suited to the information type, access frequency, and timing and situation of use.	**Design Objective:** Minimize all reliance on short-term memory and long-term memory for seldom-used information. Instead, store this information in a form that can be intuitively accessed with little cognitive and physical effort.

Note: For supporting information, see Appendices K.3, K.4, and K.5 (Communication, Human, and Information Workspace Design).

prototype lifecycle model look for opportunities to employ a Meta SA approach that offloads computation and information storage where it improves the barrier task execution and performance and does not introduce potentially unacceptable new risks. Examples opportunities to consider its use occur in detailed design step C8-4a and the cognitive assessment and mitigation process in Appendix I (performed in the detailed design processes C-3 through C-8).

C.1.6 Team SA Application

Table C.1 provides an overview of the basic design objectives when using the presented team SA model.

C.2 NON-TECHNICAL SKILLS

Non-technical skills (NTS) are those considered essential to effective team functioning and performance. According to Flin et al. (2008, p. 1), NTS include the following:

- Situational awareness
- Decision-making
- Communication
- Teamwork
- Leadership
- Identifying and managing stress
- Managing fatigue
  Each is addressed in the lifecycle model.

C.3 IMPLICIT AND EXPLICIT COORDINATION

C.3.1 Implicit Coordination

For Hollnagel (2003, Ch. 31, p. 758),

> One of the most important team processes that is affected by shared mental models is communication. It is hypothesized that shared mental models allow team members to explain and predict the informational needs of teammates. Because team members rely on their shared mental models, communication can take place efficiently and effectively.

Coordinating communications can be less frequent and of shorter duration. The term used for this collective set of results is **implicit coordination** (IC). Over time, IC develops as an automatic, skill-based capability.

HEs working in a barrier team must develop the mental models, schemas, and action scripts needed for effective and efficient team interactions and functioning. Implicit coordination improves over time as individuals gain team experience and develop the skills and long-term memories that enable and create this emergent behavior and capability. Through repetitive, time-pressured team exercises and drills, team members develop the ability to anticipate what others need and when they need it. IC can reduce the communication frequency and duration needed to maintain team functioning and coordination, but to a limited extent. Explicit coordination is also required, as discussed in C.3.2. (See Appendix F.5 for information on mental models.)

C.3.2   EXPLICIT COORDINATION

From Hollnagel (2003, Ch. 31 pp. 758),

> Communication during task execution refines team members' shared mental model with contextual cues. This may result in more accurate explanations and predictions of the teamwork demands… For maintenance purposes, communication is needed to keep the shared mental model up to date with the changes that occur during task execution. Especially in dynamic or novel situations, when teams must deal with unanticipated disturbances, communication is needed to preserve an up-to-date shared mental model of the situation and to adjust strategies or develop new ones to deal with the situation.

The term used for this type of communication is **explicit coordination**. Explicit coordination may be needed:

- If a barrier leader provides information on plans, priorities, progress, and changes.
- When a team member responds to an information request from another team member. (The timing and expectation of the request may be included in a procedure or triggered by an event.)
- To activate or coordinate an additional resource such as a helicopter evacuation service or telemedicine resource. (For background, see the preliminary design step B20-4 and the detailed design step C14-5.)

C.4   TRUST

The book *Teams That Work: The Seven Drivers of Team Effectiveness* (Tannenbaum and Salas 2021) provides an up-to-date and well-researched review of the essential traits, behaviors, and attributes of a team. The book identifies the following seven drivers of team performance:

- Capabilities: Includes task work and teamwork competencies
- Cooperation: Includes trust, psychological safety, collective efficacy, and cohesion
- Coordination: Includes monitoring, providing backup/support, adapting, and managing team emotions and conflict
- Communications
- Cognitions: Shared mental models, transactive memory systems

- Conditions: The context and environment in which the team operates, for example, organizational policies and practice, senior leadership support, resources, time availability, autonomy/decision-making authority, and team mission/purpose
- Coaching and leadership

These topics are addressed to varying degrees throughout this book and the life-cycle model. The focus of this section is "trust." Tannenbaum and Salas (2021, p. 61) state that "When team members trust one another (intrateam trust), they consistently demonstrate better team performance." Also from this source, researchers confirmed that "trust was even more strongly related to performance in highly interdependent teams…, when authority is more centralized, and when team members skills are more dissimilar." (This describes the typical emergency response barrier team for an offshore O&G facility.)

Tannenbaum and Salas (2021, p. 61) further state,

> In a more centralized team, people with the authority to make decisions must trust the rest of the team to feed them accurate information, and individuals without that authority must trust that those in power will make decisions that benefit the team. And when team members possess dissimilar expertise, they must trust that their teammates know what they are doing and are making contributions, since they lack the expertise to readily assess if that is true.

By necessity, active human barriers with several to many team members tend to be highly interdependent.

Tannenbaum and Salas (2021, pp. 61–62) elaborate that, like co-located teams, trust is also important when team members are in physically separate locations, a condition that occurs with a remote barrier support (RBS) task performed from an ROC. (For background, see Appendix L.) In this case, trust plays a role in the willingness to accept and consider high-consequence information and guidance from personnel who are not at the facility. Developing and maintaining an elevated level of trust may be facilitated if rotational practices swap the barrier leader and RBS roles with each rotation. Their shared mutual experience working in each role should enhance the degree of overlap and similarities in their shared mental models. Providing the RBS-assigned HE with access to live CCTV video, radio communications, Incident Command Board, and other information contributes to maintaining the shared situation awareness required to perform the assigned task in coordination with the HEs located on the protected facility.

Based on the abovementioned book (p. 63), trust develops between individuals when the other person:

- Demonstrates their capability to do the assigned work (competent),
- Takes action that considers my interest (has my back), and
- Demonstrates personal integrity (follows acceptable values and principles).

Trust development is aided by team-based drills and exercises, and the actual experience of working together on an activated barrier. (The team culture and provided interpersonal tools should provide for and demonstrate the ability to resolve differences and conflicts, which are essential to maintaining trust.) Trust is earned and develops over time but can also change in response to a single event. The operate and maintain phase PIF assessment (process E-9) provides one possible means to assess the level of trust within barrier teams and among the barrier-assigned maintenance personnel.

Appendix D
Performance Standards

This appendix is organized as follows:

- D1 Introduction
- D2 Differences in Global Standards and Practice
- D3 Suggested Performance Standards
- D4 CIEHF (2016) Suggestions for a Human Performance Standard

D.1 INTRODUCTION

Appendix G defines the verification and validation scope and schemes adopted in the prototype lifecycle model. This appendix defines the performance standards that provide the acceptance criteria for these processes.

According to CCPS (2018, p. xix), a performance standard is a

> measurable statement, expressed in qualitative or quantitative terms, of the performance required of a system, equipment item, person, or procedure (that may be part or all of a barrier), and that is relied upon as the basis for managing a hazard. The term includes aspects of functionality, reliability, availability, and survivability.

NORSOK S-001 (2008, cl. 4.3) considers that a

> Safety performance standard shall be the verifiable standard to which safety system elements are to perform. The objective of the specific safety performance standard is to add any supplemental safety requirements other that those specified by authority requirements and standards.

Several countries' regulatory regimes and associated practice standards define performance standards to include the applicable regulatory requirements and applicable industry standards, and good engineering practice. The standards discussed here address other types of performance standards applicable to active human barriers.

Note: EI (2020b, pp. 67, 74) uses the terms "assurance" and "verification" to differentiate between two different verification processes. Assurance applies to verification activities performed by the owner/operator or assignee, for example, an inspection or Factory Acceptance Test. Verification applies to verification schemes and activities performed by an independent assessor. The suggested scheme includes observing an assurance activity, verifying an assurance activity is performed according to a published procedure and plan, and reviewing sample design documents to verify that they address and correctly apply the requirements listed in a PS. Appendix G addresses assurance and verification processes. The suggested Barrier Safety Management Plan (Appendix B.2) should define the management system for both.

Figure G.1 (Appendix G.1) shows the timing of the proposed verifications at various stages of the barrier lifecycle.

D.2 DIFFERENCES IN GLOBAL STANDARDS AND PRACTICE

D.2.1 PERFORMANCE STANDARD TYPES AND APPLICATION

CCPS (2018) and EI (2020b) recognize two different performance standard types, design-focused and operations-focused. The CCPS (2018, p. 112) provides the following examples:

- Design: "The fire water system shall have two independent water supplies with two fire water pumps and with diverse power supplies delivered through a ring main, which can isolation any leak."
- Operations: "The fire water system shall deliver, for example, 5000 gpm at 120 psi at the most remote location, within 3 minutes of actuation."

CCPS (2018, p. 110) also notes,

Acceptance criteria are measurable standards that are used to establish whether a barrier is functioning properly or is degraded to some degree, or in the design phase meets the original design specifications. Acceptance criteria are also called performance standards. Acceptance standards during the operations phase can be simpler than full confirmation of the detailed design specification.

D.2.2 GLOBAL DIFFERENCES IN PS PROPERTIES

As with other areas of active human barrier design and lifecycle management, global regulatory requirements and practice standards vary in the properties addressed in a performance standard. Figure D.1 compares the performance standard properties in different documents and regulatory regimes, and those proposed in the performance standards developed via the prototype lifecycle model (see Section D.3).

PSA (2013, 18) contends that "performance means the properties which a barrier element must possess in order to ensure that the individual barrier and its function will be effective."

Material differences in the properties and requirements of performance standards could contribute to different verification outcomes such as findings, effectiveness, and comprehensiveness.

For PSA (2013, p. 3), performance requirements are "verifiable requirements related to barrier element properties to ensure the barrier is effective."

To this, EI guidance adds two additional properties, interactions and dependencies. They are described in EI (2020b, p. 31) as follows: "Interactions' refers to the interfaces of an SCE with other systems, which usually are also SCEs. 'Dependency'

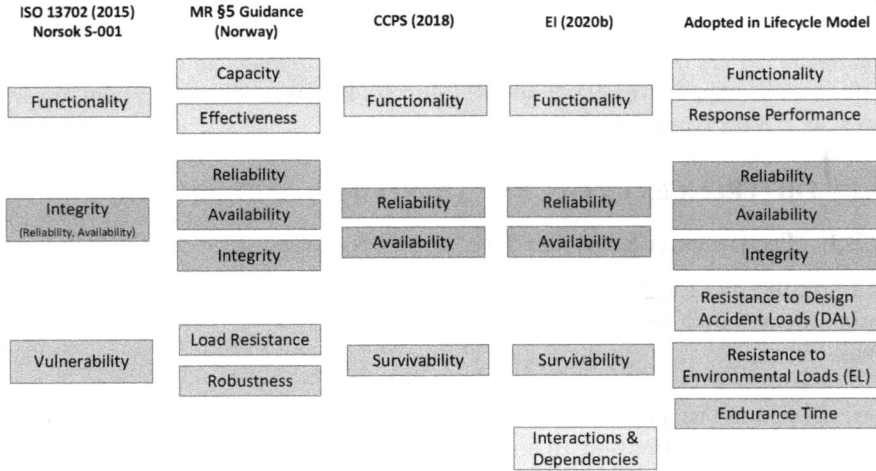

FIGURE D.1 Comparison of Performance Standards Requirements; Modified and Adapted from Figure 12, PSA (2013); "MR §5" – Petroleum Safety Authority Norway, Management Regulation §5.

refs to the degree of reliance of the SCE on other systems (usually also SCEs)." In the lifecycle model, example dependencies include external support systems (e.g., lighting, electrical power) and external protective barriers (e.g., passive and active). Figure D.2 provides a visual representation of active human barrier dependencies and interactions that are common in an offshore O&G facility and other facility types.

The performance standards for an external support system or external protective barrier that supports one or more active human barriers should be guided by the most onerous applicable requirements among those barriers. This includes reliability, endurance times/durations, and load capacities. (For example, that load demand should be considered when specifying and sizing a battery backed power supply system.) In addition, those requirements should consider all plausible scenarios and their associated endurance times, durations, and capacity/load demands. An example (and likely) scenario is the simultaneous activation of multiple barriers, and each of those barriers rely on other ESSs, in-place PEs, workspaces, and external protective barriers that themselves may rely on the same ESS or EB to provide that supporting function. (Figure D.2 provides a visual representation of this complex interdependency. It may be helpful to create a drawing like this for each barrier that shows its specified endurance time/duration, and the required ESSs and EBs that must be in place (and functioning) to realize that requirement.)

D.2.3 PROPERTIES SELECTED IN THE PERFORMANCE STANDARDS IN THE PROTOTYPE LIFECYCLE MODEL

Figure D.1 suggest the model-adopted performance standard properties to address in performance standards that support the verification processes described in

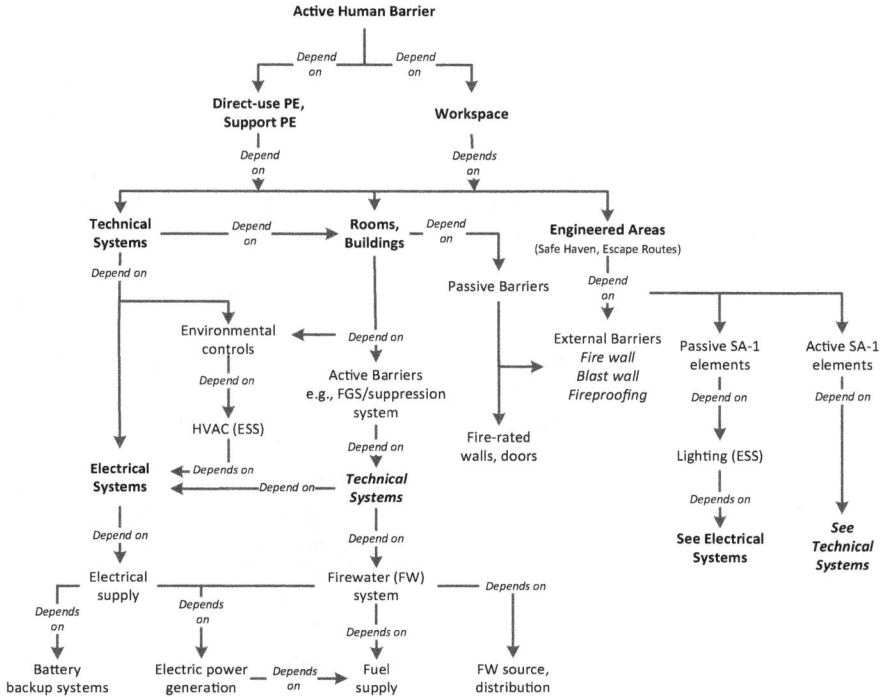

FIGURE D.2 Interdependencies: Barriers, Technical Systems, External Support Systems, and Protective Barriers; FGS – fire and gas detection system; HVAC – Heating, Ventilation, and Air Conditioning.

Appendix G. The deviations from other performance standards reflect the differences and unique properties in active human barriers. The tables in Appendix B.7 identify the source requirements and information developed in the lifecycle model that may aid and guide the development of performance standards. The selected model properties indicated in Figure D.1 are elaborated below.

D.2.3.1 Reliability

Reliability is defined by CCPS (2007b, p. 259) as *"The probability that equipment operates according to its specification for a specified period of time under all relevant conditions."*

D.2.3.2 Availability

Availability is the fractional time a barrier is online and ready to perform its specified safety function with the required performance.

D.2.3.3 Integrity

For CCPS (2015, p. 47), "Integrity is a property of the IPL that is a measure of its capability to satisfy its specified requirements. For an IPL to be dependable, it must be available when needed."

The integrity requirements for performance standards vary based on their application (e.g., use in verifying or accepting a PE, an HE, or the barrier system as a whole). A PE may have an assigned Safety Integrity Level (SIL) target that requires periodic proof testing to verify it is achieved. A performance standard for an HE may define integrity in terms of error rate or the number of deviations from procedures, whereas one for a barrier system may define it in terms of the success (or failure) to achieve the required barrier safe state within the defined barrier safety time (BST).

Note: Active human barriers often include elements that have a fail-to-danger design. For example, a siren that activates a muster barrier does not function if its electrical activation circuit fails or power to the circuit is lost. Depending on the design, the failure may be detected and announced immediately or remain hidden. Humans can also exhibit a fail-to-danger behavior. The performance of a barrier-assigned person (HE) can significantly degrade in cases of excessive fatigue or stress and other internal and external conditions that degrade physical and cognitive capabilities. The timing, frequency, and scope of a verification process applied to personnel should consider the likelihood and frequency of barrier activation and the consequences of barrier failure.

D.2.3.4 Survivability

Depending on the performance standard type, the included survivability requirements may apply to a barrier PE or HE. Both must survive the event that activated the barrier and do so for the barrier activation period. Further, the specified design accident and environmental load levels should be addressed in the desgin so their effects cannot degrade PE and HE performance to a level that places the barrier (or an unprotected person) at risk. For example, a continued unprotected HE exposure to excessive heat, cold, vibration, or noise during barrier activation can significantly degrade HE performance and potentially place the barrier (and personnel) at risk.

D.2.3.5 Time

Figure D.1 includes two time-related properties: response performance (functionality) and endurance time (survivability). In some cases, endurance time may be constrained by the capacity of an external support system, for example, a battery-backed power supply system.

As a constrained resource, the available barrier safety time (BST) is allocated (parsed) to physical and human elements. The maximum possible time should be allocated to the HE as the element with the higher degree of variability. (The HE is also the only element in the barrier system that can respond to unexpected situations not considered in the barrier design.) Because time can adversely affect humans (e.g., potential negative effects on performance and behaviors), the verification process and performance standards should focus on activities that are the primary time consumers and on metrics that track the inappropriate use of time.

Typically, the hazard that activates the barrier fixes/controls the time available to respond to that hazard, for example, the time available to activate and implement barrier functions to regain control and prevent hazard escalation. If not responded to quickly enough, the size and scale of the hazard event can increase at an exponential rate. A countering human response cannot match this change rate. In response, the barrier design process should seek to respond in a way that limits the opportunities for escalation and escalation pathways. An appropriate performance standard metric can address HE and team responsiveness, i.e., the reliable achievement of the target task and phase safety times defined in Table 3.27. Another could examine the total elapsed time consumed versus the specified barrier safety time (BST). A different metric may verify the quality (accuracy and usefulness) of an SA-3 projection or how that information is actually used in a pre-emptive action like completing the mental (or team) preparation and planning should the action be required.

Refer to Figure 3.8 (preliminary design step B-18). If the abandon-facility barrier is activated, the endurance time of the firewall may become one of the primary constraints limiting the time available to initiate and safely evacuate all personnel. Such scenarios warrant ongoing status monitoring and alerts of the barrier-dependent external protective barriers and support systems (e.g., remaining time before failure or capacity depletion) given their potential to constrain time that remains to achieve the barrier function and safe state. As such, it may be appropriate to identify dependencies of this type on bow tie diagrams.

Consider an offshore production facility with a barrier that addresses a potential ship collision. The available response time depends on the ship speed, direction, and attributes (size, type, and capability to change speed and direction), the time the potential hazard is first detected, and communication between the ship and the protected facility (timeliness and effectiveness). The barrier safety time is unique to the scenario. The procedures should address these varied situations and also define the decisions, actions and timings that ensure time remains (with a defined safety margin) to initiate and safely complete an abandon facility if required to do so.

For additional background, see Appendix B.5.3 and preliminary design step B-16.

D.3 SUGGESTED PERFORMANCE STANDARDS

This section provides example performance standards that support the verification and validation processes defined in Appendix G.

D.3.1 DESIGN PERFORMANCE STANDARDS FOR A PHYSICAL ELEMENT – EXAMPLE CONTENT AND REQUIREMENTS

Figure D.3 outlines the suggested properties and requirements of a performance standard for a barrier physical element. As discussed in the introduction, different performance standards (and standard types) may be needed to support the design phase and the operate and maintain phase verification processes.

Note: EI (2020b, Annex E) provides an example of a performance standard for a barrier system physical element. A PS developed in the prototype lifecycle model may benefit from using the document approach described in EI (2020b, pp. 90–91),

Property	Subject	Requirements
Functionality	Functionality	Define the functions to be performed or achieved by the physical element.
	Response Performance	Define the maximum permitted response time to achieve the specified PE functionality. (See the response targets in Table 3.27.)
Integrity	Reliability	Define the minimum required PE reliability, e.g., mean time between failure. (PE reliability may include the contribution from Its dependent ESS and EB.)
	Availability	Define the minimum required PE availability to perform the specified PE function upon barrier activation and do so for the full activation duration, e.g., mean time between failure. (PE availability may include the contribution from its dependent ESS and EB.)
	Integrity	Define the minimum PE integrity metric, i.e., a combined reliability/availability metric, a maximum accepted system error or deviation rate/value, a specified Safety Integrity Level target, etc.
Survivability (Vulnerability)	Resistance to Design Accident Loads (DAL)	Define the minimum acceptable resistance to defined DALs so as to not effect functionality or compliance to other performance requirements. (This may apply to the PE and dependent ESS and EB.)
	Resistance to Environmental Loads (EL)	Define the minimum acceptable resistance to defined ELs so as to not effect functionality or compliance to other performance requirements. (This may apply to the PE and dependent ESS and EB.)
	Endurance Time	Define the required duration and period of resistance to each identified DAL and EL, e.g., the resistance or capacity that must be achieved or maintained for the duration of barrier activation.

FIGURE D.3 Performance Standard – Physical Elements.

which identifies (lists) the SCE interactions. The list may provide the means to identify a dependency on a cable system or tray that carries the barrier-required signal, power, and network cables. (For background, see step CX-0.1g in Appendix B.6). The approach may also support activities that seek to identify the most onerous requirements (among all barriers) that should be applied to the dependent element, for example, an external support system or external protect barrier.

D.3.2 DESIGN/OPERATIONAL PERFORMANCE STANDARDS FOR HUMAN ELEMENTS – EXAMPLE CONTENT AND REQUIREMENTS

Figure D.4 provides the suggested properties and requirements of a performance standard for barrier-assigned personnel (HE). As discussed in the introduction, different performance standard types may be needed for a design phase versus an operate and maintain phase verification process.

Consider: An additional performance standard may be warranted for maintenance personnel assigned to perform sensitive maintenance (a safety critical task) on a high-consequence safety system. For background, see the hypothetical case study example in Chapter 9, Section 9.5, Application to Safety Critical Tasks.

Example integrity requirements may include the following:

- Communication:
  - Conveys the correct information at the correct time 95% of the time (e.g., correct use of terms, correct and appropriate content, and timing).

Property	Subject	Requirements
Functionality	**Functionality**	Define the functions to be performed or achieved by the human element.
	Response Performance	Define the required response performance achieved by the HE in the performance of an assigned barrier task or task activity.
Integrity	**Reliability**	A quantitative value that defines the minimum required reliability of a HE performed physical action or cognitive response meaning it is reliably performed with sufficient accuracy at the required/ appropriate time.
	Availability	A quantitative value that defines the minimum required availability of the barrier assigned HE to perform assigned tasks and task activities (physical and cognitive) at the required/appropriate time. (Improved availability may be achieved by adding one or more HE-backup roles.)
	Integrity	A quantitative value that defines the maximum permitted HE errors (number or rate) when performing assigned barrier tasks and task activities. Errors may be physical, cognitive or temporal. (Errors may result from gaps in skills, knowledge or mental models, poor HMI display design, deficient procedures, etc. **(See example below.)**
Survivability (Vulnerability)	**Resistance to Design Accident Loads (DAL)**	Define the minimum resistance to defined DALs needed to maintain the HE capability to meet the other performance requirements.
	Resistance to Environmental Loads (EL)	Define the minimum resistance to defined Els needed to maintain the HE capability to meet the other performance requirements.
	Endurance Time	Define the minimum duration and period of resistance to each identified DAL and EL, e.g., the resistance must be achieved for the duration of barrier activation.

FIGURE D.4   Performance Standard – Human Elements.

- Message duration does not exceed x seconds.
- Responds to a valid request within x seconds of request (e.g., requests identified in procedures and/or training).

- Detect: SA-1:
  - SA-1 information (e.g., a task or barrier activator) is noticed and detected 95% of the time within the specified "notice + detect" time. (See Table 3.27.)
  - Change – detects material change to SA-1 information within x minutes.

- Detect: SA-2/3:
  - Achieves the minimum specified SA-2 comprehension 90% of the time.
  - In any instance, the inaccuracy of an SA-2 comprehension result is not sufficient to materially contribute to task or barrier failure.
  - SA-2 comprehension result is achieved within the specified target time 95% of the time. (See Table 3.27.)

- Detect: SA-3:
  - Achieves the minimum specified SA-3 projection 80% of the time.
  - In any instance, the inaccuracy of an SA-3 projection (nature and degree) is not sufficient to materially contribute to task or barrier failure.
  - SA-3 projection result is achieved within the specified target time 95% of the time. (See Table 3.27.)

- Decide:
    - Makes the appropriate decision 95% of the time.
    - All task decisions and the associated act phase response planning are completed within the target phase safety time 95% of the time. (See Table 3.27.)

- Act Response:

    - Initiates and completes the act phase action (the action that achieves the barrier safe state) within the barrier safety time (BST) 100% of the time. (See Table 3.27.)
    - Initiates the appropriate act phase response 95% of the time.
    - Once the decision is made to initiate the selected act phase response, promptly initiates the response 100% of the time.
    - Self-identifies an incorrect act phase response and takes the appropriate corrective action to recover at least 60% of the time.

- Shared Situation Awareness (SSA):
    - Achieves and maintains the minimum required SSA 95% of the time. (This includes monitoring for and maintaining the required SSA in response to changes, for example, a change in the barrier status or priorities.)

- Coordination:
    - Maintains timely coordination of actions with others 90% of the time.

- Leadership
    - Provides the required and appropriate leadership 95% of the time for the duration of barrier activation.

- Overall
    - Achieves the barrier safe state within a period that is 90% of the specified barrier safety time (BST), i.e., maintains a 10% safety margin for time (a time contingency).

D.3.3 Operational-Based Performance Standards for Barrier Systems

Appendix G presents the suggested verification processes and schemes performed in the operate and maintain phase, such as the Barrier Site Acceptance Test (B-SAT). A dedicated barrier system performance standard (one per barrier) may provide the appropriate acceptance basis for the verification activity. The performance standard should be guided by the SRS, a document that includes every barrier system requirement (PE, OE, and HE). As an operational-based performance standard, examples of acceptance criteria include the following:

- Achieving the specified barrier function and safety state within the barrier safety time.
- Achieving the specified individual task functions and safe states within the task and task phase safety times. (See Tables 3.6 and 3.27 for times.)

- Acceptance metrics on barrier team communications, coordination, interactions, and behaviors.
- Acceptance metrics on errors/actions that may contribute to barrier failure, HE injury, etc.
- Acceptance metrics on the appropriate use of the available and additional resources.
- Acceptance metrics on the usability/interference of Support PE in the performance of assigned tasks.
- Acceptance metrics of the types shown in the integrity example in D.3.2.
- Acceptance metrics on the HE response performance in the presence of different environmental loads.

D.4 CIEHF (2016) SUGGESTIONS FOR A HUMAN PERFORMANCE STANDARD

The CIEHF white paper (2016, Section 5.4, p. 51) "contains recommendations for the content and structure of a Human Performance Standard associated with human barrier elements." The following are thoughts and comments on this recommendation.

The white paper includes example requirements for tasks, organizational elements, performance standards, and barrier management systems. It appears to suggest that this information be packaged into a single document with the title "Human Performance Standard." If so, this author does not recommend the title, or the document as proposed.

As an alternative, the lifecycle model suggests an approach that captures the above and additional information in several documents having the titles barrier safety management plan (BSMP) and safety requirements specification (SRS). The model adopted purpose and content of both documents are guided by the existing, globally accepted and widely used consensus standard, IEC 61511-1 (2016). Safety management system information is captured in the BSMP. Appendix B.2 summarizes its suggested contents and organization. The barrier system requirements and information are captured in the SRS. Appendix B.3 summarizes its suggested contents and organizations.

Appendix D.3.2 proposes a performance standard that is specific to the operations personnel (HEs) assigned to the barrier system. The BSMP, SRS, and performance standard are inputs to the verification and validation schemes and activities described in Appendices G.5 and G.6. This suite of documents appears to be more aligned and consistent with current practice.

Note: The CIEHF white paper appears to recognize the need for a document (and processes) that more fully defines and specifies the human activities in an active human barrier system. As there is no comprehensive lifecycle model to draw from, their suggested document is a step in that direction. However, the lifecycle model (described in Chapters 2 to 7) goes well beyond that suggestion. The model suggests methods for organizing and structuring the information in forms that may be better suited to the information type and its likely uses and users and do so using an approach that is consistent with current practice and project document expectations.

Appendix E
Procurement

This appendix is organized as follows:

- E.1 Input to Requisitions, Purchase Orders, and Contracts
- E.2 Proposal Review
- E.3 Input to Purchase Order or Contract
- E.4 Purchase Order/Contract Kick-off Meeting
- E.5 Vendor Data Review
- E.6 Periodic Inspection
- E.7 Assessments
- E.8 Verification #3, Validation #2
- E.9 Factory Acceptance Testing and Validation #3
- E.10 Site Services
- E.11 Site Acceptance Test, Verification #4, Validation #3
- E.12 Purchase Order or Contract Closeout
- E.13 Procurement Information: Reference Tables

This appendix provides guidance on the barrier and barrier-related procurement and contract activities. Figure E.1 provides an overview of the possible activities that may occur at the indicated lifecycle model phases.

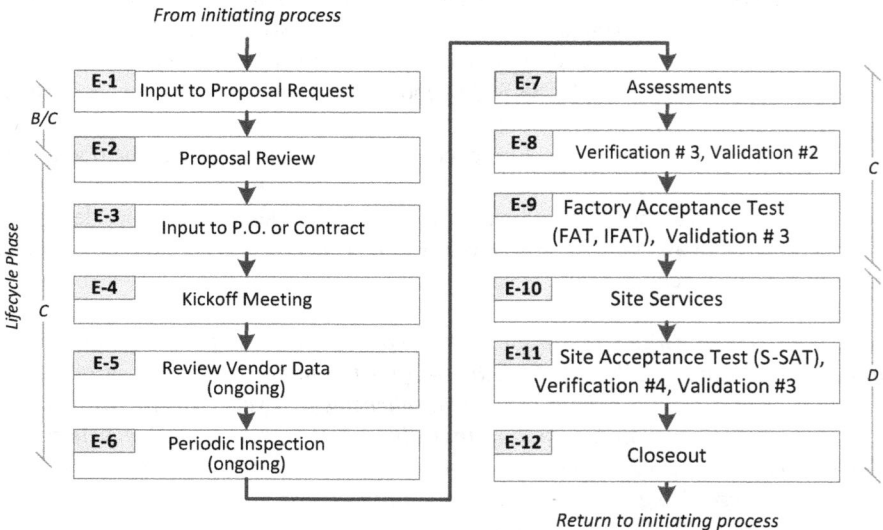

FIGURE E.1 Procurement Activities.

422

E.1 INPUT TO REQUISITIONS, PURCHASE ORDERS, AND CONTRACTS

This step suggests the content and process to integrate barrier requirements into a proposal request, purchase order, or contract. Example procurement package documents may this document should identify and define every purchaser-requested document include the following:

- Scope of supply (text added to scope section)
- Technical package
- Vendor data requirements
- Inspection and test requirements (may include verification and validation requirements)

Scope of Work: Procurement packages typically include a scope of work (SOW) section. It outlines the general scope of the order. Beyond the requested equipment, the additional scope may include training, assessments, verification and validation activities, and a requirement to address findings and recommendations. For less common requirements or those easily overlooked, it may be helpful to add a few sentences or a single paragraph to frame or summarize this scope.

Note: Given the high cost of responding to proposal requests, responders do not always read the entire package, which can include many hundreds (or more) of pages. A barrier system requirement buried in this often very large volume of information is easily missed. Identifying an atypical requirement in the SOW improves its visibility to the bidder.

Technical Package: The technical package should include the barrier/task requirements, specifications, design documents, and drawings needed to fully define and correctly understand the requirements.

Vendor Data Requirements (VDR): Procured systems commonly include a list of the vendor/contractor-provided documents such as design documents, drawings, parts lists, quality reports, inspection, and test records, operating and maintenance procedures, deficiency and corrective action records, and spare parts list.

Inspection and Test Requirements Specification (ITRS): This document should identify and define every purchaser-requested, barrier-specific quality activity such as inspection, testing, verification, and validation. The document should identify the party responsible for planning, tracking, leading/participating, providing activity reports, and responding to corrective actions identified by these activities. (The required plans, procedures, and reports should be identified in the VDR.)

The development and management of a procurement or contract package may be assigned to a technical or project individual. Timely communication with this person is essential to knowing when information is needed to gain early agreement on what information to add. Resources are then needed to perform these activities.

Note: A suggested approach is to proactively gather and maintain a database of the procurement/contract packages. This could include the package owner and contact information, a list of the barrier elements in the package, and key milestone dates. Review the package before issue to verify the SOW, technical package, VDR and ITRS documents are included and appropriately integrated.

The following tables (included in step E.13) identify the sources of information that could be included in a requisition, purchase order, or contract package. They also suggest where to include this information in the package.

- Table E.1. Commercial Off-the-Shelf Components
- Table E.2. Control Consoles and Panels
- Table E.3. HMI Displays (VDU-Based)
- Table E.4. Technical Systems
- Table E.5. Buildings, Rooms, and Walk-in Enclosures
- Table E.6. Packaged Equipment Systems

E.1.1 Discussion

The term "packaged equipment system" applies to an equipment or process package delivered as a fully designed, fabricated, integrated, and tested unit or package. The package may include a walk-in building and/or an indoor or outdoor mounted control panel. For example, a turbine-driven power generation package may be provided as a complete and fully integrated unit that includes the driver and generation equipment, fuel receivers, electrical switchgear, a walk-in enclosure, and control and safety systems that monitor and control the entire package. An instrument air or nitrogen generation package could be provided as a skid-mounted unit that includes the process equipment and a freestanding control panel designed for an outdoor installation.

Equipment providers/vendors may limit the range of changes that can be made to their "standard" product. A requested change to meet a barrier system requirement may not be permitted as proposed. For example, the procured item or package may have a regulatory "type" approval that does not permit changes to any of its components or basic design. (Should this occur, this may require a return to an earlier design process to seek an alternative approach.) Equipment providers / vendors often limit the type and range of changes permitted to their "standard" product. Examples may include the following:

- Selective and limited changes to the vendor's standard control or display equipment (e.g., the make or model of a preferred programmable logic controller).
- Selective and limited changes to the vendor's standard HMI displays (e.g., styles, elements, or layout).
- A selective change that enables receiving a command signal from an external control system (used to coordinate package actions with other barrier functions).
- Adding a new instrument or an associated instrument function.
- A change to comply with a barrier system performance requirement (e.g., reliability, availability, or integrity requirement).

E.2 PROPOSAL REVIEW

Review the vendor's proposal to verify that the barrier scope and requirements are acknowledged and accepted without deviation, including the requirements in the VDR and ITRS. Confirm that this full scope is included in the vendor's quoted pricing.

Identify exceptions to and deviations from the requirements in the bid review package. Review both with the bidder to seek options to address the requirements or determine the options available.

To the extent possible and practicable, attend bid clarification meetings to verify the abovementioned requirements. The vendor's acceptance of requirements related to barriers may be a challenge for various reasons:

- The vendor relies on others to implement programmed changes to the control and display systems.
- Functional and specialty testing may be costly to conduct and stage, especially if the testing is atypical.
- The added requirements present an added risk (schedule, internal costs) to the vendor and project.
- The bid response to a lump sum or fixed fee purchase order or contract is not explicit. The vendor proposal does not explicitly acknowledge acceptance of and compliance with every requirement. Not doing so requires further bid clarification discussions and exchanges.

E.3 INPUT TO PURCHASE ORDER OR CONTRACT

To prepare a purchase order or contract for issue, update the barrier information to ensure it reflects all agreements and clarifications identified and accepted during the bid review process. Review the fully prepared purchase order or contract to confirm that all barrier system requirements are included, accurate, and appropriately integrated into the package. (See Appendix E-1 for common and supporting information.)

E.4 PURCHASE ORDER/CONTRACT KICK-OFF MEETING

Soon after a purchase order or contract is issued and accepted, a meeting typically occurs to review the scope and agreements, confirm the understanding of the requirements, and identify and resolve potential areas of ambiguity or disagreement.

To the extent possible and practicable, attempt to attend the vendor kick-off meeting. The objectives are to:

- Confirm the vendor's intention to comply with the included requirements.
- Confirm that the vendor accurately understands each requirement. Clarify areas of potential misunderstanding.

The results of the meeting are typically recorded and published in a Minutes of Meeting (MOM) document. The MOM often becomes part of the purchase order or contract package. A *change in understanding* captured in this document can (and often does) result in an intended or unintended material change to the purchase order or contract and the delivered goods. Carefully review the MOM for potential changes or any misunderstanding of barrier requirements.

E.5   VENDOR DATA REVIEW

See Figure G.1 and Appendix G.2 for suggested guidance on the vendor design and data review process.

E.6   PERIODIC INSPECTION

The ITRS may include requirements that allow the owner/operator (or assignee) to periodically inspect the goods during fabrication, assembly, and construction. See Figure G.1 and Appendix G.3 for guidance on this activity.

E.7   ASSESSMENTS

Based on the prototype lifecycle model, the scope may include assessments performed by the vendor, owner/operator, or assignee. Example assessments may include the following:

- Phase safety time assessment (See detailed design processes C-3 to C-8 for detail.)
- Cognitive assessment (See detailed design processes C-3 to C-8 and Appendix I.2 for detail.)
- Performance influencing factors – working environment/physical ergonomics assessment (See sub-process CX-0.1e in Appendix B.6 and Appendix H.3 for detail. The CX sub-process is performed in detailed design processes C-3 to C-8.)
- Functional analysis/assessment – control consoles and panels, rooms, and engineered areas, buildings (See processes C-9 through C-12 for detail.)

The purchase order or contract should identify the scope and suggested execution plans for each assessment. It should also define the agreed-to contractual approach to addressing the assessment findings and assign responsibilities for approving and implementing corrective actions and recommendations (e.g., who, when, and how much).

Note: Working environment/physical ergonomics assessments are becoming increasingly common on major capital projects such as large-scale, offshore O&G facilities. Vendors may have little to no experience with the other assessments.

E.8   VERIFICATION #3 AND VALIDATION #2

See Figure G.1 and Appendices G.5 and G.6 for the suggested verification and validation scope, process, timing, and execution model that may apply to the scope of the purchase order or contract.

E.9   FACTORY ACCEPTANCE TESTING AND VALIDATION #3

See Figure G.1 and Appendix G.4 for the suggested Factory Acceptance Testing (FAT) and integrated FAT (IFAT) scope, process, timing, and execution model that may apply to a barrier-required or barrier-dependent component, control console or panel, technical system, building, or packaged equipment system.

Note: Barrier system elements may reside in equipment and systems procured under different purchase orders or contracts. The element(s) may be a small part of the overall order or contract; for example, it may be a single display element on an HMI display or a single signal to a package-controlled control valve. A project manager may not approve the participation of a cognitive ergonomics specialist in a Factory Acceptance Test when this scope is limited. Thus, confirming if the barrier element is correctly implemented and validated may rely on those authorized to attend the test.

See Figure G.1 and Appendix G.6 for the suggested validation scope, process, timing, and execution model that may apply to the purchase order or contract.

Note: Validation of an HMI display, a control console or panel, or prefabricated building (e.g., a room) should take place at the same time as the FAT. If deferred to a later time (e.g., during the Site Acceptance Test), any findings identified at this time will be significantly more disruptive in terms of cost, schedule, and resource availability.

E.10   SITE SERVICES

Example site services requested from the vendor may include equipment pre-checks (e.g., before introducing power), support/perform basic checks after power-up, and provide onsite support for commissioning and start-up.

E.11   SITE ACCEPTANCE TEST, VERIFICATION #4, VALIDATION #3

See Figure G.1 and Appendix G.4 for the suggested System Site Acceptance Test (S-SAT) scope, process, timing, and execution model that applies to barrier system and barrier dependent elements that lie within the now fully installed control console or panel, technical system, building, or packaged equipment system.

See Figure G.1 and Appendices G.5 and G.6 for the suggested verification and validation scope, process, timing, and the execution model that may apply to the barrier-related scope included in the purchase order or contract.

The notes in process E-9 apply to the S-SAT.

E.12 PURCHASE ORDER OR CONTRACT CLOSEOUT

As a precondition to closing out the purchase order or contract, the following barrier-specific items should be checked to confirm completion:

- All barrier-specified scope was met and delivered.
- All barrier-related and requested documents have been reviewed, approved, and formally issued in the required forms and the requested number of copies.
- All recommendations and pending issues from assessment studies, reviews, inspections, and tests are resolved and closed.

E.13 PROCUREMENT INFORMATION: REFERENCE TABLES

The following tables summarize the sources of information that may be included in a requisition, purchase order, or contract package, and give suggestions on where to include that information in the package.

- Table E.1, Procurement Input: Commercial Off-the-Shelf Components
- Table E.2, Procurement Input: Control Consoles and Panels
- Table E.3, Procurement Input: HMI Displays (VDU-Based)
- Table E.4, Procurement Input: Technical Systems
- Table E.5, Procurement Input: Buildings, Rooms and Walk-in Enclosures
- Table E.6, Procurement Input: Packaged Equipment Systems

TABLE E.1
Procurement Input: Commercial-Off-the-Shelf Components

Requirement Type	Process	Table	Document/Topic	Scope of Work	Technical Docs.	Vendor Data	Quality, Testing
			Requirement Source				
Base requirements	—	B.8	Applicable base documents/requirements	✓	✓		
Component/device requirements	C-3, C-7, C-8 (See sub-process CX-0.1 in App. B.6)	B.9	Device specifications, data sheets, equipment list, performance requirements, etc.	✓	✓		
General arrangements	C-3, C-7, C-8 (see sub-process CX-0.1 in App. B.6)	B.15	Preliminary drawings, typical drawings, and details: mounting and installation, signal, electrical and utility connections, etc.	✓	✓		
Performance requirements	C-3, C-7, C-8 (see sub-process CX-0.1 in App. B.6)	3.7, 3.27, 3.32, 4.24, 4.25, B.12, B.13, B.14	See component/device requirements	✓	✓		
Procedures (by provider)	C-3, C-7, C-8 (see sub-process CX-0.2 in App. B.6)	B.5	Operating – normal, start-up, shutdown, abnormal, etc.	✓		✓	
		B.6	Maintenance – inspection, preventive, periodic, calibrate, etc.	✓		✓	
Training (by provider)	C-3, C-7, C-8 (see sub-process CX-0.2 in App. B.6)	B.7	Training syllabus	✓		✓	
	See Figure G.1 and App. G.3	G.1, G.2, BSMP	Inspection: ongoing	✓	✓	✓	✓
Quality processes	See Figure G.1 and App. G.4	G.3, G.4, BSMP	Factory Acceptance Test (assembled/complex component, e.g., control valve or lifeboat) *See App. G.3 (Tables G.1, G.2) for acceptance inspection*	✓	✓	✓	✓
	See Figure G.1 and App. G.5	G.5, G.6, BSMP	Verification	✓	✓	✓	✓

TABLE E.2
Procurement Input: Control Consoles and Panels

Requirement Type	Process	Table	Document/Topic	Scope of Work	Technical Docs.	Vendor Data	Quality, Testing
Component/device requirements	See Table E.1, Commercial Off-the-Shelf Equipment						
HMI displays (VDU-based)	See Table E.3, HMI Displays (VDU-Based)						
Technical systems	See Table E.4, Technical Systems						
Base requirements	—	B.8	Applicable base requirements	✓	✓		
General arrangements	C-3, C-7, C-8 (see sub-process CX-0.1 in App. B.6)	B.15	Preliminary sketches to show relative sizing, layout, equipment placement, components, etc.	✓	✓		
Vendor design, assembly, and fabrication documents	—	—	Detailed design, layout, assembly, fabrication, and interconnection drawings and documents, etc.	✓		✓	
Assessments	C9-1	4.18	Functional analysis – Control console and panels	✓	✓		
Performance requirements	—	3.7, 3.27, 3.32, 4.24, 4.25, B.12, B.13, B.14	Performance standards and other sources of performance requirements	✓	✓		
	C-3, C-7, C-8 (see sub-process CX-0.2 in App. B.6)	B.5	Operating – normal, start-up, shutdown, abnormal, etc.	✓		✓	
		B.6	Maintenance – inspection, preventive, periodic, calibrate, etc.	✓		✓	
Procedures (by provider)	See Figure G.1 and App. G.3	G.1, G.2, BSMP	Inspection – ongoing	✓	✓		✓
	See Figure G.1 and App. G.4	G.3, G.4, BSMP	FAT (acceptance test)	✓	✓		✓
			See App. G.3 (Tables G.1 and G.2) for acceptance inspection				
Quality processes (*see Tables E.3 and E.4 for processes that apply to console/panel HMI displays and technical systems*)	See App. G.2	BSMP	Design reviews: ongoing, formal	✓	✓	✓	✓
	See Figure G.1 and App. G.4	G.3, G.4, BSMP	Factory Acceptance Test	✓	✓	✓	✓
			See App. G.3 (Tables G.1 and G.2) for acceptance inspection				
	See Figure G.1 and App. G.5	G.5, G.6, BSMP	Verification	✓	✓	✓	✓
	See Figure G.1 and App. G.4	G.3, G.4, BSMP	Site Acceptance (functional) Test	✓	✓	✓	✓
	See Figure G.1 and App. G.6	G.7, G.8, BSMP	Validation	✓	✓	✓	✓

Requirement Source (spanning Process, Table, Document/Topic columns)

TABLE E.3
Procurement Input: HMI Displays (VDU-Based)

Requirement Type	Process	Table	Document/Topic	Requirement Source			
				Scope of Work	Technical Docs.	Vendor Data	Quality, Testing
HMI display design/style guidelines	—	—	HMI display guidelines, style guides, etc.	✓	✓		
HMI display – elements, integrated	C8-4, C8-5, C8-6	4.12, 4.13, 4.14, 4.15	Display element and integrated display requirements	✓	✓		
Display sketch	C8-4, C8-5, C8-6	—	Display sketch	✓	✓		
Functional requirements	C8-4, C8-5, C8-6	B.10	Functional specification, logic diagrams, cause-and-effect chart, control narratives	✓	✓		
Performance requirements	B-4, C-13	3.7, 3.27, 3.32, 4.24, 4.25, B.11–B.14	Performance standards and other sources of performance requirements	✓	✓		
Vendor functional specifications	—	—	HMI display functional design specifications	✓		✓	
Vendor HMI display development	—	—	HMI display development and coding	✓		✓	
Assessments	C8-9	3.27	Safety time assessment	✓	✓	✓	
	C8-10	I.1	Cognitive assessment	✓	✓	✓	
Quality processes	See Figure G.1 and App. G.2	BSMP	Design reviews: ongoing	✓	✓	✓	✓
	See Figure G.1 and App. G.4	G.3, G.4, BSMP	Factory Acceptance Test *See App. G.3 (Tables G.1 and G.2) for acceptance inspection*	✓	✓	✓	✓
	See Figure G.1 and App. G.5	G.5, G.6, BSMP	Verification	✓	✓	✓	✓
	See Figure G.1 and App. G.6	G.7, G.8, BSMP	Validation	✓	✓	✓	✓

TABLE E.4

Procurement Input: Technical Systems

Requirement Type	Requirement Source			Scope of Work	Technical Docs.	Vendor Data	Quality, Testing
	Process	Table	Document/Topic				
Component/device requirements	See Table E.1, Commercial Off-the-Shelf Equipment						
Control consoles and panels	See Table E.2, Control Consoles and Panels						
HMI displays (VDU-based)	See Table E.3, HMI Displays (VDU-Based)						
Base requirements	—	B.8	Applicable base requirements	✓			
General requirements	—	B.21	Source requirements	✓	✓		
Barrier block diagrams	B15-3/C-1	—	Barrier block diagram		✓	✓	
Communication systems	B-7, C3-3, C7-9	3.9, 3.12	System requirements	✓		✓	
Public address/public alarm systems	C7-6	—	System requirements	✓	✓	✓	
CCTV systems	C8-7	4.16	System requirements	✓	✓	✓	
Off-console VDU displays	C8-8	4.17	Placement and implementation requirements	✓	✓	✓	
Support PE (complex systems)	B-8	3.13	Equipment requirements	✓	✓	✓	
Control and display system requirements	C14-2, C14-6, App. L.2.2	4.27	System requirements	✓	✓	✓	
Functional requirements	C-3 to C-8	B.10	Functional specification, logic diagrams, cause-and-effect chart, control narratives, etc.	✓	✓	✓	
Vendor design, assembly, and fabrication documents	—	—	Detailed design, layout, assembly, fabrication, and interconnection drawings and documents, etc.	✓	✓	✓	
Performance requirements	B-4, C-13	3.7, 3.27, 3.32, 4.24, 4.25, B.11-B.14, B-21	Performance standards and other sources of performance requirements	✓	✓		

Category	Reference	Code	Item				
Procedures (by provider)	C-15	4.5	Operating – normal, start-up, shutdown, abnormal, etc.		✓	✓	
		4.30	Maintenance – inspection, preventive, periodic, calibrate, etc.		✓	✓	
	See Figure G.1 and App. G.3	G.1, G.2	Inspection – ongoing		✓	✓	
	See Figure G.1 and App. G.4	G.3, G.4	FAT, IFAT, S-SAT (acceptance)		✓	✓	
			See App. G.3 (Tables G.1 and G.2) for acceptance inspection		✓	✓	
Training (by provider)	C-16, C32-4	4.32, 4.33, 4.34, 5.7	Training syllabus		✓	✓	
	C3-9, C7-10, C8-9	3.27	Phase safety time		✓	✓	
	C3-10, C7-11, C8-10	I.2	Cognitive assessments		✓	✓	
Assessments	C-3, C-7, C-8 (App. H.3 and sub-process CX-0.1e in App. B.6)	H.3	PIF assessments: working environment, physical ergonomics assessment			✓	
Quality Processes	See App. G.2	BSMP	Design reviews: ongoing, formal	✓	✓	✓	✓
	See Figure G.1 and App. G.3	G.1, G.2, BSMP	Inspection: ongoing, acceptance	✓	✓	✓	✓
	See Figure G.1 and App. G.4	G.3, G.4, BSMP	Factory Acceptance Test (FAT, IFAT), Site Acceptance Test (S-SAT)		✓	✓	✓
	See Figure G.1 and App. G.5	G.5, G.6, BSMP	Verification	✓	✓	✓	✓

TABLE E.5

Procurement Input: Buildings, Rooms, and Walk-in Enclosures

Requirement Type	Requirement Source			Scope of Work	Technical Docs.	Vendor Data	Quality, Testing
	Process	Table	Document/Topic				
Component/device requirements	See Table E.1, Commercial Off-the-Shelf Equipment						
Control consoles and panels	See Table E.2, Controls Consoles and Panels						
Base requirements	—	B.8	Applicable base requirements	✓	✓		
Room design/requirements	—	B.17	Source requirements	✓	✓	✓	
Engineered Area Design/Requirements	—	B.18	Source requirements	✓	✓	✓	
Building Design/Requirements	—	B.22	Source requirements	✓	✓	✓	
Off-console VDU displays	C8-8	4.17	Placement and implementation requirements	✓	✓	✓	
Functional analysis/assessments	C10-1, C11-1, C12-1	4.18, B.10	Functional analysis/assessment	✓	✓	✓	
Technical systems: dedicated to building functions	C12-3	4.20, 4.22, 4.23	Overall – see Table E.4, Technical Systems, for guidance	✓		✓	
Other technical systems in building	C9-3, C10-3, C11-3, C12-4	3.16	In-Place technical system requirements (not included in the building-dedicated systems)	✓		✓	
	C10-2	4.20	Rooms	✓	✓	✓	
General layout and design (barrier-required elements)	C11-2	4.22	Engineered areas	✓	✓	✓	
	C12-2	4.23	Overall – general arrangements, layouts, locations, etc.	✓		✓	

External support system requirements	C9-4, C10-4, C11-4, C12-5	3.29, 4.19–4.23, B.19	Identify all barrier-specific ESS requirements (building-provided), e.g., lighting, electrical power supply and distribution, HVAC, environmental controls.	√	√	√
External protective barrier requirements	C10-5, C11-5, C12-6	4.21, 4.23	Identify all barrier-specific external protective barriers (building provided), e.g., passive and active	√	√	√
Detailed design documents	—	—	Detailed design specifications and documents, general arrangement, layout, and location drawings	√	√	√
Fabrication and construction drawings and documents	—	—	Building fabrication, construction, reports, certifications, etc.	√	√	
Performance requirements	C-13	3.7, 3.27, 3.32, 4.24, 4.25, B.12, B.13	Performance standards and other sources of performance requirements	√	√	
Assessments	C10-1	4.20	Functional analysis – room	√	√	√
	C11-1	4.22	Functional analysis – engineered area (egress/escape route, safe haven)	√	√	√
	C12-1	4.23	Functional assessment – building	√	√	√

(Continued)

TABLE E.5
(Continued)

Requirement Type	Process	Table	Document/Topic	Scope of Work	Technical Docs.	Vendor Data	Quality, Testing
Procedures (by provider)	C-15	4.5	Operating – normal, start-up, shutdown, abnormal, etc.	✓	✓	✓	
		4.30	Maintenance – inspection, preventive, periodic, calibrate, etc.	✓	✓	✓	
	See Figure G.1 and App. G.3	G.1, G.2, BSMP	Inspection – ongoing	✓	✓	✓	
	See Figure G.1 and App. G.4	G.3, G.4, BSMP	FAT, S-SAT (acceptance) *See App. G.3 and Tables G.1/3.7 for acceptance inspection*	✓	✓	✓	
Training (by provider)	CX-0.2, C-16	4.32, 4.33, 4.34	Training syllabus, training delivery	✓	✓	✓	✓
	See App. G.2	BSMP	Design reviews: Ongoing, formal	✓	✓	✓	✓
	See Figure G.1 and App. G.3	G.1, G.2, BSMP	Inspection: ongoing, acceptance	✓	✓	✓	✓
Quality processes	See Figure G.1 and App. G.4	G.3, G.4, BSMP	Factory Acceptance Test, Site Acceptance Test	✓	✓	✓	✓
	See Figure G.1 and App. G.5	G.5, G.6, BSMP	Verification	✓	✓	✓	✓
	See Figure G.1 and App. G.6	G.7, G.8, BSMP	Validation	✓	✓	✓	✓

Requirement Source

TABLE E.6
Procurement Input: Packaged Equipment Systems

Requirement Type	Requirement Source		
	Process	**Table**	**Document/Topic**
Component/device requirements	See Table E.1, Commercial Off-the-Shelf Equipment		
Control consoles and panels	See Table E.2, Control Consoles and Panels		
HMI displays (VDU-based)	See Table E.3, HMI Displays (VDU-Based)		
Technical systems	See Table E.4, Technical Systems		
Buildings, rooms, and walk-in enclosures	See Table E.5, Buildings, Rooms, and Walk-in Enclosures		

Appendix F
Cognition: Baseline Science and Application

This appendix is organized as follows:

- F.1 Automatic and Conscious Processes – Overview and Comparison
- F.2 Cognitive Capabilities and Limitations (Design Considerations)
- F.3 Cognitive Access to Visual Information
- F.4 Acute Stress – Physical and Cognitive Effects
- F.5 Understanding Mental Models and Long-Term Memory

Note: The roles and responsibility tables in Appendix A.2.3 identify the suggested participant in the identified lifecycle model activities. Appendices C and F provide the foundational information on human cognition that underpins the prototype lifecycle model processes, activities, and tools. The participant expected to have expertise in this material is identified by the term "cognitive system engineer" (CSE). Those identified as process safety and human factors engineers should have a general understanding of this material and know when CSE input, guidance, and support are needed. It may be helpful to provide the task analysis participants (and others) with an overview of the primary material to gain sufficient awareness and understanding of how it affects barrier system design and lifecycle processes.

F.1 AUTOMATIC AND CONSCIOUS PROCESSES – OVERVIEW AND COMPARISON

Human cognition is the collective product of several interdependent and very different cognitive processes:

- Sensory receptors and pre-processing
- Autonomic (amygdala and hypothalamus; e.g., freeze, fight, flight (FFF))
- Automatic processes also known as System 1, subconscious, unconscious processes
- Conscious processes also known as System 2, attention/working memory

The relative response time of these different cognitive processes based on a single perception cycle and indicative times is as follows:

- Autonomic process: **Very Fast** (e.g., 20 milliseconds (ms))
- Automatic process: **Fast** (e.g., 70 ms, 25–170 ms range; Sträter 2005, p. 85)
- Conscious process: **Slow** (e.g., 285 ms; Carter 2014, p. 121)

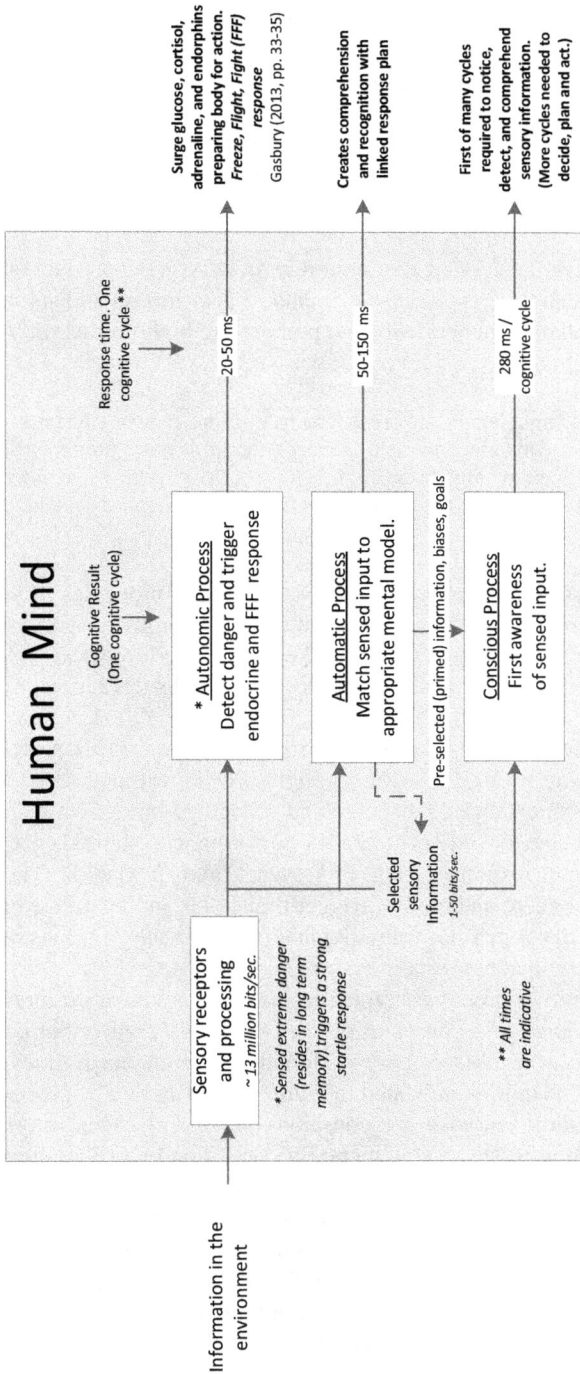

FIGURE F.1 Overview of Common Cognitive Processes and Example Response Times.

Figure F.1 provides a generalized view of these processes and interactions. Each creates a different perception and response to sensed information in the environment. The physical response depends on which process drives and controls perceptions, decisions, and physical responses from moment to moment.

Note: The term "automatic" is commonly used in the research and academic world. Other terms applied to automatic processes include subconscious, unconscious, or System 1.

A startle or fear-driven response, an autonomic process, is the product of human evolution. Once activated, the response may include a near-instant and automatic release of performance enhancing chemicals that prepare the body for action. According to Kelly et al. (2023, p. 8),

> Activation of the 'fight' or 'flight' response to a minor degree can improve performance, but further activation can lead to prioritisation of gross motor skills over fine motor skills, reducing manual dexterity. In addition, prioritisation over hearing, along with narrowing of the visual field, contribute to poor communication and loss of situation awareness...

Further, a strong activation tends to suppress conscious processes in favor of automatic processes. An epinephrine-fueled automatic response (e.g., instinctual and impulsive) may be a lifesaving strategy in the natural world, for example, chased by a saber tooth tiger. However, in the technological world such actions often prove to be ill-advised.

Individuals differ in their autonomic startle/fear response to a perceived danger. Response forms may be freeze, fight, or flight. For emergency responders, it may be of life-critical importance to discover and recognize one's own response behavior and how that behavior aids or degrades performance. Methods are available to identify one's default response, triggering events, and thresholds. The results can then be used to recognize and modify trigger thresholds and modulate behaviors and responses in ways that may improve individual performance. (This is common training for U.S. military medical responders.)

Automatic cognitive processes typically dominate moment-to-moment perceptions and actions. However, a conscious process can modify one or both. Summarized in Table F.1, these two processes have profoundly different capabilities, limitations, behaviors, biases, and quirks. Indicated in Figure F.1, an automatic process may select the initial information provided for conscious processing. This information (pre-priming) may include biases, goals, mental models, and other long-term memories. Thus, the starting point for a conscious cognitive process is biased by this information.

Humans tend to believe their decisions and actions are the product of deliberate and thoughtful (conscious) processes. This is not the case in most instances. Instead, humans commonly default to automatic processes that are not deliberative or reflective by nature. If not modified by a conscious process, automatic processes and long-term memories provide the recognition of what the sensed information means and the plans and actions that may be associated with or linked to those memories. A conscious process is required to verify if that plan and response is appropriate to the current situation.

TABLE F.1

Comparative Overview of Automatic and Conscious Processes

Trait, Function, or Behavior	Automatic Processes	Conscious Processes
Span of control	**Always active:** as a seemingly unlimited resource. Responsible for 95% of daily cognitive activities (Kahneman 2011, p. 20; Mlodinow 2012, p. 34).	**A limited resource selectively allocated.** On average, overrides automatic processes to directly affect up to 5% of daily cognitive activities (Kahneman 2011, pp. 23, 34–35; Mlodinow 2012, p. 34)
Normal operation	Automatic, continuous, and effortless. (Reason 1990, p. 98; Kahneman 2011, p. 20) Open loop, positive feedback only. (Sträter 2005, p. 118).	**Capacity-limited resource.** Employs a least-effort behavior that maintains reserve capacity. – Highly effortful. – Lazy tendencies: seeks *cognitive ease* (Kahneman 2011, p. 21, Ch. 3). – Closed loop, negative feedback (Sträter 2005, p. 118). – Runs concurrent to automatic processes (Reason 1990, pp. 132–134). – Has a natural speed. Going beyond that speed consumes working memory, thereby reducing what remains for other conscious processing (Kahneman 2011, pp. 40–43).
Executive mode	A **recognition engine** that continuously compares input stimulus to long-term memories and mental models seeking a match. If a "match" is found, it automatically selects the associated memory. If not found or recognized, it may call a conscious process to resolve (Kahneman 2011, p. 24) *Note: A "match" does not necessarily mean the mental model is appropriate or applicable to the current information and situation. Heuristics seek a MM that seems to match. If not found, model does not exist or insufficient cues to find a match), heuristics may switch to the "most recently accessed" or "most frequently accessed" calling condition. The MM may be somewhat useful or unsuitable for the situation.*	**Slow, linear, sequential** processing cycles (Reason 2008, p. 12) **Realized by Working Memory (WM)**, essential to all conscious processes (Reason 2008, pp. 12–13; Wickens et al. 2013, pp. 197–201) Working memory comprises a central executive that controls WM activities, and three short-term memory stores. (Wickens et al. 2013, pp. 198–207) The three stores are as follows: – *Episodic buffer*, a passive store supports the interaction between WM components and information. – *Phonological store* for information in linguistic form. – *Visual-spatial* store for analog, spatial, and visual information.

(Continued)

TABLE F.1
(Continued)

Trait, Function, or Behavior	Automatic Processes	Conscious Processes
Response time	**Recognition Time: Fast,** e.g., 70 milliseconds (indicative), 25–150 millisecond range for a single cognitive cycle (Sträter 2005, pp. 85, 126–128) **A recognition callup (which may include an embedded response action/plan): Fast,** e.g., 70–250 milliseconds for clear response to recognized input (Sträter 2005, pp. 119, 126–127; Carter 2014, p. 121)	**Slow:** Fractional to many minutes (Carter 2014, p. 121)
Attention	**Attention/Working memory:** To support a habituated skill or action (e.g., driving) captures for brief periods that go unnoticed by conscious processes. An overly focused conscious process may prevent this brief access, which also goes unnoticed, creating a potentially dangerous condition.	– A conscious process call. Continuous focus possible for limited periods of time. (Reason 1990, pp. 132–134; Sutcliffe 2002, p. 31; Kahneman 2011, pp. 22, 105) – An automatic process call that overrides a conscious effort to direct attention, for example, re-direct attention in response to a highly salient sensed source or in response to a situation is not recognized/no mental model available (Sutcliffe 2002, p. 31; Kahneman 2011, p. 24).
Observability (able to monitor by conscious processes)	**Mostly hidden from conscious mind:** Recognition may be experienced as an intuition or gut feel.	**Partial:** General visibility into the object of one's directed attention, decisions, results, and some conscious processes (Endsley 1995; Reason 2008, p. 12) **Hidden:** Bias and priming effects from automatic processes (aka, System 1). System 1 "generates impressions, feelings, and inclinations; when endorsed by System 2 these become beliefs, attitudes and intentions" (Kahneman 2011, p. 105. Also see Reason 1990, pp. 11–12).
Decision and Analytical Capabilities	**Decision Capability: None** (see "Executive Mode") **Analytical Capability: Limited and Selective:** Some intuitive ability to estimate averages, but not sums. Highly prone to error when responding to a statistical query. (Kahneman 2011, pp. 92–93).	**Yes:** Powerful analytical and computational capability (Reason 2008, p. 12). Maximum throughput of 10 bits/second (binary decision) **Caveat:** Results may be degraded by inappropriate biases/ priming effects from automatic process inputs (Kahneman 2011, p. 86). Also, subject to memory/ execution-induced errors discussed in Section F.2. Limitations in the WM/short-term memory store capacity are problematic when processing negative facts because they require more short-term memory capacity to hold and process. This may contribute to the bias pattern to seek positive facts for use in analyses and confirmations. Reasoning attempts using negative facts may be illogical.

Ability to Detect Danger	**Fast, continuous, automatic** (Sylvestre 2017, pp. 66–71). Detected dangers are those learned from prior experience and training. (This information resides in long-term memory).	**Yes, though limited.** Occurs only when tasked to do so. Slow detection/response relative to automatic processes. Ability to maintain persistent monitoring tends to be limited to 20 minutes common to all vigilance tasks. (Sylvestre 2017, pp. 66–71). Freeze, fight, flight (FFF) response may temporarily suppress conscious process activation.
Ability to Detect Risk	**None** (Sylvestre 2017, pp. 68, 70).	**Yes,** but only if activated and tasked (Sylvestre 2017, pp. 66–71).
Memory Capture, Storage, Recall, and Retrieval	"Memory function is an attribute of System 1" (Kahneman 2011, p. 46). Automatic processes capture, store, consolidate, modify, recall, and retrieve long-term memories. These activities are often affected by one's emotions. "Emotion facilitates memory consolidation as well as (a) increases attention to emotional aspects of events, (b) makes events more distinct, and (c) results in more information organization." (Radvansky 2021, p. 23) "If there is an emotional reaction to an event, that reaction would be encoded into the memory trace via the amygdala." (Radvansky 2021, p. 41) From Radvansky (2021, p. 399), emotions can affect LTM recall. "The more emotional the event, the more likely it will be remembered." Further, "the emotional intensity of negative events is tempered more so than positive events." These processes may enable hidden cognitive biases. Automatic processes use search heuristics and cues embedded in stored long-term memories. (See Appendix F.5 for details.) A common heuristic uses a similarity (like-with-like) match. If not found, the criteria may switch to frequency gambling by searching for a frequently or recently accessed memory (Reason 1990, pp. 98, 127–147; Reason 2008, pp. 12–25). Memory retrieval time is affected by the nature, applicability, and number of the available cues in the environment, how often the memory is recalled, the interval from its prior recall, and other factors.	Current theories posit that working memory has separate short-term memory stores (buffers) to hold transactionally used episodic, phenological, and visual-spatial information. The store functioning in WM is analogous to RAM in a desktop computer. Unlike RAM, the WM short-term memory stores have several safety critical attributes. – The store capacity is limited and a primary constraint on conscious cognitive processes (see below). This capacity degrades when the person experiences high stress, excessive cognitive workload, fear, fatigue, etc. – The stored information quickly and automatically fades if not continually refreshed (an ongoing attentional demand). – Distractions, interruptions, and the above-noted conditions can cause store data loss or integrity failure (e.g., an unintended/undetected data change, misremembering). (These attributes and behaviors are primary challenges to barrier and safety critical task design.) The commonly stated capacity limit for a WM STM store is 7 ± 2 chunks. (Some believe the limit may be closer to 5. Others believe it becomes less problematic for individuals with unique aptitudes or increased domain expertise.) If the person is exposed to one or more of the above-discussed conditions, the store capacity can degrade to a low of 1 chunk. Each successive drop degrades the cognitive processing capability.

(Continued)

TABLE F.1
(Continued)

Trait, Function, or Behavior	Automatic Processes	Conscious Processes
Long-Term Memory (LTM) Access	Automatic access to all recalled/retrievable long-term memories. *Note: Be aware of the recall criteria noted in "Memory Capture, Recall, Storage and Retrieval" above.*	**Default:** Automatic processes control access to long-term memories. With effort and focus, can: – Accept or reject the selected long-term memory prompted by automatic processes (Reason 1990, p. 131; 2008, pp. 12–25) – Modify LTM recall criteria to seek a different memory (Reason 1990, pp. 130–131; 2008, pp. 12–25).
Memory Retention	**The retention of long-term memories is subject to varying degrees of "forgetting."** This process depends on the memory type, frequency of access, and other factors (Radvansky 2021, pp. 82–83, 406–410). A long-term memory can change with every recall, a change that typically goes unnoticed. (This attribute is a likely contributor/enabler to drift and normalization of deviance. See Appendix J.7.2 for background.) Skill-based memories are also subject to fade (decay). See Appendix J.7.1 for further details.	**Information in short-term working memory stores/buffers can rapidly fade,** e.g., 30 to 120 seconds, if not refreshed (Guastello 2023, p. 165). See Radvansky (2021, pp. 118–121) for additional detail. Dedicated sensory information stores provide temporary storage for raw sensory information. The retention duration for visual (iconic) information may range from 40 to 600 milliseconds, depending on how long the information is sensed and available to sense (Radvansky 2021, pp. 106–108). For auditory (echoic) information, the period may be 3–4 seconds (Wickens et al. 2013, p. 202). Longer retention requires an intended act that moves the information to the WM information processing buffer.
Validate Response Before Acting	**No** – impulsive behavior (Sträter 2005, pp. 118–119; Kahneman 2011, pp. 85–86).	**Yes,** but only with focused effort. Otherwise: 1. Does not automatically check input data or decision validity, or if essential info is missing (Kahneman 2011, pp. 85–86, 99) 2. Tends to limit validity checks to confirm information only, i.e., confirmation bias (Kahneman 2011, pp. 80–82, 105).

Ability to Self-Monitor; Self-Correct, Control emotions	**None** (Sträter 2005, p. 118; Kahneman 2011, pp. 41–43, 47–48) *A new situation assessment may result in a different decision or action response. Its utility depends on the timing of the new assessment and if the process allows a timely recovery from the prior action.*	Yes, but only when activated and tasked, and working memory capacity is available to do so. Provides the only means to monitor and modify one's performance, decisions, emotional state, and behavior. With cognitive overload, behavior becomes more representative of automatic processes (e.g., less tempered, less controlled, unfiltered demonstration of beliefs and prejudices; Kahneman 2011, pp. 24, 41–42).
Access to Sensory Data	Sensory data (11–13 million bits/second) Yes, all senses, high bandwidth, some latency in accessed data.	**Yes**, if the data is transferred from the dedicated sensory store to working memory/short-term memory before it fades or is replaced by new sensory data (Carter et al. 2014, pp. 78–107). **Max input data rate:** 1–50 bits/second processed in working /short-term memory. Data latency may be several times longer than the automatic process. *The automatic process may pre-select sensory data based on what is expected or looked for (mental model, top-down driven). Attention may automatically shift to a high-salience object like flashing lights or movement (bottom-up driven).*
Habituated Actions and Skills	Once fully learned, has full automatic control of habituated skills and routines (Reason 2008, p. 14) *Note: Skills and habits become automatic with continued repetition over time (e.g., 2–6 months). Prior to that, the action is a sliding mix of conscious and automatic control (Radvansky 2021, pp. 182–189).* *Note: All skills are subject to varying degrees of skill fade (forgetting) over time if not used or refreshed. See Appendix J.7.1 for detail.*	Initially, a skill or habit begins as a consciously controlled action that is often clumsy and effortful. Errors in this initial state are the most likely (Reason 2008, pp. 13–14, Kahneman 2011, p. 35; Radvansky 2021, pp. 182–189).

(Continued)

TABLE F.1
(Continued)

Trait, Function, or Behavior	Automatic Processes	Conscious Processes
		Event sequence: Yes, subject to the limitations of the working memory.* **Reliably track clock time:** Conditional and limited. Possible for short periods, subject to the limitations in attention and working memory. "Our perception of time, however, is intimately related to the level of attention given to the passage of time. When attention is diverted, a systematic shortening of the subjective duration occurs... Representations of time are reflected in the pulse count accumulated over a particular period, which critically depends on the degree of attentional engagement. The perception of time also relies on our stored representations of intervals in working memory" (Rao et al. 2001, pp. 317, 322).
Time	Temporal capabilities: Optimal for 10–20 seconds Track event sequence: Yes Reliably track clock time: No *Note:* Humans often demonstrate highly accurate clock-time tracking under ideal conditions. However, common situations and conditions cause this tracking to become increasingly inaccurate and unreliable and therefore unsuitable for safety critical tasks.	

*Tracking event sequences and clock time consumes attention and working memory resources. Both tracking capabilities degrade with conditions that degrade working memory/ short-term memory capacity and retention duration.

Our understanding of automatic processes has increased dramatically since the mid-1990s when functional magnetic resonance imaging (MRI) systems started providing a real-time look at the human brain under dynamic conditions. This research also increased our understanding of how automatic processes function and interact with conscious processes.

Consideration: The information in Table F.1 provides many indications and evidence that human behavior is systematic and, despite appearances, is seldom random. Behavior and actions are unique to each person because of their unique aptitudes, skills, long-term memories, mental models, attitudes, and beliefs. Different situational and environmental factors also affect behavior and actions. *Why is this important?* Existing practices often attempt to calculate safety system reliability and integrity. Those calculations may include the human contribution. If those methods or the employed factors assume that the incorrect human actions are random, the results may be incorrect.

Consideration: In a time-pressured active barrier environment, situational and environmental conditions commonly exist that can lead to attention capture and degrade WM/STM capacity and retention duration. These conditions degrade the ability to accurately track and remain internally aware of clock time. This capability is often needed to recall and perform a required action at the appropriate time. For these reasons, the barrier system design should employ methods and provide support aids and displays that support time-based task tracking and alerts. These functions and displays should be provided in all locations where a barrier person (HE) may be expected to perform a timed or a time-critical future task. (See Chapter 3, step B-10 for additional discussions and examples on this topic.)

A unique aspect of the human memory system, namely, the capture and storage of information, is influenced by an area of the brain tied to human emotions, the amygdala. Internal and external events that trigger increased emotions may selectively change or increase the event information (type, nature, extent) that is stored in long-term memory, information that may include one's emotional reaction. This implies an emotion-based bias regarding what information is captured and recorded in the memory in terms of type, context, breadth, and fidelity. (This is a major deviation from a computer-based concept that takes a mechanistic and dispassionate approach to capturing and storing information in memory.)

F.2 COGNITIVE CAPABILITIES AND LIMITATIONS (DESIGN CONSIDERATIONS)

Figure F.2 provides an overview of the cognitive and physical contributors that affect human behavior and performance. The affects are situation-specific and vary according to internal and external factors, conditions that create a near-infinite range of situational and transient affects and cognitive outcomes. Each can positively or negatively affect active human barrier and safety critical task reliability and performance.

The remainder of this section provides examples of cognitive traits and issues that can contribute to various cognitive challenges and affect humans' response in different situations and environments.

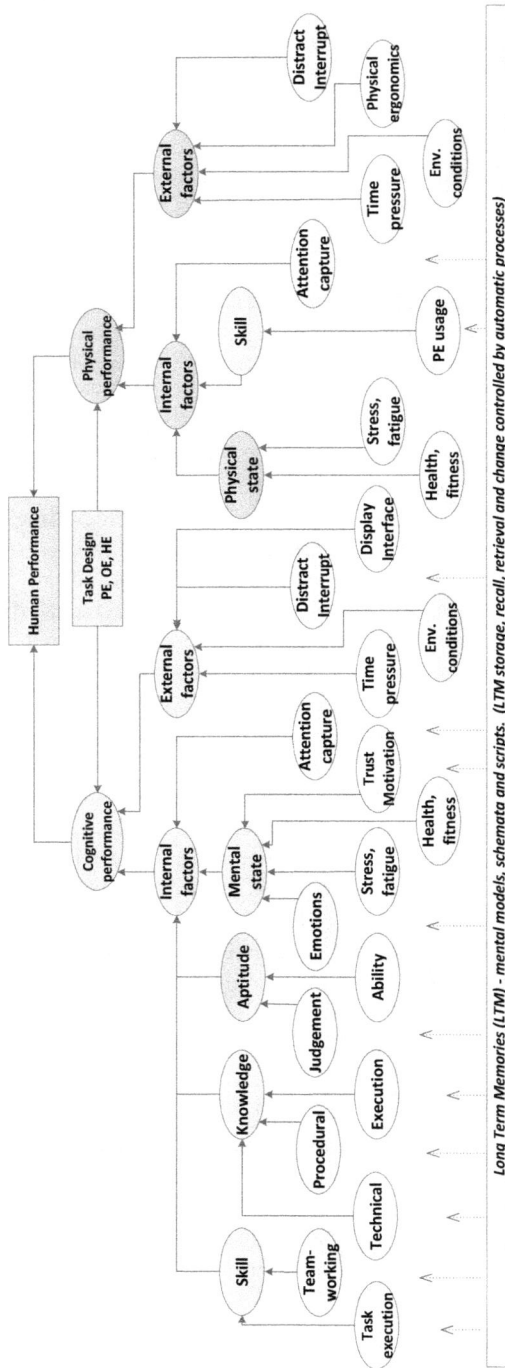

FIGURE F.2 Contributors to Human Performance and Behaviors.

F.2.1 ATTENTION

Attention is the cognitive aspect that directs automatic processes and working memory to an object and brings it into the mind's eye.

Primary Design Principle: A primary constraint on the barrier system design, as presented and assumed in the lifecycle model, is that a human can reliably attend to one and only one object at a time. (As with many aspects of human cognition, attention is not this simple. Humans appear to have the ability to attend to several objects simultaneously, as seems to occur when one drives a car while simultaneously conversing with a passenger. However, the mechanisms at work in this example are complex and not adequately addressed by the information provided in this book. For that reason, the guidance in the lifecycle model is limited to barrier designs and processes that conform to this principle.)

Terms used to differentiate attention types include selective, focused, divided, and sustained. A default state, attention is commonly guided by automatic processes identified in Appendix B.5.1 as *top-down processes*. Attention is also directed by a conscious process, and can thus be selective, focused, or sustained. When scanning or looking for something, we tend to scan *for what we expect* and *in the expected locations*, a process guided by mental models and memories. An automatic process can also automatically re-direct one's attention to an object of high salience (e.g., movement or a flashing light) or in response to an interruption or distraction. The unintended attentional shift can have an undesirable effect if it interrupts a complex sequence of actions or the flow of cognitive processes to make a safety critical decision.

Sustained attention refers to an intentional effort to maintain one's focus on an object or activity over a sustained, contiguous period, i.e., a vigilance task. The ability to do so tends to be limited to 15–20 minutes or less. A task design that relies on sustained vigilance should be avoided given this limitation. Doing so also places a persistent demand on working memory/short-term memory, which may be required elsewhere. Focused attention narrows the attentional focus to a specific object or area, which may cause a person to ignore or not notice other information in the environment or available from other sources. This has an undesirable effect if the undetected information is safety critical or time-sensitive. Divided attention occurs when one shifts attention/working memory between concurrent tasks and task activities. Thus, task design creates attentional demands and challenges that should be considered in the design process. Understanding the positive and negative effects of attention and attention management on barrier system function and performance must therefore be considered.

Attention and attention management can degrade in the presence of conditions and situations that degrade working memory. Example conditions and situations:

- All conditions that degrade working memory and short-term memory span (e.g., acute stress or fatigue)
- Workload exceeds capacity (increases error opportunities, likelihood, and the range of error types)
- Poor attention management (misdirected attention)

- Attention capture (see below for more information)
- A continuing effort to maintain a physical or mental work pace above one's "normal" pace
- A continuing effort to maintain one's emotional state or behavior in the presence of an internal or external stress-inducing condition (Kahneman 2011, pp. 39–42)
- Any need to continuously monitor any object, information source, and so on, i.e., a sustained vigilance task

Consider two types of attention capture: *internal preoccupation* and *external distractions*. Examples of *internal preoccupation* (Reason 1990, 2008) include the following:

- Excessive workload-induced tunnel vision (ignores information)
- Intended intense focus (lose awareness of surroundings, sensory information in the scanned workspace)
- Problems at home (misdirected attention, difficult to maintain attentional focus)
- Internally driven fear may trigger a freeze/flight/fight response (unintended re-direction of attention, loss of focus, temporarily disables/disrupts conscious cognitive processes)

Examples of external distractions (Reason 1990, 2008) include the following:

- Interruptions: two-way radio call, ambient conversations
- Sudden distractions: explosion, panicky voices, smell of toxic gas
- Sudden external event (startle event that triggers a freeze, flight fright response – an unintended re-direction of attention, loss of focus, temporarily disables/disrupts conscious cognitive processes)

Attention capture can lead to:

- The unintentional blocking out of other sensory or information inputs. (The missed information may be the barrier/task activator or required by a barrier task.)
- Automatic withdrawal of attention from a current task, which may be a top priority.
- *Execution errors*: For example, place losing, forgetting, or misremembering information once held in working memory (Reason 2008, p. 33).
- *Change blindness*: Failure to see what is not looked for, i.e., tunnel vision/ attention tunneling (Wickens 2013, pp. 53–54).
- *Strong habit intrusion*: Automatically perform a familiar skill-based task sequence that is not appropriate to the current task. Here, the sequence is similar but has a safety critical difference. This represents 40% of all absent-minded slips (Reason 2008, p. 42).

Recommended Reading: see Wickens et al. (2023)

F.2.2 Short-Term Memory (Separate Stores for Sensory Input and Information Processing)

See Table F.1, *Executive Mode, Memory Retention,* and *Memory Capture, Storage, Recall, and Retrieval.* Working memory (WM) includes an executive function and separate short-term memory (WM/STM) stores for episodic, phonological, and visual-spatial information. These stores provide the temporary buffers for holding information used in a WM process. WM/STM store capacities are severely limited to seven chunks of information, plus or minus two chunks. The store duration is brief (e.g., 30 seconds) if not regularly refreshed/attended to. If not refreshed, the stored information fades and is therefore forgotten.

Separate from WM/STM, the sensory system has dedicated stores that briefly hold sensory input. There are dedicated stores for visual, audible, smell, and haptic (touch) sensory inputs. If needed in a WM process, an automatic cognitive process must move the information from the sensory store before it fades. (This automatically occurs if the process is not interrupted.) The sensor store durations vary for each sense with the visual storage duration being the shortest, for example, 600 milliseconds. The store duration for audible information can be several seconds before it fades. Though not advised, the barrier design may require the task assignee to recognize and capture rapidly changing sensory information (e.g., from an audible conversation) by moving it to WM/STM and holding it there for its duration of use. As such, the sensory store duration and the WM/STM store duration and capacity must be considered and addressed in the barrier task and display design.

Note: The functioning of the STM sensory store for visual information can create a false perception of how long the sensed information is present. The duration of a lightning flash may be 200 milliseconds. However, the observer perceives the event as lasting longer. What the perceiver experiences is an echo of the event that results from its temporary retention in the dedicated visual STM.

Table F.1 indicates that conscious cognitive processes are slower than automatic processes. They also tend to commonly progress in a stepwise and sequential way. Errors in the working or short-term memory can occur that negatively affect the results of the detect, decide, and act phase; non-technical skills; and team communication and coordination. This can lead to a wide range of potential human errors that contribute to a wide range of barrier/task degradation and failure scenarios. Examples of types of errors attributed to working memory and short-term memory include the following:

- Forgetting or misremembering the information held in the working or short-term memory
- Forgetting to remember a pending future task, i.e., prospective memory (Reason 1990, p. 107)
- Losing track of time or poor time management
- Losing place: What step am I in?
- Losing track of barrier/task goals or priorities

F.2.3   TASK-SWITCH ERROR

Figure F.3 shows what actually occurs when an individual is attempting to progress two (or more) task simultaneously.

Attention must frequently shift between tasks and activities within each task. Task-switch error refers to a cognitive error mechanism or behavior that is seldom discussed. An example, the human (HE) fails to switch to a higher priority task (or task activity) when it is appropriate and warranted. Common issues that can delay or prevent a required task switch can result from personal motivation, barrier/task priority ambiguities not addressed in procedures or training, negative influences from a problematic safety culture and other hidden biases, and non-rational behaviors (Wickens and Gutzwiller 2015).

A timely switch to a higher priority task may fail to occur or be delayed 30% of the time (Wickens and Gutzwiller 2015). Interestingly, approximately 300 milliseconds are needed to switch one's attention between objects. Timewise, a single switch may have a negligible effect on barrier performance. However, a behavior that leads to an indecisive dithering between many objects with no meaningful result may cause a detrimental delay in the barrier response and performance.

Task-switch errors include the following:

- A reluctance to switch to a task that is unknown or is known to have a drastic effect (Sträter 2005, p. 50).
- Plan continuation error: a strong resistance to switch from a task that is nearing completion to a task has a higher priority. (Sträter 2005, p. 50).
- Under a high mental load, a switch may fail to occur due to attention capture and cognitive tunneling (Wickens and Gutzwiller 2015).

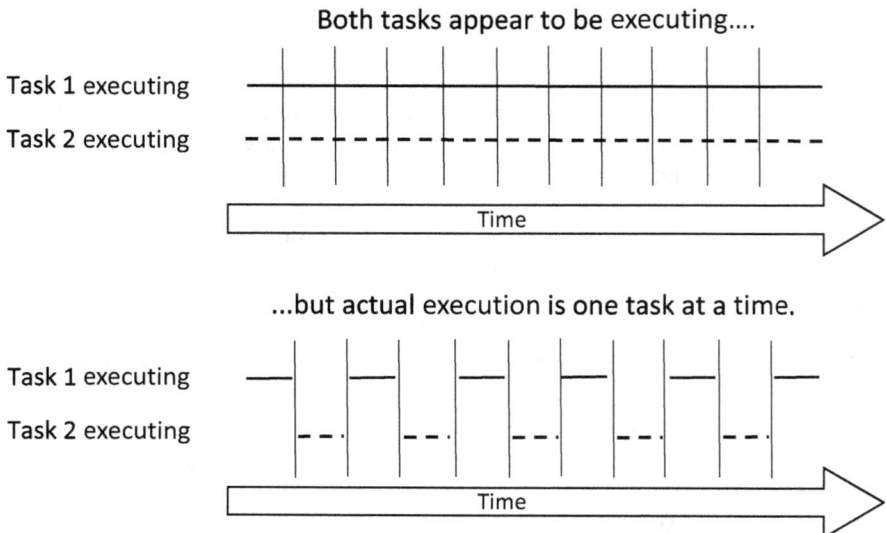

FIGURE F.3   Understanding Concurrent Task Execution.

- The presence of a goal conflict (real or perceived) reduces the certainty that the higher priority, safety critical task is selected (Sträter 2005, p. 50).
- If attempting to progress two tasks simultaneously (a high mental load), the more cognitively demanding task may be dropped even though it may have a higher priority (Wickens and Gutzwiller 2015).

For additional information on this topic, see Wickens and Gutzwiller (2015).

F.2.4 Change Blindness

Change blindness is a potential source of human error rooted in attention management, working memory, and mental models. For Kahneman (2011, pp. 23–24), attentional focus allows one to select specific sensory information to perceive, while other critical information may go unnoticed. Sensory perception may be inhibited if one's attention is fully focused on a difficult mental task. A form of change blindness, inattention blindness, can occur if the remaining working memory/short-term memory approaches its capacity limits or the person does not expect to see the missed object, as can occur with top-down processing (Wickens et al. 2013, pp. 55–56). In each case, we may easily miss important sensory information directly in front of us, i.e., we look but do not see. Furthermore, detecting change often relies on noticing and remembering a prior condition. From there, we need to compare the past and current information, understand what the differences mean, and know the expected response. Detecting change (a tracking task) places a non-trivial reliance and demand on WM/STM because one must remember the prior state (or know where to find it) to perform the compare function. Kahneman (2011, p. 24) also points out that individuals tend to be unaware that these and other forms of perceptual blindness are common and apply to everyone.

F.2.5 Cognitive Ease and Non-Rational Biases

Working memory and short-term memory, essential to a conscious cognitive process, are limited-capacity resources. A conscious process feels effortful and consumes a great deal of energy (e.g., sugars). Unconsciously, humans tend to seek a "least-effort" approach that limits its use, a behavior termed *cognitive ease* (Kahneman 2011, pp. 59–78). The following are examples of the causes and consequences of this behavior (Kahneman 2011, Figure 5, p. 60):

Causes of cognitive ease:

- Repeated experience
- Clear display
- Primed idea
- Good mood

Consequences of achieving cognitive ease:

- Feels familiar
- Feels true

- Feels good
- Feels effortless

These causes and conditions do not reflect the proactive, rational behavior expected from those charged with performing safety critical functions. The design process should recognize and respond to the resulting behaviors and non-rational biases that are individual adaptations of cognitive ease. These causes provide insight and basic guidance on how to present barrier essential information. The presented information must be clear and fit within a primed idea such as an appropriate mental model. Adapting the "chronic unease" behavior achieved by a high reliability organization (HRO) culture may offset the "feels good" consequence that can degrade the motivation to assess, decide, and implement an appropriate response to a complex barrier situation and time-critical decision.

Biases and behaviors attributable to cognitive ease include the following:

- **Confirmation bias:** The tendency to validate one's own understanding by seeking confirming information but ignoring and not seeking information that may contradict this understanding (Kahneman 2011, pp. 80–81; Wickens et al. 2013, pp. 261–262).
- No effort is made to assess if essential information is missing.
- If a prior alarm was caused by a spurious and non-hazardous condition, the tendency is to assume the same is true when this alarm activates in the future (Kahneman 2011, pp. 70–75).

Negative logic and word statements that have negating or double negative words require more working memory/short-term memory to process. Without knowing it, over time humans attempt to minimize the more effortful cognitive processes by seeking and focusing on confirming statements, while ignoring and failing to seek information that has negative logic or negating words.

More recent responses to major accidents, such as the white papers discussed in the introduction, identified confirmation bias as a persistent contributor to major accidents, warranting further attention in when designing barriers and safety critical tasks. Other common non-rational biases (discussed below) include loss aversion, availability bias, affect bias, and first theory bias.

F.2.5.1 Loss Aversion Bias

According to Kahneman (2011, pp. 302–303), "Loss aversion refers to the relative strength of two motives: we are driven more strongly to avoid losses than to achieve gains." A decision to activate a plant emergency shutdown function or abandon an offshore O&G facility may face stiff internal resistance if not counterbalanced against an equally compelling consequence, such as an acute realization of potential bodily harm to oneself, friends, and co-workers. The counterbalance may lie in having full confidence that one's management expects and fully supports such a decision.

F.2.5.2 Availability Bias/Heuristic

Kahneman (2011, pp. G.2-130) "defined the availability heuristic as the process of judging frequency by 'the ease with which instances come to mind.'" This heuristic applies to automatic and conscious processes. "The availability heuristic, like other heuristics of judgement, substitutes one question for another... Substitutions of questions inevitably produce systematic errors." This bias may be the initial calling criteria that call up the initial long-term memory or mental models in response to new sensory information. The recalled memory may or may not be appropriate to the actual situation and information context.

F.2.5.3 Affect Bias

Kahneman (2011, pp. 138–139) notes that "the affect heuristic is an instance of substitution, in which the answer to an easy question (How do I feel about it?) serves as an answer to the harder question (What do I think about it?)." "An inability to be guided by a 'healthy fear' of bad consequences is a disastrous flaw."

F.2.5.4 First Theory Bias

First theory bias is a common behavior in which the initial understanding of the information (i.e., first theory) is compelling. The default behavior creates a biased behavior that tends to resist changing this perception. Confirmation bias may be a contributor to that behavior. To offset this bias in complex scenarios and barrier types, it may be justified to pre-load procedures and training to establish the desired "first theory." Given the challenges with barriers activated by flammable gas alarms, it may be prudent to train personnel that the alarms are valid until unambiguous evidence proves otherwise.

F.2.6 Bounded Rationality/Keyhole Effect

Bounded rationality and the keyhole effect can lead to or cause comprehension (SA-2), projection (SA-3), and decision errors. A critical gap in one's understanding and experience can limit the range of explanations that may be considered when first attempting to understand a situation or condition. Similar gaps may also limit the range of decision options and response plans that are considered. These gaps may occur because essential input (SA-1) information needed to comprehend the event or situation is missing. The cause may be the information is no longer available, was never available, or was not presented in a form that innately contributes to comprehension. On the latter, refer to the following example for details.

Example Display that does not Convey Understanding. The operator's view of a flammable gas release and the potential risk if the gas ignites may be constrained, biased, or misled by the information presented on an HMI display. Widespread practice often presents this information as point measurements that indicate flammable gas concentration (e.g., a percentage of the lower explosive limit – LEL). This form is inherently abstract because it is an inferred indication of the true hazard, referring to the effects if the gas ignites. If ignition occurs, the cloud concentration, size, location, dilution rate, and area congestion determine the resulting thermal and overpressure loads on human and facility receptors at varying distances. Under the right conditions and environment, the gas cloud may detonate, creating an explosive force that may markedly increase these loads and the likelihood that high-speed shrapnel is ejected in many directions. The facility may have insufficient detector quantities, locations, and types needed to quickly detect the leak, identify its source, and assess its potential cloud size, location, and movement. If the HMI display does not adequately convey meaning, the operator may be unable to accurately understand the incident, its threat potential, and the time available to respond. An insufficient comprehension can lead to incorrect decisions, actions, and response timing. (For additional insight as it relates to display design, see Chapter 4, process step C5-1c.)

Yet another behavior may occur when the current state or situation is novel and not adequately addressed by procedures, experience, or training. A person may be missing the long-term memories and mental models needed to support recognition and comprehension. If so, the person may spend too much time searching for reasons that explain why the available information does not support what one believes to be the situation. The stronger the belief, the greater the resistance to pursuing alternate explanations or accepting that one's understanding may be wrong, misleading, or incomplete. As such, decisions and actions are based on an incorrect or incomplete understanding. At this point, the barrier task decision space may be limited to the options provided in procedures and training or guided by one's experience. Without a time-out and a team environment that allows for dissenting views, this situation may persist.

Consider: On accepting that the event is novel, a new ad hoc plan is needed. The plan may start with taking a fresh look at the available information to see if that creates a different and, perhaps, a more accurate understanding. Decision-making, no longer guided by procedures and prior experience, must adopt the more effortful and error-prone conscious cognitive processes. This is a shift to a knowledge-based approach (defined in the preliminary design step B-10). The cognitive assessment process in Appendix I attempts to address this situation in Table I.8, Item MM-1, solution OE2.

F.3 COGNITIVE ACCESS TO VISUAL INFORMATION

Figure F.4 summarizes the visual system acuity and access to information detail as one moves away from the eye's line of sight. The ability to read text or see fine detail is limited to a few degrees of the visual angle from the eye's line of sight. Beyond that angle, the capability to see detail quickly decreases. People vary in their individual capabilities. Subject to the wide variations possible in visual capability and acuity, a person may detect an object but cannot discern its details, for example, read text or fully discern a symbolic image. It may be possible to discern the presence of text but not the individual letters. In Zone C, visual acuity and access to information may be limited to detecting movement but unable to detect the presence of a smaller, less salient object.

This information should be considered when seeking the optimal placement of a barrier display object in a workspace. Example objects may provide visual, audible, tactile, or other sensory information types. An example workspace may be an integrated HMI display, a control console, a VDU display wall, a control room, an engineered area, or a process/utility area. (Appendix K provides guidance on placement within different workspace types.)

Consider displays sources (e.g., visual or audible) that must be reliably detected in a large workspace (e.g., at various elevations and locations along an egress/escape route). The design should consider how each display user may need to move their eyes, head, or body to gain access to a visual display positioned in various locations within the workspace. The workspace may be relatively small and provide access

Zone A - Foveal Vision
Able to discern:
Text (read)
Fine detail
Object / details
Movement / Trajectory
Color variation/change

Zone B – Central Vision
Able to discern:
Objects (limited detail)
Movement / Trajectory
Color variation /change

Zone C – Peripheral Vision (Mid)
Able to discern: Movement
Object presence (light emitting/reflecting)
Color: limited to black, white, variations

Visual receptors in eye – cones (10%) and rods (90%). (Proctor and Van Zandt, 2018, pp. 108-132)
Cones – Detect fine detail and color. Degrades in low light, e.g., unable to see color in low light.   Rapid light>dark visual adjustment.
Rods – Primary vision in very low light. Does not detect color. Low acuity, i.e., displays must be larger.
Cone density – Max in zone A. Progressively declines through zone B. None exist in Zone C or beyond.
Rod density – Exist in all zones, including far peripheral.
Eye movement - Normally expectancy driven (LTM/MM). May be triggered by conspicuous object, e.g., flashing light

Zone C: To 70° from LOS for older persons

Zone B: +/- 15° from LOS optimal for primary displays. +/- 35° for secondary displays.

Zone A: +/- 3° from LOS

Notes (Example design information)
Zone B: Up 15 /40 °, down 15/20 ° LOS for optimal/maximum view. (All data for eye movement only.)

For further detail (eye, optimal/max head movements) see Karawowski et al. (2021, pp. 516) and Guastello (2023, p. 108)

Line of sight

A B
C

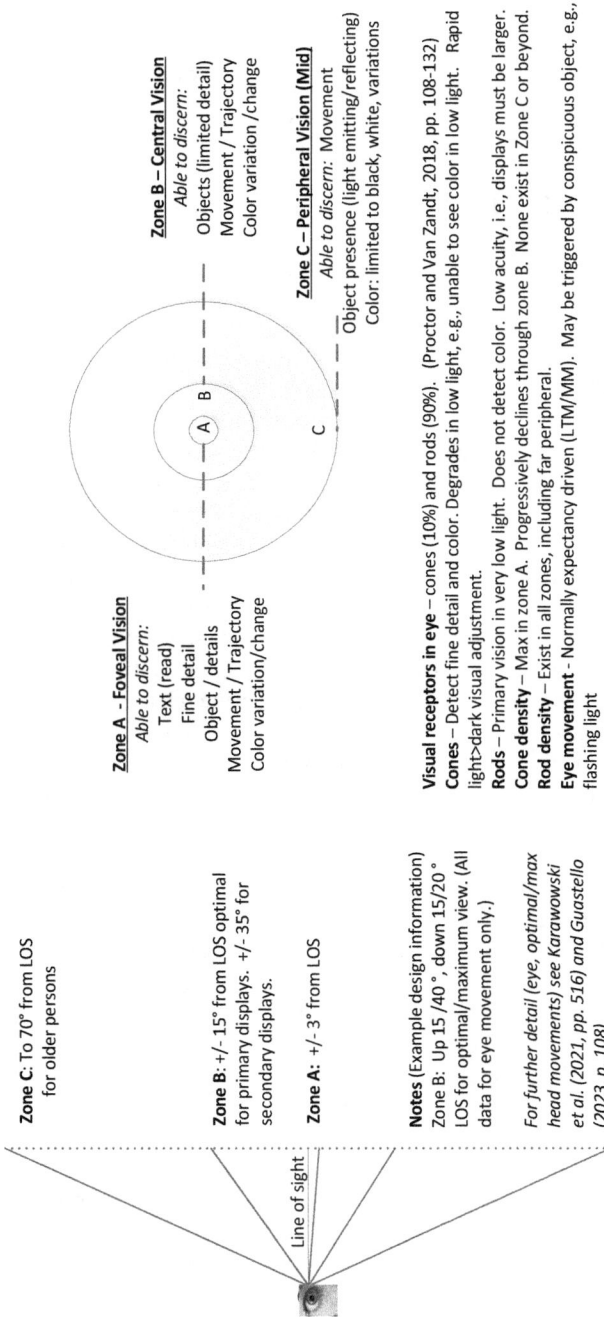

FIGURE F.4   Human Visual System, Capabilities, and Limitations. LOS – Line of Sight.

from a fixed location (e.g., access a VDU-based display on a control console) or large and provides access from a variety of locations (e.g., visual and audible objects accessed while transiting an egress/escape route during an emergency). The significantly degraded capability to detect and accurately perceive an object in Zone C requires additional consideration to better ensure the object in this zone is detected and accessible. For example, an object that gives the appearance of movement, optical blinking object (e.g., a flashing lamp), or an audible tone (non-directional) increases the likelihood that the object is noticed and accessed.

The design effort should consider and recognize situations (physical and cognitive) that can affect the noticing, accessing, and comprehending the meaning of display objects, which include assessing environmental conditions that may impede access. Making the display information available does not ensure it will be promptly and reliably detected or understood. The raw information provided by the human sensor systems generates a vast amount of information provided at a rate of 11–13 million bits per second. Roughly 80% of the information is visual, and many areas of the brain contribute to the processing of that information. Variations in individual visual and audible acuity, peripheral vision, object salience, task design (e.g., excessive workload or distractions), and experience (e.g., mental models) affect the reliability, timeliness, and accuracy of the abovementioned processes, thereby affecting barrier functions and performance.

Refer to Figure F.4. Within the visual system, dedicated neurons are believed to enable an early warning system that near-instantly detects a potential danger, such as a large and rapidly approaching object seen in one's peripheral vision. The sensed object may have been previously experienced and perceived as an acute source of danger. When a similar image is viewed and matched to that long-term memory, the autonomic process may trigger a near-instant endocrine release response that prepares the body for action. A moment later, automatic cognitive processes (followed several moments later by a conscious cognitive process) may finally "see" the image though only after the endocrine response is initiated. The viewed image may or may not be an actual source of danger. Regardless, it contributes to chemically induced stress, a starting point for automatic and conscious processes. (See Section F.1 for additional details on autonomic processes and the startle/freeze, fight, flight response.)

Known limitations and variations (person-to-person) exist in the human visual system. The actual raw information sensed by the human visual system is not the continuous and seamless image that we all experience. Instead, the raw information contains gaps and missing information. Moment-to-moment, automatic processes fill in these gaps and smooth the image based on what is expected (information recalled from long-term memories). Automatic processes have access to visual information from all three visual zones, as discussed in Figure F.4. The likelihood for noticing and detecting this information depends on its salience (a bottom-up, attention-driven process) and prior experience (a top-down, LTM-driven process). See Appendix B.5.1 for additional details.

Given the stated "slowness," relatively low bandwidth, and other limitations in conscious cognitive processes, its ability to capture, process, and make sense of rapidly changing information may be limited if an expected action places an unrealistic demand on the working memory's short-term memory stores. If an external means is not available to capture and later replay rapidly changing information, that information may be lost before it can be captured and used. As such, a task activity (one dependent on conscious cognitive processes) may be unreliable if the cognitive

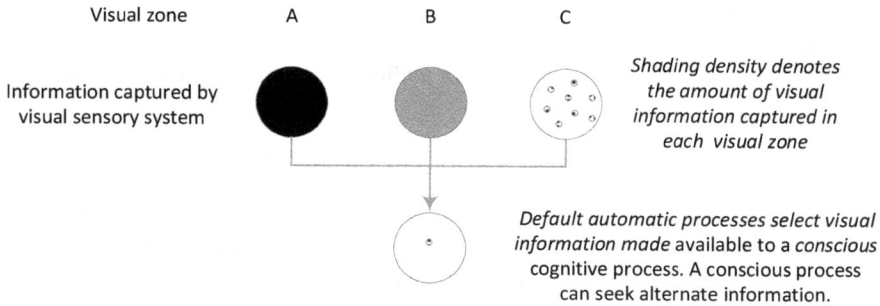

FIGURE F.5   Visual Information Available to Conscious Cognitive Processes.

demand is not reduced through the use of support aids, improved displays, or a task design change that shifts one or more activities to one that is skill-based (automatic and effortless). See Table J.3 (Appendix J.6) for other possible options.

Automatic processes continuously match sensory and other information to long-term memories as a means of deriving comprehension from that information and the plan for what to do next. What is "seen" and acted on depends on what resides in mental models and other long-term memories. (The latter is identified as a top-down process.) If the design requires the barrier assignee to quickly and reliably detect information in a peripheral zone, the applicable model processes should be employed to achieve that end. One option may rely on both skill-based and knowledge-based training. The knowledge-based training may explicitly identify and discuss the available information (including source and location) in each visual zone and how that information contributes to the barrier/task safety function. It may include a discussion on the situations and visual limitations that may impede that access. Skill-based training may employ drills or simulations that develop a desired scanning behavior that automatically looks for the required information in the expected locations. Doing so allows the trainee to learn when they may need to move their body or head to gain visual access to that information.

What happens if the design requires the conscious processing of specific information? As indicated in Figure F.1, conscious cognitive processes are significantly slower than automatic cognitive processes. Conscious processes may lack the resources needed (e.g., short-term memory) to acquire and process high-bandwidth sensory information. As mentioned, conscious processes are limited to processing 1–50 information bits per second, typically closer to the former. As symbolically indicated in Figure F.5, a small fraction of the available visual information is *selected* for conscious cognitive processing. Selection may be driven by a rational, conscious process or by a heuristic-based automation process that may or may not be appropriate to the situation. If the workload exceeds one's capacity, the person may default to attention tunneling, a behavior that ignores information (regardless of criticality) as a means of balancing the workload to the available WM/STM resources. This raises an important question to consider in design. Is the ignored information essential to the barrier/task function and performance? If so, what barrier system elements should address that challenge and how?

As a final comment, Figure F.4 is a simplification of the human visual system. This system has additional constraints and idiosyncrasies to consider in the design process.

For example, the discernible information in the horizontal direction (referenced to eye's line of sight) differs from that viewed in the vertical direction. See ISO 11064-3 (1999), ISO 11064-4 (2013), and similar sources for additional information and guidance.

F.4   ACUTE STRESS – PHYSICAL AND COGNITIVE EFFECTS

Performing safety critical and complex activities under time pressure, especially if a highly dangerous incident is underway, is a source of acute stress. Another, discussed in Figure F.1, is the occurrence of a sudden and unexpected event that triggers a startle response. The response is a near-instant release of stress-inducing hormones and stimulants that prepare the body for action. For an injured person, endorphins may mask a serious or potentially fatal injury. Epinephrine and cortisol can positively and negatively affect cognitive processes. Gasbury (2013, p. 38) explains,

> As your stress level increases and your body triggers the physiological reactions discussed earlier, it becomes more difficult to comprehend complex and detailed information. You may even struggle to understand things that, under normal stress levels, you would find easy to comprehend.

Gasbury (2013, p. 39) advises that

> Stress drives a person to use a primitive, instinct-driven form of decision-making. Intuition, triggered by stress, can lead to the often-referenced intuitive *gut feelings*, a primal decision process that uses subconsciously stored training and experiences supported by prewired evolutionary programming in your brain.

This indicates the importance of recognizing and understanding how and why long-term memories, mental models, and skilled actions ("mindware") drive automatic behaviors and responses when a dangerous situation is suddenly detected. Revealing one's automatic response and behaviors may be accomplished using time-pressured exercises that do not allow time for conscious cognitive processes to affect and temper recognitions, decisions, actions, and behaviors. The results provide insight into how a person might respond in a time pressured, emergency response barrier environment.

Case Study, Deepwater Horizon Accident: *Personnel positioned at one of the life-boats were asked to count off before boarding commenced. With repeated attempts, personnel were unable to perform this simple and basic task. Another person reached the lifeboat but did not wait to board (Skogdalen et al. 2011). Instead, the person jumped from height to the sea, an act that has a 5–20% chance of ending in a fatality.*

Table F.1 describes an automatic cognitive process that is able to quickly detect and automatically respond to a detected danger. (A detectable danger condition is one that is previously experienced or learned and therefore captured in long-term memory). It does not process risk. In the above case study, it is plausible the person's action to jump (rather than wait for the lifeboat) was driven by an automatic process that had no capability to consider the potential risk of injury or fatality when jumping from height; a conscious process is required to do so. Unfortunately, acute fear or being startled often suppress conscious processes when needed most, contributing to observed reactive and contextually irrational behaviors. In the above case study the jumper survived; an outcome partially attributed to a prompt response from a nearby ship. In a post-event interview, the person was questioned on that decision. The person did not appear to have a conscious rationalization for the act, which contributes to a view that the act was driven by a reactive, automatic cognitive process. Because these processes are mostly hidden from our conscious mind, humans commonly (and unwittingly) attempt to create a story and reasoned logic that explains their actions. However, humans remain unaware the story is partially or fully "made up". (When interviewing accident victims and witnesses, accident investigators should be aware of this normal human behavior.)

F.4.1 Sources and Effects of Acute Stress

The sources of acute stress may be environmental (intense noise or heat), exposure to novel sudden or dangerous events or situations (not previously experienced), and challenging task environments that may include high workloads, intense time pressures, exposure to danger, and performance anxiety (Flin et al. 2008, Figure 7.4).

Flin et al. (2008, pp. 177–178) point out that the effects of acute stress in a team reduce its ability to work together and coordinate activities. In individuals, stress can lead to:

- Memory impairment: Example presentations and forms include easily distracted, an increased tendency to employ confirmation bias, a degraded ability to manage and process information, or abandoning tasks (regardless of priority) that are contributing to cognitive overload)
- Reduced ability to manage one's attention and focus. As examples, poor attention management may impede one's efforts to prioritize work, maintain one's attention on complex information and activities, or lead to attention narrowing that reduces the ability to notice highly salient/highly conspicuous objects.
- Difficulty making decisions. Examples include stalled thinking (mind blanking), and an increased tendency toward availability bias (e.g., default to routine ideas and solutions with little to no consideration of other possibilities). Acute stress can also contribute to bounded rationality, where the person is not aware of and therefore does not consider explanations or courses of action of which he/she has no prior knowledge. Such limitations have been identified as a failure of imagination, a conclusion that may or may not be appropriate when applied to a high-consequence, time-pressured emergency response barrier.

F.4.2 Solutions to Manage and Mitigate Stress and Its Effects

Flin et al. (2008, pp. 180–183) give training solutions and methods to manage and limit the effects of acute stress over three progressive stages:

- First prevention phase: Provide training to gain knowledge on the causes, effects, and signs of acute stress.
- Second prevention phase: Provide training to develop the skills needed to actively recognize and manage the effects of acute stress in individual and team environments. For example, learned skills may increase one's ability to

regulate emotions, breathing, and heart rate. Another may seek to overtrain (overlearn) an assigned task or seek to improve attention management under stress. See Flin et al. (2008, pp. 182–183) for a more comprehensive list.

- Third prevention phase: Practice learned skills and behaviors in settings closer to what may be experienced in actual conditions. Conditions are more closely matched (to the degree possible and practicable) to those that may occur.

Note: The prototype lifecycle model does not explicitly address chronic stress, a condition that can have detrimental effects like those caused by acute stress. It develops over time in response to persistent issues or conditions such as chronic overwork, a poor safety culture, and poor physical working environment. The model assumes that chronic stress is addressed by corporate-level programs such as health and wellness management programs or continuous improvement in leadership and management culture. Process E includes monitoring processes to monitor for signs of chronic stress (e.g., the fitness for service monitor processes E3-2 and E6-2).

F.5   UNDERSTANDING MENTAL MODELS AND LONG-TERM MEMORY

Note: All references in this section are from Radvansky (2021) unless noted otherwise. See this reference for a more complete introduction to this critical topic.

The ability to understand why humans do what they do and think what they think should begin with a basic understanding of long-term memory. This includes understanding its nature, content, recall, update/retention/stability, idiosyncrasies, and differences in how it contributes to automatic and conscious cognitive processes. This section reviews a few facts that provide glimpses into how mental models and other long-term memories affect human performance and barrier task execution.

A common source of latent (designed-in) error in active human barriers begins with an inadequate understanding of the design's acute reliance (an assumption) that the target person possesses the required long-term memories and mental models. The design is at risk if the designer does not have an adequate understanding and knowledge of these memory types, including their development, unique strengths, and sources of fallibility. As a human moves through a physical workspace, the sensory information available to automatic and conscious cognitive processes constantly changes. In a near-instant, automatic processes search long-term memories using the available cues. On finding a match (a relative term), the recalled memory creates a recognition experience as affected by the detected information and situation. Each person's unique long-term memories create and contextually guide perceptions, comprehension, decision-making, and action responses. The memory match and recall process is subject to a range of variability that may include insufficient or misleading cues, imprecise search heuristics, and the influences from non-rational biases.

From page 22, the types of long-term memories include non-declarative and declarative memory. Non-declarative memory includes procedural (motor) and implicit memory. Procedural memory applies to how activities are done, like riding a bike. (This long-term memory type includes skills.) Implicit memories are employed in automatic cognitive processes that tend to be hidden from our conscious awareness. Declarative memory includes semantic and episodic memory. Semantic memory records our understanding of things and the relationships between them. An episodic memory is information from or attributed to an event or experience. (A mental model is a form of episodic memory.)

F.5.1   Long-term Memory Recall: Importance of Cues and Search Heuristics (pp. 210–211)

From Wickens et al. (2013, p. 235),

> We have long known that information is not stored in LTM as a random collection of facts. Rather, that information has specific structure and organization, defining the ways in which items are associated with one another. In particular, systems designed to allow the operator to use knowledge from a domain will be well served if their features are congruent with the operator's organization of that knowledge.... we store different types of information for broader than for narrow instances.

Automatic processes continuously match sensory and other available information cues with one's store of mental models and other long-term memories. This process continuously seeks to match incoming cues to long-term memories to create a continuous stream of perception and comprehension. A sensory input such as a viewed image or smelling hydrogen sulfide gas automatically triggers a memory search. If successful, an instant later (e.g., 100 milliseconds), a recalled memory may identify the chemical from its smell and provide a recognition-based awareness of its potential danger. This search/recognition process relies on cues from our sensory system and other cues that may be generated by a conscious cognitive process.

Automatic processes that attempt to match cues to long-term memories may use a similar or like-for-like match (Reason 1990, pp. 130–131). If insufficient cues are available, other heuristics guide the search, for example, revert to seeking a memory that is readily available (most recently called) or most frequently called (Reason 1990, p. 131). A recalled memory is not necessarily one that applies or is helpful. Understanding this process should encourage the need to examine the recalled memory for validity and actively seek another if needed to find one that fits (Reason 1990, p. 135).

"Long-term memory is content addressable, i.e., we can access the information using the components that make it up" (p. 210). The employed search cue may be feature- or context-based. A feature cue resides within the memory, such as a personal experience, a date, or smell. "Feature cues involve components of the memory itself" (p. 210). A sensed odor often tends to be one of the strongest. In addition, a cue that is easily associated with oneself tends to be stronger. On the contrary, "Context is important for episodic memory because context changes can indicate new events..., such as a shift in special location or temporal framework (e.g., a day later). Context is a powerful memory cue" (p. 213). A stronger contextual cue may duplicate the contextual information included in the memory.

The nature of the available cues (number, similarities, and differences) affects the speed and accuracy of the memory recall process. Too few cues may recall no memories, leaving one with no understanding. If cues are common to many memories, the first recalled memory may not actually apply to the situation. Having several strong and different cues improves the memory search and search discrimination (assuming the memory exists).

If a recalled memory does not provide the information needed to recognize or comprehend an attended object, a conscious cognitive process is needed to change the memory calling conditions and seek a different memory. This process is repeated until an adequate long-term memory is recovered or we begin to realize that we may have no memory applicable to this cue or situation (meaning the cue or event or situation is novel).

Thus, how can this information be used in barrier design? Every display, support aid, display, communication, procedure, and training module should consider the inclusion of cues (e.g., terms, organization, color) that supports the memory recall of the appropriate memory. This creates a design linkage between the following:

- A barrier/task activity that specifies the required skills and knowledge.
- Long-term memories and mental models that may contain the required knowledge (context applicable).
- The training and on-the-job experience needed to develop and maintain the abovementioned long-term memories and mental models.
- The cues needed to support the search, recall, and activation of the above-mentioned memory. These cues may be provided or available:
  - In procedures, support aids, displays, and communications.
  - In the environment, for example, sensed visual, audible, tactile, or olfactory information.

Note: The abovementioned concepts tend to be missed in current practice.

F.5.2   MENTAL MODELS

This book makes extensive references to mental models. Mental models

> often contain the who, what, when, where, why and how of an experience. The binding of information that makes up episodic memories occurs in the hippocampus. Thus, a mental model can include some perceptual or other experiential details. In general, mental models are remembered over long periods. People use knowledge at this level to make memory decisions about what was encountered before.

> *(pp. 208–209)*

The term is also used to describe the product of a conscious cognitive process that integrates sensory information with mental models and other long-term memories to develop an understanding of an event or situation.

Two additional important LTMs are schemata and scripts. A schema is a proto-typical state of the mental model that allows its information to be quickly found and recalled, that is, it functions like a shortcut. Schemata develop through direct experience but may also develop through indirect means (e.g., reading about it or learning about it in a training class). Once accessed, the schema may provide comprehension and projection/anticipation information in the activated memory. Scripts "are a set sequence of actions on what to do in each case that a schema represents" (Endsley and Jones 2012, p. 23). Each contributes to appropriate and timely task execution and performance.

Note: For an advanced discussion on mental models and applications see Borders (2024).

F.5.3 Remembering Event Order (pp. 370–373)

Humans tend to accurately remember the order of events, information that may be important in an SA-3 projection or anticipation (event escalation or sequence) or when attempting to recognize a situation or simulate options for solutions when using the recognition-primed decision model. Using this recall helps to play a sequence of events in the order in which they occurred. As a caveat, humans have a greater problem if attempting to work backward in that sequence, for example, given event A, what happened immediately prior to that event?

F5.4 Remembering Event Duration

From page 377, "Memory for time can be distorted forward or backward in time." Recalling the duration of a remembered event is subject to time compression. As an example, the recalled duration of an event may be remembered as a much shorter period (clock time) than occurred with the actual event. Recalling a memory of a similar (but different) event may provide misleading guidance. This becomes important when deciding how much time is needed to complete a barrier task activity, information that can be used to shift between tasks or to determine how long a decision can be delayed.

Essentially,

> Memory for time is worse than memory for space. In addition to memory errors due to age and serial position, temporal memory shows scale, telescoping and compression effects. Despite these errors, people are often able to remember the sequence or order in which events occur, even if they cannot place them properly in time.
>
> *(p. 377)*

This type of information can be used to guide an SA-3 result (time projection) or the use of the recognition-primed decision model. A typical requirement of an HE who is assigned an SA-3 requirement or employs the recognition-primed decision model is that he/she be an expert in that domain. Domain expertise developed over 10 or 20 years encompasses a wide range of knowledge and hands-on experience. (The experience may also be developed using a high-fidelity simulator.) Experts should have the experience of discovering if and when their memorized duration estimates could be incorrect and develop ways to adjust that data over time.

Appendix G
Verification and Validation

This appendix is organized as follows:

- G.1 Introduction
- G.2 Design Reviews
- G.3 Inspection: Processes and Execution
- G.4 Testing: Processes and Execution
- G.5 Verification: Processes and Execution
- G.6 Validation: Processes and Execution

G.1 INTRODUCTION

Figure G.1 summarizes the inspection, testing, verification, and validation activities in the prototype lifecycle model. The approach is guided by ISO 11064-1 (2000), IEC 61511-1 (2016), and EI (2020b). These activities should be identified in the barrier safety management plan (BSMP), described in Appendix B.2.

Note: Figure G.1 encompasses all project phases defined in ISO 11064-1 (2000). The preliminary design phase indicated in the figure encompasses the ISO phases A (clarification), B (analysis and definition), and C (conceptual design). Clause 7.6 in the ISO standard provides high-level verification and validation guidance that apply to the ISO phases A and B.

As shown in Figure G.1:

- Design reviews (discussed in G.2) that may have different scopes and follow different procedures and execution plan requirements (e.g., timing, independence, and competency requirements). The review verifies that the phase output deliverables conform to the input phase requirements.
- Inspection activities (discussed in G.3) that may have different scopes, occur at different locations, follow different procedures, and have different execution plan requirements. The inspection visually verifies that the item conforms to requirements and to the approved documents, drawings, and installation details.
  - **Periodic (ongoing) inspections** that occur during construction, fabrication, assembly, installation, and software/HMI display development.

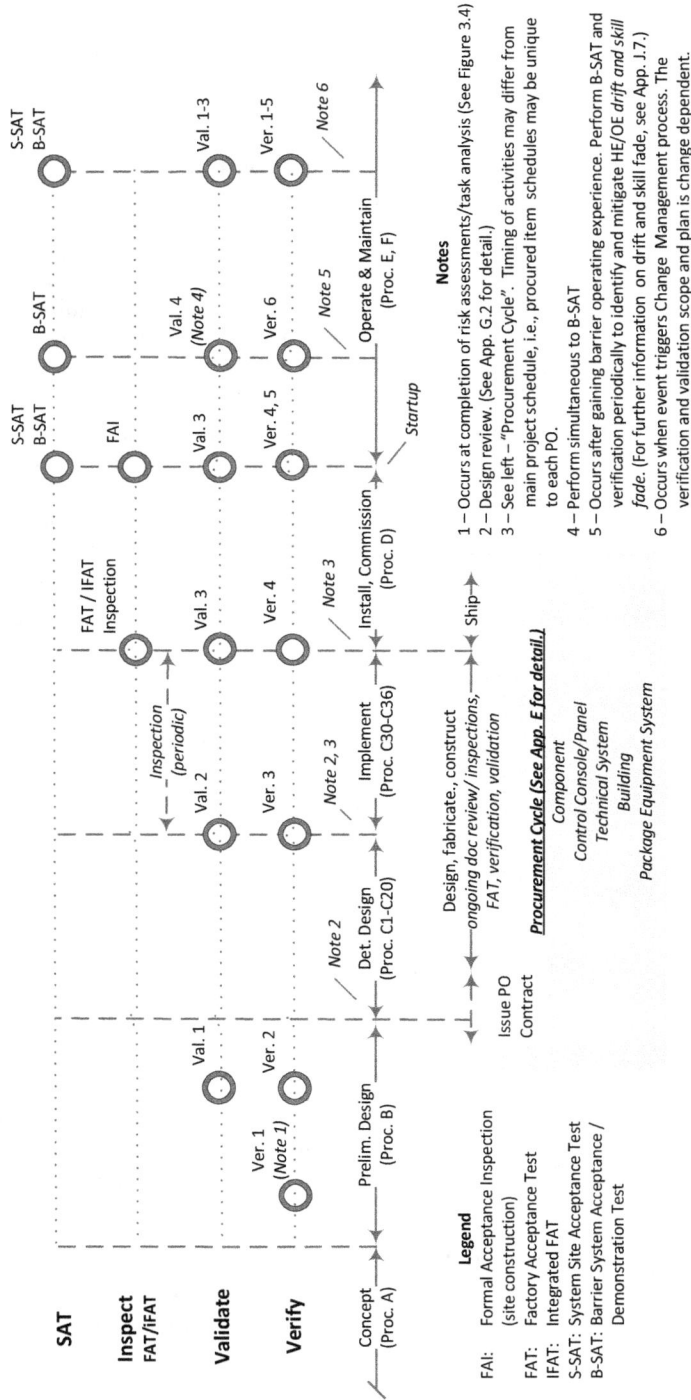

FIGURE G.1 Verification and Validation Activities and Timing.

- **Formal Acceptance inspection** (FAI) may be a pre-condition for acceptance and shipment of the procured item. For field fabrication, installation, and construction, it provides a means to verify the works (goods) conform to requirements and are mechanically complete.

- Testing activities (discussed in G.4) that may have different objectives and scopes, occur at different locations, and follow different procedures and execution plan requirements:

  - **Factory Acceptance Test (FAT)** – Formal testing verifies that barrier elements residing in the tested equipment or system conform to the specified system, function, and performance requirements. Examples of "equipment" are HMI displays, control consoles or panels, technical systems, packaged equipment systems, and shop-fabricated buildings.
  - **Integrated FAT (IFAT)** – Formal testing verifies that a communication/functional interface between technical systems conforms to requirements (e.g., the automatic interchange of display and control information and functions between a technical and packaged equipment interface).
  - **System Site Acceptance Test (S-SAT)** – Formal testing conducted onsite verifies that the barrier elements that reside in equipment (e.g., a technical or packaged equipment system) conform to requirements once fully installed, commissioned, and made operational.

    Note: The FAT typically seeks to test every system function and verify performance. The test may include connected devices (typically one each) and simulated interface connections to other systems. As such, it does not test every connected device and interface or the effect of using the site infrastructure (e.g., field network and cabling systems, site-supplied power, or external environmental controls or lighting systems). The site environment also differs (e.g., electrical noise, heat, light). Rather than repeating the FAT, the S-SAT should perform tests that could not be performed or fully demonstrated in the FAT.

  - **Barrier System Site Acceptance Test (B-SAT)** – A comprehensive, live activation and demonstration of the fully installed, commissioned and operational barrier system (PE, HE, and OE) under representative conditions (e.g., time pressured). *This is the only test that demonstrates (verifies) the barrier system achieves the specified safety function and safe state at the required performance and within the specified safety times.*

- Six verification schemes (discussed in G.5), which may have different objectives and scopes, occur at various locations, and follow different procedures and execution plan requirements. These schemes verify if the barrier system conforms to requirements and includes assurance activities.

- Four validation schemes (discussed in G.6), which may have different focus areas, occur at different locations, and follow different procedures and execution plan requirements. These schemes focus on the adequacy of the barrier system from the end-user perspective.

Active human barrier element design (e.g., elements common to emergency response barriers) often employs a fail-to-danger design. The design may lack the capability to automatically detect element failure. Inspection and testing provide a means to detect this failure type. Proposed processes in the lifecycle model work to detect drift, a failure condition that tends to remain hidden. (For information on drift, see Appendix J.7.)

G.1.1 VERIFICATION

The suggested verification schemes described in Section G.5 are guided by the following adopted definition.

ISO 11064-1 (2000, cl. 3.16) defines verification as a

confirmation by a systematic examination and tangible evidence that specified requirements have been fulfilled. Note 1: In design and development, verification concerns the process of examining the result of a given activity to determine conformity with the stated requirements for that activity. Note 2: Tangible evidence is regarded as being information that can be proven to be true, based on facts obtained through observation, measurement, test or any other means.

The following excerpts indicate similar industry perspectives on verification schemes and activities.

According to EI (2020b, p. 52),

A verification scheme is a management system defined and implemented by the operating company to confirm whether SCEs will be, or are, suitable … A verification scheme comprises the verification process and processes for its implementation, reporting requirements, record keeping and review… The operating company should be responsible for owning, establishing, implementing, *reviewing and revising the verification scheme, whereas the verifier should be responsible for performing the verification activities and reporting the results.*

EI (2020b, p. 56) also notes that

the verifier should not repeat the work of the designer – the intent is to review a sample of the design documentation to get confidence that the design will deliver suitable SCEs with appropriate PSs, that will meet the defined risk criterion (e.g., ALARP).

(For the abbreviations: SCEs – safety critical elements. PSs – performance standards.)

IEC 61511-1 (2016, cl. 3.2.87) defines verification as a "confirmation by examination and provision of objective evidence that the requirements have been fulfilled." Note 1 in this clause states that verification is "the activity of demonstrating for each phase of the relevant SIS safety lifecycle by analysis and/or tests, that, for specific inputs, the outputs meet in all respects the objectives and requirements set for the specific phase." Note 2 from this clause states, "example verification activities include:

- reviews of outputs (documents from all phases of the safety life cycle) to ensure compliance with the objectives and requirements of the phase taking into account the specific inputs to the phase;
- design reviews;
- tests performed on the designed products to ensure they perform according to their specification;
- integration tests performed where different parts of a system are put together in a step-by-step manner and by the performance of environmental tests to ensure that all the parts work together in the specified manner."

The suggested verification schemes presented in Section G.5 introduce additional verifications and criteria to better address the unique nature and challenges of an active human barrier. An example is verification #1, which seeks to verify that errors did not occur in the barrier origination, definition, translation, transfer, or recording activities. Later schemes include the development and application of performance standards that focus on the HEs and overall barrier system performance. Adding these new acceptance criteria enhances and expands the focus and results of the verification process.

G.1.2 VALIDATION

The suggested validation schemes, described in Section G.6, are guided by the following, adopted definition:

ISO 11064-1 (2000, cl. 3.15) defines validation as a "confirmation by examination and tangible evidence that the particular requirements for a specific intended use are fulfilled. Note: In design and development, validation concerns the process of examining a product to determine conformity with user needs."

Because an active human barrier relies on humans, the design of the human-system interface shares similarities with the commercial software domain. Chemuturi (2013, pp. 104–105) describes the following validation approaches used in this domain:

- **Expert Review:** "Expert in this context is a person who is well versed in the functional domain of the proposed project or product. The expert may be a single individual or a set of individuals. The method may be an independent review or a guided review. Expert review uncovers gaps in the requirements..."
- **End-User Review:** "End users are the individuals who are likely to use the resultant software product...These end users ought to be knowledgeable about the document conventions and be able to understand the documents and provide intelligent feedback. The review may be independent or guided, but in most cases, a guided review is preferred....Arranging

a meeting and get them all together and present them with the requirements for review would be better as their feedback can be collected in one go. It is also effective on timeline considerations. A guided review also facilitates in ensuring the end users understand the requirements as intended without bringing in their individual experiences to interpret the requirements."

The validation schemes suggested in Appendix G.6 (Tables G.7 and G.8) use both approaches. The *end-user review* approach may contribute to closing the gap between work-as-imagined and work-as-done. (Closing this gap is a stated objective for the prototype lifecycle model. See Appendix 9.6 for further information and background.)

Note: The end-user review should provide reliable and insightful information and a primary means to close the WAI-WAD gap. As a caveat, it can also be a source of design bias if not guided and tempered. Based on personal experience, operations personnel assigned to a large capital project tend to be (1) among the most experienced and skilled. (2) The base tendency is toward the "normal day" when considering the time needed to reliably complete a barrier task. (3) Less consideration may be given to the "off-normal day," which is certain to occur. On these days, the person may be fatigued, distracted by events at home or a looming layoff, or recovering from an injury or illness. The net bias effect may be overly optimistic belief on what personnel can actually and reliably do (a common human behavior). From the perspective of response performance, the difference between the normal day/optimistic view and the plausible bad day view can be significant. The review should consider the full range of experience and competency that is likely to be realized in the pool of persons assigned to barrier roles. A procedure well suited to a highly experienced and knowledgeable person (one end of the spectrum) will likely be inadequate for a person recently transferred into a role to gain needed experience (the other end of the spectrum).

As an observation, the adopted ISO definition aligns with the definition for validation provided in IEC-61511-1(2016). IEC 61511-1 (2016, cl. 3.2.86) defines validation as "confirmation by examination and provision of objective evidence that the particular requirements for a specified intended use are fulfilled." However, it appears to deviate from the ISO standard when it identifies an activity as a validation process, a process that ISO identifies as verification. For example, IEC 61511-1 (2016, cl. 15.2.1, see Note) identifies example validation activities that include "loop testing, logic testing, calibration procedures, simulation of application program." *Site Acceptance Tests (SAT) are also described as a validation activity.* As is often the case in current practice, the scope of the SAT is to verify the functionality and performance of the tested system in the operational environment. This test seldom seeks to demonstrate the adequacy of the system (end-user perspective) given the plausible range of operating conditions, situations, and environments.

G.1.3   COMPETENCE AND INDEPENDENCE

ISO 11064-1 (2000), IEC 61511-1 (2016), and similar industry documents state that personnel assigned to perform verification and validation activities should have the required competencies and be fully independent of (not part of) the design and implementation team. (The competency and independence requirements should be documented in the barrier safety management plan described in Appendix B.2).

G.1.3.1   Competence

EI (2020b, p. 53) asserts that "verifiers should have sufficient numbers of competent staff in all relevant disciplines (e.g., process, electrical, etc.)."

Furthermore, as per IEC 61511-2 (2003, cl. 5.2.2.2),

> The skill and knowledge required to implement any of the activities of the safety life cycle relating to the safety instrumented systems should be identified, and for each skill, the required competency levels should be defined. Resources should be assessed against each skill for competency and also the number of people per skill requirement.

G.1.3.2   Independence

Finally, EI (2020b, p. 53) emphasizes that

> verifiers should be sufficiently independent as to be impartial and objective in their judgement…Practically this means the verifier should be independent of the operating company's management, should not have been involved been involved in any aspect of anything likely to be examined…The role of the verifier may either be undertaken by a single organization or by several different individuals or organisations considering separate aspects of the operation.

G.1.4   COMPARISON TO IEC 61511-1 (2016)

Figure G.2 provides an overview of the verification, validation, and functional safety assessments identified in the IEC 61511 standard.

Some activities in Figure G.1 overlap with those shown in Figure G.2, such as design reviews and acceptance testing. There are also differences. For example, the FSA activities are integrated into the verification processes.

G.1.4.1   Discussion – Functional Safety Assessment

IEC 61511-1 (2016, cl. 5.2.6.1) defines the Functional Safety Assessment (FSA) performed in each lifecycle phase. These activities are integrated into the suggested verification schemes.

The FSA includes activities that may be more common to verification, with at least one exception. The FSA does not limit its scope to a single lifecycle phase. Examples of FSA activities cited from IEC 61511-1 (2016, cl. 5.2.6.1.4) are as follows:

- the H&RA has been carried out (see 8.1).
- the recommendations arising from the H&RA that apply to the SIS have been implemented.

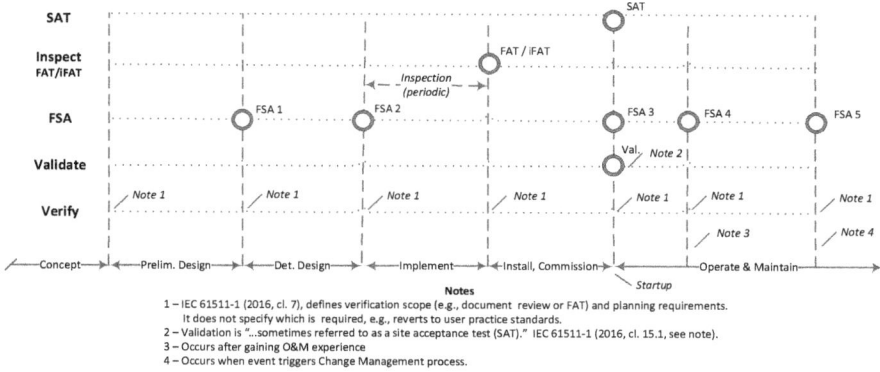

FIGURE G.2 IEC 61511-1 (2016): Verification, Validation, and Functional Safety Assessment.

- project design change procedures are in place and have been properly implemented.
- the recommendations arising from any FSA have been resolved.
- the SIS is designed, constructed and installed in accordance with the SRS, any differences have been identified and resolved.
- the safety, operating, maintenance and emergency procedures pertaining to the SIS are in place.
- the SIS validation planning is appropriate and the validation activities have been completed.
- the employee training has been completed and appropriate information about the SIS has been provided to the maintenance and operating personnel.
- plans or strategies for implementation further FSAs are in place.

Furthermore, as per cl. 5.2.6.1.10,

A FSA shall also be carried out periodically during the operations and maintenance phase to ensure that maintenance and operations are being carried out according to the assumptions made during design and that the requirements within IEC 61511 for safety management and verifications are being met.

Finally, cl. 5.2.6.2.1 highlights that "the purpose of the audit is to review information documents and records to determine whether the functional safety management system (FSMS) is in place, up to date, and being followed."

Note: In the prototype lifecycle model, the term for the abovementioned FSMS is "Barrier Safety Management Plan" (BSMP).

G.2 DESIGN REVIEWS

Figure G.1 suggests the timing of the design reviews included in the lifecycle model. The BSMP should define the requirements for formal design reviews, including the scope, objectives, execution plan requirements, and timing. It should also define requirements for the assessor/reviewer competencies and independence, the expected response to identified deficiencies, and so on. The review should verify that all requested phase deliverables are complete and delivered, and the phase input requirements are correctly integrated and applied. The review may be part of or separate from the verification activities. The scope may be a spot-check of randomly or purposely selected documents or a more comprehensive review.

Note: If the scope of the design review in step B-27 (preliminary design) includes a barrier-dependent Remote Operations Center; see ISO 11064-1 (2000, clause 8.3) for additional guidance.

G.3 INSPECTION: PROCESSES AND EXECUTION

Table G.1 suggests inspections to verify that the inspected item or element conforms to requirements. The criteria for conformance or acceptance reflect the inspected item type, inspection scope and objectives, and the lifecycle phase when the inspection occurs. Figure G.1 indicates the suggested timing of each activity. Table G.2 suggests example execution topics to address in the inspection plan. The BSMP should define the basic elements in the plan including the inspection scope, type, participants, and timing.

Ongoing inspections commonly occur during the fabrication and construction phase to prevent late detection. Acceptance inspections (e.g., an inspection included in a factory acceptance test) visually verify if the provided PE complies with approved documents and drawings. Periodic O&M phase inspections should confirm whether an element's observable design and integrity conforms to requirements included in the Safety Requirements Specification.

The scope and objectives of the inspection should be defined in the BSMP. Inspection activities may seek to visually verify that:

- Barrier element sizing, locations, dimensions, and configurations are consistent with approved documents and drawings.
- Placement and mounting of a barrier PE conform to drawings and installation details.
- Supplied components conform to specifications and data sheets.
- The design of an engineered area or room, including design, supply and placement of equipment, lighting, and Support PE storage, conforms to approved specifications and drawings.
- Control consoles and panels conform to specifications and general arrangement drawings and plans.
- Specialty material conforms to defined material requirements as confirmed by a review of vendor data and positive material identification reports.
- All requested documents are provided, such as procedures, device and equipment certifications, reports, and performance test results.
- The installation of equipment conforms to requirements (e.g., covers installed, tags and labeling in place and readable, and provider recommendations).

TABLE G.1

Inspection: Examination Form and Tangible Evidence

Inspection Type and Timing	Process/Step	Applies to	Systematic Examination Inspection Form	Tangible Evidence
Periodic — *Implementation phase*	C30-6 C31-6	Procured equipment, technical systems, packaged equipment systems, and building	1. Verify inspection planning and execution performed according to BSMP requirements (e.g., planning, procedures, participation, addressing deviations). 2. Verify inspected items conform to requirements. If applicable, verify the inspected elements conform to performance standard requirements (See D.3.1 for a PE example.) 3. Verify identified deficiencies are recorded and resolved.	– Inspection activities conform to requirements in the BSMP. – Inspections verify if inspected elements conform to requirements. – All requested documents are provided and conform to requirements. – Identified deviations and corrective actions are recorded and resolved.
Formal Acceptance — *Performed with FAT, prior to shipment*	C30-8 C31-8		– See activities 1–3 in the "Periodic" inspection above. – Verify all requested documents are provided and comply with requirements (e.g., Factory Acceptance Test procedures, operating procedures, maintenance procedures, inspection records, deficiency records, equipment certificates).	
Periodic — *Site/facility construction and installation*	D1-4 D2-3 D4-3	Site-constructed, engineered areas, buildings, barrier	– See activities 1–3 in the "Periodic" inspection above. – If applicable, verify inspected elements conform to performance standard requirements. (See D.3.1 for PE example.)	– See tangible evidence from Periodic and Formal Acceptance Inspections above. – Visual inspection verifies if construction and installations conform to approved documents and drawings.
Formal Acceptance Inspection (FAI) — *(mechanical completion)*	D1-6 D2-5 D4-5	element installations, dependent external support systems and protective barriers	– Verify construction and installation conform to approved documents and drawings.	

TABLE G.2

Inspection: Example Execution Plan Elements

Inspection Type	Inspection Staffing				Inspection Procedure Source	Inspection Locations
	Barrier Lead	Other Barrier HEs	PE Provider	Others		
Periodic Purchased item (Engineered equipment, systems, and buildings)	Typically not	Typically not	Yes (support)	Often by EPC contractor	By PE provider if vendor data requirement	PE provider's staging location
Formal Acceptance (*before shipment*)	TBD	TBD		EPC contractor Barrier element	May be by EPC contractor guided by owner/operator standards	Onsite
Ongoing Inspection (onsite construction)	TBD	TBD	TBD	purchaser, other technical disciplines	May be by EPC contractor guided by owner/operator standards	Onsite
Formal Acceptance Inspection (FAI) (mechanical completion)	Possibly yes, or designee	TBD	Yes (support)	Owner/operator onsite team or designee	May be by EPC contractor guided by owner/operator standards	Onsite

Note: The scope of the inspection is limited to active human barrier elements and the dependent external support systems and external barriers.

G.4 TESTING: PROCESSES AND EXECUTION

Table G.3 suggests the testing scope and acceptance criteria. Test activities seek to verify the tested items and systems conform to the barrier-specified equipment, system, function, and performance requirements. Figure G.1 indicates the suggested timing of each activity. Table G.4 suggests example execution topics to address in the test execution plans. The BSMP should define the basic elements for these plans, such as the test scope, type, participants, and timing.

Performing the suggested Factory Acceptance Test (FAT), integrated FAT (IFAT), and System Site Acceptance Test (S-SAT) are common in current practice. Barrier system elements that reside in the tested system, may represent a small part of the overall test scope.

The suggested Barrier System Acceptance Test (B-SAT) is unique given the acceptance criteria in the suggested operations-based performance standard defined in Appendix D.3.3. This is a live, comprehensive test of the complete barrier system which is now fully installed and commissioned. The test may occur when the protected facility is operating under normal operating conditions.

Note: The B-SAT provides the only means to verify the barrier system achieves and conforms to the barrier-specified function, performance, and performance standards (e.g., error rates, recovery actions). By nature, the test includes the effects of the external support systems and the installed environment.

Note: Consider a barrier element (e.g., a single HMI display or display element) that resides in a technical system that contain functions from many applications. The time needed to verify a barrier element may be brief (e.g., minutes or hours) relative to the overall testing period (e.g., days or weeks).

G.4.1 DISCUSSION

Testing an active barrier type that resides fully within PEs (a barrier type that does not rely on a human to achieve its safety function) can reveal hidden deficiencies resulting from design errors, component and system degradation, undocumented hardware, or software change. The same is true when testing physical elements that reside within an active human barrier. However, the active human barrier is more complex given its reliance on one or more humans to achieve its safety function and safe state. Design induced error/failure mechanisms may exist at the point of interaction/interface between HE's (person-to-person communication or observations) and between an HE and a direct-use physical element, procedure, or support aid. Cognitive and physical failure situations tend to be situational and transient, a reflection of constantly changing task loads, physical and cognitive activities, and environmental conditions. Figure F.2 (Appendix F.2) provides insight into the wider range of the human related/affected aspects, conditions, and situations that can passively and negatively affect human performance. This performance contributes to barrier degradation, failure, or success.

TABLE G.3

Testing: Examination Form and Tangible Evidence

Test Type	Process or Step	Applies to	Systematic Examination Form	Tangible Evidence
FAT Procured equipment or systems	C30-8 C31-9	Procured, offsite-fabricated items: – engineered equipment – technical systems – control console/panels – packaged equipment systems (PES) – buildings	1. Verify the FAT planning and execution are performed according to BSMP requirements (e.g., planning, procedures, participation, deviations). 2. Through FAT, verify the tested item conforms to the barrier system, function, and requirements specified or referenced in the safety requirements specification (SRS). 3. Verify the tested item conforms to approved performance standard requirements. (See D.3.1 for the suggested performance standard for physical elements.) 4. Verify all test deviations were recorded, corrected/implemented, and closed.	1. Test performed according to BSMP requirements 2. FAT verifies equipment conforms to system, function, and requirements in the SRS. 3. FAT verifies the system conforms to performance standard requirements. 4. Records identified deficiencies and required corrective actions (CAs). No pending deficiencies or CAs remain.
IFAT Procured equipment or systems	C30-8 C31-9		See the examination form from the FAT above. (Modify for IFAT scope and purpose.)	See the FAT above.
S-SAT Procured equipment or systems	D2-8 D3-3 D4-8		1. Verify the S-SAT planning and execution is performed according to BSMP requirements (e.g., planning, procedures, participation, deviations). 2. Through S-SAT, verify the tested item conforms to the barrier system, function, and performance requirements specified or referenced in the safety requirements specification (SRS). 3. Verify the tested item conforms to approved performance standard requirements. (See D.3.1 for the suggested performance standard for physical elements.)	1. S-SAT performed according to BSMP requirements. 2. S-SAT verifies equipment conforms to system, function, and requirements specified in the SRS. 3. S-SAT verifies equipment conforms to performance standard requirements. 4. Records identified deficiencies and required corrective actions. No pending deficiencies or CAs remain.

			4. Verify all test deviations were recorded, corrected/implemented, and closed. *Note: Depending on the equipment tested, the S-SAT is often performed before the overall site becomes operational. In some cases, it may occur after facility start-up to demonstrate the system achieves contractual operational performance specifications.*	

B-SAT
Barrier system
(tests full barrier system, installed and operating)
	D-8
	E-8
		Fully installed and operational barrier system	– See examination form in FAT, items 1 and 2. Exception: activities specific to the scope and purpose of the B-SAT.
			– Verify the barrier system achieves the stated safety function and safe state within the specified task and barrier safety times.
			– Verify the barrier systems conform to all performance standard (operations-based) requirements:
			  • HE-based performance standard requirements (See Appendix D.3.2 for detail.)
			  • Barrier system performance standard requirements (See Appendix D.3.3 for detail.)
				– See tangible evidence from FAT above, items 1 and 2.
				– Testing verifies the barrier system achieves the specified safety function and safe state within the specified task and barrier safety time (TTST and BST)
				– Testing verifies the barrier system conforms to all performance standard requirements

TABLE G.4
Testing: Example of Execution Plan Elements

Test	Staffing				Test Procedure Source	Example Test Locations
	Barrier Lead	Other Barrier HEs	PE Provider	Others		
FAT	TBD	TBD	Yes Typically conducts test witnessed by others	Barrier designer or designee Other disciplines as required	Often by PE provider. (approved by owner/operator)	Vendor staging or fabrication site
IFAT	TBD	TBD	Yes	Yes (TBD)	TBD (approved by owner/operator)	TBD May be at PE provider site or other locations
S-SAT	TBD	TBD	May witness or support	EPC contractor Operations	Often by PE provider. (approved by owner/operator)	Onsite
B-SAT	Yes	Yes	—	Expert assessor: monitors conformance to HE and barrier system performance standards. Others may monitor other aspects of test	Owner/operator or designee	Onsite

Identified aspects and conditions are detected by monitoring processes included in the prototype lifecycle model. (See Chapter 7 for detail.) In design, the challenge lies in having and applying the right processes that guide the holistic selection, application, and integration of the barrier system PE, OE, and HE elements. The proposed B-SAT is the only means to verify this complex system of interacting and interdependent elements actually achieves the barrier safety function at the specified performance.

Note: A common practice with a safety instrumented function (a different active barrier type), its function and equipment are tested multiple times before the barrier is placed into service. This is often not the case for active human barriers. The full B-SAT can only be performed after completing the testing and verification of the individual OEs, HEs, and PE subsystems and elements. (Additional time may be allocated after the initial startup so HEs can gain additional experience with system elements. This may place the timing of the test after the barrier is placed into service. A similar practice is not permitted for safety instrumented functions.) Once the plant is operational, additional safeguards should be in place pending completion of the B-SAT and the test results demonstrate the barrier system conforms to all requirements. Some active human barrier types may be assigned a specified level of risk reduction (i.e., a Safety Integrity Level). To achieve and maintain that reduction, the B-SAT, maintenance, and other verification activities must be performed at the frequencies specified in the SRS.

G.5   VERIFICATION: PROCESSES AND EXECUTION

Table G.5 suggests verification processes that include the systematic examination form, tangible evidence of compliance. Figure G.1 indicates the suggested timing of each activity. Table G.6 suggests example execution topics to address in the various verification schemes. The BSMP should define the verification scope, objectives, performance standards, timing, recommended participants, and examination form and tangible evidence.

Depending on the regulatory regime, regional practice, project type, and company practice, the proposed verification activities may include formal design reviews (G.2), inspections (G.3), and testing (G.4). EI (2020b) uses the terms assurance and *verification* to differentiate between a formal verification scheme and other quality activities. Here, the term *verification* is reserved for the more formal verification schemes. *Assurance* collectively refers to the other quality activities. Requirements and differences in the various verification activities should be defined in the BSMP.

Table G.5 seemingly shows that verifications #4 and #5 occur at the same time. In practice, the procured items that may be delivered at various times, such as those delivered by different purchase orders or contracts. Verification #4 applies to individual items that include engineered and fabricated control consoles, technical systems, packaged equipment systems, and buildings. Verification #5 verifies the readiness of the full barrier system as a pre-condition to placing the barrier into service.

Note: A different (though potentially less rigorous) verification option may include CRIOP (SINTEF 2011). This, according to SINTEF (2011, Section 1.1), "is a methodology that contributes to the verification and validation of the ability of a control centre to safely and efficiently handle all modes of operation including start-up,

TABLE G.5

Verification: Examination Form and Tangible Evidence

Ver. No.	Process/ Step	Applies	Systematic Examination Form	Tangible Evidence
1	B3-3	Completion of risk/ task analysis, and data transfer to Tables 3.5 and 3.6 **Objective:** verify if the information in Tables 3.5 and 3.6 is complete and correct.	1. Verify the risk assessments (RAs) are complete and performed according to approved procedures and include the required participants. 2. Review the active human barrier origination requirements in the RA (e.g., HAZOP, LOPA, other processes). Verify the validity of the scenarios, assumptions, and recommendations. 3. Verify the RAs captured the required information (e.g., assumptions, process safety time, confirm the proposed barrier will not create a new hazard). 4. Verify the task analysis (TA) is complete, performed according to the approved procedures and the BSMP, and includes the required participants. 5. Verify the TA results: does it achieve the barrier function and have assigned goals? Verify tasks are correctly framed to encompass cognitive challenges, have defined task assignees, etc. 6. Verify the RA/TA reports accurately capture and migrate information from the assessment/analysis worksheets and tools. 7. Verify the TA activity conforms to the BSMP requirements. 8. Verify the RA/TA action items are resolved and closed. 9. Verify the selection and recording of RA and TA information in Tables 3.5 and 3.6 is complete and accurate. 10. Verify all pending issues from prior phases are resolved and closed	1. RAs completed and performed according to published requirements and plans. The required personnel attended. 2. RA analysis and recommendations appear to be clear and valid. 3. RAs captured the required additional information, and it appears to be valid. 4. TA completed and performed according to BSMP requirements. The required personnel attended. 5. Tasks appear to be correctly framed to the appropriate level of detail with actionable and appropriate goals. 6. RA and TA reports accurately capture/migrate information from the assessment worksheets and tools. 7. TA activities conform to the BSMP requirements. 8. All RA and TA action items (those applicable to the barrier system) are completed/implemented and closed. 9. Tables 3.5 and 3.6 accurately capture the barrier and task requirements from the RA and TA. 10. No pending recommendations/actions (those affecting active human barriers)

			Verifications	Results
2	B-21	Completion of barrier definition/ preliminary design phase **Objective:** Verify if phase deliverables conform to the input requirements with no pending actions.	1. Verify the phase verification activities (e.g., methods and processes) are performed according to published plans, procedures, and competency requirements in the BSMP. 2. Verify all required phase deliverables are published, approved, and conform to phase input requirements. 3. Verify updates to the SRS fully reflect barrier requirements and design until this stage. 4. Verify phase deliverables conform to base performance standard requirements (Table 3.7). 5. Verify a formal design review (App. G.2) was performed according to BSMP requirements. 6. Verify the employed design / development tools do not negatively affect those processes or products. 7. Verify that no pending action items remain from this or a prior phase. *See ISO 11064-1 (2000, clause 7.6) for additional verifications applicable to barrier task design, function allocations, etc.*	1. Phase verifications performed according to all requirements and BSMP. 2. Deliverables from this phase are complete and conform to phase input requirements. 3. SRS fully updated and accurately captures barrier requirements. 4. Phase deliverables conform to the base performance standard requirements. 5. Formal design reviews conform to BSMP requirements. 6. Design/development tools had no negative effect on design/development process or products. 7. No pending action items remain from this or prior phase. *See ISO 11064-1 (2000, clause 7.6) for additional requirements.*
3	C-19 C30-4 C31-4	Completion of detailed design phase **Objective:** Verify phase deliverables conform to input requirements with no pending actions.	1. See examination form from verification #2. 2. Verify requisitions correctly include barrier element requirements (PE and OE). 3. Verify phase deliverables conform to the PE-based performance standard requirements (See App. D.3.1). 4. Verify assessments were performed as per requirements and the approved results appropriately implemented.	1. See verification #2. 2. Requisitions include the required scope, technical documents, vendor data, and inspection/test requirements. 3. Phase deliverables conform to approved performance standards. 4. Assessments performed as per requirements, approved results are appropriately implemented, and no actions remain.

(Continued)

TABLE G.5
(Continued)

Ver. No.	Process/Step	Applies	Systematic Examination Form	Tangible Evidence
		PE and OE elements fabricated/constructed/quality-checked to confirm readiness for commissioning activities.	1. See examination form from verification #2. 2. See examination form from verification #3, item 3. 3. Verify periodic and formal acceptance inspections (G.3) are complete and performed as per approved procedures and BSMP requirements. Verify if deviations are recorded and resolved. 4. Verify all FAT and IFAT activities (G.3) are complete and performed as per approved procedures and BSMP requirements. Verify all deviations are recorded and resolved. 5. Verify all barrier HE positions are filled (e.g., internal transfers and hiring completed. See process C-33). 6. Verify that all required additional resources were acquired (See process C-34).	1. See verification #2. 2. See verification #3, item 2. 3. Periodic/formal acceptance inspections completed and performed as per approved procedures and BSMP requirements. 4. All FAT and IFAT activities are complete and performed, and all deviations recorded and resolved according to approved procedures and BSMP requirements. 5. All HE staffing positions filled and available for training. 6. Acquisition of all *additional resources* completed and verified.
4	C-35 D1-7 D2-6 D4-6	**Objective:** Verify fabricated/constructed elements and OE elements completed and conform to requirements with no pending actions.	7. Verify all supplied, constructed, and installed PEs (barrier-required/dependent) conform to approved documents and drawings. No action remains. 8. Verify all supplied, constructed, and installed PEs (barrier-required/dependent) conform to performance standards. (See Appendix D.3.1.) 9. Verify all barrier-required procedures are completed and approved as per PDMS requirements. 10. Verify training plan and modules are developed and approved according to TDMS requirements. 11. Verify no actions remain for any of the above activities.	7. Supplied, constructed, and installed PEs conform to all approved documents and drawings. 8. Supplied, constructed, and installed PEs conform to performance standards. 9. All barrier-required procedures completed and approved as per PDMS requirements. 10. All barrier training planning/module development conforms to TDMS requirements. 11. No pending/open deviations remain from the above evidence sources.

#	ID	Objective	Verification	Criteria
5	D-7	Input to Startup Readiness Check. **Objective:** Verify inputs to pre-start-up readiness check completed with no actions pending. (See process D-10, in Figure 6.1).	1. See examination form from verification #2. 2. Verify all verification #4 activities are complete with no remaining deviations or pending action items. 3. Verify all validation #3 recommendations are addressed and closed (See processes C30-7, C31-7, D1-5, D2-4, D4-4). 4. Verify that procured items are installed as per approved drawings and procedures (See process D3-1). 5. Verify commissioning activities (pre-S-SAT) are completed as per approved documents and procedures (See processes D2-7, D3-2, D4-7). 6. Verify all S-SATs are complete and performed as per approved procedures and BSMP requirements (See processes D2-8, D3-3, D4-8). 7. Verify all required training is complete (See process D6-1). 8. Verify no deviations or pending actions remain from any of the above activities.	1. See verification #2. 2. All verification #4 activities complete with no remaining deviations or pending actions. 3. All validation #4 recommendations addressed and closed. 4. Procured items installed as per approved drawings and procedures. 5. Commissioning activities complete with no pending deviations or action items. 6. All S-SATs completed and performed according to approved procedures and BSMP requirements with no pending deviations or action items. 7. All required training is complete. 8. No pending/open deviations remain from the above evidence sources.
6	E-10	O&M Phase Periodic **Objective:** Verify barrier system and activities conform to the BSMP and SRS requirements.	1. Verify the physical and operational status of the barrier system complies with and conforms to requirements in the safety requirements system. 2. Verify the barrier system conforms to operational performance standards (See Appendices D.3.2 and D.3.3) 3. Based on the B-SAT, verify the barrier system meets all function and performance requirements. Verify all deviations are recorded, corrected, and closed. 4. Verify all O&M phase monitoring activities are performed as per the BSMP. Verify all findings and deviations are recorded, corrected, and closed.	1. The barrier system conforms to all SRS requirements. 2. Barrier system conforms to performance standards. No action items remain. 3. B-SAT verifies the barrier system meets all barrier system function and performance requirements. No open deviations or action items remain. 4. O&M monitoring functions performed per BSMP requirements. All identified findings and deviations are recorded, corrected/implemented, and closed.

(Continued)

TABLE G.5
(Continued)

Ver. No.	Process/ Step	Applies	Systematic Examination Form	Tangible Evidence
	E-10 (continued)		5. Verify reviews completed for incidents and barrier to identify potential deficiencies in the barrier system. If so, verify all deficiencies are appropriately addressed, corrected, and closed. Verify that the change management process was used.	5. Review of incidents and barrier activations identified no system deficiencies. Findings and corrective actions are recorded, corrected, implemented, and closed. The change management process was used to address a change that affects an SRS stated requirement.
			6. Verify that barrier-dependent ESS and EBs contain no deficiencies or action items that can degrade a barrier system.	6. No deficiencies or action items remain for barrier-dependent ESS and EB.
			7. Verify the process E-9 PIF assessment was performed as scheduled, and a report issued with all findings and recommendations. Verify all actions and recommendations were addressed through closure.	7. A review confirms the timely completion of the process E-9 PIF assessment. A report was issued. No pending actions or recommendations remain.
			8. Verify no changes were made to owner/operator (o/o) programs and policies employed in the barrier lifecycle. Verify changes (if any) are reported. Verify all subsequent actions and recommendations were addressed through closure.	8. A review confirms no changes made to o/o programs/policies occurred. If changes, no pending recommendations, or actions remain.
			9. Verify there are no open/pending recommendations or actions that affect a barrier. (These may result from a required assessment, change management process, or a change in staffing, programs, or policies.)	9. A review of all sources of barrier and barrier related changes confirms there are no pending actions or recommendations.
			10. Verify all changes from item 9 above used the change management process and conform to the applicable plans and processes in the BSMP.	10. A review of records from step 9 above verifies all changes employed the change management process and conformed to the plans and processes in the BSMP.

TABLE G.6
Verification: Example Execution Plan Elements

		Staffing				
Ver. No.	Barrier Lead	Other Barrier HEs	PE Provider	Others	Verification Process Requirements	Verification Locations
1	No	No	–	Independent assessor with expertise in risk assessments and task analysis (see BSMP)	Barrier safety management plan (BSMP), other sources when required by regulation or statute, or defined by owner/ operator	Design office
2	No	No	Support function	Independent Assessor (verification lead) (See BSMP for detail.)		Design office
3	No	No	Support function			Design office
4	Yes, or designee	TBD	Support function			Onsite
5	Yes, or designee	TBD	–			Onsite
6	TBD	TBD	–		BSMP, SRS	Onsite

normal operations, maintenance and revision maintenance, process disturbances, safety critical situations and shut down."

Note: In current practice, performance standards often include the regulatory, statutory, industry, and owner/operator standards. As this is common practice in many existing regions and regulatory regimes, it is not shown in the prototype lifecycle model.

G.5.1 Discussion

Verification #1 – Like other barriers identified in a risk assessment (RA), active human barrier identification and origination is the product of an RA. The information from the RA is an essential input to the task analysis (TA). Errors in this process (analysis, developed information, results) may be random and/or systematic. Verification #1 seeks to verify, to the extent possible and practicable, that errors did not occur in these processes. (The scope of the RA review process is limited to active human barriers.) Example error sources are as follows:

- The RA (preliminary design step B-2) may incorrectly identify, define, or analyze hazard scenarios and err in the selection of the active human barrier as a viable and achievable means to prevent or mitigate the hazard or its consequences. Errors may occur when defining the process safety time or confirming the proposed barrier cannot create a new hazard. Perhaps, a required participant (or role) was missing or other deviations from the published RA plan or procedures occurred.
- The TA (preliminary design step B3-1) may be incorrectly performed. Defined task goals may be vague. The barrier tasks may be unable or

unsuited to reliably achieve the barrier function within the barrier safety time. One or more tasks may fail to capture and encompass all likely decisions and the associated SA-1 information requirements. Perhaps a required participant (or role) was missing or other deviations from the published TA place or procedures occurred. Finally, assumptions may be unrealistic or not documented.

- Errors may (and often do) occur when migrating the results of the RA/TA to their respective reports. Common types of errors include translation errors, cut-and-paste errors, and missing information. Errors can also occur when the RA recommendations are considered and addressed in post-RA processes (e.g., a layer of protection analyses).
- Errors may occur when information is migrated from the RA/TA reports to Barrier and Task Origination Tables 3.5 and 3.6 (preliminary design steps B2-5 and B3-2).

EI (2020b) presents an increasingly common approach that assesses physical elements to determine which should be classified as safety critical elements (SCEs). This designation invokes a more rigorous set of verification processes and activities. In contrast, the prototype lifecycle model defines (and views) every PE, HE, and OE component in the barrier system and its dependent external support systems and protective barriers as critical meaning its failure (or critical deficiency) can prevent the barrier from achieving its specified safety function at the required performance. The failure may be acute (e.g., the capability to achieve the barrier function is immediately lost). Other failures may reduce the duration of time the barrier remains available or degrades a barrier element or task function (e.g., loss of area lighting degrades access to labels, physical signs, and paint markings).

Note: The prototype lifecycle model does not prevent applying the SCE approach when developing verification schemes and performance standards. Appendix D.3 suggests different performance standards to consider given the added verification requirements that may apply to barrier organizational and human elements.

G.6 VALIDATION: PROCESSES AND EXECUTION

Table G.7 suggests validation processes that include the systematic examination form and tangible evidence of compliance. Figure G.1 indicates the suggested timing of each activity.) Table G.8 provides example execution topics to address in the various validation schemes. The BSMP should define the validation scope, objectives, timing, recommended participants, and example examination form and tangible evidence.

Note: CRIOP (SINTEF 2011) may be an alternative validation option to supplement or replace the suggested schemes.

TABLE G.7
Validation: Examination Form and Tangible Evidence

Val. No.	Process/ Step	Applies to	Systematic Examination Form	Tangible Evidence
1	B22	End of basic barrier requirements definition/design, before control room activities	See ISO 11064-1 (2000, cl. 7.6)	
2	C-20	End of detailed design phase **Objective:** To validate the adequacy of the barrier system based on the detailed design deliverables.	Validate the adequacy of the barrier design. Examples may include validating the adequacy of the following based on the intended use, and use locations and environments: – Direct Use detect and act phase physical elements. – Support PE (carried, worn) – Engineered workspaces (e.g., VDU displays, control consoles and panels, rooms and engineered areas). – Identified barrier and task procedures and support aids – Specified competency requirements (input to training and required competencies) – Defined/required decisions – Defined/required (and plausible) skill-based actions – Performance standards (e.g., PE, HE, and barrier system)	For each intended (and plausible) user, the assessor confirms the assessed item is adequate for the intended (and plausible) uses/ purposes and use locations and conditions.
3	C30-7 C31-7 D1-5 D2-4 D4-4	Completion of implementation phase and prior to commissioning and the S-SAT **Objective:** To validate the adequacy of the fabricated or constructed equipment and workspaces (e.g., engineered area, remote operations center, room).	Validate the adequacy of each fabricated, constructed, and implemented barrier system elements and their dependent external support systems and protective barriers. See Validation #2 for the suggested adequacy assessment areas.	

(Continued)

TABLE G.7
(Continued)

Val. No.	Process/ Step	Applies to	Systematic Examination Form	Tangible Evidence
4	D-9	O&M Phase – concurrent with B-SAT **Objective:** To validate the barrier system in live (full demonstration) conditions	Validate the adequacy of the fully installed and operational barrier system. The assessment scope should encompass the entire barrier system (PE, OE, and HE). Examples adequacy assessments may include the following: – Barrier system capability to achieve the specified barrier, task, and task phase safety times. – Timely detection of detect phase information and changes. – Team design, interactions, coordinated performance – Defined competencies (those defined as new skills and knowledge, or others) – Defined/required decisions – Defined/required (and plausible) skilled actions – Workspace design and support for efficient and interference free movements with limited interruptions – Workspace environment (e.g., lighting, temperature, humidity, air quality, noise) – Provided procedures and support aids. – Provided training. – Competency verification process. – Direct-use PE and Support PE – External support systems – External protective barriers.	*See above for tangible evidence*

TABLE G.8
Validation: Example Execution Plan Elements

		Staffing		Validation	
Val. No.	Barrier Lead	Other Barrier HEs	Others	Validation Process Requirements	Validation Locations
1	TBD	TBD	Expert reviewer (*see note*)	See ISO 11064-1 (2000, cl. 7.6)	TBD
2	* TBD	* TBD	** Validation performed by expert reviewer or expert end-user (defined in BSMP)*	Barrier safety management plan (BSMP)	TBD
3	* TBD	* TBD			Vendor fabrication site Onsite
4	** TBD	** TBD	*** Validation performed by external expert end-user with post-B-SAT input from barrier HEs (defined in BSMP)*		Onsite

Note: Barrier HEs may be participants in the phase design activities.

Appendix H
Performance Influencing Factors

H.1 INTRODUCTION

This appendix is organized as follows:

- H.2 PIF Assessment for Barriers, Tasks, and Task Phases
- H.3 PIF Assessment (Working Environment) for Direct-Use Components and Support PE
- H.4 Basis for Selecting Performance Influencing Factors

H.4 evaluates the performance influencing factors (PIFs) and performance shaping factors identified in four industry guidance standards. Using the described process, a short-list is developed from these sources and one new PIF is added. The resulting list, presented in Table H.1, is adopted for use in the lifecycle model.

H.2 PIF ASSESSMENT FOR BARRIERS, TASKS, AND TASK PHASES

Figure H.1 summarizes the PIF assessment process performed in the detailed design phase (process C-2) and the operate and maintain phase (process E-9). This process evaluates the effects of each PIF (positive and negative) on barrier-assigned personnel. Table H.1 identifies the suggested PIFs addressed in this assessment. The results are captured in Table H.2.

Tables A.2 and A.4 (Appendix A.2.3) identify the suggested participants and roles. Participants should have the appropriate barrier and facility experience, knowledge, and expertise to provide input to guide the assessment.

Note: If the PIF assessment described in an existing guidance or practice standard is sufficiently similar to the process in Figure H.1, it can be substituted in the lifecycle model if it evaluates all of the PIFs included in Table H.1.

H.2.1 Step H2-1, Select Barrier, Task, or Task Phase for Evaluation

This step selects the barrier, task, or workspace activities for evaluation.

1. Select the barrier, barrier task, and task phase (if applicable) for assessment. (The HE or maintenance role is inherent to the task.) Record this information in Table H.2

From detailed design process C-2 /
operate and maintain process E-9

Tables: 3.5, 3.6, 4.4, 4.15, 4.18, 4.19,
4.20, 4.22, 4.23
Docs: Task analysis, barrier roster
and org. chart, SOPs, specifications,
general arrangement drawings
Process E-9: Also, tables 4.5, 4.30,
B.5, B.6 (O&M procedures).

H2-1	Select barrier, task, and task phase for evaluation	→ Table H.2
H2-2	Select PIF for evaluation	→ Table H.2
H2-3	Evaluate PIF for positive / negative effects	→ Table H.2
H2-4	Develop recommendations	→ Table H.2

Table H.1 → (to H2-2)

Table H.2 → (to H2-3)

Table H.2 → (to H2-4)

To process C-3 / E-10

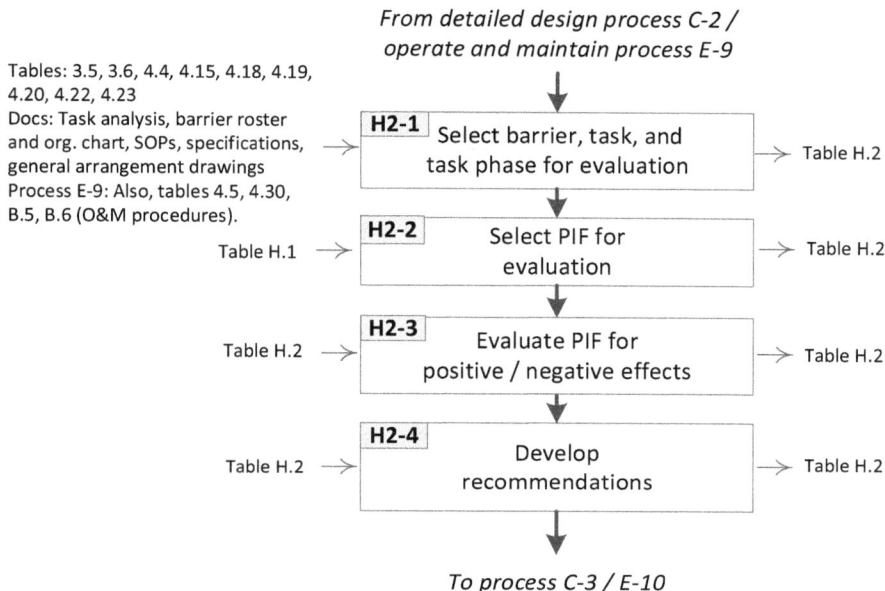

FIGURE H.1   PIF Assessment: Organization, Barrier, Task, and Workspace.

2. If a task phase is assessed, enter the task phase. Example entries are noted in the **Task Phase** field.

Note: A different PIF assessment, described in Appendix H.3, performs the WE assessment for direct-use PE components and Support PE.

H.2.2   STEP H2-2, SELECT THE PERFORMANCE INFLUENCING FACTOR FOR EVALUATION

1. Select the PIF from Table H.1 for evaluation. (Perform this step for each applicable PIF in Table H.1.)
2. Record the selected PIF in Table H.2.

Note: The process described in Appendix H.3 is limited to a working environment assessment for direct-use PE and Support PE. The WE assessment in the H.2 process is performed at the barrier, task, and task phase level. The assessed WE effects may contribute (positively or negatively) to personnel performance, behaviors, attitude, beliefs, and injuries, among others.

H.2.3   STEP H2-3, EVALUATE ITEM TO IDENTIFY THE PIF EFFECTS

1. For the selected item and PIF, evaluate the positive or negative effects of the selected PIFs on personnel (HEs and maintenance). As applicable, consider the following information:

TABLE H.1

Suggested Performance Influencing Factors for Assessment

PIF	Source Ref.	PIF Description	Application	Process
Threat stress (IFE 2022, pp. 56–57)	IFE (2022)	1. Event is uncommon or novel. 2. Increased emotive local environment. 3. Threat to one's own life or others. 4. Threat to self-esteem, professional status, or legal liability if performing a wrong decision or action.	HE performance of barrier/task/task phase activities Maintenance personnel performance of barrier system maintenance activities	C-2 E-9
Task complexity (IFE 2022, pp. 57–58)	IFE (2022)	Barrier/Task Complexity Types: 1. **Goals:** Competing or parallel, have multiple path options. 2. **Size:** Task size. Many SA-1 info sources and cues (high HE demands on information processing). 3. **Step:** "The number of mental or physical acts, steps or actions that are qualitatively different from other steps in the task." This increases with an increasing step count (IFE 2022, p. 57). 4. **Connections:** Interdependent elements or components with uncertain interactions. 5. **Dynamic:** Unpredictable or changing environment wherein the task is performed. "This includes the change, instability, or inconsistency of task elements." (IFE 2022, p. 57) 6. **Structure:** Task sequence order and branching possibilities affected by number, availability, and mutual compatibility of rules.	HE performance of barrier/task/task phase activities Maintenance personnel performance of barrier system maintenance activities	C-2 E-9
Physical working environment (WE) IFE, 2022, pp. 65–66)	IFE (2022)	Effects of the physical working environment personnel, e.g., potential to cause degraded performance, fatigue, injury, or exposure to danger.	Barrier/task HE and maintenance personnel activities in a workspace Workspace: control console or panel, room, engineered area, or process area containing barrier PE	C-2 E-9

See NORSOK S-002 (2018) for additional guidance and examples	IFE (2022)	Effects of WE (IFE 2022, pp. 65–66): 1. Potential to aid/inhibit/interfere in HE interactions with direct-use PE. (The scope may be expanded to include Support PE.) 2. Potential effects on personnel maintaining barrier PE.	HE interactions with direct-use components: detect and act phase PE; Support PE (worn/carried); Maintenance personnel performance of barrier system maintenance activities C-3, C-7, C-8 (See sub-process step CX-0.1e. in App. B.6)
Morale, attitude, motivation	NUREG (2016). See applications in Appendix A tables.	Refers to a person's state of mind, temperament, personality, and other characteristics that affect the effort and willingness to complete a task.	HE performance of barrier/task/task phase activities; Maintenance personnel performance of barrier system maintenance activities C-2 E-9
Other load	NUREG (2016)	Refers to incidental workload that, while not specific to the task, still consumes time and takes effort, such as non-barrier activities, interference, and distractions. The loads may be actual or perceived.	HE performance of barrier/task/task phase activities; Maintenance personnel performance of barrier system maintenance activities C-2 E-9
Safety culture	NUREG (2016)	Refers to the organization's policies, practice, beliefs, efforts, and attitudes as these apply to personnel, facility, and public safety. (Includes management support for those performing barrier tasks. This is an addition from IFE 2022, p. 63.)	HE performance of barrier/task/task phase activities; Maintenance personnel performance of barrier system maintenance activities C-2 E-9
Corrective action program	NUREG (2016)	Refers to the organization's policies and practice regarding correcting deficiencies and the willingness, level of effort, and priority for doing so.	Operate and maintain practices affecting barrier system integrity C-2 E-9
Societal influencers	Bea et al. (2009)	Influencing effects stem from: 1. Regulatory regimes/environment 2. Business climate/environment 3. Local operating practices and culture 4. Legal/litigation environment 5. National culture and norms	Persons and organizations responsible for barrier system design and implementation; HE performance of barrier/task/task phase activities; Maintenance personnel performance of barrier system maintenance activities C-2 E-9

TABLE H.2

PIF Assessment: Barrier, Task, Task Phase

Barrier ID	Task ID	Task Phase	PIF Name	PIF Assessment				Recommendations	
				Describe PIF effects	PIF Consequence	Potential to Recover		Describe Recommendations	Status
Enter barrier ID	Enter task ID	If applicable, enter task phase: – **SSA** – **Detect** – **Decide** – **Act** *Include ref. to source table.*	Enter PIF name (from Table H.1)	Describe PIF effects	Enter potential consequence of the identified effect	Enter potential to recover from a negative effect		Enter proposed recommendation to mitigate a negative effect or exploit a positive effect	Enter status: – Pending – Modify – Accept – Reject – ALARP

- Task phase requirements in Tables 3.5, 3.6, 3.8, 3.9, 3.11–3.20, 3.22, 3.24, 3.25, 3.27–3.29, 3.33, and 3.34
- Remote monitoring task requirements in Table 4.6
- New skill and competency requirements identified in the above tables
- HMI displays (Table 4.15)
- Direct-use PE and Support PE used in the performance of an assessed task
- Procedures. For the detailed design process C-2, consider the applicable standard operating procedures (SOPs). For the E-9 process, consider the new operating and maintenance procedures identified in Tables 4.5, 4.31, B.5, B.6, and others as may be listed in Table 4.30
- Activities performed in each identified workspace (indicated in Tables 4.18–4.20, 4.22, and 4.23)
    As applicable, also consider the following information:
- Risk assessment findings (step B2-x)
- Reliability study report and recommendations (Tables 3.29 and 3.30)

2. Identify the PIF consequence/effects. (As an option, consider the frequency and consequence magnitude, i.e., the potential risk/benefit.)
3. For a negative effect, identify the opportunities for recovery.
4. Record the findings in Table H.2.

H.2.4 STEP H2-4, DEVELOP RECOMMENDATIONS

1. For each entry in Table H.2, develop recommendations to mitigate each negative effect and exploit each positive effect, as appropriate.
2. Confirm that the proposed recommendation cannot create a new hazard or undesirable effect.
3. Record the recommendation in Table H.2.

Note: For the process C-2 assessment (shown in Figure 4.1*), the product of the assessment and preliminary design phase update (process C1) provides design input to the detailed design phase processes. For the process E-9 assessment (shown in* Figure 7.1*), the product of the assessment may be an input to the verification process, E-10.*

H.3 PIF ASSESSMENT (WORKING ENVIRONMENT) FOR DIRECT-USE COMPONENTS AND SUPPORT PE

The assessment of WE PIFs (from Table H.1) applies to barrier system *components* that are in direct use (detect and act phase PE). This assessment occurs in the CX-0.1 sub-process described in Appendix B.6. The sub-process is performed in the detailed design processes C-3, C-7, and C-8 presented in Chapter 4. The results are captured in Table H.3. Refer to Figure H.2 for an overview of this process. (This assessment may be modified to include Support PE.)

From subprocess CX-0.1d, Response Time
(See Figure B.3, App. B.6.1)

See listed tables →

CX-0.1e1 Select component for evaluation → Table H.3

Table H.1 → **CX-0.1e2** Select WE PIF aspect for evaluation → Table H.3

Table H.3 → **CX-0.1e3** Evaluate WE aspect → Table H.3

Table H.3 → **CX-0.1e4** Develop recommendations → Table H.3

Table H.3 → **CX-0.1e5** Implement recommendations

To subprocess CX-0.1f
External Support Systems

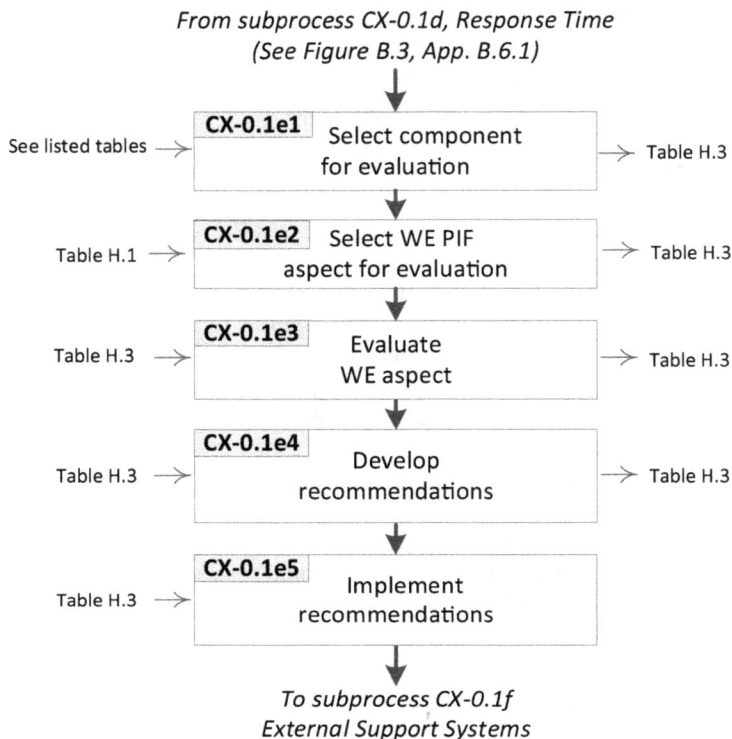

FIGURE H.2 PIF Assessment (WE): Direct-Use PE Components.

H.3.1 STEP CX-0.1E1, IDENTIFY COMPONENT FOR EVALUATION

Record the evaluated component in Table H.3. (The components were selected in step CX-0.1).

H.3.2 STEP CX-0.1E2, SELECT WE AREA FOR EVALUATION

1. Select the PIF WE aspect (from Table H.1) for evaluation.
2. Record the selected WE aspect in Table H.3.

H.3.3 STEP CX-0.1E3, EVALUATE THE EFFECTS OF PIFS IN THE WE

1. For the selected component and WE PIF aspect, evaluate the positive or negative effects on personnel (HEs, maintenance).
2. Identify the potential consequences of the effect. Consider its frequency and potential consequence magnitude (potential risk/benefit).
3. For a negative effect, identify the opportunities for recovery.
4. Record the above information in Table H.3.

TABLE H.3
PIF Assessment (WE): Direct-Use PE Component

Direct-Use or Support PE ID and Function	PE Source Table	ID-affected Person	Describe WE Issue	PIF Assessment Describe PIF Effects	PIF Consequence	Potential to Recover	Recommendations Describe Recommendations	Status
Enter PE ID and function detect or act PE (direct use) or Support PE	Enter PE source table	**Enter** – HE ID – Maint. ID/ role	Describe WE factor of concern (e.g., low light, obstructed by smoke, awkward position)	Describe: – Obstruction mechanism – Effect on usability – Effect on user access – Risk to user	Enter potential consequence of the identified effect	Enter potential to recover from a negative effect	Enter proposed recommendation to mitigate a negative effect or exploit a positive effect	Enter status: – Pending – Modify – Accept – Reject – ALARP

H.3.4   Step CX-0.1e4, Develop Recommendations

1. For each finding from step CX-0.1e3 (and recorded in Table H.3), develop recommendations to mitigate a negative effect or exploit a positive effect.
2. Confirm that the proposed recommendation cannot create a new hazard or undesirable effect.
3. Record the recommendation in Table H.3.

H.3.5   Step CX-0.1e5, Review and Implement Recommendations

Note: Steps 1–4 are performed by those listed in the BSMP. Step 3 is performed by those responsible for component design.

1. Review each recommendation in Table H.3 to determine whether it should be implemented as is, modified, rejected, an alternative recommendation developed, or whether the design be accepted as is (e.g., the design is ALARP).
2. Define the implementation approach. Have the approach approved.
3. Implement the recommendation. This may require changes to the prior CX-0.1 activities.
4. Update the information and status in Table H.3.

H.4   BASIS FOR SELECTING PERFORMANCE INFLUENCING FACTORS

The PIF assessment processes described in Sections H.2 and H.3 differ from current practice in the assessment timing, the assessed PIFs, and other aspects of the assessment process. This discussion provides the rationale for selecting the PIFs in Table H.1 and the reasons to exclude other PIFs and performance shaping factors (PSFs).

There is little global consensus on which PIFs to address in an assessment and on a shared and well-grounded basis for their selection. Tables H.4–H.7 summarize the PIFs and PSFs included in the following published industry documents:

- IEE/SINTEF (2011) – CRIOP: checklist-guided deficiency identification of scenarios, evaluation, mitigation
- EI (2020a) – safety critical tasks: guideword-based, human error identification, evaluation, mitigation
- IFE (2022) – Petro-HRA: human reliability analysis (scenarios, tasks), error quantification, mitigation (optional)
- NUREG (2016) – HRA: human reliability analysis (scenario, events), cognitive ergonomics focused (e.g., micro/macro cognition, teams, situations). Calculates the probability of human error.

Below are a few observations from a simple review of these lists:

- They share areas of overlap and areas of significant divergence.
- A PIF assessment using the different PIF lists will likely produce different results and levels of specificity.
- An assessment using the different lists may require different levels of assessor/participant knowledge, experience, and expertise.

TABLE H.4
Performance Shaping Factors (SINTEF, CRIOP)

Performance Shaping Factors

Competency and training
Procedures
Human-system interface
Teamwork
Goal conflicts
Time of day
Time available
Work environment
Emergency response
Interventions

Source: SINTEF (CRIOP) 2011 para 5.2.6.

TABLE H.5
Performance Influencing Factors (Energy Institute, Safety Critical Tasks)

Performance Influencing Factors

Control/display design
Equipment/tool design
Memory aids
Training
Work design
Procedures
Supervision
Reducing distractions
Environment
Communications
Decision aids
Behavior safety

Source: EI 2020a, Table 5.

TABLE H.6
Performance Shaping Factors (Petro-HRA)

Performance Shaping Factors

Time
Threat stress
Task complexity
Experience and training

(Continued)

TABLE H.6

(Continued)

Performance Shaping Factors

Procedures and supporting documents

Human-machine interface

Attitudes to safety, work, and management support

Teamwork

Physical working environment

Source: IFE (2022), Petro-HRA, p. 49.

TABLE H.7

Performance Influencing Factors (NUREG)

Category	Performance Influencing Factors
Organization-Based	Training program: availability, quality
	Corrective action program: availability, quality
	Other programs
	Safety culture
	Management activities:
	Staffing: number, qualifications, team composition
	Scheduling: prioritization, frequency
	Workplace adequacy
	Resources: procedures, tools, necessary information
Team-Based	Communication: availability, quality
	Direct supervision: leadership, team member
	Team coordination
	Team cohesion
	Role awareness
Person-Based	Attention: to task, to surroundings
	Physical and psychological abilities: alertness, fatigue, impairment, sensory limits, physical attributes, other
	Knowledge/experience
	Skills
	Familiarity with situation
	Bias
	Morale, motivation, attitude
Situation/ Stressor-Based	External environment
	Conditioning events
	Task load
	Time load
	Other loads: non-task, passive information
	Task complexity: cognitive, task execution

TABLE H.7
(Continued)

Category	Performance Influencing Factors
	Stress
	Perceived situation: severity, urgency
	Perceived decision: responsibility, impact (personal, plant, society)
Machine-Based	HMI display: input, output
	System responses

Source: NUREG 2016, Table 2-2.

Based on a simple inspection process, the PIFs and PSF in the four tables appear to fit into four categories:

1. Cognitive (e.g., attention, bias, task load)
2. HMI displays (PE)
3. Organizational elements (e.g., procedures, experience, skills, staffing, non-technical skills, leadership, and supervision)
4. Other (e.g., working environment, attitudes)

Group 1 (Cognitive Aspects): The need to assess this group is replaced by the more detailed, cognitive-focused activities in the prototype lifecycle model. The cognitive assessment (suggested in Appendix I) exceeds what is possible with the conventional PIF assessment. *In response, this author believes the results of these activities should exceed what is achievable using conventional PIF/PSF assessment processes.*

Group 2 (HMI Displays): The need to assess the HMI display is replaced by a more thorough, first-principles–based design process applied to HMI displays, display elements, and multi-VDU display systems. Each display is then assessed using the more advanced cognitive assessment described in Appendix I and guided by the applicable workspace design guidance provided in Appendix K. *In response this author believes the results of these activities should exceed what is achievable using conventional PIF/PSF assessment processes.*

Group 3 (Organizational Elements): The need to assess organization elements is replaced by a more rigorous and detailed set of process that reduce the potential for latent errors in OEs. *In response, this author believes the results of these activities should exceed what is achievable using conventional PIF/PSF assessment processes.*

A Caveat: *A systematic type error can be introduced into barrier organizational elements when an error exists in an owner/operator policy or program that directs or guides a barrier design or management activity. For example, this can occur if the error resides in the organization's Procedure Development and Management System. Example error forms may include a faulty procedure template, or procedure update practice, or deficiencies in the minimum competencies expected for procedure developers, updaters, and approvers. Analogous errors may also exist in the owner/operator*

Training Development and Management System or Competency Management System. A separate assessment of these programs may be needed to identify these potential sources of systematic error that can affect all active human barriers.

Group 4 (Everything Else): This group comprises the PIFs that did not fit into the other three categories and adds the additional PIF, societal influencers. Use of the added PIF appears to be appropriate in the detailed design phase assessment (process C-2) and the suggested periodic assessment in the operate and maintain phase (process E-9). The reduced list may improve the assessment team's focus on these more complex PIFs.

Societal Influencer PIF: The last item in Table H.1, Societal Influencers, is generally not included in a PIF assessment. Bea et al. (2009) posits that risk assessment consistently underestimate risk by one to two orders of magnitude if the influencing effects of this PIF is not considered. To understand the potential effect mechanism, consider how societal influencers can and often do affect decision-makers and leaders (typically hidden), for example, the effects on their management and leadership style, beliefs, priorities and focus, motivations, decision-framing, timing, sense of urgency, chosen words, and actions. The definitive case and justification for adding it to the PIF assessment is beyond the scope of this book. However, this author believes its addition aligns with one of the key objectives of the prototype lifecycle model, that of identifying and addressing all cognitive ergonomics considerations that affect barrier function and performance. (Ignoring the ever-present, pervasive, and changing societal influencers may be a form of bounded rationality.)

As an exercise, use this new lens to re-examine the Deepwater Horizon and Boeing 737 Max accidents. Is it possible (perhaps likely?) that the confluence of a permissive regulatory regime, changes in the external business environment, and "flexible" organizational risk attitudes contributed to each accident? Is it possible that the ever-present effects of societal influencers contributed, in small ways or large, to many if not most incidents and major accidents? Personnel on the Deepwater Horizon intentionally deviated from a published safety procedure by lining up the mud gas separator (MGS) in response to a detected well-kick. (See Appendix M, Note 2 for background and detail.) Using the MGS captured oil that may have otherwise flowed to the sea, a regulatory reportable incident. It also captured costly mud that would have been dumped into the sea. However, in this case it became a causal contributor to the accident. Did the hazard risk assessment teams consider this possibility? If they had, was the event assessed to be unlikely or implausible? Would that change if the assessment teams were more aware of the slow, silent, pervasive, and persistent influence of a confusing safety culture, increasing workloads, or the unintended effects of a company recognition/award program?

Regarding the Boeing 737 Max accident, did the risk assessment teams consider the possibility that the initial design would become increasingly untenable? If they had, would that assessment team consider it possible that these events and societal influencers would move the Boeing management and team to an unsafe "in-for-a penny, in-for-a-pound" behavior? Admittedly, in hindsight it appears that the collective Boeing team fell into a downward drift-to-failure spiral. Perhaps these are sufficient arguments for adding "societal influencers" to risk assessments.

Appendix I
Cognitive Assessment and Mitigation Process

This appendix is organized as follows:

- I.1 Introduction
- I.2 Cognitive Assessment and Mitigation Process
- I.3 Cognitive Assessment Tables
- I.4 Execution Considerations

I.1 INTRODUCTION

Section I.2 presents a prototype, first-of-kind process to address the typically hidden design errors (cognitive attributed) that tend to be persistent contributors to barrier system and safety critical task degradation and failure. The primary error type is a designed-in cognitive demand (expectation) that is inconsistent with the human capability to respond to that demand reliably and promptly when faced with deficient designs and the changing situations and conditions possible with these barrier and task types. This is a *first principles-based* process because it is directly underpinned by the science and knowledge from the fields of human factors, cognitive and behavioral psychology, and neuropsychology. As such, it has capabilities that are not available in existing assessment methodologies, for example, a performance influencing factor assessment. A primary capability is the identification and prevention or mitigation of cognitive-type mismatches and other cognitive error types. Its first-principles attributes should contribute to reduced variability in the results and an increased likelihood that the identified corrective actions (solutions) achieve the intended results without introducing an unintended new error.

Note: Though derived and developed from a well-researched foundation, this process is a prototype. As such, the above statements remain posits until proven through repeated use and verification.

Regarding use, the lifecycle model integrates this process into the detailed design phase activities for team and task phase design elements, summarized in Figure 4.1 (Chapter 4 introduction). It assesses the product of the detailed design phase processes. Section I.2 describes the suggested process. Section I.3 provides the tables (assessment tools) used in the process. Section I.4 provides selective execution guidance for those responsible for planning and implementing this process.

Note: This assessment is complementary to the performance influencing factor assessment described in Appendix H.2.

Sträter (2005, Ch. 6) describes the challenges when using an HRA as an input to a risk assessment. In the process industries, similar processes are used (e.g., Petro-HRA, IFE 2022). A different variant is applied to safety critical tasks (EI 2020a). The outputs from these methods can vary widely, most likely because they are not first-principles based and not adequately underpinned by cognitive ergonomics and cognitive science. For this reason, their use as a design tool may unintentionally introduce errors into the design while providing a potentially unwarranted confidence that the design is free of such errors.

From Sträter (2005, cover sleeve),

Safety suffers from the variety of methods and models used to assess human performance. For example, operation is concerned about human error while design is aligning the system to workload or situational awareness. This gap decouples safety assessment from design. As a result, system design creates constraints for the Human working at the sharp end, which eventually leads to errors. Accidents and incidents throughout all industries demonstrate the safety relevance of this gap.

A major challenge when designing a complex sociotechnical system is to prevent the introduction of design errors of that type that misdirect or fails to support the responding human in a manner that contributes to barrier degradation of failure. Such errors commonly occur in detection, decision-making, and problematic interactions with other barrier system elements (PEs, OEs, and other HEs). Continually changing situations and challenging work environments can also have delegatory effects that degrade human performance. Design errors that cause unintended human actions (incorrectly identified as human error) remain persistent and common contributors to incidents and major accidents.

A key statement from Reason that should inform and guide a cognitive assessment process (1990, p. 238),

...most operator errors arise from a mismatch between the properties of the system as a whole and the characteristic of human information processing. System designers have unwittingly created a work situation in which many of the normally adaptive characteristics of human cognition (its natural heuristics and biases) are transformed into dangerous liabilities.

CIEHF (2016, Section 4.3.6, p. 38) asserts: "There is often a lack of understanding of the nature or complexity of the tasks – and especially the cognitive elements of those tasks – that need to be carried out for barriers to function as intended."

In addition, the SPE (2014, p. 11) cautions that "...most people in the industry lack awareness of the realities and limitations of human cognition and the 'tricks' the brain uses to be able to function in the complex modern world."

The nuclear industry recognized these challenges. NUREG (2016) is a solution response to the historical problem of inconsistencies and inaccuracies in HRAs. (The HRA is one of many inputs to a larger risk assessment and quantification process to develop a more complete and accurate risk picture. The product of an extensive literature search for cognitive information and failure mechanisms, experts compiled and organized cognitive type information into the organized form presented in a NUREG document, a document that may rank among the most comprehensive if its type.)

NUREG (2016, para 1.1.1, p. 3) explains, "Research in the behavioral sciences has accumulated knowledge about the mechanism underlying human performance, including how human performance may be affected by situational factors." "Such a cognitive basis can...elucidate the effects of PIFs on human failure, and define the strength of PIFs with respect cognitive mechanisms and human vulnerabilities." It further states, "our task was to mine current state-of-the-art behavioral sciences research for information to establish direct links (causal relationships) between performance influencing factors, mechanisms of human cognition, and human performance."

The cognitive assessment tables in Section I.3 are inspired by the NUREG (2016) document. The assessment process in I.2 provides the opportunity for a deeper and greater level of specificity when applied to the more detailed information created in the lifecycle model processes defined in Chapters 3 and 4. The following are potential attributes and results that seem achievable through this process:

- First-principles (cognitive)–based
- Fully traceable processes
- Potential improvements in the results repeatability
- Greater level of detail and comprehensiveness
- Executable in a project or operating environment
- Applicable as a design or development tool in a project environment
- Does not rely on abstractions, translations, or manipulations to generate results and outcomes

I.2   COGNITIVE ASSESSMENT AND MITIGATION PROCESS

To understand this new process, the following describes the basic activities, tools, and data capture tables employed in this process. Its objective is to identify and prevent or mitigate cognitive-attributed (designed-in and system) error types.

- **Main Process Flowchart ("CI" Process):** Figure I.1 is a simplified overview of the "CI" process which encompasses the sub-process II, summarized in Figure I.2. The CI process performs a cognitive assessment that analyzes each task phase requirement (object) and the direct use elements designed and provided to realize phase functions and requirements.

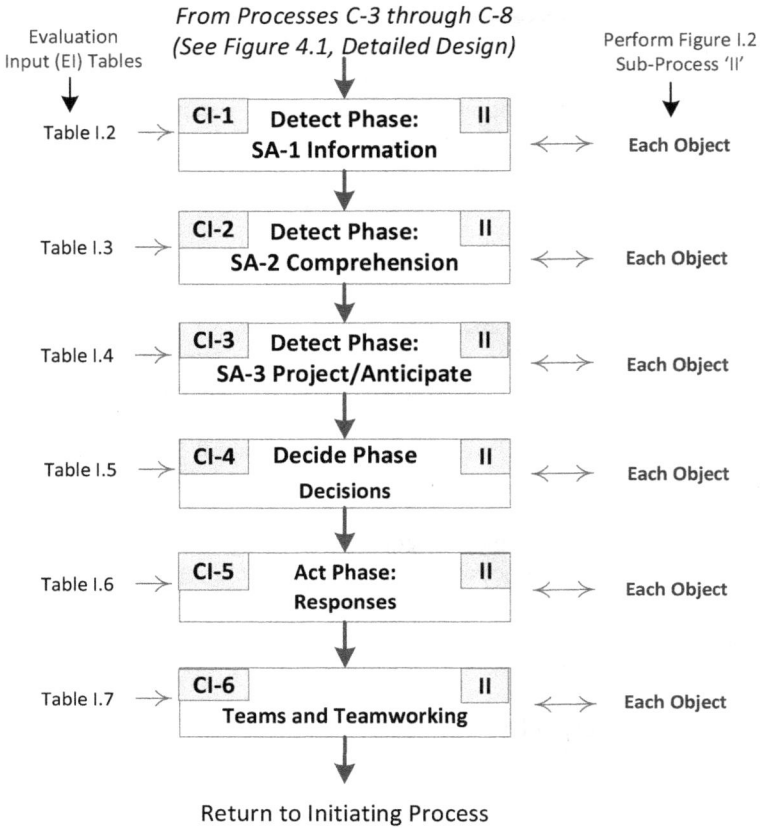

FIGURE I.1   Cognitive Assessment and Mitigation Process (CI).

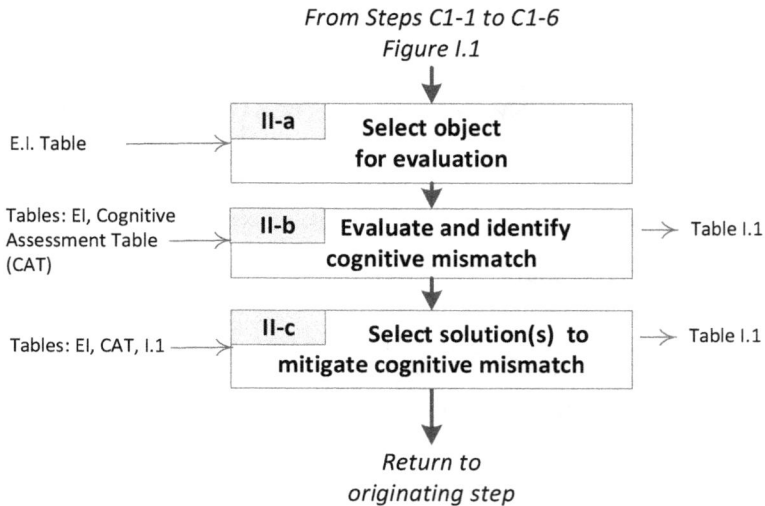

FIGURE I.2   Cognitive Assessment Sub-Process (II).

Note: This process is applied to the task phase design outputs from the detailed design processes C-3 to C-8, described in Chapter 4.

- **Sub-process Flowchart ("II" Process):** Section I.2.2 describes the "II" sub-process. Figure I.2 summarizes the steps performed in this sub-process.
- **Assessment Results and Recommendations Table:** Table I.1 captures the input information, findings, and the corrective actions proposed to eliminate, prevent, or mitigate an undesirable finding. (With most findings, the focus is on cognitive mismatch errors of the type discussed in the introduction.)
- **Evaluation Input (EI) Tables:** the six (6) task phase specific EIG tables, included in Section I.2.3, identifies the recommended supporting input information, identifies the object (requirement) and the associated direct use element to examine in the assessment, and the Cognitive Assessment Table that applies to the task phase type.
- **Cognitive Assessment Table:** the seven (7) tables in Section I.3 identify the full range of potential cognitive issues (and plausible failure mechanisms) to consider in the assessment. If a mismatch/cognitive issue is identified, the tables provide a starting list of matched corrective solutions to consider when seeking possible corrective solutions to prevent or eliminate source of the mismatch/issue or mitigate its effects.
- **Execution:** Section I.4 provides execution guidance and information to support planning and personnel resourcing.

Note: The CI process and tools may also be used (adapted) to assess a safety critical task or a task component, for example, a procedure, display, or hand-held device.

I.2.1 PROCESS "CI" (SEE FIGURE I.1)

Each of the detailed design processes C-3–C-8 (Chapter 4) contains a step that calls this process. The timing of the assessment is indicated in the process overview figure included in the introduction to each of these processes. For example, Figure 4.9 shows the process C-8 steps that design the detect phase (VDU-based) elements. Step C8-10 identifies (calls) the CI process.

I.2.1.1 Select/Enter the Task Phase

In Table I.1, enter the following information into the **Barrier Task ID** and **Task Phase** fields.

Barrier Task ID: enter unique ID for the selected barrier task.
Task Phase: enter the applicable descriptor for the task phase under evaluation:

- Detect SA-1
- Detect SA-2

TABLE I.1

Cognitive Assessment Evaluation and Recommendations

| Barrier Task ID | Identified Object | | | Cognitive Mismatch or Issue
(From Cognitive Assessment Table or CAT) | | | | Corrective Action (CA) Recommendation | |
	Source Table	Object ID	Task Phase	ID	Issue	Mismatch/ Issue Source	Failure Mechanism (s)	CA ID and Descr.	Implement Req.
Enter barrier task ID	Enter object source table number	Enter object ID and description	Enter task phase or team	Enter ID for this issue	Enter issue description	Enter mismatch/ issue source	Enter the failure mechanism (s) *All that apply*	Enter CA ID and description *(From CAT)*	Specify CA requirements *All that apply*

- Detect SA-3
- Decide
- Act
- Team

I.2.1.2 Perform Sub-process "II"

For the selected task phase, perform the sub-process II steps described in Section I.2.2 below.

Note: This sub-process repeats to assess each of the objects and associated elements identified in the respective EI table. On completion, the process returns to the first step to commence the assessment for the next barrier task. When that is complete, progress to the next barrier.

I.2.2 SUB-PROCESS "II" (SEE FIGURE I.2)

I.2.2.1 Step II-a, Select Object for Evaluation

Refer to the six (6) Evaluation Input (EI) guidance tables (located in Section 1.2.3), one for each task phase and team/teamworking.

Guided by the applicable EI table, select the first object for evaluation.

In Table I.1, enter the following information for the selected object. **Object Id** and **Source Table**.

Object ID: enter the unique ID and brief description for the assessed object, available from the object's source origination table.

Source Table: if the object is a task phase, enter the task phase description. If the object is a task phase requirement, enter the originating table for that requirement. Example origination tables include the following:

- Table 3.8. Shared Situation Awareness Requirements
- Table 3.12. Act Phase Response: Direct-Use PE Requirements (for background, see Table 3.9)
- Table 3.13. Act Phase Response: Support PE Requirement (for background, see Table 3.9)
- Table 3.17. Decide Phase Requirements
- Table 3.18. Detect Phase: SA-2 Comprehension Requirements
- Table 3.19. Detect Phase: SA-3 Projection Requirements
- Table 3.20. NTS Requirements: Teamworking
- Table 3.23. NTS Requirements: Leadership
- Table 3.24. NTS Requirements: Monitor and Manage Acute Stress
- Table 3.25. Detect Phase: SA-1 Requirements. (This table integrates the SA-1 requirements from other tables. For a full listing of the "other" tables, see Tables 3.2 and 4.1.)

I.2.2.2 Step II-b, Evaluate Object for Cognitive Issues or Mismatches

1. Guided by the respective EIG table and listed references, assess each task phase object (requirement) and its associate user interface/implementation element (direct use) against the plausible cognitive issues indicated in the EIG indicated Cognitive Assessment Table. Identify the certain and likely cognitive mismatches / cognitive failure issues.

   - **SA-1 Objects:** guided by EI Table I.2, assess each object identified in the **Assessed Object** field against the potential cognitive issues in Cognitive Assessment Table I.9.

     Note: For this and all remaining evaluations, consider the effects of all plausible situational and environment conditions that can inhibit or interfere with the object in a way that may cause it not to achieve the intended barrier/task function and performance.

   - **SA-2 Objects:** guided by EI Table I.3, assess each object identified in the **Assessed Object** field against the potential cognitive issues in Cognitive Assessment Table I.10.
   - **SA-3 Objects:** guided by EI Table I.4, assess each object identified in the **Assessed Object** field against the potential cognitive issues in Cognitive Assessment Table I.11.
   - **Decision Objects:** guided by EI Table I.5, assess each object identified in the **Assessed Object** field against the potential cognitive issues in Cognitive Assessment Table I.12.

     Note: The process assesses rule-based and recognition-primed decision model processes.

   - **Act Response Objects:** Guided by EI Table I.6, assess each object identified in the **Assessed Object** field. Assess the object against the potential cognitive issues in Cognitive Assessment Table I.13.
   - **Team and Teamworking Objects:** Guided by EI Table I.7, assess each object identified in the **Assessed Object** field. Assess the object against the potential cognitive issues in Cognitive Assessment Table I.14.

2. Record the identified fault or mismatch in Table I.1. Enter the information in the cognitive **ID, Issue, Mismatch/Issue Source**, and **Failure Mechanism** fields.

   ID: Enter the identifier for the cognitive mismatch or issue. (This is the **ID** field in the applicable Cognitive Assessment Table.)

   Issue: Enter the issue assigned to the abovementioned ID. (This is the **Issue** field in the applicable Cognitive Assessment Table.)

Mismatch/Issue Source: Enter the mismatch or issue source. (This is the **Mismatch/Issue Source** in the applicable Cognitive Assessment Table.)

Failure Mechanism: Enter the identified failure mechanism for the identified mismatch or issue. (Use the description (or similar) from the applicable Cognitive Assessment Table.)

I.2.2.3 Step II-c, Recommend Corrective Action

1. For the identified mismatch/issue recorded in Table I.1, evaluate and select an effective corrective action(s) (CA) that prevents or eliminates the entry or mitigates is consequences. The applicable cognitive assessment table provides example solutions that are matched to the identified mismatch/issue type.

2. Record the selected CA(s) in Table I.1. Enter the information in the **CA ID & Descr.** and **Implement Req** fields. (Repeat as necessary. When complete, move to step 3 below.)

   CA ID & Descr: Enter the selected CA(s) ID from the Cognitive Assessment Table.

   Implement Req.: Enter the design and implementation requirements (if any) for further clarifications regarding this CA.

   Note: Recommended corrective actions (changes) identified in this process should be reviewed by the appropriate persons to determine whether the recommendations are accepted, modified, or should be replaced by an alternate recommendation. The design may also be accepted as is, which can occur if the design is assessed to be ALARP.

   Note: Concatenating IDs entered into Table I.1 may be an approach to generating a unique ID that contributes to traceability and supports cross-referencing and ready identification. For example, concatenating the Barrier ID + Task ID + task object ID + Item ID + CA ID.

3. Return to Step II-a to select the next object in this task phase for assessment.

   Note: On completing the assessment for all objects for this phase, return to process CI to begin the assessment of the next task phase. Once complete, return to select the next barrier for evaluation.

I.2.3 EVALUATION INPUT GUIDE TABLES

The following evaluation input (EI) tables I.2–I.7 provide topic-specific guidance applied to each of the assessment topics indicated in Figure I.1.

TABLE I.2

Evaluation Input Guidance: SA-1 Information

Review HE Interface Object (Direct-Use)

See Table 3.26 for descriptions of display types.

Assessed Object	Non–VDU-Based	VDU-Based	Support Aids, Alternative HMI Displays	Supporting Documents	Cog. Assess. Table
Object = inbound comms req. from Table 3.11	Communications: see Table 3.11	Communicated by live video conferencing: see Table 3.11		– SRS, BSMP	Table I.9
Object = SA-1 info req. from Table 3.25	Remote monitoring of Support PE display: see Table 4.6	HMI display element: see Tables 4.12 and	SA-1 support aids	– Task analysis report	
	Passive display: see Table 4.7	4.14 (see	SA-1 alternative HMI display types	– Tables H.2, H.3 (PIF assess. recommendations)	
Object = SA-1 info req. from Table 4.6	Simple active display: see Table 4.8	Table 4.15 for integrated display)	See Table 4.13 (see process C8-4b, for information)	– Functional analysis: Tables 4.18, 4.19 (CP), 4.20 (rooms), 4.22 (engineered areas)	
Object = Each SA-1 req. from Table 4.9	Incident scene: see Table 4.9	CCTV image: see Table 4.16		– Appendix K.1, Tables K.1, K.2	
Object = Each SA-1 req. from Table 4.10	Incident Command Board: see Table 4.10	Off-console display wall: See Table 4.17		– Procedures: SOPs, Tables 4.5, 4.30, B.5, B.6	
				– Training: standard, Table B.7	
				– Competencies: standard	
				– New skills/knowledge noted in the assessed object table	
				– Table 3.27 – safety times	
				– Performance standards, Tables 3.7, 3.32	
				– Data sheets	
				– Environmental data	
				– Lighting design	
				– Noise studies	
				– Installation details/drawings: mounting, location, layout	

Note: See detailed design processes C-7 and C-8 for design details and documents.

TABLE I.3

Evaluation Input Guidance: SA-2 Requirements

Assessed Object	Review HE Interface Object (Direct-Use)			Supporting Documents	Cog. Assess. Table
	See Table 3.26 for descriptions of display types.				
	Non–VDU-Based	VDU-Based	Support Aids, Alternative HMI Displays		
Object = SA-2 req. from Table 3.18	See SA-1 input to SA-2 object noted in Table 3.18 See Table I.2 for display source		SA-2 support aid SA-2 alternative HMI display (See Table 4.13, process C8-4b, for both)	See Table I.2 for applicable documents See SA-1 assessment recommendations	Table I.10

Note: See detailed design process C-5 for design details and documents.

TABLE I.4

Evaluation Input Guidance: SA-3 Requirements

Assessed Object	Review HE Interface Object (Direct-Use)			Supporting Documents	Cog. Assess. Table
	See Table 3.26 for descriptions of display types.				
	Non–VDU-Based	VDU-Based	Support Aids, Alternative HMI Displays		
Object = SA-3 req. from Table 3.19	See SA-1 input to SA-3 object noted in Table 3.19 *See Table I.2 for display source*		SA-3 support aid SA-3 alternative HMI display (See Table 4.13, process C8-4b, for both)	See Table I.2 for applicable documents See SA-1, SA-2, and SA-3 assessment recommendations	Table I.11

Note: See detailed design process C-5 for design details and documents.

TABLE I.5

Evaluation Input Guidance: Decisions

Assessed Object	Review HE Interface Object (Direct-Use) See Table 3.26 for descriptions of display types.			Supporting Documents	Cog. Assess. Table
	Non–VDU-Based	VDU-Based	Support Aids, Alternative HMI Displays		
Object = decision req. from Table 3.17	See SA-1 input to decision noted in Table 3.17 See SA-2 input to decision noted in Table 3.18 See SA-3 input to decision noted in Table 3.19 See Table I.2 for display source		Decision Support Aid Decision alternative HMI display (See Table 4.13, process C8-4b, for both)	See Table I.2 for applicable documents See SA-1, SA-2, and SA-3 assessment recommendations	Table I.12

Note: See detailed design process C-4 for design details and documents.

TABLE I.6

Evaluation Input Guidance: Act Phase Response PE

Review HE Interface Object (Direct-Use)

See Table 3.26 for descriptions of display types.

Assessed Object	Non–VDU-Based	VDU-Based	Support Aids, Alternative HMI Displays	Direct-Use PE	Support PE	Supporting Documents	Cog. Assess. Table
Object = outbound comms. requirement from 3.11							Table I.13
Object = act phase response requirements from Table 3.12	See SA-1 input to act response noted in Tables 3.17, 4.10		Act phase response support aid	See Tables 3.11 and 3.12 for direct-use PE info		See Table 3.9 for act phase response info.	
Object = support PE identified in Table 2.13	See SA-2 input to act response noted in Table 3.18		Decision alternative HMI display (See Table 4.13, process C8-4b, for both)	Hand/electronic marking on ICB	See Table 3.13 for Support PE info	See Table I.2 for applicable documents.	
Object = act phase remote monitor response requirements in Table 4.6	See SA-3 input to act response noted in Table 3.19					See SA-1, SA-2, SA-3, and decision assessment recommendations.	
Object = hand marking of information on Incident Command Board noted in Table 4.10.	(See Table I.2 for display sources)						

Note: See detailed design process C-3 design details and documents.

TABLE I.7
Evaluation Input Guidance: Teams and Teamworking

Review HE Interface Object (Direct-Use)
See Table 3.26 for descriptions of display types

Assessed Object	Non–VDU-Based	VDU-Based	Support Aids, Alternative HMI Displays	Act Phase PE	Supporting Documents	Cog. Assess. Table
Object = SSA requirement from Table 3.8					– See Table I.2 for applicable documents	
Object = teamworking requirement from Table 3.20	See Table I.2	See Table I.2	See Table I.2	See Tables 3.11 and 3.12	– See SA-1, SA-2, SA-3, decision, and act phase response assessment recommendations – Roster/ org chart – Design basis: active human barrier, remote barrier support – Philosophy: Emergency response – Task analysis – Standard operating procedures	Table I.14
Object = leadership requirement from Table 3.22						
Object = stress monitoring requirement from Table 3.24						

Note: See detailed design process C-6 design details and documents.

I.3　COGNITIVE ASSESSMENT TABLES

The following tables support the cognitive assessment and mitigation process. Their contents were guided by NUREG (2016) and information in Appendices C, D, F, K, and L and the companion guides to this book (listed in Section 1.4).

- Table I.8. Cognitive Assessment Table: Common Issues
- Table I.9. Cognitive Assessment Table: SA-1 Information
- Table I.10. Cognitive Assessment Table: SA-2 Comprehension
- Table I.11. Cognitive Assessment Table: SA-3 Projection
- Table I.12. Cognitive Assessment Table: Decisions
- Table I.13. Cognitive Assessment Table: Act Phase Response
- Table I.14. Cognitive Assessment Table: Teams and Teamworking

Note: These tables are intended as a starter source. The examples show the types of information and level of detail to consider in this assessment. The presented corrective actions are intended to show the range of options (OE, HE, or PE) available to address each deviation or finding. Before use, experts from the reader's organization should review these tables for completeness and accuracy based on its internal knowledge, practice, and preferences (e.g., corrective actions).

I.3.1　COMMON COGNITIVE ISSUES

Table I.8 (located in Section 1.3.8) addresses the cognitive issues common to several task phases. It includes the following aspects:

- Task goals: incorrect goal selected
- Task goals: goal ambiguity/confusion
- Task goals: efforts not guided by goals
- Task priorities: uncertain/ambiguous priority
- Task priorities: wrong priority
- Excessive workload
- Attention: not monitoring the required information
- Attention: monitoring the wrong information
- In working memory: short-term memory information fade, replacement, misremembered
- In working memory: degraded short-term memory capacity/retention
- In working memory: activity/demands exceed short-term memory capacity/capability
- Degraded psychological state
- Freeze, flight, fight response
- Degraded physiological state

- Mental model: missing mental model
- Mental model: wrong mental model selected
- Mental model: incomplete/inadequate mental model
- Mental model: errors in mental model
- Skills: skill not fully developed
- Skills: deficiencies and errors in skilled actions
- Non-rational biases and behaviors

Task-specific tables include references to this table on applicable cognitive issues, mechanisms, and possible corrective actions. It also provides an additional perspective on the importance and contribution of task goals in barrier/task performance.

According to Endsley and Garland (2000, p. 20),

> The active goals direct the selection of the mental model. ... That goal and its associated mental model are used to direct attention in selecting information from the environment. They serve to direct scan patterns and information acquisition activities. For this reason, the selection of the correct goal(s) is extremely critical for achieving SA. If the individual is pursuing the wrong goal (or a less important goal), critical information may not be attended to or may be missed.

I.3.2   Detect Phase: SA-1 Information

Table I.9 (located in Section 1.3.8) provides sources of design and situation-based deviations that may contribute to a failure to perceive SA-1 information (not detected or not attended to) or incorrectly perceive the information. It includes the following aspects:

- SA-1 information not detected: not noticed
- SA-1 information detected but forgotten, replaced, or misremembered
- SA-1 information not detected: not looked for
- SA-1 information not detected: not accurately perceived
- SA-1 information not detected: found but ignored
- SA-1 information: error in information
- SA-1 information: not available (communications)
- SA-1 information: attending to the wrong information

It may be helpful to consider the SEEV detectability equation to understand factors to improve noticeability: Salience, Effort, Expectancy, and Value. For further details and information on this model, see the *Note* in Appendix B.5.2.

Noticing and detection topics to consider include attention and attentional issues, working or short-term memory, mental models, information object salience, display placement within a visual zone, and environmental factors. Additional factors become relevant for SA-1 information received through communications in a verbal, written, or text form. Appendix B.5.1 provides an overview of the various challenges in noticing and detecting SA-1 information. B.5.2 provides information for improving display salience. Appendices F.3 and K.1.2 discuss limitations that are inherent in the human visual system.

Note: As preparation and input to this process, review the PIF assessment and recommendations from the detailed design process C-2 and the SA-1 display processes C-7 and C-8.

I.3.3 DETECT PHASE: SA-2 COMPREHENSION

Table I.10 (located in Section 1.3.8) provides sources of design and situation-based deviations that may contribute to a failure in comprehending SA-2. The following aspects are covered:

- SA-2 comprehension: no comprehension
- SA-2 comprehension: insufficient comprehension
- SA-2 comprehension: wrong comprehension

Humans commonly use *abductive logic* in the process of developing understanding and comprehension. Different from deductive logic, this process begins when we first perceive SA-1 information. Automatic cognitive processes attempt to match that information (current context) with an existing long-term memory or mental model. Once matched, that model provides (creates) context and meaning to that information. See Section 4 and Appendix A.2 in NUREG (2016) for a more advanced discussion of four competing sensemaking models that include Endsley's SA model and Klein's data/frame model.

Note: As preparation and input to this process, review the PIF assessment and recommendations from detailed design phase process C-2 and the SA-2 design activities in process C-5.

I.3.4 DETECT PHASE: SA-3 PROJECT/ANTICIPATE

Table I.11 (located in Section 1.3.8) provides sources of issues and mismatches that may contribute to an inadequate, misleading, or incorrect SA-3 projection. The following aspects are covered:

- SA3 projection/anticipation: none achieved
- SA3 projection/anticipation: inadequate results/accuracy in event evolution or time projection
- SA3 projection/anticipation: significant error in the recalled projection/anticipation memory

This capability tends to be limited to domain experts. Gasbury (2013, pp. 28–29) contends that "experts…are able to comprehend the meaning of clues and cues much quicker. In fact, experts can quickly assemble groupings of clues and cues, mentally package them together, and determine the meaning of the packaged information." Experts also have the added "ability to process and comprehend negative clues and cues. Negative means absent clues or cues" (Gasbury 2013, p. 30). For additional background, see Appendix F.5 to understand the importance of long-term memories and mental models in achieving this capability.

A SA-3 requirement may create a staffing challenge given the greater level of domain expertise that may be needed to meet that requirement. A mental model sufficient to meet SA-2 requirements is not necessarily sufficient to meet the more challenging SA-3 requirement. The requirement may include the capability to adequately estimate time (e.g., the time needed to complete a complex series of tasks in a multi-person barrier). In a time-pressured environment, this may provide an important edge, because it allows one to begin planning early for a potential future need. For more information, see the notes in preliminary design step B-12, Appendix B.5.3 and Table J.3.

Note: As preparation and input to this process, review the PIF assessment and recommendations from detailed design phase process C-2 and the SA-3 detailed design activities in process C-5.

I.3.5   DECIDE PHASE: DECISIONS AND RESPONSE PLAN

Table I.12 (located in Section 1.3.8) provides sources of design and situation-based deviations that may contribute to a failure in decisions and decision-making. The following aspects are covered:

- Decide phase (rule-based): wrong decision
- Decide phase (rule-based): wrong act phase response plan
- Decide phase (RPD): wrong decision or response plan due to errors in the mental model/long-term memory
- Decide phase: decision too late, not made

Note: The decision phase may consume a high percentage of the task and, and the barrier safety time. The HE may consistently view the safety time as an opportunity to complete other activities before attending to this barrier. If the barrier has a long safety time (e.g., 30 to 60 minutes), the HE may use that time in an attempt to correct the condition that activated the barrier. The decision to do so may be influenced by an awareness of the disruption caused by the barrier activation and the additional effort needed to return to normal operations. (The corrective attempt may or may not succeed. With poor time tracking or a task-switch error, the HE may spend too much time on this failed effort, leaving insufficient time to complete the barrier response. The scenario is a common production vs safety goal conflict example.)

Note: As preparation and input to this process, review the PIF assessment and recommendations from detailed design phase process C-2 and the decide phase process C-4.

I.3.5.1 Recognition-Primed Decision Model

The recognition-primed decision model (RPD) is a recognized decision process developed to "explain decision-making of experts in stressful situations and under time pressures" (NUREG 2016, p. 79). With highly experienced fire brigade leaders, the approach is used to guide up to 80% of decisions. This process is one of satisficing, i.e., identifying an adequate and acceptable solution that is not necessarily the best option. The RPD process tends to require little time, for example, seconds to minutes. It is best suited to complex barriers and barrier situations when a more rapid decision is essential to preventing a hazard escalation or controlling/limiting further risk to personnel or limiting conditions that lead to barrier degradation or failure.

The RPD model (Klein 1993, 2003, pp. 20–29) includes two distinct phases (recognition/evaluation and mental stimulation) and three possible scenarios.

1. Situation recognition: the decider recognizes the current situation being the same as or sufficiently like a prior recalled experience based on the available information and its changing behavior (change direction, type, or rate). If it appears to be a like-for-like match, the recalled memory may include the appropriate act phase response actions and timing (a response plan).
2. Serial option evaluation: the decider may choose to use the memory from step 1 to consider other response plan options (e.g., change the timing of an act phase action). If the first option appears viable and applicable (intuitive-based), it becomes the decision plan for the act phase response. If not, a serial process considers the next option until one is accepted for the act phase response plan.
3. Mental simulation: for a complex situation or scenario or if there is doubt about the plan from step 2, the decider may perform a mental simulation of the plan to visualize its performance and expected or possible results. The validity and accuracy of the simulation rely on the validity and accuracy of the activated mental model(s). The product of this effort may be accepted as is, modified to address a perceived deficiency, or rejected. A rejection requires a return to step 1 or 2 to seek a different decision or solution.

Each of the abovementioned activities provides insight into the potential sources of cognitive failure that may contribute to a wrong decision or RPD result. A wrong decision is possible if the recognition is based on missing, incomplete, or incorrect SA-1 information. The option evaluation and mental simulation may be negatively affected by issues with attention, working or short-term memory, mental models, or accessing/using timely and necessary information. A valid RPD result relies on having a mental model that adequately encompasses the current situation. Its use, from a design perspective, may be limited to only those with the requisite domain-specific expertise. Others, by design, should be trained to use a rule-based decision process.

For additional discussion, perspectives, and insight, see Harris (2011, Chapter 5).

I.3.6 ACT PHASE: RESPONSE

Table I.13 (located in Section 1.3.8) provides sources of design and situation-based deviations that may contribute to an act phase response error or failure. The following aspects are covered:

- Act phase response: not performed
- Act phase response: not completed within the barrier/target task safety time (BST, TTST)
- Act phase response: not performed correctly
- Act phase response: information not communicated on time
- Act phase response: error in communicated message
- Act phase response: other communication issues

Note: As preparation and input to this process, review the PIF assessment and recommendations from detailed design process C-2 and the act phase process C-3.

I.3.7 TEAMS AND TEAMWORKING

Table I.14 (located in Section 1.3.8) provides sources of design and situation-based deviations that may contribute to barrier team deficiencies that can lead to barrier/ task failure. The following aspects are covered:

- Team shared situation awareness (SSA) deficiencies
- Barrier leader deficiencies
- Barrier team deficiencies

Note: As preparation and input to this process, review the PIF assessment and recommendations from detailed design process C-2 and the teams and teamworking process C-6.

I.3.8 COGNITIVE ASSESSMENT TABLES (DETAILS)

TABLE I.8

Cognitive Assessment Table: Common Issues

| Item ID | Issue | Cognitive Mismatch/Issue Information | | Possible corrective action changes to: |
		Mismatch/Issue Source	Failure Mechanisms	Physical, Human, Organizational Elements
GOA-1	Task goals *Incorrect goal selected*	Forget or misremember a barrier/task goal	Insufficient understanding of barrier/task goal	OE1: procedures – ensure task goals are clearly stated and easy to remember. OE2: barrier leader training should include monitoring of team activities to confirm team alignment to the barrier functions, task goals, and priorities. OE3: consider procedural stop points to assess current actions, goals, and priorities. PE1: add a remote barrier support task to monitor HE alignment to the barrier function, task goals, and priorities. (See Appendix L.3.1/2 and L.4 for suggestions and considerations.)
GOA-2	Task goals *Goal ambiguity/ confusion*	Goal conflicts	Unclear which goal to select, so may default by selecting the easiest or the most interesting goal, or one that can be completed quickly. (For discussion and examples, see preliminary design step B-10).	OE1: employ training exercises that present goal conflict situations, i.e., a production vs. safety scenario. (See step B-10 for background and example.) OE2: review barriers, tasks, and procedures to identify conflicting tasks and task goals. Resolve and update procedures and training accordingly. OE3: conduct a survey to reveal potential goal confusion. PE1 – provide decision support displays that identify active barriers, goals, tasks, and priorities. *See GOA-1 for additional suggestions.*

(Continued)

TABLE I.8
(Continued)

		Cognitive Mismatch/Issue Information		Possible corrective action changes to: Physical, **Human**, Organizational Elements
Item ID	Issue	Mismatch/Issue Source	Failure Mechanisms	
GOA-3	Task Goal *Effort not guided by goals*	Deviation between goals and current actions:	Lose focus on goal caused by: – Goal ambiguity/confusion – Excessive workload (see TL-1) – Attentional failures (ATT 1 to 5) – WM/STM failures (See WM-1 to 7)	**OE1:** training to develop awareness when actions are not adequately guided by the barrier/task goal. Provide simple tools to refocus attention on the goals of the activated barriers and tasks. **OE2:** consider procedural stop points to assess current action and goals. *See GOA-2, TL-1, ATT-1 to 5, and WM-1 to 7 for additional suggestions.*
PR-1	Task Priorities *Uncertain/ambiguous priority*	Uncertain priorities when selecting barrier/task activity	Insufficient understanding of goals, priorities, or procedures Inadequate MM	See GOA-1 to GOA-3, MM-3, WM-5, and WM-6 for suggestions.
PR-2	Task priorities *Wrong priority*	Select wrong priority	Task-switch error (see App. F.2.3 for background) Wrong goal (see GOA-1) Ignores known priority	See GOA-3 and PR-1 for suggestions.
TL-1	Excessive workload (task load)	Workload demand (base, peak) greater than resource capacity/ capability *For additional information, see process C-2 PIF assessments for:* – *Task complexity* – *Working environ.*	– Poorly designed barrier system/task (e.g., excessively complex, too many tasks, poor task allocations or communications procedures, insufficient resources) – Starting a task too late creates work peak and enables new hazard escalation pathways.	**OE1:** repetitive training drills (time-pressured) creating the opportunity to convert some conscious process activities to skill-based actions to reduce activity duration and free up cognitive resources (WM/STM). **OE2:** see Table J.3, Appendix J.6 for options to reduce workload by reducing the time (effort) needed to complete an activity. **OE3:** exploit SA-3 projection capabilities to predict and plan for new work demand possibilities. **OE4:** procedures allow for early start and pre-emptive actions to "gain time."

		– Other loads – *Corrective action program*	– Many external attention capture events (e.g., distractions, interruptions, communications)	**PE1:** modify task allocation between HEs, or between HEs and PEs. (Return to preliminary design step B-15.) **PE2:** simplify excessively complex tasks or procedures. **PE3:** provide support aids designed to maintain focus on activated barrier goals and task priorities. **PE4:** review validity of barrier, task, or task phase safety times. (See Table 3.27, preliminary design step B-16.)
ATT-1	Attention *Not monitoring the required information* (ATT-1 to ATT-6) *See Appendix F.2.1/4 for background*	Change blindness *Includes* *Inattention blindness*	Failure to notice an observable change, (e.g., state, conditions, location) may result from: – A failure to notice/remember initial conditions – A failure to compare the initial and current condition – An overly intensive focus on a specific object – Performing an excessively challenging task	**OE1:** train to develop a skilled behavior that looks for changes in monitored SA-1 information. Examples include the presence/absence of personnel, personnel location (exposure to hazard, movement times), declining vital signs (injured person), and hazard escalation (actual and potential pathways). **OE2:** procedures should explicitly state a requirement to monitor/look for a specific change (SA-1 information) or change type and when that monitoring should occur. **OE3:** training simulations that demonstrate *look-but-do-not-see* situations and consequences. **PE1:** provide simple tools to track change (e.g., paper or a simple visual device).

(Continued)

TABLE I.8
(Continued)

Item ID	Issue	Cognitive Mismatch/Issue Information		Possible corrective action changes to:
		Mismatch/Issue Source	Failure Mechanisms	Physical, **Human**, **Organizational Elements**
ATT-1 (cont'd)			This may also be a failure to notice an object within foveal view Zone A (inattention blindness). This can occur when there is: – Lack of MM expectancy – Low object salience – Limited WM/STM capacity See Appendices F.2.1 and F.2.4 for additional information	**PE2:** where feasible, use technical systems to track changes in monitored information. Indicate the change (nature, scale, direction) at the required use locations. Provide alerts on change occurrence and information that may change task phase activities and outcomes. **PE3:** consider using CCTV motion detection to alert/detect change (e.g., personnel movement in an unsafe area). The event may activate an automatic pre-recorded warning message through the area public address/general alarm (PA/GA) speakers. **PE4:** enhance the salience of SA-1 information objects. *Also see FIT-1 and FIT-3 for degraded physiological and physiological contributors to reduced WM/STM capacity and retention.*
ATT-2	**External attention capture** – *Interruptions*: communications, alarms, etc. – *Distractions*: nearby conversations or movements, noise, explosion, smell of gas, etc.		See Appendix F.2.1 for example mechanisms	**OE1:** provide training to demonstrate the effects of distractions and interruptions and develop individual approaches to mitigate their effects. **PE1:** consider the trade-off between the negative distraction and interruption effects of excessively loud alarms against a standard that requires this audible level. (Excessive noise can have a highly cognitive disruptive effect.) Also see WM-1 to WM-4 for additional suggestions.
ATT-3	**Internal attention capture:** problems at home, frustration, mind wandering, etc.			

ATT-4	Excessive workload	Excessive workload can lead to attention tunneling.	See Table TL-1 for suggestions. *Note: Attention tunneling occurs when one ignores other tasks/task activities with a higher priority. (This is also an example of a task-switch error. See Appendix F.2.3.)*
ATT-5	Divided attention/task-switch errors *See Wickens et al. (2023) for additional background and CA suggestions. Also see Appendix F.2.3 for task-switch error types and details.*	Place-holding error Forget or misremember information on return to prior activity *Errors occur when switching between activities (multi-tasking). The potential for error increases when a switch occurs in the middle of a required sequence of steps or activities.*	**OE1**: develop task design and procedures that seek to minimize conditions and situations that require switching between tasks. **OE2**: optimize design to take the least time to complete the barrier/task activities. (This may reduce the switch demand/ frequency.) **OE3**: training exercises include situations that require switching between different activity stages (e.g., different steps in a required sequence). Demonstrate and develop awareness of the range of errors or failures when this occurs. (e.g., communication interrupts a multi-step set of activities.) **OE4**: procedures provide guidance on when to complete a step sequence before switching. (Some sequences may be inviolate and must be completed once started.) **OE5**: provide training to increase awareness of the diverse types of task-switch errors, conditions when each may occur, and how to recognize the signs. **OE6**: consider procedural stop points to assess current goals, priorities, and activities. **PE1** – provide support aids to reduce errors in a multi-tasking environment. Create checklists or placeholders to track incomplete activities and guide the correct return to the incomplete activity or place in a step sequence.

(Continued)

TABLE I.8
(Continued)

		Cognitive Mismatch/Issue Information		Possible corrective action changes to:
Item ID	Issue	Mismatch/Issue Source	Failure Mechanisms	Physical, Human, Organizational Elements
ATT-6		Sensory access to information obstructed	Ambient environment interferes with sensory access to SA-1 information	OE1: training to include exercises that simulate plausible sources of environmental interference when attempting to access SA-1 information from diverse sources. The training objective may be to gain experience and develop skills to manage and overcome the interference effects. PE1: select SA-1 display types and communication equipment to prevent or mitigate the effects of plausible sources of environmental interference.
ATT-7	Attention *Monitoring the wrong information*	Monitoring the wrong information	Wrong goal or priority selected Wrong MM selected Error in selected mental model	See GOA-1 to GOA-3, PR-1, PR-2, MM-2 to MM-4
WM-1	**WM/STM** *STM information fade, replaced, or misremembered* (WM-1 to WM-4)	Forget future action (STM memory fade/replacement. See WM-2.)	Prospective memory task	OE1: delete or mitigate the effects of prospective memory tasks when possible. (May require a return to the function allocation in step B-15.) *See WM-2 for additional suggestions.*
WM-2		STM memory fades if not refreshed	**STM memory fades** if not frequently refreshed. *Fade becomes more likely with prospective memory or a demand that requires retaining information in STM under demanding conditions.*	OE1: revise procedure so activities do not require holding information in STM for longer than 30 seconds. OE2: communication procedures allow for request/repeat last info. (See Appendices K.3, K.4, K.5.) PE1: provide displays and alerts to inform an HE of a pending future action. This may be in the form of a timed re-alarm function.

ID	Description	Notes / Mechanisms	Suggestions
WM-2 (cont'd)		*For background, see Table F.1 (Memory Capture, Storage, Recall and Retrieval), Appendix F.2.2., and preliminary design step B-10, discussion item 4.*	**PE2:** if the future action performs the act phase response, consider automatically performing the response if it is not started by a pre-defined time. **PE3:** if task requires remembering SA-1 information for future use, retain/store that information in a form that can be easily accessed when needed. (See Appendices K.3 and K.5.) **PE4:** add a remote barrier support task to monitor pending future actions that may be at risk of being forgotten or not started on time. *See Appendix L.3.1/2 and L.4 for suggestions and considerations.*
WM-3	FFF response replaces information in WM/STM (See FIT-2)	Instantly interferes with normal conscious cognitive process (e.g., immediate loss of STM information, reduces STM capacity and retention duration).	*See FIT-2 for suggestions.*
WM-4	**Attention capture** – External capture (*See ATT-2*). – Internal capture (*See ATT-3*)	See Appendix F.2.1 for example mechanisms	*See GOA-1 (OE2, OE3), ATT-2/3 (OE1), and WM-1 to WM-3 for suggestions*
WM/STM *Degraded STM capacity/retention* *(Insufficient to support activity)* *(WM-5 to WM-7)*			
WM-5	Degraded psychological state (*See FIT-1*)	These conditions degrade WM/STP capacity, retention, and performance	See FIT-1, GOA-1/3, WM-1, WM-2, WM-4, and WM-8 for suggestions
WM-6	Degraded physiological state (*See FIT-3*)		See FIT-3, GOA-1/3, WM-1, WM-2, WM-4, and WM-8 for suggestions
WM-7	Sustaining a well-above-normal cognitive or physical work pace consumes WM/STM resources		See TL-1, GOA-1/3, WM-1, WM-2, and WM-8 for suggestions

(Continued)

TABLE I.8
(Continued)

| Item ID | Issue | Cognitive Mismatch/Issue Information | | Possible corrective action changes to: Physical, Human, __Organizational__ Elements |
		Mismatch/Issue Source	Failure Mechanisms	
WM-8	**WM/STM** *Activity demand exceeds WM/STM capability*	Insufficient capacity to store and integrate information from several MMs, sensory stores, and other information	Insufficient skills with processing complex information streams *For additional information, see process C-2 PIF assessments for task complexity*	**OE1:** modify procedures to simplify activity. **OE2:** modify (increase) aptitude requirements of assigned personnel. **OE3:** increase experience working with complex integration tasks. (May limit who is qualified for role.) **PE1:** modify task to reduce complexity. This may require a return to the preliminary design process. **PE2:** changed the task HE/PE allocation to move complexity to PE system function. This may require a return to the preliminary design process. **PE3:** provide support aids to reduce the complexity of the integration effort and WM/STM demand. **PE4:** provide a single integrated display that provides the required information. *See SK-1 for additional suggestions*
FIT-1	Degraded *psychological* state	Acute stress, fear, degraded morale *For additional information, see process C-2 PIF assessments for:* – *Threat/stress* – *Morale, attitude, motivation*	– Persistent/long duration high-tempo workload in a hazardous environment or situation. – Low confidence in self or others – Low trust in the barrier team, leader, or facility leadership	**OE1:** train barrier leader and others to improve their capability to monitor and manage stress in self and others **OE2:** design team training to develop and maintain trust between members, learn to resolve conflicts, and disagreements. **OE3:** barrier leader monitors team and uses enhanced leadership skills (e.g., optimal use of tone/words that reduce stress and fear and maintain morale).

	See Appendix F.2 for further information (cognitive performance)	– Safety culture – Corrective Action Program – Societal Influencers	– Poor safety culture – Frequent incidents – Low morale due to poor leadership, pay/reward policies, forced layoffs, loss of benefits, etc. – Poor perception of safety systems or practice	**OE4:** repetitive training under time pressure to develop resistance to stress and improve confidence and trust. **OE5:** maintain the barrier team continuity to reduce workload (e.g., increased use of implicit coordination) and improve trust. **OE6:** consider procedural stop points to assess current actions, goals, and priorities **PE1:** add support aid to reduce cognitive demand and human error likelihood
FIT-2	Event triggers freeze, flight, flight response (FFF)	Event may trigger automatic and undesired cognitive response *See Appendix F.1 and Figure F.1 for detail.*	– Immediate effect may suppress conscious cognitive processes – Increased propensity to automatic, impulsive response – Awareness of learned or experienced dangers but no awareness of risk	**OE1:** provide training that describes the body's response to a FFF event (e.g., near instant release of corticosteroids, endorphins) and tendency to suppress conscious processes and activate an FFF response. **OE2:** include training or simulated experience to better prepare HE for sudden, violent situations. *See FIT-1 for additional suggestions*
FIT-3	Degraded physiological state *See Appendix F.2, Figure F.2 for more information (physical performance)*	Excessive fatigue or health issue	– Deficient fatigue management program/fatigue target exceedance – Unforeseen events or poor planning results in personnel shortages – Short-term or transient health issue – Inadequate tracking and management of personnel health and fitness for service	**OE1:** train the barrier leader and others to improve their capability to monitor and manage high fatigue and its effects on self and others. **OE2:** train barrier leader to maintain awareness of an HE health issue and effect, and if/how it may degrade team performance. **OE3:** consider procedural stop points to assess current actions, goals, and priorities. **PE1:** create remote barrier support task to monitor team *physiological* state and when it can threaten barrier system performance. Identify solutions to use that information without an unintended action that degrades team performance. (See Appendices L.3.3 and L.4 for suggestions and considerations.)

(Continued)

TABLE I.8
(Continued)

		Cognitive Mismatch/Issue Information		Possible corrective action changes to: Physical, Human, Organizational Elements
Item ID	**Issue**	**Mismatch/Issue Source**	**Failure Mechanisms**	
FIT-3 (cont'd)				**PE2:** create barrier support task that provides the barrier leader with options to assign preselected ad hoc tasks to the RBS HE. (See Appendices L.3 and L.4 for suggestions and considerations.) **PE3:** add support aid to reduce cognitive demand and human error likelihood
MM-1	Mental Model *Missing MM*	*No MM exists for detected SA-1 information*	Situation is unforeseen, so no prior experience, i.e., no useful MM available to create comprehension *Note: A cyber security attack (e.g., a man-in-the middle attack) presents false information to HMI displays. The information does not make sense given the sensed state of the plant.*	**OE1:** an unforeseen situation that changes requirements warrants a return to the preliminary design to update the assessments and requirements. **OE2:** include at least one training exercise for an unforeseen (not trained for) situation. Exercise purpose: 1. Develop awareness and skills to recognize a new or novel situation. 2. Establish expectations and approaches to seek additional information and understanding before responding. (Knowledge-based decision-making significantly increases the potential for human error. See preliminary design step B-10 for background)
MM-2	Mental Model *Wrong model selected*	*Access wrong MM* (See Appendix F.5, cues)	– Directed by wrong SA-1 information (*See Note in MM-1.*) – Directed by wrong goals or priorities – Cues insufficient to discriminate between MMs – Failure to examine/reject an incorrect MM	**OE1:** include MM cues in training, procedures, HMI displays, and support aids to improve MM recall speed and accuracy. (See Appendix F.5 for background on cues.) **OE2:** provide training on MMs: – Understand contents (e.g., topics, level of detail) – Understand automatic search (e.g., like-for-like, employed heuristics (traits, consequences))

MM-2 (cont'd)		– Failure to change calling conditions to seek a more appropriate MM (Reason, 1990, pp. 130/1)	– Examine recalled memory for applicability – Seek a new memory by changing calling conditions *See GOA-1, PR-2 for additional suggestions.* *See Table 1.9 for SA-1 type errors/issues that may call the wrong MM*
MM-3	Mental Model *Incomplete/ inadequate MM*	MM is incomplete or not adequate	– Missing experience and training needed to develop the required MM content and detail. – Experience/training not adequately integrated into the MM.	**OE1:** provide targeted training in the areas in which the MM is deficient. Include testing to enhance the learning results (Radvansky, 2021, p. 221). Perform a competency assessment to verify the required competency is achieved. **OE2:** until the HE acquires the necessary MM, provide additional supervision by a fully competent person. *If an advanced MM (domain expertise) is required:* **OE3:** select an alternate person that has the required MM. **PE1:** revise the requirement. (Return to preliminary design phase.) **PE2:** provide the necessary support aid to meet the requirement.
MM-4	Mental model *Errors in MM*	MM contains errors that may degrade or cause barrier/task failure	– MM contains knowledge-based errors in procedures, technical information (systems, processes, the hazard that activated the barrier), or task execution. – MM drift. Slow changes over time introduce errors in this memory. (See SK-2.)	**OE1:** perform ongoing competency verification and re-training to detect and correct MM drift. (See operate and maintain processes E2-2, E2-3, E3-1, E5-2, E5-3, and E6-1 for detail.) *Also see MM-3 (OE1) for an additional suggestion*
SK-1	Skills *Not fully developed* *See Harris (2011,* *Table 5.2) for* *example skill types.*	Skill not adequately developed to achieve reliable performance and reduce WM/STM load	Due to inadequate drills or poorly timed repetitions, performance of the partially developed skill is effortful (increases WM/STM load) and more prone to error.	**OE1:** add repetitive training exercises and additional experience to fully develop a required skill. **OE2:** the HE/barrier leader should remain aware a team member competency is not yet met. Time is needed to do so. Additional supervision may be warranted.

(Continued)

TABLE I.8
(Continued)

Item ID	Issue	Mismatch/Issue Source	Failure Mechanisms	Possible corrective action changes to: Physical, **Human**, **O**rganizational **E**lements
			Cognitive Mismatch/Issue Information	
	Skills			
	Deficiencies and errors in skilled actions			
SK-2	(SK-2 to SK-4)	Drift Unsafe deviation in behaviors, beliefs, or actions. *(See Appendix J.7.2 for background.)*	The initial deviation may be unnoticed or intentional (e.g., a necessary adaptation or an unsafe violation taken to reduce effort). Repetition over time may habituate the deviation in the individual (MM) and organization (culture, discipline).	OE1: monitor personnel for deviation from procedures. (See operate and maintain processes E2-3, E2-5, E5-2, and E5-5.) OE2: provide periodic training updates and competency assessments to detect and correct skill-based errors. (See operate and maintain processes E2-2, E3-1, E5-2, and E6-1.) OE3: train barrier leader to monitor team for compliance to procedures. OE4: consider stop points (included in procedures) to assess current action and goals.
SK-3		Skill fade Includes skilled use of equipment, procedures, task execution *(See Appendix J.7.1 for background.)*	Complex skills (e.g., team interaction, RPD) not practiced in 3–6 months may begin to fade. A person may be unaware of the skill loss and the potential task performance degradations and risk attributed to that loss.	OE1: training department tracks the use frequency of a required skill (e.g., training events, barrier activation). Automatically schedule training and/or competency verification for a skill not used in the last 3–6 months. (For additional information, see Appendix J.7.1.)
SK-4		Strong habit intrusion (Action slip) Reason (2008, pp. 42–43)	Two skills share similarities and one or more safety critical differences. One is frequently used, and the other is rarely used. A demand occurs that requires the rarely used skill. Instead, the other is performed and goes unnoticed (can also occur with similar locations or movements).	OE1: where possible, design the barrier system to quickly reveal a missed/incorrect step, and fully recover if the error is promptly detected and corrected. OE2: review procedures that may contribute to this failure mechanism. Identify and discuss these procedures in training to develop awareness. Provide suggestions to prevent this error type. PE1: provide and use a checklist to track multi-step actions.

| NR-1 | Non-rational bias and behaviors | Non-rational biases and behaviors contribute to task activity errors | Non-rational biases may affect decisions, priorities, etc. See Appendices F.2.5 and F.2.6 for details. | **OE1:** develop training to recognize and combat confirmation bias in oneself and others.
OE2: barrier leader training should include monitoring of team activities for non-rational behaviors and making corrections as noted in procedures.
OE3: consider procedural stop points to assess current action and priorities.
OE4: procedures define the accepted/expected approach to identify and mitigate non-rational bias/behavior observed in others such as a barrier leader and other HEs.
PE1: add a remote barrier support task to monitor key barrier team behaviors (e.g., decisions, priorities) for confirmation bias and other non-rational biases and behaviors and respond according to procedures. (See Appendices L.3.3 and L.4 for suggestions and considerations.) |

TABLE I.9

Cognitive Assessment Table: Detect Phase SA-1 Information

ID	Cognitive Mismatch/Issue Information			Possible corrective action changes to: Physical, Human, Organizational Elements
	Issue	Mismatch/Issue Source	Failure Mechanisms	
SA1-1	SA-1 information not detected: *Not noticed* (SA1-1 to SA1-6) *For background see App. B.5.1.*	"Bottom-up" scan not activated	Low salience of SA-1 display/information object Noisy sensory environment interferes with sensory detection or discrimination (See App. B.5.2 for detail.)	**OE1:** establish minimal HE visual requirements: visual acuity, peripheral vision capability, low light capability, time needed to achieve minimal levels of vision when transitioning from high>low light environment. **OE2:** establish minimal hearing requirements: minimum hearing capability at defined frequencies, minimum ability to discriminate sounds with high background noise, etc. **OE3:** add simulation training to increase awareness of missed objects in environment due to low salience. **PE1:** enhance salience of SA-1 implementation, e.g., enhance display attributes (color, shape, size, behavior). Reduce clutter and salience of other elements in visual zone.
SA1-2		SA-1 object not located in the optimal visual zone (See Figure F.4, and Appendices F.3, F.4 and K.1.2 for detail)	Placement outside optimal visual zone decreases likelihood of being noticed Reduced salience (relative) of objects in non-optimal visual zones Reduced visual/hearing capabilities to notice and detect a display in the visual peripheral zone	**OE1:** training reviews SA-1 displays located in visual periphery. Train to scan in that direction, e.g., develop mental model (expected locations and information value). **OE2:** consider training that uses a visual simulator to demonstrate degraded noticeability and detection as one moves away from eye's line of sight. *The following may already be implemented if following the prototype lifecycle model. See Appendix F.3, K.1, and K.2 for detail.* **PE1:** for target personnel, place visual displays in zones A and B (where possible) assuming the expected position of the target viewer.

SA1-2 (cont'd)			**PE2:** for peripherally located displays, the display type should be selected based on what the eye can reliably detect in that zone. (See Appendix F.3 for background.) **PE3:** add other means to increase the noticeability (salience) of displays in the visual peripheral zones (e.g., add an audible sound at/near the display, employ flashing or movement).
SA1-3	"Top-down" scan driven by mental model (e.g., expectancy driven)	Incomplete MM *See Table I.8 MM-3* (Not looking for items that are not expected. See Endsley and Jones 2012, pp. 27)	See Table I.8, MM-3 for suggestions.
SA1-4	Change blindness	See Table I.8, ATT-1	See Table I.8, ATT-1 for suggestions.
SA1-5	Sensory obstruction	Environmental conditions may interfere with noticing and accurately perceiving SA-1 information. Example conditions include (visual) smoke, low light, rain, distance, blocked by personnel movements. Audible: high background noise, noise in voice frequencies, nearby conversations, etc.	**PE1:** select SA-1 displays and solutions that overcome or mitigate the effects of environmental interference (e.g., sun/rain shields, higher lamp intensity, automatic display intensity adjustments, larger signs/text, use of flash sequences, enhanced speaker systems, redundant audible messaging). **PE2:** consider display redundancy (e.g., convey information visually and audibly). **PE3:** select communication equipment designed for noisy environments (e.g., noise canceling, dynamic volume control, wearable options).
SA1-6	Missed due to distraction or interruption	A distraction or interruption may cause an HE to miss SA-1 information that is transient (e.g., a communicated message, a one-time alert, or transient information in the environment). Also see Table I.8, ATT-2/3	**OE1:** employ communication protocols and procedures that allow for request/repeat last info. (See Appendix K.3 and K.5 for supporting information.) **PE1:** convey information using methods that support/allows for later access/recall if the information was missed due to a distraction or interruption (e.g., use text, email, video replay). (See Appendices K.3 and K.5 for supporting information.) Also see *Table I.8, ATT-2/3 for additional suggestions.*

(Continued)

TABLE I.9
(Continued)

Cognitive Mismatch/Issue Information

ID	Issue	Mismatch/Issue Source	Failure Mechanisms	Possible corrective action changes to: Physical, Human, Organizational Elements
SA1-7	SA-1 information detected but forgotten, replaced or misremembered (SA1-7 to SA1-11)		The result of a distraction or interruption or divided attention (multi-tasking error)	*See Table 1.8, ATT-2/3, ATT-5, and WM-1 to WM-4 for suggestions.*
SA1-8		Loss of SA-1 information temporarily held in WM/STM	Degraded WM/STM capacity/capability caused by degraded *psychological state* *See Table 1.8, FIT-1*	*See Table 1.8, TL-1, FIT-1, WM-1, WM-2, and WM-4 for suggestions.* *(See Appendix F.2 Figure F.2 (cognitive performance) for further information.)*
SA1-9		(See Appendix F.2)	Freeze, fight, flight response causes the loss or replacement of STM information *See Table 1.8, FIT-2*	*See Table 1.8, FIT-2 for suggestions.*
SA1-10			Degraded WM/STM capacity/capability caused by degraded *physiological state* *See Table 1.8, FIT-3*	*See Table 1.8, FIT-3 for suggestions.* *(See Appendix F.2 Figure F.2 (physical performance) for further information.)*
SA1-11		WM/STM memory fade	– Prospective memory task, the need to remember a future required action (see Table 1.8, WM-1) – Required SA-1 information is a tracked change value (e.g., hold prior value to compare to future value). – Unable to reliably retain information in STM while gathering other information for integration and processing.	*See Table 1.8, WM-1, and WM-2 for suggestions.*

SA1-12	SA-1 information not detected: *Not looked for* (SA1-12 to SA1-17)	Attention capture *External events:* distractions or interruptions *Internal events:* problems at home, mind wandering, etc.	Attention capture stops scanning of new SA-1 information, performance of other barrier/ task activities. See Appendix F.2.1 for examples of mechanisms.	*See Table I.8, ATT-2/3 for suggestions.*
SA1-13		Attention capture *Communications too long, too frequent*	Attention capture stops scanning of new SA-1 information, performance of other barrier/ task activities.	**OE1:** training exercises demonstrate the effect of inappropriate communications (too long, too frequent). **OE2:** procedures define communication requirements including the appropriate timing, content, and target maximum durations. *See Table I.8, ATT-2/3 for additional suggestions.*
SA1-14		Excessive workload halts scanning for new SA-1 information	Excessive workload can lead to attention tunneling, task fixation. See Appendix F.2.1 for examples of mechanisms.	See Table I.8, TL-1 for suggestions.
SA1-15		Inattention blindness	Failure to notice information in foveal view (Zone A). (See Table I.8, ATT-1.)	*See Table I.8, ATT-1 for suggestions.*
SA1-16		Not attending to SA1 info due to MM issue	– Missing MM – Accessed wrong MM – Incomplete MM – Error in MM Selection and scanning for SA-1 information not guided by expectations in a MM (see Table I.8, MM-1 to MM-4.)	*See Table I.8, MM-1, MM-2, MM-3, and MM-4 for suggestions.*

(Continued)

TABLE I.9
(Continued)

Cognitive Mismatch/Issue Information

ID	Issue	Mismatch/Issue Source	Failure Mechanisms	Possible corrective action changes to: Physical, Human, Organizational Elements
SA1-17		SA-1 info not attended to due to incorrect goal or priority	Not guided by the correct goal or priority. *See Table I.8, GOA-1 to GOA-3, PR-1 to PR-2*	*See Table I.8, GOA-1 to GOA-3, PR-1, PR-2 for suggestions.*
SA1-18	SA-1 information not detected: *not accurately perceived* (SA1-18 to SA1-21)	Sensory access to information obstructed	Ambient environment interferes with sensory access to SA-1 information.	*See Table I.8, ATT-6 for suggestions.*
SA1-19		Misleading display presentation	*Gestalt theory* Display of SA-1 information creates erroneous perception. (See compatible SA, App. C.1.3)	**PE1:** present information in a form/layout/approach consistent with its intended use and user. (See compatible SA, App. C.1.3 for detail.)
SA1-20		Communicated message not understood	– Message content not clear (e.g., ambiguous, conflicting, poor use of terms, ambient interference). – Message delivery not clear (e.g., talking too fast, poor diction).	*See Table I.8, SK-1, ATT-6, and Table I.13, ACT-35 for suggestions.*
SA1-21		Inconsistent use of coding system	Misinterpretation caused by inconsistent/ incorrect coding conventions in the presented information (e.g., color, shape, location).	**PE1:** use consistent coding in the presentation of all SA-1 information.

SA1-22	SA-1 Information not detected: *Found but ignored/ rejected*	Confirmation bias and other non-rational biases	Limit SA-1 search to information that is confirming. Ignore other information. (*See App. F.2.5 for further detail*)	See Table I.8, NR-1 for suggestions.
SA1-23	SA-1 information *Error in the Information* (SA1-23 to SA1-25)	Error in received communication	– Errors in the received (conveyed) communications – Verbal conveyance of correct information is misheard (e.g., poor diction, environmental interference muffles or blanks out key words, divided attention).	**OE1:** provide a training exercise on the potential hazard of using information that may have been misheard or garbled. (What I thought I heard was information I expected, e.g., incorrectly *filled in the blanks.*) **OE2:** train barrier leader to monitor team communications for possible errors **PE1:** add support barrier role to monitor critical communications for accuracy, potential to misunderstand, signs of confusion, etc. (See Appendix L.3.1 and L.4 for suggestions and considerations.) *See Table I.8, SK-1, and WM-2 for additional suggestions.* *See Table I.13, ACT-29 to ACT-32 for background.*
SA1-24		Incorrect information on Incident Command Board (See detailed design process C7-8)	Information: – Forgotten or misremembered (See SA1-7 to SA1-11) – Not detected/looked for (See SA1-12 to SA1-17) – Not accurately perceived (See SA1-18 to SA1-21) – Contains errors (See SA1-23 to SA1-25)	See SA1-7 to SA1-21, and SA1-23 to SA1-25 for suggestions.

(Continued)

TABLE I.9
(Continued)

Cognitive Mismatch/Issue Information

ID	Issue	Mismatch/Issue Source	Failure Mechanisms	Possible corrective action changes to: Physical, **Human**, <u>Organizational</u> Elements
SA1-25		VDU-displayed SA-1 information not consistent with plant state, values, operation, etc.	A type of cyber-attack is a man-in-the-middle attack that presents false information on VDU-based displays. Other attack types may take hidden control actions or disable safety functions or equipment.	**OE1**: provide training on potential types of cyber-attacks (e.g., the potential false information on displays and how to proceed when this occurs). **OE2**: provide procedures to support detection and response to a possible cyber-attack. **PE1**: create remote barrier support tasks that provide procedure-defined support (e.g., assist with diagnoses or remote support management). See Appendices L.3 and L.4 for potential opportunities and considerations.
SA1-26	SA-1 information *Not available – Communications*	Received (communicated) message missing information or arrived too late (See Apps. K.3, K.5)	– Information request sent too late – Communication procedures not followed (timing, content) – Message sender issues (e.g., attention tunneling, goal or priority selection issue)	**OE1**: based on procedure guidance, make a second request for information if not received within the defined period. **OE2**: barrier team leader monitors communications with a focus on communication errors. Learn procedurally defined actions to correct or mitigate the error.
SA1-27	SA-1 Information *Attending to the wrong information*	Attending to the wrong information	See Table I.8, ATT-7.	See Table I.8, ATT-7 for suggestions.

TABLE I.10

Cognitive Assessment Table: Detect Phase SA-2 Comprehension

Item ID	Issue	Cognitive Mismatch/Issue Information		Possible corrective action changes to: Physical, Human, Organizational Elements
		Mismatch/Issue Source	Failure Mechanisms	
SA2-1	SA-2	Missing MM	See Table I.8, MM-1 for detail.	See Table I.8, MM-1 for suggestions.
SA2-2	Comprehension	Change blindness (change or aspects of change not noticed)	See Table I.8 ATT-1 for detail.	See Table I.8 ATT-1 for suggestions.
SA2-3	*No comprehension* (SA2-1 to SA2-3)	SA1 information not detected or noticed (no comprehension from the missing information)	See Table I.9, SA1-1 to SA1-17.	See Table I.9, SA1-1 to SA1-6 and SA1-12 to SA1-17 for suggestions.
SA2-4	SA-2	Incomplete/inadequate MM	See Table I.8, MM-3.	See Table I.8, MM-3 for suggestions.
SA2-5	Comprehension	Failure to accurately comprehend changing conditions	Failure to monitor, notice, or track changing information. Incomplete or inadequate MM	See Table I.8, ATT-1, and MM-3 for suggestions.
SA2-6	*Insufficient comprehension* (SA2-4 to SA2-8)	Missing SA-1 information leads to incomplete comprehension	See Table I.9, SA1-1 to SA1-22. *Note: An insufficient number of sensors may contribute to a keyhole effect. See App. F.2.6 for detail and examples.*	See Table I.9, SA1-1 to SA1-22 for suggestions.
SA2-7		Inability to correctly integrate SA-1 information with MM/ other LTMs to gain sufficient comprehension	Degraded WM/STM capability Aptitude insufficient for task type/complexity	See Table I.8, TL-1, FIT-1, FIT-3, WM-5 to WM-8 for suggestions.

(Continued)

TABLE I.10
(Continued)

Item ID	Issue	Cognitive Mismatch/Issue Information		Possible corrective action changes to: Physical, Human, Organizational Elements
		Mismatch/Issue Source	Failure Mechanisms	
SA2-8		HMI Display element does not directly convey meaning.	Unrealistic WM/STM effort needed to interpret/ extract the actual meaning from displayed information.	**PE1:** consider using an alternative display type if the SA-1 source/display does not directly convey the meaning needed to achieve the SA-2 requirement. (See step C5-1c, *Alternative Display Type* note for example.)
SA2-9	SA-2 Comprehension / *Wrong comprehension* (SA2-9 to SA2-15)	Error in MM	See Table I.8, MM-4.	See Table I.8, MM-4 for suggestions.
SA2-10		Incorrect comprehension of changing conditions	Error in SA-1 information; Failure to monitor, notice, or track changing information; Wrong MM selected; Error in MM	See Table I.8, ATT-1, MM-2, MM-3, MM-4 and Table I.9, SA1-23 to SA1-25 for suggestions.
SA2-11		Access wrong MM (see Appendix F.5, cues)	See Table I.8, MM-2.	See Table I.8, MM-2 for suggestions.
SA2-12		Monitoring the wrong SA-1 information.	See Table I.8, ATT-7.	See Table I.8, GOA-3, PR-2, MM-2 to MM-4 for suggestions.
SA2-13		Error in the accessed SA-1 information	See Table I.9, SA1-23 to SA1-25.	See Table I.9, SA1-23 to SA1-25 for suggestions.
SA2-14		SA-1 information ignored or not sought (*Confirmation bias*)	See Table I.9, SA1-22.	See Table I.8, NR-1 for suggestions.
SA2-15		Inability to correctly integrate SA-1 information with MM/ other LTMs to gain sufficient comprehension	See SA2-7.	See Table I.8, TL-1, FIT-1 to FIT-3, WM-5 to WM-8 for suggestions.

TABLE I.11
Cognitive Assessment Table: Detect Phase SA-3 Project/Anticipate

Item ID	Issue	Cognitive Mismatch/Issue Information		Possible corrective action changes to: Physical, Human, Organizational Elements
		Mismatch/Issue Source	Failure Mechanisms	
SA3-1	SA-3 projection/ anticipation *None achieved (Unable to develop event evolution or time estimate/projection) (SA3-1 to SA3-2)*	Missing MM	Unforeseen situation or experience (see Table I.8, MM-1) Insufficient experience or training needed to develop the required advanced MM/LTM content	**OE1** – Select a different HE who meets this competency requirement. (Meeting this objective requires domain-specific expertise, a state that is only achieved over many years.) **OE2:** change the SA3 requirement. Requires return to preliminary design. **PE1:** provide support aids that provide the required SA-3 project information. *Also see Table I.8, MM-1 for additional suggestions.*
SA3-2		Failure/unable to detect changes to SA-1 information (e.g., magnitude, direction, and rate)	See Table I.8, ATT-1, and WM-8.	See Table I.8, ATT-1, and WM-8 for suggestions.

(Continued)

TABLE I.11
Continued

Item ID	Issue	Cognitive Mismatch/Issue Information		Possible corrective action changes to: Physical, <u>Human</u>, <u>Organizational</u> Elements
		Mismatch/Issue Source	Failure Mechanisms	
SA3-3		Incomplete/inadequate mental model	See Table I.8, MM-3.	See SA3-1 and Table I.8, MM-3 for suggestions.
SA3-4	SA-3 projection/ anticipation	Projection/anticipation not consistent with changing SA-1 information (e.g., rate, magnitude, or direction)	Failure to monitor/notice/track changing information Incomplete/inadequate MM	See Table I.8, ATT-1, and MM-3 for suggestions.
SA3-5	*Inadequate results/ accuracy in event evolution or time projection*	Missing SA-1 Information needed to develop required projections	See Table I.9, SA1-1 to SA1-17. *Note: An insufficient number of sensors may contribute to a keyhole effect.* *See App. F.2.6 for detail and examples.*	See Table I.9, SA1-1 to SA1-17 for suggestions.
SA3-6	(SA3-3 to SA3-6)	Inability to correctly integrate SA-1 information with MM/other LTMs to gain adequate projections	Degraded WM/STM capability Aptitude insufficient for task type	See Table I.8, TL-1, FIT-1 to FIT-3, WM-5 to WM-8 for suggestions.
SA3-7		Error in MM	See Table I.8, MM-4.	See Table I.8, MM-4 for suggestions.
SA3-8	SA-3 projection/ anticipation	Projection/anticipation not consistent with changing SA-1 information (e.g., rate, magnitude, or direction)	Error in SA-1 information Failure to monitor/notice/track changing information Wrong MM selected Error in MM	See Table I.8, ATT-1, WM-8, MM-2 to MM-4, SK-1, and Table I.9, SA1-6, and SA1-23 to SA1-25 for suggestions.
SA3-9	*Significant projection/ anticipation error of a magnitude that can place the barrier/task at risk*	Access wrong MM (see Appendix F.5, cues)	See Table I.8, MM-2.	See Table I.8, MM-2 for suggestions.
SA3-10	(SA3-7 to SA3-10)	Inability to correctly integrate SA-1 information with MM/other LTMs to gain adequate projections	See SA3-6.	See SA3-6 for suggestions.

TABLE I.12

Cognitive Assessment Table: Decisions

Cognitive Assessment and Mitigation: Decide Phase and Response Plan

Item ID	Issue	Cognitive Mismatch/Issue Information		Possible corrective action changes to:
		Mismatch/Issue Source	Failure Mechanisms	Physical, _Human_, <u>Organizational</u> Elements
DEC-1	Decision: _Rule-based process_	Wrong goal or priority	See Table I.8, GOA-1, GOA-3, and PR-2.	See Table I.8, GOA-1, GOA-2, PR-1, and PR-2 for suggestions.
DEC-2	_Wrong decision Review PIF assessment C-2 process (all PIFs)_	No comprehension of situation	See Table I.10, SA2-1 to SA2-3.	See Table I.8, ATT-1 and MM-1, and Table I.9, SA1- to SA1-6 and SA1-12 to SA1-17 for suggestions.
DEC-3	(DEC-1 to DEC-9)	Wrong comprehension of situation	See Table I.10, SA2-9 to SA2-15.	See Table I.8, GOA-3, PR-2, ATT-1, MM-2 to MM-4, and Table I.9, SA1-23 to SA1-25 for suggestions.
DEC-4		Insufficient comprehension of situation	See Table I.10, SA2-4 to SA2-8.	See Table I.8, ATT-1, MM-3, TL-1, FIT-1, FIT-3, WM-5 to WM-8, Table I.9, SA1-1 to SA1-22, Table I.11, SA2-8 for suggestions.
DEC-5		Divided attention	See Table I.8, ATT-5.	See Table I.8, ATT-5 for suggestions.
DEC-6		Significant error in SA3 projection	See Table I.11, SA3-7 to SA3-11.	See Table I.8, ATT-1, MM-2, MM-4, TL-1, FIT-1, FIT-3, WM-5 to WM-8, Table I.9, SA1-6, and SA1-23 to SA1-25 for suggestions.
DEC-7		Personnel using RPD process without having sufficient expertise	Lacks the necessary expertise (e.g., insufficient mental model/RPD skills)	**HE1:** select candidates who have and demonstrate the required expertise **OE1:** procedures provide rule-based decision requirements.
DEC-8		Errors introduced by non-rational biases and bounded rationality	Decision errors attributed to non-rational biases (Appendix F.2.5) or limited by bounded rationality (Appendix F.2.6)	See Table I.8, NR-1 for suggestions.

(Continued)

TABLE I.12
(Continued)

Cognitive Assessment and Mitigation: Decide Phase and Response Plan

Item ID	Issue	Cognitive Mismatch/Issue Information		Possible corrective action changes to: Physical, Human, Organizational Elements
		Mismatch/Issue Source	Failure Mechanisms	
DEC-9		Decision capability degraded due to insufficient WM/STM capacity/resources	See Table I.8, WM1 to WM-8.	See Table I.8, WM1 to WM-8 for suggestions.
DEC-10	Decision: *Rule-based process* *Decision included wrong act phase plan* (DEC-10 to DEC-11)	Wrong goal or priorities	See Table I.8, GOA-1 to GOA-3, PR-1, and PR-2.	See Table I.8, GOA-1 to GOA-3, PR-1, and PR-2 for suggestions.
DEC-11		Incomplete/inaccurate recall of act phase responses and response sequences	Insufficient MM/memory recall containing this information Unreasonable or inappropriate reliance on MMs/LTMs	**OE1:** provide periodic barrier re-training on infrequently activated barriers. Include testing to enhance learning results (Radvansky 2021, p. 221). **OE2:** procedures define act phase responses, options, and steps in easy-to-recall/remember forms. **PE1:** provide quickly and widely accessible access to procedures noted in OE2 above.
DEC-12	Decision: RPD process *Wrong decision due to incorrect recognition or serial option selection process* (DEC-12 to DEC-14)	Incorrect recognition	Not monitoring the correct information None or inadequate MM Incorrect MM selected	**HE1:** for roles requiring the use of RPD, limit candidates to only those having the required domain expertise and fully developed/proven RPD skills.
DEC-13	*RPD = Recognition-Primed Decision model*	Error in serial selection process	Attentional issues interrupt process Working memory limitations introduce errors in process.	**HE2:** those assigned to a barrier backup role (a role expected to use RPD) should have the same RPD capability and skill use as the primary role. **OE1:** perform more frequent competency assessments of emergency response barrier leaders. *See Table I.8, ATT-1 to ATT-5, WM-1 to WM-8, FIT-1, to FIT-3, SK-1 for additional suggestions.*

DEC-14		RPD skill use not fully developed	Person has the required domain expertise but insufficient/unproven RPD expertise for assigned role	**HE1:** if this is an emergency response barrier leader role, select a different candidate. **OE1:** for other barrier types and roles, see Table I.8, SK-1 for suggestions. *See Harris (2011, Table 5.2) for example RPD skill types.*
DEC-15	Decision: RPD process *Wrong decision due to error in mental simulation* (DEC-15 to DEC-17)	No mental simulation performed	Simulation not performed (e.g., forgotten, skipped, excessive workload)	See Table I.8, FIT-1 to FIT-3, WM-5 to WM-8, SK-1 to SK-3 for suggestions.
DEC-16		RPD skilled use not fully developed	Person has the required domain expertise, but insufficient/unproven RPD expertise for assigned role	See DEC – 14
DEC-17		Mental simulation error	Non-viable/inadequate simulation: – Attention management issues – Mid-simulation interruption introduces undetected memory error leading to inaccurate results – Insufficient WM/STM resources to successfully complete simulation	See Table I.8, ATT-2 to ATT-5, FIT-1 to FIT-3, MM-3, MM-4, WM-5 to WM-8, and SK-1 to SK-3 for suggestions.
DEC-18	Decision: *Decision too late, not made (Unable to complete barrier/ task act phase response within the specified safety time, BST/ TTST)* (DEC-18 to DEC-24) *Review PIF assessment C-2 process (all PIFs)*	Lose track of time	Prospective memory task (see Table I.8, WM-1) Reliance on WM/STM to track time (see Table I.8, WM-2)	See Table I.8, WM-1, and WM-2 for suggestions.

(Continued)

TABLE I.12
(Continued)

Cognitive Assessment and Mitigation: Decide Phase and Response Plan

Item ID	Issue	Cognitive Mismatch/Issue Information		Possible corrective action changes to:
		Mismatch/Issue Source	Failure Mechanisms	Physical, **Human**, **Organizational Elements**
DEC-19		Inappropriate delay in decision or start of act phase response because of inaccurate knowledge of the time needed to complete an act phase response	Incorrect knowledge of the longest plausible time needed to complete the response Failure to include contingency time in the planned delay time *Past experience (time) may be an inadequate guide to plausible situations that have not yet occurred*	**OE1:** procedures provide guidance on the minimum time contingency needed when choosing to delay a decision or act phase response. **OE2:** provide tabletop training sessions to show/explore the effects of late decisions. Include scenarios where the act phase takes much longer to complete than planned. **OE3:** applied to the emergency response barrier leader role, see Table I.11 (SA-3) for suggestions. For others, see Table I.8, MM-3 for suggestions.
DEC-20		Divided attention/task-switch errors	See Table I.8, ATT-5.	**PE1:** add alerts at key decision points. **PE2:** provide count up/down timers at shared displays to enhance time awareness. Indicate timed events, target, demanding time limits, time remaining, and the information needed for planning and coordination. **PE3:** add a remote barrier support task to monitor timing for critical decisions and provide reminders as defined in procedures. (See Appendices L.3.2 and L.4 for suggestions and considerations.) *See Table I.8, ATT-5 for additional suggestions.*

DEC-21	Incorrect barrier/task priority	See Table I.8, PR-1, and PR-2.	See Table I.8, PR-1, and PR-2 for suggestions.
DEC-22	Incorrect SA-3-time projection	See Table I.11, SA3-7 to SA3-11.	See Table I.8, ATF-1, TL-1, MM-2 to MM-4, FIT-1 to FIT-3, WM-5 to WM-8, SK-1, Table I.9, SA1-6, and SA1-23 to SA1-25 for suggestions.
DEC-23	Degraded decision-making caused by degraded *psychological* state	See Table I.8, FIT-1, and FIT-2.	See Table I.8, FIT-1, and FIT-2 for suggestions.
DEC24	Degraded decision-making caused by degraded *physiological* state	See Table I.8, FIT-3.	See Table I.8, FIT-3 for suggestions

TABLE I.13
Cognitive Assessment Table: Act Phase Response

Item ID	Issue	Cognitive Mismatch/Issue Information		Possible corrective action changes to: Physical, Human, Organizational Elements
		Mismatch/Issue Source	Failure Mechanisms	
ACT-1	Act phase response	Incorrect goal selection	See Table I.8, GOA-1.	See Table I.8, GOA-1 for suggestions.
ACT-2	*Not performed*	Goal conflicts	See Table I.8, GOA-2.	See Table I.8, GOA-2 for suggestions.
ACT-3	(ACT-1 to ACT-7)	Effort not guided by barrier/ task goal	See Table I.8, GOA-3.	See Table I.8, GOA-3 for suggestions.
ACT-4		Uncertain/ ambiguous priority	See Table I.8, PR-1.	See Table I.8, PR-1 for suggestions.
ACT-5		Wrong priority	See Table I.8, PR-2.	See Table I.8, PR-2 for suggestions.
ACT-6		Forget to perform action (delayed or not started)	Prospective memory task Reliance on WM/STM to track time	See Table I.8, WM-1, and WM-2 for suggestions.
ACT-7		Personnel or equipment not available to perform actions	1. HE cannot reach required location 2. HE injured or fatality 3. Direct-use PE failed or not accessible 4. PE failed or not accessible	**OE1:** provide role backups for HE roles (e.g., barrier leader, others). **OE2:** procedures define specific roles, tasks, or task activities that the barrier leader may assign (on-the-spot) to an HE located in a remote location like a remote operations center. (See Appendices L.3.1/2, L.4 and C.4 for options and considerations.) **PE1:** provide redundancy/backups for direct-use and Support PE.
ACT-8	Act phase response *Not completed within the barrier/ task safety time (BST/TST)* (ACT-8 to ACT-13)	Decision made too late to complete action within BST/ TTRT	See Table I.12, DEC-18 to DEC-24.	See Table I.12, DEC-18 to DEC-24 for suggestions.

ID		Description		
ACT-9		Incorrect time information used when deciding to delay a future action (e.g., start the act phase response too late)	See Table I.12, DEC-19.	See Table I.12, DEC-19 for suggestions.
ACT-10		Uncertain or wrong priority	See Table I.8, PR-1, and PR-2.	See Table I.8, PR-1, and PR-2 for suggestions.
ACT-11		Forget to perform action (delayed or not started)	See Table I.12, DEC-19.	See Table I.12, DEC-19 for suggestions.
ACT-12		Internal or external attention capture	See Table I.8, ATT-2, and ATT-3.	See Table I.8, ATT-2, and ATT-3 for suggestions.
ACT-13		Skills needed to achieve response time not fully developed	See Table I.8, SK-1.	See Table I.8, SK-1 for suggestions.
ACT-14	Act phase response *Not performed correctly* (ACT-14 to ACT-26)	Targeting errors	Visual blockage or restricted movement contributes to the potential mis-selection or incorrect manipulation of a targeted device or object.	**PE1:** eliminate situations where the movement to reach the target device blocks the view or movement needed to concurrently monitor and select/use the act phase (direct-use) object. **PE2:** ensure Support PE does not inhibit/interfere with response object targeting/manipulation.
ACT-15		Degraded WM/STM capability/capacity causes response errors	See Table I.8, WM-5 to WM-8.	See Table I.8, WM-5 to WM-8 for suggestions.
ACT-16		Strong habit intrusion *Slip*	Error made in a required sequence	See Table I.8, SK-4 for suggestions.
ACT-17		Skill not fully developed	See Table I.8, SK-1.	See Table I.8, SK-1 for suggestions.

(Continued)

TABLE I.13
(Continued)

| Item ID | Issue | Cognitive Mismatch/Issue Information | | Possible corrective action changes to: |
		Mismatch/Issue Source	Failure Mechanisms	Physical, Human, Organizational Elements
ACT-18		Procedure use or selection deviation	See Table I.8, SK-2.	See Table I.8, SK-2 for suggestions.
ACT-19		Skill-based error *Includes skill fade*	See Table I.8, SK-3.	See Table I.8, SK-3 for suggestions.
ACT-20		Divided attention/task-switch error	See Table I.8, ATT-5.	See Table I.8, ATT-5 for suggestions.
ACT-21		Dual task interference	"Execution of one task may interfere with aspects of a second task due to "crosstalk" between the two tasks. This interference can occur in cognition or in coding a motor response" (NUREG 2016, A-51).	**OE1:** design tasks, functional allocations, and procedures to minimize the need for concurrent response actions that may be subject to crosstalk.
ACT-22		Support PE inhibits or interferes with response actions	Support PE physically or cognitively interferes with response actions	**PE1:** create remote monitoring task to monitor complex Support PE functions. (See detailed design step C3-8.) **PE2:** ensure Support PE does not obstruct essential sensory information or the performance of a response action.

Drift (appears between ACT-18 and ACT-19 rows, under Mismatch/Issue Source column)

ACT-23	Inadequate act phase monitoring, adjustment, modulation or correction *Act phase action requires real-time feedback and modulation/adjustment to respond to achieve the required result and response performance. (e.g., monitor the effect of a manually applied/directed fire suppressant.)*	Detect/decide/act phase failures in the monitoring/ modulating actions	**OE1:** provide the repetitive training and experience needed to develop an automatic skill. This skill provides a closed-loop, real-time monitoring, and modulated control action to achieve a required result. **OE2:** Consider opportunities to reduce the duration of the monitoring/corrective effort and action duration, or degree of precision required in those actions. **PE1:** seek a design that requires minimal physical movement (preferably limited to eye movement only) when monitoring response feedback or accessing/manipulating the response device/object during execution. (See Appendix K.2 for guidance and examples.) See Tables I.9, I.10, I.11, and I.12 for additional suggestions.
ACT-24	Wrong response to current mode	Did not check or unaware of current mode. Incorrect response action that does not apply to current mode.	**OE1:** training exercises to develop and maintain mode awareness and describe the consequence of actions that are not mode-correct/applicable. **OE2:** procedures define required response in each mode, with recommendations on how to remain aware of current mode. (Problems occur if actions are similar, though one is performed more often than others.) **PE1:** Provide shared VDU displays that constantly show current modes. **PE2:** design displays to query action before entry acceptance. Perhaps show the required action in this mode.

(Continued)

TABLE I.13
(Continued)

Item ID	Issue	Cognitive Mismatch/Issue Information		Possible corrective action changes to: Physical, **Human**, <u>Organizational</u> Elements
		Mismatch/Issue Source	Failure Mechanisms	
ACT-25		Incorrect information recorded on Incident Command Board (ICB) (See detailed design process C7-8.)	– Incorrect information conveyed or retrieved – Forget or misremember information before it is recorded – Error in the process of recording the information (e.g., interruption, distraction, crosstalk with local discussion) – Information recorded in the wrong location	**OE1:** develop and validate the skills of the scribe assigned to gather and record information on the ICB. (Recognize this activity as a uniquely challenging skill with many possible error entry points.) **OE2:** develop procedural standards on board layout, content, use of color, etc. (For background see the detailed design process C7-8, and Tables K.1, K.2, and K.5) **PE1:** add remote support barrier task to monitor information on ICB for correctness, accuracy, updates to date. (See Appendices L.3.1 and L.4 for options and considerations.) Also see Table I.8, SK-1 to SK-4 for suggestions.
ACT-26		Inappropriate entry device	Inconsistent coding of device contributes to use/entry errors. Coding does not aid in correct use/functioning of the device.	**OE1:** apply coding standards to data or user command entry devices/displays defined in the facility HF standards and HMI display standard/style guide. **PE1:** select the control entry device that best aligns with the entry type or function.
ACT-27	Act phase response	Message not conveyed	See ACT-1 to ACT-7.	See Table I.8, GOA-1 to GOA-3, PR-1, PR-2, WM-1, WM-2, and ACT-7 (this table) for suggestions.
ACT-28	*Information not communicated on time*	Message sent too late.	See ACT-8 to ACT-13.	See ACT-8 to ACT-13 for suggestions.

ACT-29	Act phase response *Error in communicated message* (ACT-29 to ACT-32) Deliver wrong information *Misunderstood request*	Attentional and other error sources may result in misunderstanding a written or verbal request for information.	**OE1:** communication procedures include a read-back response to all received information. **OE2:** procedures and training include a challenge/repeat request element if received information "does not sound right." **OE3:** provide training that demonstrates the consequences of errors in a conveyed message. **OE4:** barrier leader task monitors communications to detect and correct errors. **OE5:** review prior incidents where communications errors contributed to the incident or degraded the response. Assess possible changes to procedures or training. **PE1:** add a remote barrier support task to monitor communications for potential errors. (See Appendices L.3.1 and L.4 for options and considerations.)
ACT-30	Deliver wrong information *Unaware that conveyed information is incorrect*	Error in gathered or developed information to be conveyed. (This error may occur in the detect or decide phase. See the applicable error topics in Tables I.9, I.10, I.11, and I.12.)	
ACT-31	Deliver wrong information *Verbal Slip*	Slip-based error in conveyed message (e.g., similar sounding words or word segments) (Reason, 1990, p. 107).	
ACT-32	Dichotic listening introduces incorrect information in verbally conveyed message	Error occurs when relaying (repeating) verbally received information from a primary source and a secondary conversation occurs nearby. *Information/words from the secondary source are accidentally inserted into the conveyed message.*	**OE1:** provide training that demonstrates this potential error source and the recommended methods to reduce its likelihood. See ACT-31 for additional suggestions.

(Continued)

TABLE I.13
(Continued)

| Item ID | Cognitive Mismatch/Issue Information | | | Possible corrective action changes to: |
	Issue	Mismatch/Issue Source	Failure Mechanisms	Physical, **H**uman, **O**rganizational Elements
ACT-33	Act phase response *Other communications issues* (ACT-33 to ACT-35)	Communications message or duration too long	Captures attention of senders/receivers and prevents/limits work on other barrier/task activities. Attempts to multi-task (divided attention/ task-switch) may introduce a wide range of possible errors. (See Table I.8, ATT-5.)	**OE1:** communications procedures to define protocols and terms to minimize communication duration and content. **OE2:** repetitive training designed to develop a skilled action that delivers a concise message that is appropriately timed and delivered.
ACT-34		Inappropriate communication from a non-valid source	Communications from non-authorized (e.g., not barrier assigned) persons. *Non-barrier personnel using barrier team radio communication channels introduce misinformation, disrupt team coordination, etc.*	**OE1:** procedures create communication discipline that limits this occurrence. **OE2:** train barrier leader to monitor and control/halt unauthorized communications.
ACT-35		Diction, pace, use of terms not clear	Verbal diction not clear. Speaking too fast. Unclear use of terms.	**OE1:** perform training exercises to practice communications in time-pressured environment. Purpose: improve communication clarity, diction, pacing, tone, and correct use of terms. *See Table I.8, SK-1 for additional suggestions.*

TABLE I.14

Cognitive Assessment Table: Teams and Teamworking

ID	Issue	Cognitive Mismatch/Issue Information		Possible corrective action changes to: Physical, Human, Organizational Elements
		Mismatch/Issue Source	Failure Mechanisms	
SSA-1	Team SSA deficiencies (SSA-1 to SSA-5)	Team members do not have a shared understanding of the roles and interactions	One or more team members (HE) missing a minimum shared MM	**OE1:** conduct team drills to develop the required shared SA mental models/LTMs **OE2:** perform training drills that demonstrate the roles and needs of each member and the required interactions.
SSA-2		Errors in communications	See Table I.9, SA1-23, Table I.13, ACT-29 to ACT-32.	See Table I.9, SA1-23, Table I.13, ACT-29 to ACT-32 for suggestions.
SSA-3		Poorly timed communications	See Table I.9, SA1-26, Table I.13, ACT-27 to ACT-28.	See Table I.8, GOA-1 to GOA-3, PR-1, PR-2, ATF-1 to ATT-3, SK-1, WM-1 to WM-8, FIT-1 to FIT-3 Table I.9, SA1-6, SA1-23 to SA1-25, Table I.12, DEC-19, and DEC-20, Table I.13, ACT-7 for suggestions.
SSA-4		Unauthorized communications	See Table I.13, ACT-34	See Table I.13, ACT-34 for suggestions.
SSA-5		Errors in other shared information sources (e.g., the Incident Command Board)	See Table I.9, SA1-24 and SA1-25, and Table I.13, ACT-25, and ACT-26	See Table I.9, SA1-24 and SA1-25, and Table I.13, ACT-25, and ACT-26 for suggestions.
TE-1	Barrier leader deficiencies (TE-1 to TE-3)	Monitor communications *Communication errors*	Inadequate monitoring of deficient communications and its effects Not taking timely and necessary actions to correct communication issues and the effects of those communications	See SSA-1 for suggestions (this table). Also see Table I.9, SA1-14, SA1-23, and Table I.13, ACT-27 to ACT-35 for additional suggestions.

(Continued)

TABLE I.14
Continued

ID	Issue	Cognitive Mismatch/Issue Information		Possible corrective action changes to: Physical, Human, Organizational Elements
		Mismatch/Issue Source	Failure Mechanisms	
TE-2		Barrier leader-to-team communication *One or more HEs not aligned to barrier goal, priorities, requests, or instruction based on communicated information*	**Barrier leader monitors communication for team members who:** – Do not appear to know correct goal or priority – Conveys incorrect information – Conveys information at the wrong time (e.g., not often enough, too early to be remembered) – Indicates the message was misheard or misunderstood information	**OE1:** for emergency response and an exceptionally high workload environment, employ advanced training (e.g., see OPITO 2014). **OE2:** provide team exercises and drills to develop and demonstrate conveyance and alignment to conveyed goals and priorities. **OE3:** barrier leader – Develop and demonstrate advanced skills for monitoring and responding to errors in team communications.
TE-3		Monitor resources *Excessive workload or degrading or soon-to-be-depleted resources places the barrier/ task at risk*	Inadequate awareness of barrier system resource status and utilization rate Failure to make timely and necessary adjustments in resources when needed	**OE1:** train barrier leader to monitor resources (HE and PE) for excessive workload that may place the barrier at risk. Include exercises to practice/demonstrate leader changes to team activities when needed to complete work within the remaining available time.

			Mitigation
TE-3 (cont'd)			**OE2:** procedures define barrier leader options to shift selected tasks or task activities to other HEs who may have a lower workload. (The other HEs should be trained and verified competent on that task.) **PE1:** task remote support task to monitor barrier system resources and dependent systems when approaching depletion of capacity limits. (See Appendices L.3.1 and L.4 for suggestions and considerations.)
TE-4	Monitor team for excessive fatigue and stress *Excessive fatigue/stress degrades cognitive capabilities/capacity*	Inadequate monitoring of excessive stresses and fatigue in self and others. Failure to mitigate excessive fatigue or stress when options are available	See Table I.8, FIT-1 to FIT-3 for suggestions.
TE-5	Barrier team (HE) deficiencies (TE-4 to TE-6) Excessive communications create unnecessary workload, degrade capability	Inadequate use/development of implicit coordination	**OE1:** procedures define communications and target reliance on implicit coordination. *See Table I.8, SK-1 to SK3 and TL-1, SSA-1 (this table), and Table I.9, SA1-13 for additional suggestions.*
TE-6	HE activities not adequately coordinated, contributing to degraded barrier/task performance	Inadequate communications to achieve the necessary coordination Shared displays used to coordinate actions missing information, or not available	**OE1:** procedures define communications and target level of reliance on explicit coordination (e.g., function, timing, and content). *See Table I.8, SK-1 to SK3 and TL-1, SSA-1 to SSA-3 (this table), Table I.9, SA1-6, SA1-18 to SA1-27, and Table I.13, ACT-27, and ACT-28 for additional suggestions.*

I.4 EXECUTION CONSIDERATIONS

I.4.1 ASSESSMENT PARTICIPANTS

Table A.2 (Appendix A.2.3) identifies the suggested participants in this process.

I.4.2 EXPERTISE REQUIREMENTS

At least one participant in the assessment and corrective action selection process should have the necessary cognitive system engineering/human factor experience and execution skills needed to support and guide these activities. This person should have a high or expert level of knowledge of the information in Appendix F and in the cognitive assessment tables provided in Section I.3.

I.4.3 POTENTIAL EXECUTION CHALLENGES

If applied to a major capital project, the execution challenges in implementing these processes may be significant in terms of the following:

1. Requirements for new expertise (e.g., cognitive system engineer or equivalent).
2. Budget increase to address the new work.
3. Schedule challenges to integrate these new activities into an often-compressed project schedule.
4. Support may be needed from the appropriate operations personnel if the team seeks additional information or a clarification on a task phase activity or physical or organizational element (e.g., a procedure, training detail, Support PE, HMI display, alternative display type, or support aid).
5. A process to review, approve (or modify/change), design, and monitor the follow-up and implementation of the approved corrective actions.

Appendix J
Human and Organizational Elements

This appendix is organized as follows:

- J.1 Overview and Introduction
- J.2 Procedures
- J.3 Training
- J.4 Competency Requirements and Management Systems
- J.5 HE Staff and Staffing Policies
- J.6 Optimizing Human Performance (Minimize Task Execution Time)
- J.7 Skill Fade and Drift

J.1 OVERVIEW AND INTRODUCTION

Figure J.1 shows the relationship between the organizational elements addressed in the lifecycle model presented in this book. The steps and processes are noted in the upper left-hand corner. The figure highlights the traceable linkage between the indicated barrier activities and human and organizational elements. The tasks defined in the task analysis provide the basis for defining the procedures, skills, and knowledge requirements. The requirements then drive procedure and training development. Owner/operator programs provide consistent framing and template inputs to procedure and training development and competency management. The hashed box (drift, skill fade) indicates the slow-developing (and typically undetected) changes in skills, mental models, and procedure usage that degrade barrier performance over time. Operate and maintain phase processes (monitoring, periodic refresh training, and competency assessments) identify and correct these deviations before they contribute to barrier failure.

Note: The prototype lifecycle model does not provide comprehensive guidance on many aspects of procedures, training, and competency management systems. Instead, it shows where and how these elements fit into a lifecycle model and how processes can work to prevent or minimize barrier system deficiencies that can lead to failure. The inclusion of end-users in these activities (e.g., the barrier leader) contributes to closing the gap between work-as-imagined and work-as-done.

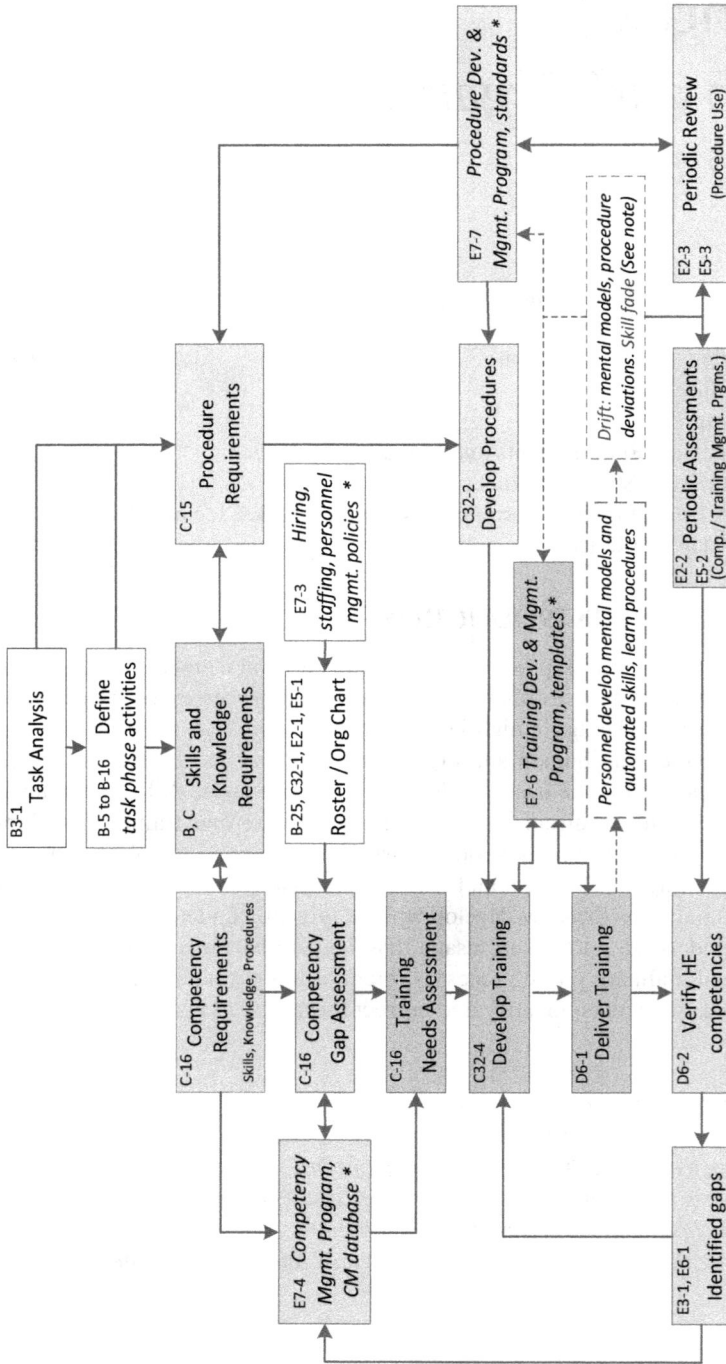

FIGURE J.1   Relationships: Skills, Knowledge, Procedures, Training, and Competencies.

*Owner/operator programs, policies, standards, and templates
Note: For information on drift and skill fade, see Appendix J.7.

TABLE J.1
Organizational Elements: Procedures

Procedure			Identification and Requirements		Write and Approve			
Type	Application	Source	Required Procedure	Purpose, Content, Format (Note 2)	Select Author/ Provider (Note 2)	Write and Review (Note 2)	Approve	Monitor Procedure
Operate barrier system	Task Start-up, Normal/ abnormal operation, Shutdown	Table 4.5	C15-2a	C32-2a	C32-2b	C32-2c/d		E2-3
	Component (direct-use, Support PE)	Table B.5	CX-0.2a	CX-0.2a	CX-0.2a	CX-0.2a C32-2c/d	C32-3 (all)	
Maintain task	Preventive proof test Repair & restore	Table 4.30	C15-2b (see Note 1)	C32-2a	C32-2b	C32-2c/d		E5-3
	Component (direct-use, Support PE)	Table B.6	CX-0.2b (see Note 1)	CX-0.2a	CX-0.2b	CX-0.2b C32-2c/d		

Note 1: Use a task analysis to identify safety critical tasks that may require dedicated procedures.

Note 2: Procedures may be written and provided by the Owner/operator or others, for example, included in vendor purchase order or contract for a barrier system element.

J.2 PROCEDURES

J.2.1 DEVELOPMENT OVERVIEW

The lifecycle model processes address procedure requirements, development, content, approval, and monitoring. The following are the included activities based on Stanton et al. (2010, Ch. 4), CCPS (2022, Part 2), and CCPS (2007a, Ch. 22):

1. Identify tasks that require procedures
2. Determine and select the procedure content, format/type, format, and level of detail
3. Write and review the procedure
4. Approve the procedure
5. Provide training on the procedure (see J.3 for detail)
6. Monitor procedures for effectiveness and correct use. Update when an issue is found. (Consider: procedures that do not change should still be periodically re-published with the new date. This notifies the user that the procedure is up to date.)

Procedure-related activities are guided by the owner/operator's Procedure Development and Management System (PDMS), templates, and standards. Additional requirements may be included in the barrier safety management plan (BSMP). Table J.1 provides an overview of the lifecycle processes that address each activity.

Regarding *Item 1* above, the required procedures are captured in Tables 4.5, 4.30, B.5, and B.6, and the procedure plan information is compiled into Table 4.31. Procedures define barrier task actions and how those actions should be performed.

Note: The prototype model identifies maintenance procedures. The industry guidance standard EI (20202a) presents a different approach that may be used to identify and define safety critical maintenance tasks, a prerequisite to defining the procedures needed to support these tasks.

Item 2 examines the task activities and target task safety time to identify the appropriate content and level of detail. From that information, select the procedure template from the standard templates in the PDMS. Emergency response barriers tend to be time-of-the-essence, suggesting a procedure type that can be quickly scanned and applied. For further guidance on procedure types and selection criteria, see CCPS (2022, Part 2).

Note: Task phase and team detailed design processes C-3 to C-8 define support aid requirements. Review these requirements to understand the relationship and allocation of information between procedures and support aids. (See Appendix B.8 for further information on support aids.)

Item 3 identifies the procedure writer who develops the draft procedure for review. End-users and other competent (verified) reviewers review draft procedures. Each activity is guided by the requirements in the PDMS and BSMP.

For *Item 4*, the final procedure is approved by the designated (competent) persons based on the guidance provided in the PDMS and BSMP.

For *Item 5*, as noted in Table J.1, barrier-assigned personnel receive the appropriate (purposely developed) training on each procedure.

For *Item 6*, monitoring processes in the operate and maintain phase seek to identify deviations in procedure use or a procedure that may need to be revised. Identified in Figure J.1, drift can result from seemingly minor (or expedient) unsafe deviations occur, continue and become normal practice over time. Drift in the appropriate use of procedures may be isolated (the product of individual behaviors, attitudes, or training problem) or chronic (more widespread). The later may reflect a broader issue with procedure validity or usability, trust, safety culture, procedure discipline, or insufficient time is available to use procedures.

Suggested Reading: for additional guidance on procedures and procedure development, see CCPS (2007a, Ch. 22), Stanton et al. (2010, Ch. 4), and Edmonds (2016, Ch. 17). Additional resources include governmental and regulatory authorities, industry organizations (e.g., COS 2020 and HPOG 2021), professional societies, and academic institutions (e.g., Hendricks et al. 2021).

J.2.2 Execution Planning and Guidance

J.2.2.1 Personnel and Competency Requirements

The owner/operator PDMS and CMS, and the BSMP may define competency requirements for the personnel assigned to each procedure activity. Personnel having the required competencies should be available to perform these roles.

J.2.2.2 Scheduling

The activities in Table J.1 should be scheduled to meet key procedure development, training, and the affected project milestones. Input considerations include the following:

- Example milestone dates:
  - Target date to receive procedures provided by others
  - Target procedure approval dates (the approved procedure may be a prerequisite to training module development, training delivery, etc.)
  - Target date for placing the barrier into service
- Schedule activity duration: procedure development, review, and approval

J.3 TRAINING

J.3.1 Development Overview

Training is one of the essential tools used to develop and achieve a barrier-required skill or knowledge competency. (Others include on-the-job experience, supervision with timely feedback, and incident discussions.) The training methods used to develop and achieve a skill competency typically differ from those used to develop and achieve a knowledge competency. As shown in Figure F.2 (Appendix F.2), skill competencies include task execution, teamworking (non-technical skills), and the skilled use of equipment or a device. A skill develops as an automatic (cognitive) process through repetition and kinetic/interactive training. Knowledge competencies include knowledge of procedures, task execution, and technical information and systems. This competency relies on developing and storing this knowledge in one's mental models and other long-term memories. Knowledge-based training (different training methods), experience (varies by knowledge types and depth), and time are needed to develop the necessary mental models and other long-term memories. This knowledge must be sufficiently complete and accurate to adequately guide barrier activities such as detection, recognition, comprehension, anticipation, and decision-making. The selected training approach should be appropriate to the competency type and training objectives.

Note: As detailed in Section J.7.1, infrequently used complex skills like task execution and teamworking tend to fade over time. (The term for this process is skill fade.) As detailed in Section J.7.2, drift is a different process that also occurs over time. As an example, a verified competent person develops an incorrect understanding or behavior that is an unsafe deviation from a required knowledge competency.

As discussed in Appendix F.5, long-term memory can change with every recall. A potential change attributed to that recall, which tends to occur slowly over time, may have a positive or negative effect on the barrier function and performance. Complex skills (e.g., teamworking or complex communication interactions), information stored in long-term memories (e.g., knowledge and beliefs), and memory-driven behaviors are subject to this memory drift effect. Thus, periodic competency re-assessments and refresher training (included in the lifecycle model) are required to identify and address skill fade and drift.

The training activities included in the lifecycle model are guided by Stanton et al. (2010, pp. 43–45), CCPS (2007a, pp. 78–81), and Edmonds (2016, pp. 388–394). Table J.2 provides an overview of the lifecycle model processes that address each activity. For additional information and guidance, see Salvendy (2012, Part 3, Chapter 17), Design, Delivery, Evaluation, and Transfer of Training Systems.

The *design* of this organizational element (training) begins with a *competency gap assessment*. The assessment seeks to identify a gap between a barrier-required competency and the current competency of the barrier-assigned HE or maintenance personnel. An identified gap provides input to a *training needs assessment* process. This process identifies if/how training can close a competency gap. The results are then used to guide the remaining training activities in the lifecycle model. These activities provide the required input to the following training activities:

1. *Training design:* review the training options to address an identified competency gap. Review the practical considerations (e.g., available time and resources, trainee characteristics).
   The CCPS (2007a, p. 79) advises, "The actions required of trainees when performing a task should be specified, including the conditions under which the task is performed and the standards that must be met for successfully performance."

2. *Training content:* Identify the training content that addresses the skill or knowledge competency gap.
   To this end, CCPS (2007a, p. 79) recommends the selection of the "best available subject matter experts who are doers rather than supervisors should be consulted and managers should help determine content."

3. *Training delivery methods:* select the appropriate training method (e.g., an interactive exercise, drill, simulation, or lecture). Refer to the owner/operator's Training Development and Management System (TDMS), training examples and templates, and the BSMP for potential guidance.
   Regarding methods, Edmonds (2016, p. 393) notes,

   There are different levels of learning: individual, team and organizational … Learning is most effective when the learning method is active, (i.e., one that gets people involved) and when the type of experience is direct (i.e., the person is involved in the experience). However, indirect methods can be effective if participants are engaged with interpreting and understanding cause and effect relationships.

IOGP (2019a/b) provides example training drills, exercises, and content for developing the technical and non-technical skills of drilling rig personnel in the O&G industry. Also see Radvansky (2021, pp. 182–189) for additional information on skill development training.

4. *Resource planning and activity scheduling:* develop a schedule for all training activities that include key milestone dates and realistic activity durations. Select the training developer and the person(s) who will deliver the training. Schedule these and other resources (e.g., the trainees, and the training venue, equipment, and materials.)
5. *Develop the training modules and material:* The TDMS and BSMP may provide guidance and the requirements.
6. *Review and approve the training module:* The TDMS may provide guidance and the requirements such as who is authorized to review and approve training modules and their associated required competencies.
7. *Training results:* Gain feedback from those taking the training to determine whether changes may be needed. The TDMS may provide a training validation process and timing, and how to address the findings.
8. *Ongoing training validation:* according to CCPS (2007a, p. 80), "The performance of representative trainees on the job should be observed and any performance problems identified. Training should be modified accordingly."

Note: The above is an abbreviated list of training activities. For a more complete description and discussion, see the previously mentioned references.

J.3.2   EXECUTION PLANNING

J.3.2.1 Personnel and Competency Requirements

The owner/operator PDMS and BSMP should define the competency requirements for personnel assigned to each training activity. Personnel having those competencies must also be available to fill these roles.

J.3.2.2 Scheduling

Schedule considerations for training based on the activities indicated in Table J.2 include the following:

Key milestone dates:
- Complete the training design, identify its content, and select the training method
- Complete the training module and material development
- Complete the training module review and approval processes
- Target training completion date
- Complete the required competency verifications

Schedule duration:
- Training development, review, and approval
- Time required to complete all training for barrier-assigned HEs (each barrier)

TABLE J.2
Implementation of organizational elements: Training, Owner/Operator Provided

		Training Identification and Requirements				Develop/Delivery Training			
Type	Applies	Training Req. and Objectives	Training Design	Select Method	ID Trainees	Develop Training	Review/ Approve	Deliver Training	Refresher Training
Operate Train HEs on barrier-assigned tasks	Task	C16-2	C32-4	C32-4		C32-4		D6-1	
	Component	CX-0.2c *See Notes 1, 2*	CX-0.2c *See Note 2*	CX-0.2c *See Note 2*		CX-0.2c *See Note 2*		D6-1 *See Note 2*	E2-2
Maintain Train maint. Personnel on barrier safety critical tasks	Task, System, Equip.	C16-2 *See Note 1*	C32-4	C32-4	C32-4	C32-4	C32-4g (all)	D6-1	E5-2
	Component	CX-0.2c *See Notes 1, 2*	CX-0.2c *See Note 2*	CX-0.2c *See Note 2*		CX-0.2c *See Note 2*		D6-1 *See Note 2*	

Note 1: See Note 1 in the process C-16 introduction. Note 2: Training may be provided by others, for example, included in a purchase order for vendor-provided equipment or systems. See CX-0.2c (App. B-6) for additional information.

- Time required to complete training for barrier-assigned maintenance personnel
- Time required to complete the verification process
- Known or likely delays in gaining access/participation to trainees, trainers, training materials (including the new/approved procedures), and other required training resources
- Planning should achieve the timing (with contingencies) that ensures these activities are aligned with the overall project schedule milestones.

J.4 COMPETENCY REQUIREMENTS AND MANAGEMENT SYSTEMS

Figure J.1 identifies some of the competency-related activities in the lifecycle model. Tables 3.2 and 4.1 identify documents that capture new skill and knowledge requirements (if any), and the steps and processes that defined those requirements.

Note: This aspect of the prototype lifecycle model may set it apart from current practice. Defined competencies are specific to a task activity, a marked difference from assigning a person to a barrier task based on a job classification or supervisor recommendation. This new and specific information is available to guide staff selection, procedures, support aid content, training content and type, and competency assessments.

Note: The competency processes in the prototype model align with HSE 2007a. See that reference for additional detail.

As Figure J.1 shows, the prototype model assumes the owner/operator has an active and viable competency management system (CMS). The following are the expected features of a CMS:

- Maintains an active and managed database of the competencies required for each barrier task/role.
- Tracks and maintains complete records of the competency verification assessment completions and results (each barrier-assigned HE).
- Generates alerts and reports that identify each required competency that is not achieved and verified. For each unmet/unverified competency, the report should identify who, which barrier, and the length of time in this state. (The status may provide important input to a real-time barrier health status indicator.)
- For periodic competency updates, the CMS should generate alerts and reports on all required assessments including planned, update required, completed, results, and overdue assessments and duration.

Note: Operating a barrier when a competency requirement is not met is a latent deficiency that places the barrier at risk. This can occur if a required skill or knowledge competency is not identified, tracked, and managed in a competency management system. (The latter indicates the importance of a functioning CMS.) For additional guidance on competency management and CMS, see CCPS (2022, Part 4), Edmonds (2016, Chapter 20.2), HSE (2007 a/b), COS (2013 Section 8), and Stanton et al. (2010, Ch. 2).

J.5   HE STAFF AND STAFFING POLICIES

The barrier roster and organization chart should be a controlled document. The information in J.2 and J.3 highlights the reasons for doing so. A proposed roster change should first seek to verify that the proposed person meets the competency requirements for the proposed role(s), and preferably has obtained the required competencies before the change is made. The same applies if a named designated backup is changed. (This becomes more difficult in facilities with lean staffing or events (planned or unplanned) are underway that places additional demands on existing staff.) If the proposed person lacks required competencies, identify the contingencies or mitigation solutions that should be in place pending achievement of those competencies. A change in staff rotations or overtime may lead to conditions that exceed the maximum fatigue target identified in step B-26 and recorded in Table 3.34. The BSMP should address how to manage these and similar changes and identify conditions that require using the change management process.

J.6   OPTIMIZING HUMAN PERFORMANCE (MINIMIZE TASK EXECUTION TIME)

As discussed in Appendix B.5.3, time is the resource that often presents the greatest challenges. Table J.3 suggests tools and methods to reduce the time needed

TABLE J.3

Example Methods to Optimize (Reduce) Human Response Time

Barrier/Task Phase	Optimization Approach	Solution Areas	Comments
Detect SA-1	Minimize time to detect and accurately perceive the barrier/task activator, other SA-1 information sources.	• Enhance MMs to improve top-down scanning (automatic process) to reduce the time needed to find/detect SA-1 information. • Enhance SA-1 object salience. Reduce salience of less-important information. • Reduce workload if excessive (e.g., re-allocated task functions, simplify tasks, develop procedures that enhance the use of skill-based actions, reduce sources of distractions and interruptions, increase reliance on implicit coordination (if appropriate)). • Prevent excess fatigue.	See Appendices B.5.1, B.5.2, and K for additional information and guidance.
Detect SA-2	Minimize the time to comprehend the SA-1 information and situation.	• Use alternative display types that provide direct comprehension with minimal effort. • Provide additional training to develop a more advanced mental model. • Provide support aids.	See the notes in detailed design step C5-1c on alternative display design. See App. B.8 for support aids.
Detect SA-3	Exploit projection/anticipation capabilities that support proactive readiness and planning for potential events, decisions, etc.	• Increase the experience/expertise requirements in the task-specific domain (develop more advanced MMs). • Increase the frequency of team exercises to increase the capability to anticipate the needs of others. • Add support aids to support improved projection/anticipation results (e.g., time-displays, troubleshooting charts). • Modify procedures to take pre-emptive actions rather than wait (e.g., for a muster barrier, initiate emergency response actions begin to preemptive actions before everyone reports in).	See App. B.8 for support aids.

Decide	Minimize the time to complete the decision phase.	• Training and drills that focus on complex decision-making • Staff selection (e.g., increase the aptitude requirement) • Modify task to simplify decision and decision-making • Add support aids (e.g., a decision tree or diagnostic flowchart) • Define an increased experience and expertise requirement allowing for increased use of the recognition-primed decision (RPD) model. See Appendix I.3.5 for additional information. • Minimize conditions and effects of performance influencing factors (e.g., threat stress and social influences; see Appendix H.2, Table H.1)	See App. B.8 for support aids.
Team SSA	Minimize the time required to maintain adequate individual and collective SSA.	• Procedures should eliminate non-essential communication. • Minimize communication exchange durations by increasing short-cut code words that increase accuracy and brevity. (Employ standardized terms with clear meanings, employ communication discipline, etc.) • Enhance the use of implicit coordination while the non-essential reliance on explicit coordination. Limit 2-way communication exchanges to only those required to maintain coordination, make changes, etc. (See Appendix C.3 for information on implicit and explicit coordination.)	See Apps. K.3, K.5 for communications, information guidance.
Team coordination & cohesion	Minimize time needed to maintain team coordination and cohesion.	See Team SSA	—
All	Minimize time needed to maintain team morale and manage stress in self and others.	• Increase barrier leadership training and skills to enhance and manage morale and motivation. • Add exercises to improve HE capability to monitor stress in self and others through increased awareness (skill). • Examine means to improve team trust.	See Table I.14 (TE-1), and Appendices C.4 and F.4 for additional information.

(Continued)

TABLE J.3
(Continued)

Barrier/Task Phase	Optimization Approach	Solution Areas	Comments
All	Pre-position supplementary resources to ensure they are immediately available if needed.	• Exploit opportunities that may be possible by adding a remote barrier support task. • Exploit options that may be available as an *Additional Resource*.	See App. L.3, preliminary design step B20-4 and detailed design step C14-5 for background information.
All	Minimize task execution by eliminating unnecessary linkages and coordination points between HEs and HE tasks.	• For the muster barrier, begin follow-up tasks and additional barrier activations before all personnel report it. • Procedures: pre-approve actions after a fixed time. • Task design and procedures: minimize the number of tasks that must be executed in sequence. • Task design and procedures: minimize the number of coordination points affecting multiple HEs and tasks. (Decouple tasks to the extent possible. The time needed to perform linked or coordinated actions may exceed the time needed to perform individual actions.)	See App. H.2 and Table H.1 (task complexity PIF) for additional information.
Action	Minimize time needed to perform an action response.	• Minimize travel distances and other unnecessary time consumers that delay starting/performing a required time-sensitive action. • Minimize task steps (e.g., number, complexity). • Minimize the physical effort required to perform the action. • Maximize physical ergonomics, i.e., the optimal selection and application or use of a direct-use physical element. • Add support aids (e.g., use a checklist to prevent errors that require time-consuming recovery effort). • Modify tasks to increase the use of skilled (automatic) actions. • Exploit automation where it can be selectively and reliably applied to reduce barrier, task, and task phase response times.	For additional detail, see App. H.2 and Table H.1 (task complexity PIF), App B.8 for support aids, and Apps. K.1 and K.2 (workspace design).

to progress a task or task activity or improve the utilization of the available time. (Additional examples are included throughout this book.)

J.7 SKILL FADE AND DRIFT

Many factors affect human performance in the operate and maintain phase, the longest period in the active human barrier lifecycle. A barrier change (positive or negative) can result from changes to owner/operator policies and programs, personnel, personnel management, budgets, operations, and maintenance (O&M) practices, and modifications to facility systems and processes. Changes to barrier performance can also result from **skill fade** (Wickens et al. 2013, pp. 242–243; CCPS 2022, p. 159 and para 22.3.2) and **drift**.

J.7.1 SKILL FADE

From CAA (2021):

> *'Skill fade' means that skills have decayed over time because they have not been used. Personnel may experience this as: finding it harder to remember and carry out some tasks, carrying them out in the wrong sequence, doing the wrong task at the right time or the right task at the wrong time, forgetting whole or parts of tasks and processes, or failing to realise that they have carried out a task incorrectly. If personnel realise that their skills have decayed this may make them feel frustrated or anxious.*

CCPS (2022, p. 150) recognizes skill fade as a known source of competency degradation. "Even with acquired learning, if there is not enough opportunity to put the new competency into practice, an individual's ability to do the task to the required standards decrease over time."

According to Wickens et al. (2013, pp. 242–243),

> *sometimes the problem of skill forgetting is a substantial one, in particular when the person did not thoroughly learn the skill in the first place, or when the person has only limited opportunity to practice it (e.g., first aid procedures).*

Refer to Chapter 2.1 for the employed definition of the term *skill fade*. A skill is a memorized activity that may be psychomotor only, cognitive only, or a combination of the two. Other sections in this book identify the value of modifying a task where it becomes possible to frame activities in forms that can be learned as a skill-based approach. Doing so reduces the cognitive demand on working memory/short-term memory. Now look at this from a different perspective. In current practice, a degradation in physical element performance is viewed as a risk to the barrier function and/or performance. The same view should be taken for a specified skill that is required to achieve the barrier function at the specified performance. Industry standards and practice methodically address the prior (PE issues) but less so address the latter (skill fade).

Every skill is subject to fade (decay) over time. How quickly this occurs depends on the skill type, the training type, and the frequency of actual use. From NATO (2023, Table 4-1), psychomotor skills that do not have a complex procedural element (riding a bike) are commonly retained for long periods of time. Conversely, the skill required to perform a challenging procedure or task execution activity can begin to degrade in 5 months or less if not used or refreshed. The degradation process can begin as early as 2 months with complex skills such as those common to complex procedures or barrier team communications and coordination. An example of the former is a procedure that has many sequential steps (e.g., ten steps) that must be performed in the correct sequence or at a specific time. Attempting to do so from memory, given the many possible sources of interference (situational), increases the likelihood of a procedural error. Highly experienced personnel are more likely to perform the procedure from memory, especially if it is a very detailed, step-by-step, check-the-box sequence (a type well suited to a lesser experienced person). For this reason, some organizations are exploring procedure management systems that print or display the procedure version that is more optimal for the identified user (e.g., a detailed version if less experienced and an abbreviated version for a highly experienced person.)

From NATO (2023, Annex B), as potential predictors of skill fade, fade may occur sooner in individuals given the following influencing factors:

- Increased age
- Increased interval from the last time the skill was used, or refresher training occurred
- Lower inherent cognitive capabilities (lower innate WM/STM span and capability)
- Lower innate motor skills
- Less developed belief in one's ability
- Reduced motivation
- Lower quality of the initial learning
- Reduced levels of emotional control and stability
- Increased levels of uncontrolled stress
- Less experience and levels of expertise
- Increased number of tasks or steps within a task (procedural knowledge)
- Learning events are closely spaced (short intervals between skill development training events)
- Addition of team interaction/interdependency tasks
- Little to no use of mental rehearsal
- Training provides little or no constructive, timely, and meaningful feedback
- Training exercise provides few environmental cues (cues common to the actual skill use environment)
- Training exercise performed at a slower pace than occurs in actual use
- Training material is printed (as opposed to audiovisual or VR – virtual reality materials)

Note: For a comprehensive and advanced source of information on this topic, see NATO (2023).

J.7.2 DRIFT

CCPS (2022, p. 150) identifies drift as a known source of competency degradation. This source states, "Even individuals who are experienced and knowledgeable start to 'do things their own way' – over time. This behavior can drift into being unsafe and need to be recognized when carrying out reviews."

Refer to Chapter 2.1 for the model employed definition for the term *drift*. With physical devices (e.g., an instrument transmitter or control valve), industry practice and standards commonly address the monitoring and corrective actions needed to detect drift. However, there is less awareness and effort allocated to detecting drift in the actions and behaviors of the personnel assigned to perform a barrier system function or maintenance activity. Drift may begin with a deviation (perhaps seemingly minor or perceived to be a necessary expedient at that moment) that initially takes place (and nothing bad happens) but then is repeated to become normal practice. Common examples include a deviation in the use of a procedure or using a safety system bypass/inhibit feature for an operational purpose (though the basis used to guide its design was limited to defined maintenance activities). Drift can also occur when one's in-the-head knowledge (long-term memories and mental models) is changed by an evolving deficiency or error in understanding (technical, procedural, task execution) or belief. With procedures, drift begins when a person first deviates from a documented procedure. Left unchecked, that deviation may be adopted by others. Repeated over time, it becomes normalized and is no longer noticed. Commonly a slowly evolving process, drift can be difficult to detect if not intentionally looked for using the appropriate methods. (From Chapter 7, the O&M monitoring processes E2-3, E2-5, E5-3, and E5-5 should be designed to address each of these scenarios.)

Humans possess an alert system that triggers an awareness of a change if that change is a significant. An example is an observed and obvious deviation from a trained-in safety practice or expected behavior. If not a high reliability organization (an HRO), the normal alert sensitivity tends to be more limited and less acute meaning the awareness occurs only in response to an obvious or significant change. A problem also occurs when a significant behavior or deviation (originally detected but not acted on) repeats over time without consequence. An inherent trait in the human memory system, this repetition becomes normalized within the long-term memory system, a hidden process that desensitizes the alert capability over time until it no longer triggers an awareness of that particular event or behavior. The deviation becomes normalized (Decker 2011, p. 106). As others experience the same events (continued deviation without consequence), this dampening effect may develop in others resulting in an organization-level drift. (Another term for this process and behavior is normalization of deviance.)

Appendix K
Workspace Development and Design

This appendix is organized as follows:

- K.1 Common Workspace Design Elements and Principles
- K.2 Physical Element Workspace Design
- K.3 Communication Workspace Design
- K.4 Human Element Workspace Design
- K.5 Information Workspace Design

This appendix provides guidance on workspace design, a topic addressed in many published guidance and practice standards. It places an increased focus on selective areas of cognitive and human factors and provides additional guidance specific to the more nuanced limitations in the human visual system. The modified processes are intended to supplement and complement current practice.

A workspace may be physical, communications, human or information. Being unique, each workspace needs different design methods and tools, and creates a different vantage point for examining the design and design information. The design of each workspace type affects the barrier function and performance in terms of its reliability, availability, integrity, and survivability. It may also affect personal safety, morale, behavior, and job satisfaction.

- *Physical Workspace:* in this workspace, personnel (HEs and maintenance) performing barrier activities interact with other HEs and barrier physical elements. Examples of the latter include a single backlit pushbutton, an HMI display, a control console or panel, a VDU-based display wall, a room, or an engineered area.
- *Communications Workspace:* this workspace is bounded by the communication purpose, message content, structure, medium, timing, and the intended senders and receivers. Procedures, communication skills, and non-technical skills (e.g., teamworking) define and create the workspace. Here, the design affects the coordination, control, cohesion, morale, and trust within the barrier team.
- *Human Workspace:* the boundary of this workspace is the barrier-assigned person. The barrier-required capabilities developed in this workspace take the form of skills, mental models, other long-term memories, and behaviors. (Much of the lifecycle model focuses on this workspace.)

- *Information Workspace:* the boundary of this workspace is the information that exists in the physical, communications, and human workspaces. Some of this information is required by the design of the barrier system. Other information exists, some of which is unhelpful and may contribute to distractions and slow access to the required information. The focus is on information properties, namely, its validity, accuracy, completeness, latency, timeliness, and value over time, and what sense provides access to that information. These properties provide insight to consider in the design process, such as a need to retain information for later, on-demand recall. Much of the information has one or more dynamic attributes, for example, how frequently it changes, the nature of that change, or how quickly it may become obsolete. In addition, the information flow within a physical, communications, or human workspace may also vary depending on how quickly it is created, captured, conveyed, updated, and presented at the required access point.

Note: Information management is a primary barrier challenge, one that is unique for each HE and maintenance person. Information workspace design seeks to address this challenge.

K.1 COMMON WORKSPACE DESIGN ELEMENTS AND PRINCIPLES

Industry practice standards (e.g., the ISO 11064 standard series) provide primary design guidance that applies to control centers, control rooms, control consoles and panels, and displays. This section presents supplemental tools and methods that integrate additional human factors and cognitive ergonomics elements into the design process.

Note: As a suggested pre-read to gain important insights and background, consider reviewing Appendix B.5.2 (Salience – Design Applications and Considerations) and Appendix F.3 (Cognitive Access to Visual Information). Also review the SEEV model in Appendix B.5.2 to develop an understanding of normal human behavior expressed as a subconscious resistance (motivation) to making seemingly minor movements to access visual information.

K.1.1 MINIMIZE MOVEMENTS NEEDED TO ACCESS DIRECT-USE PE

Tables K.1 and K.2 provide simple tools that may benefit many model processes and applications, for example display design (detailed design processes C-7 and C-8) and functional analysis (detailed processes C-9, C-10, and C-11). In the latter, the tables are used to assess and optimize personnel access and physical movements to access physical elements (visual or touch) from a fixed location (e.g., a control console) or while transiting through a workspace.

A key design objective seeks a least effort movement requirement to access direct-use PE with a preference for only eye movements, followed by head and hand, and, if required, body movement. The wider range of objectives aims to:

- Design the workspace to minimize the range of movements needed (body, hand, head, and eye) to access functionally linked elements located in the workspace.

TABLE K.1

Functional Analysis: Sensory Access to Object While Transiting through a Workspace

Loc.	HE Posn.	Display Object			Potential Access Movement				Causes of Blocked or Obscured Access
		Type	Location	Mode/Function *Note 1*	Body	Hand	Head	Eye *Note 2*	*Note 3*
Varies: Frequent/continuous change. Examples, transit through a room, escape route, or process area	In Transit (moving)	Passive visual displays (e.g., signage, paint markings)	Various locations in workspace	Visual: Status	✓	—	✓	✓	Inadequate lighting. View distance. Small text size, inappropriate font. Poor display contrast.
		Active visual displays (e.g., lamp, beacon, etc.)		Visual: Status	✓	—	✓	✓	Environmental interference (e.g., rain, fog, smoke, steam mist, etc.)
		Active audible discrete displays (e.g., horn, siren, voice message)		Audible: notification Status	✓	—	✓	—	High ambient noise. Insufficient volume intensity. Ambiguous tone or message (mode conventions), garbled/difficult-to-understand voice
		Public address system (1-way)		Audible voice msg.: notification status request	—	—	—	—	High ambient noise. Insufficient volume intensity. Ambiguous tone or message (mode conventions), garbled/difficult-to-understand voice
		Access to worn/carried Support PE		Visual: notification status	✓	✓	✓	✓	Difficult to access while moving. Attention diverted from workspace info.

Note 1: See Table K.7 in Appendix K.5 for information types and attributes. Note 2: Eye movement up to 30 degrees from line of sight. (See Figure F.4 in App. F.3.) Note 3: For guidance on improving display salience, see Appendix B.5.2.

TABLE K.2

Functional Analysis: Object Access from Seated Location at Control Console

Loc.	HE Posn.	Accessed Object			Potential Access Movement				Example Reasons for Blocked or Degraded Access
		Type	Location	Mode/Function *Note 2*	Body	Hand	Head	Eye	*Notes 3, 4*
A purposely designed room or workspace, e.g., a control room, ICC, ROC, electrical room, or machinery space. *Note 1*	Seated at control console or workstation	HMI display info	Display front	Visual/audible: notification status request tracking	—	✓	—	✓	Small text. Busy/cluttered display. Similarity/proximity to other elements on display. (Manual display call-up action)
		HMI display info	Angle from seated position	tracking	—	✓	✓	✓	Visibility, readability, and discrimination may be limited without head movement. (Manual display call-up action)
		Simple, active display device on console (e.g., lamp)	Angle from seated position	Visual: Status	—	—	✓	✓	See all of the above. Non-optimal movement from normal seated position. (Manual display call-up action)
		HMI display and other text/symbolic display types	Off-console VDU-display wall	See HMI display	—	✓	✓	✓	
		VDU display-CCTV view	Control console / Off-console display	Visual/audible: notification status request tracking	✓	✓	✓	✓	Image size/detail too small to see required information. Poor contrast. Limited precision in pan/tilt/zoom control changes, or no control possible. (Manual display call-up action)
		Person-to-person communication *Device-enabled Console-mounted comms. equip*	Telephone/radio *hand-held* Telephone/radio *with headset*	Visual/audible: notification status request tracking	—	✓	—	—	Background noise, distractions, interruptions. Heavy or unfamiliar accent, poor diction. Cognitive complexity of multi-tasking. (See K.3 for background.)
		Person-to-person communication *Video conferencing*	Wall-mount VDU display, room microphone	Visual/verbal / audible: notification	—	—	✓	✓	

(Continued)

TABLE K.2
(Continued)

Loc.	HE Posn.	Accessed Object — Type	Accessed Object — Location	Mode/Function Note 2	Potential Access Movement — Body	Hand	Head	Eye	Example Reasons for Blocked or Degraded Access Notes 3, 4
		Face-to-face communication	Position of other party can vary	status request	✓	—	✓	✓	
		Public address system (1-way)	Non-directional	Verbal /audible: notification status request	—	—	—	—	Background noise, distractions, interruptions. Poor diction, voice/msg. clarity. Inadequate volume. Occurs simultaneous to other notifications.
		HMI display – access control entry target	Display front		—	✓	—	✓	Small target size. Target text blocked by hand.
		HMI display – access control entry target	Angle from display		—	✓	✓	✓	Small target size. Target text blocked by hand. Angle degrades accurate targeting.
		Simple control device, manually activated (e.g., switch or pushbutton)	Control console	Visual/tactile/hand: initiate act phase response action	✓	✓	✓	✓	Label/feedback status blocked by hand. Poor label (e.g., location, text, contrast). Device similarity to a nearby device.
		Off-console-mounted control device (e.g., switch)	Off-console/ workstation		✓	✓	✓	✓	Non-optimal movement from normal seated position. Also see 'HMI Display Info' above.

Note 1: Hazard conditions such as smoke from a fire may contribute to blocked/obscured access. Note 2: See Table K.7 in Appendix K.5 for information types and attributes. Note 3: For guidance on improving display salience, see Appendix B.5.2. Note 4: Inadequate design may require simultaneous physical handling or manipulation of two different PEs (e.g., handling a telephone or radio while switching HMI displays or entering an act phase response command).

- Design to access and maintain *visual* access (e.g., read or view SA-1 information) while moving through a workspace (e.g., transiting an egress/escape route or through a process area).
- Design to access and maintain *visual* access (e.g., read or view SA-1 information) from a fixed location (e.g., seated at a control console/workstation).
- Design to access and maintain *audible* access to information from an expected fixed location and while transiting through or performing required activities within a physical workspace.
- Design for optimal physical access to controls, i.e., least movement, optimal body positions.
- Design for efficient (and least awkward) body movements with the workspace.
- Design to minimize the causes of blocked, obscured, or visual access limitations.

Note: To understand the four factors that drive visual scanning, see the Note in Appendix B.5.2 (SEEV model). The factors are salience, effort (e.g., required movements), expectancy (e.g., mental model), and the perceived value of the information.

K.1.2 Optimal Visual Sightlines and Off-Angle Object Detection

This section provides the reasons and basic guidance for placing barrier information at locations that reside within a narrow zone from the eye's line of sight position. Guidance is also provided to address placement at non-optimal locations, i.e., in the visual near or far peripheral zones. To understand the material and guidance in this section, Appendix F.3 provides information on the capabilities and limitations of the human visual system and the cognitive system access to this sensory information. For additional information on off-console displays, see ISO 11064-3 (1999, cl. 4.5).

Note: Selecting a display placement/location assumes one knows where the viewer will most likely, may, or seldom direct their eyes (line of sight). The design efforts may be less complex if the viewer is located at a fixed location, such as being seated at a control console or workstation. The challenge increases when a design constraint limits available design options to non-optimal view angles when moving through a workspace or from a fixed position. In these cases, the design inputs may draw from assumptions on the most likely scanning and viewing patterns, the results of an eye scanning exercise, or an assumed behavior developed in purposely developed training modules or simulation exercises. (Item 2 below provides additional detail.)

Note: Optimally designed displays and display placement does not ensure it will be noticed and detected. As discussed in Appendix B.5.1, issues attributed to attention, mental models, or excess fatigue can cause one to "look but not see."

The following offers suggested guidance and consideration:

1. *Foveal and parafoveal vision field (Zone A)*: these narrow fields of vision (0–1 and 1–5 degrees, respectively, from the line of sight) provide the maximum level of visual acuity, i.e., the ability to see fine detail (Karawowski

et al. 2021, p. 516). Where possible, position display information items and functionally linked information items within this zone. (K.1.3 explains the meaning of functionally linked information.)

To improve targeting accuracy (e.g., making a command entry on an HMI display), place the target in this zone. Locate the response feedback to this action (e.g., feedback confirmation) in this same zone. One's finger or hand should not block visual access to the feedback display.

2. *Central Vision (Zone B):* this zone is the area from Zone A to ~30 degrees from the eye's line-of-site. Visual acuity is greatly reduced and could be limited to a capability to discern specific objects or movement. To enhance the likelihood that a display in this zone is noticed, make the display/display object more conspicuous or salient. (Appendix B.5.2 provides guidance on improving display salience.) Once detected, the person must still be motivated to move their eyes or head to read a display from his/her current location and distance. The addition of an audible element may improve detection. (Hearing the element is more likely, regardless of head position.)

3. *Mid-Peripheral (Zone C) and Far Peripheral Vision:* placement of primary information in this zone should be prevented or minimized, as the visual capability may be limited to movement only or sensing a flashing type display (e.g., blinking lamp or rotating beacon). The addition of the audible tone, as discussed above, is an option if changes to the placement are not possible.

K.1.3   CONSIDER FUNCTIONAL RELATIONSHIPS WHEN LOCATING/PLACING DIRECT-USE PEs

This section provides guidance on the co-location of functionally linked display and control elements; i.e., the elements are accessed concurrently or sequentially when performing a task. Table K.3 provides examples of linked display and control (direct-use) objects.

This is important because as Edmonds (2016, pp. 224–225) explains,

> The way that information is presented to the user primarily requires consideration of the visual, auditory, and tactile elements of the display to ensure that the design is matched to the capabilities of the user. This needs to include how the person detects information on the display, and how they interpret and use the information.... The focus also needs to include how the displays and controls integrate with each other to form the complete interface.

Whenever possible, position the linked objects to reside within the foveal view in Zone A. If this is not possible, position the linked objects to require the least movement to access them both (e.g., eye > head > body where access by eye movement alone is preferred). See K.1.1 for detail. If the desired placement is not possible, seek the best available alternative option. Visual linkage may be achieved if each object shares/employs the same color, shape, or symbol. As a caveat, considered solutions may be challenged by other principles or conditions such as the presence of smoke, rain, or degraded lighting.

TABLE K.3

Functional Relationship between PE Objects

Functional Relationship between PE Objects (Direct-Use – Detect and Act Phase)

Object	Source	"Linked to" Object	"Linked to" Locations
SA-1 information (essential)	HMI display	Other essential SA-1 information (both needed to achieve SA-2 comprehension and SA-3 projection)	HMI display
		SA-2/3 support aid	
SA-1 information (essential)	HMI display	Decision support aid	HMI display
SA-1 information *Status feedback to Act Phase Response*		Act phase response target (e.g., soft command target)	
SA-1 information that indicates the presence of a nearby support PE cabinet	Local sign (passive) / Active indicator (beacon or backlit sign)	Support PE storage cabinet	Support PE storage cabinet
SA-1 information (act phase device activation feedback)	Console-located indicator	Act phase response entry device (e.g., ESD pushbutton)	Console-located command entry device
SA-1 information	Incident Command Board	Related information on common topic (e.g., status of a specific resource)	Incident Command Board
SA-3 information	HMI display	Act phase response target (e.g., soft command target)	HMI display
SA-2 or SA-3 information	HMI display	SA-2 or SA-3 support aid	HMI display or paper
SA-3 information	HMI display	Decision support aid	Paper or HMI display
Decision support aid	HMI display popup image of decision support checklist or flowchart	Act phase response target (e.g., soft command target)	HMI display

K.2 PHYSICAL ELEMENT WORKSPACE DESIGN

This section suggests supplemental workspace design elements applied to the physical workspace. Table K.4 shows their potential applicability. (The listed industry guidance standards are example documents that may provide the primary design guidance.)

The following supplemental principles may be applied to the design of the physical workspace (e.g., sizing, layout, and general arrangement), and the design, location, and placement of the direct-use PE (displays and controls) in that workspace.

1. **Workspace Sizing and Layout:** The overall dimensions and configuration of a workspace should be appropriate considering the normal and maximum number of personnel within it under all plausible situations. Scenarios should consider the likely simultaneous activation of multiple barriers, and the associated number and locations of personnel (e.g., HEs) requiring access and movement within the workspace.

   Applicable workplace elements include the following:

   - K.1.1 – Minimize Movements to Access Direct-Use PE
   - K.1.2 – Optimal Visual Sightlines and Off-Angle Object Detection
   - K.1.3 – Consider Functional Relationships when Locating/Placing Direct-Use PEs

2. **Position HEs to Support Required Communications and Coordination:** consider HE placement to enhance (and not impede or degrade) HE communications and coordination. Examples include the following:

   - Place HEs with specified face-to-face communications in proximal locations that readily accommodate those communication exchanges.
   - Where required by assigned tasks, place HEs in positions that support an implicit coordination objective (e.g., the ability to see what others are doing reduces the need for explicit communications).
   - Consider placing the barrier leader to optimize the performance of assigned functions while minimizing possible exposure to disruptive disturbances and interruptions.

   Applicable workplace elements include the following:
   - K.1.1 – Minimize Movements to Access Direct-Use PE
   - K.1.2 – Optimal Visual Sightlines and Off-Angle Object Detection

3. **Physical Element Placement within the Workspace**
   - The placement of elements within the workspace should functionally support on-demand physical and sensory access to all designated direct-use PE (detect and act response phase).
   - The placement of direct-use elements should be chosen to achieve optimal sightlines and visual access, and ease of reach to act phase response objects.

   Applicable workplace elements include the following:
   - K.1.1 – Minimize Movements to Access Direct-Use PE
   - K.1.2 – Optimal Visual Sightlines and Off-Angle Object Detection

TABLE K.4

Application of Workspace Principles: Physical Elements

Physical Workplace – Application of Workspace Design Principles: Physical Elements

Workspace	Room	Engineered Area	Control Console or Panel	Displays	Off-Console/Shared Displays
Applicable Workspace Design Principles (see below)	All	All	1 to 8	3, 6, 7	3, 6, 7
Example Industry References	ISO 11064-3 (1999) ISO 11064-4 (2013, Sec. 4.4) ISO 11064-6 (2005) NORSOK S-002 (2018) NUREG (2020, Ch. 12)	ISO 11064-3 (1999) ISO 13702 (2015) NORSOK S-001 (2021, Ch. 22) NORSOK S-002 (2018)	ISO 11064-3 (1999, sec. 4.4) ISO 11064-4 (2013) NORSOK S-002 (2018) NUREG 0700 (2020, Ch. 11)	ISO 11064-5 (2008), Annex A) (HMI Displays), Annex A.5.4 (CCTV guidance) NUREG (2020, multiple chapters)	ISO 11064-3 (1999, Sec. 4.5, 4.6.2) ISO 11064-5 (2008, Annex A.5.2-shared overview displays)

4. **Minimize Shared Workspace Conflicts, Distractions, and Interruptions:** when HEs and others share a common workspace, minimize the source of distractions, interruptions, and interference that may degrade HE performance. Examples are as follows:
   - Design the physical space to minimize movement interference between HEs and others.
   - Considering the expected personnel movements and activities within the workspace, place equipment to minimize the potential for blocking visual access to direct-use PE (e.g., maintain unobstructed sightlines to off-console displays).
   - Provide adequate distance or other means between personnel to limit the potential for distractions, interruptions, or dichotic listening from nearby discussions.
   - Limit the congestion attributed to the movements of HE and others in the workspace, which considers the potential for simultaneous access (multiple HEs) to the same PE device or space.
   - Consider physical restrictions and interference that may be attributed to support PE worn/used in all expected and plausible locations.

   Applicable workplace elements include the following:

   - K.1.1 – Minimize Movements to Access Direct-Use PE
   - K.1.2 – Optimal Visual Sightlines and Off-Angle Object Detection

5. **Design for Efficient Movement:** Position direct-use PEs to limit movement distance and the potential to cross movement paths with others. Movement effort increases as it progresses from the eye to head/hand, and to physically walking to a barrier display or control. (To the extent possible and practicable, limit the requirements to move between rooms and differentiated workspaces. Research verifies that crossing a room boundary increases the likelihood that a person may lose (forget) information held in the working or short-term memory.)
   Applicable workplace elements include the following:

   - K.1.1 – Minimize Movements to Access Direct-Use PE
   - K.1.2 – Optimal Visual Sightlines and Off-Angle Object Detection

6. **Proximally Position Functionally Linked Objects:** optimally position functionally linked objects to support its expected access and use.
   Applicable workplace elements include the following:

   - K.1.3 – Consider Functional Relationships when Locating/Placing Direct-Use PEs

7. **Design for Accurate Targeting with Integral Feedback:** when accurate and reliable targeting of an act phase response (direct-use) object is required, place/locate the object to support and achieve that objective. Examples include the following:

   - Place the target (control entry) object within easy reach and an optimal access angle from the normal expected position (e.g., seated at a control console). The targeted object should be within the not-to-exceed visual distance and angle that provides the required visual acuity needed to accurately target and manipulate the object.

- Design the targeted device (type and placement) to limit the potential for mistargeting. For an HMI display, select a target size that is easily accessed and accurately targeted from all plausible access and view angles.
- To the extent possible and practicable, integrate the response feedback indication into the targeted/manipulated object when there is a need to view both simultaneously. (Optimally, both should reside in visual Zone A.) This becomes more important when a stepwise or continually adjusted (modulating) action is needed to achieve the desired effect or result.

Applicable workplace elements include the following:

- K.1.2 – Optimal Visual Sightlines and Off-Angle Object Detection
- K.1.3 – Consider Functional Relationships when Locating/Placing Direct-Use PEs

8. **Minimize Awkward PE Interactions:** through device selection design and other means, limit situations that require awkward simultaneous handling or manipulation of multiple PE objects. For example, use hands-free communication devices if the task requires the user to talk or listen on a communications device (e.g., a radio) while accessing or interacting with a display or different hand-held device.

9. **Design Workspace Lighting and Light Level Controls to Maintain Optimal Visual Access to Displays:**
   - Select lighting to support visual acuity requirements (e.g., see fine detail or read text)
   - Select lighting to support visual information access (e.g., discern objects and colors, or enhance display conspicuousness (e.g., flashing effect))
   - Select lighting and light control to prevent visual obstruction to display information (e.g., glare on a VDU or washout effects from direct sunlight)

   Applicable workplace elements include the following:
   - K.1.2 – Optimal Visual Sightlines and Off-Angle Object Detection

10. **Workspace Habitability:**

   - Maintain the appropriate environmental controls for the working environment (e.g., maintain temperature and humidity within an optimal range, and reduce airborne contaminants).
   - Eliminate or minimize air-borne contaminants and irritants that can degrade cognitive and physical activities, accelerate fatigue, or become significant distractions.

K.3   COMMUNICATION WORKSPACE DESIGN

The communication workspace is purposely designed to achieve the specified barrier task function at the required performance. Communication encompasses the communication content/message, sender and target receivers, and communication media. (The conveyance may also include unintended embedded content such as tone-of-voice, fear, background sounds, etc.) Examples of this workspace includes the following procedure-defined exchanges:

- Face-to-face communication exchange
- Device-enabled communication exchange (e.g., verbal only, verbal-visual, or text)

Selected principles applied to the physical workspace address the communication workspace issues not covered in this section, such as where to physically locate personnel requiring face-to-face communications. (Inherent to communications, the sender and receiver have a functional relationship.)

Note: Communication is not normally viewed as a workspace, although it should be. It is an essential and powerful means for the barrier team members to interact, coordinate actions, and maintain team cohesion, morale, and focus. It allows the barrier leader to convey instructions that shift resources when a workload spike negatively affects a team member in a manner that increases the likelihood of barrier failure. It also creates an attention capture event for the sender and receivers who may have other equally (or possibly more) critical activities that do not progress during the exchange. As such, it should be judiciously applied and designed to be efficient, timely, selective, and brief. When developing procedures that guide communication requirements (frequency, timing, participants, etc.), the cost and benefits should be are carefully considered and balanced. For advanced discussions on communications, see Flin et al. (2008, Ch. 4) and Hollnagel (2003, pp. 557–563).

Barriers that employ two or more HEs are highly dependent on internal team communications to achieve the barrier function and response performance. As discussed in Appendices C, I.3.7, K.3, and L, and preliminary design process B-7 and detailed design processes C3-3, C-6 and C7-9, teamwork and team performance rely on communications to:

- Activate barriers (e.g., activate the search and rescue barrier on completion of the muster barrier)
- Provide notification of a resource, plan, or task change (e.g., an instruction to delay, start, or stop a barrier task or task phase)
- Provide status information to support coordination, demonstrate leadership and support, etc.
- Request time-sensitive task information
- Maintain team cohesion, common focus, and shared situation awareness (e.g., provide status information on active barriers, barrier progress, success or failures, activity timelines, and potential hazard escalation pathways and timelines)

TABLE K.5
Selecting Communication Methods

Message Form	Target HE(s)	Replay possible? Note 1	Potential Interruption Source?	Information richness (See Note 2) Verbal	Visual	Confirm accurate message receipt?
Face-to-face	1-to-1 1-to-many	Conditional yes See Note 1	Yes Sustained attention activity	High Adds tone of voice, background sounds, etc.	High Adds HE visual state and body language, etc.	Yes, if procedures and training exercises promote and develop this action
Email text	1-to-1 1-to-many	Yes	No	None	Limited, i.e., send attached media (e.g., static information)	No
Local intercom systems	1-to-1 1-to-many	Conditional yes See Note 1	Yes, but may ignore	Moderate	No	No
Live msg. over public address system	1-to-many	No	Yes, but may ignore	Moderate	No	No
Radio telephone	1-to-1 1-to-many	Conditional yes See Note 1	Yes Sustained attention activity	High Adds tone of voice, background sounds, etc.	No	Yes, if procedures and training exercises promote, enable, and develop this action
Video conference	1-to-1 1-to-many Many-to-many	Conditional yes See Note 1	Yes Sustained attention activity	High Adds tone of voice, background sounds, body language, etc.	High Adds HE visual state and body language, etc.	

Note 1: When included in procedures, the method supports a repeat request and may ALLOW a later (separate) request to repeat/confirm the information. Note 2: INFORMATION-RICH sources can be materially important. For those with advanced mental models and experience, this additional information provides cues and clues that may be essential to understanding complex situations and events that were not considered in the design process.

- Contribute to the "being there" inclusiveness needed to maximize the effectiveness and timeliness of remote barrier support functions provided by an off-facility location. (The capability to monitor ongoing communications and live CCTV images is an essential contributor to this objective.)

Table K.5 provides example communication options to consider when designing the communication workspace.

The following provides example guidance to consider in the communications workspaces design:

1. Select the communication methods best suited to the message content and function, sender and target receivers, use situations, and environments. See Table K.5 for guidance and information in this regard.
2. Barrier and task procedures should evaluate and identify the approach for implementing each communication requirement included in Table 3.11. Validate and update this information as the barrier design develops.

    Furthermore, Gasbury (2013, p. 114) states that

    Communications can also be improved when the sender and receiver use commonly understood terminology, words, and phrases. Under stress, responders may struggle with their words. If communications are winded or in a hurry, they may also use shortcut phrases. If you know this is going to happen, take pre-emptive steps to address it. Develop words and phrases for the most common tasks, assignments, and updates.

3. Develop and monitor HE skills in using the communication and communication protocols identified for procedures. When appropriate, use a read-back or response-back process to confirm the conveyed information was received and correctly understood. Provide training drills and exercises designed to develop, maintain, and demonstrate these and other skills, for example, the automatic use of approved terms, message timing, tone of voice, and brevity. (Accident investigations demonstrate that a single communication error can have catastrophic consequences. The entire exchange may take place in a few seconds.)
4. Communication assessments and methods should consider how long the conveyed information is valid and how long it should be remembered. Communication-driven reliance on short-term memory is also problematic. See Table K.7 for further information on these topics.
5. Non-verbal visual and audible information provides vital information to the barrier leader and may have a positive or negative effect on other HEs. Non-verbal, visual, and audible cues include tone of voice (stress, fear, panic), visual information (body language, signs of fatigue, wrong location), and background noise (hazard event generated, local conditions, etc.). The embedded information may cause the barrier leader to take action. For example, that information may indicate signs of fear, high duress, over-confidence, or the potential for a hasty or inappropriate action.
6. For industry guidance (examples) that may apply to the communication workspace, see Table K.6.

TABLE K.6
Example Industry Standards: Communication Workspace Design

Topic	Type	References	Topic
Communication system	Speech-based	NUREG (2020, Ch. 10.1 and 10.2) NUREG (2016, cl. 7.3) IOGP (2018), Report 503	NUREG (2020) – general guidance, design, and procedures NUREG (2106) – communication errors IOGP – recommended communication behaviors
(device-enabled)	Computer-based messaging	NUREG (2020, Ch. 10.1 and 10.3) NUREG (2016, cl. 7.3) IOGP (2018), Report 503	NUREG (2020) – general guidance, design, and procedures NUREG (2016) – communication errors IOGP – recommended communication behaviors
Communication (face-to-face)	Live exchange	IOGP (2018), Report 503	IOGP – Recommended communication behaviors

K.4 HUMAN ELEMENT WORKSPACE DESIGN

The development of the HE workspace should consider the applicable guidance and information provided throughout this book, such as in Appendices C, F, J, K.3, and K.5.

Note: The most important and flexible "workspace" is the HE assigned to the barrier task. People often become the only means available to respond to situations that were not foreseen in the assessments, task analysis, and barrier system design. Training exercises should include unforeseen events allowing the team to understand and learn from their individual and collective responses to such events. To learn from that experience, the team should conduct a post-event discussion that evaluates their individual and team performance. Identify what went well (e.g., recommend repeating) and what did not go well (e.g., evalutate and identify possible changes in the barrier system, if any). Any change to the barrier system (from either finding) should be progressed using the change management.

Examples of activities for developing and maintaining the HE workspace are as follows:

1. Develop and maintain the skills required to perform specific barrier task activities, for example:
   - Task execution
   - Equipment usage
   - Procedure and support aid usage
   - Scanning for SA-1 information
   - Advanced decisions (e.g., use of the recognition-primed decision model)
   - Coping with stress
   - Teamworking (e.g., communication, support, and coordination)
   - Leadership

2. Develop and maintain knowledge-based long-term memories and mental models that are *sufficiently* complete and accurate. The distinct types of knowledge are as follows:
   - Technical knowledge (e.g., process and system knowledge, the nature of the hazard that activated the barrier, desired and potential effects of an act phase response action, conditions required to achieve the safe state, and potential hazard escalation pathways.)
   - Execution knowledge (e.g., the required team interactions, coordination, and communications.)
   - Procedural knowledge (e.g., what procedures apply and their intended use)

3. Develop the desired behaviors, attitude, and resilience (e.g., develop trust and resistance to stress, fear, high workload, interruptions, and distractions).
4. Select personnel with the required innate aptitude and abilities.
5. Maintain fitness for service. Important here is:

- Fatigue management (e.g., this can be accomplished through rotation and staffing policies and practice)
- Health fitness (e.g., through wellness programs)

6. Monitor the above to ensure continued alignment with requirements to prevent drift and skill fade. Enhance HE awareness of the slowly evolving drift possible in using procedures or in one's mental models and beliefs. Enhance awareness of skill fade and how it contributes to degraded performance.
7. Attention: humans have one attention resource. The design of the barrier and human training should seek to optimize this resource (attention management). There are various ways in which this can be done, including via the following:

- Training: increases awareness of the nature and constraints specific to human attention. Important here is to increase awareness of attention capture situations and consequences.
- Training: seeks to exploit appropriate automatic processes that contribute to the reliability and performance of the barrier and reduce the demand for attention.
- Design communication channels to minimize communication duration and maximize transfer effectiveness and accuracy. (Active communication is an attention capture activity.)

8. Develop teamworking and non-technical skills (e.g., shared situation awareness, implicit coordination, trust, automatic teamworking skills, exploit Meta SA (Appendix C.1.5)).
9. Responsiveness: through drills and post-drill assessments, guide the HE (or maintenance personnel) to understand their individual responsiveness against the task requirements. These include the following:

- Response rate to the initial barrier or task activation
- Response to feedback information from an act phase response or support PE
- Response to plan changes from the barrier leader
- Response to requests from others
- Response to an unexpected event (e.g., a missing person, a hazard escalation, failed equipment, activation of new barriers)

10. Goal conflict and priority selection: develop the HE's capability to understand and respond to complex situations that result from actual or perceived barrier/task goals or a conflict between priorities. For example, which task should be started next (tasks perceived to have priorities), which should be delayed, or should the HE respond to a request from others now or continue working on an existing task understood as being of a higher priority. The barrier type may create an internal goal conflict if response actions can affect one's personal reputation, legal liability, or requires choosing between who to save or who to place at risk.

11. Non-rational biases and behaviors: enhance the HE's awareness and appropriate response to non-rational biases and behaviors. (See Appendices F.2.3 and F.2.5 for detail.)

K.5   INFORMATION WORKSPACE DESIGN

The information workspace encompasses displays, communications, feedback from control actions, and other information available within the environment. A review of major accidents identifies information as a common contributor to major accidents (e.g., information is missing, incomplete, out of date, conflicting, misremembered, or misunderstood). For the HE, information challenges commonly include the following:

- **Access:** gaining access to essential information is challenging when it must be accessed from many different sources, each with its own difficulties in terms of navigating to and acquiring information from each source.
- **Type:** the information exists in many forms (e.g., voice, CCTV displays, HMI displays, or a hand-written Incident Command Board).
- **Presentation:** the presented information may not be in a form that directly conveys meaning and comprehension; i.e., it requires additional cognitive resources to use and translate the presented information.
- **Quality:** information must be (sufficiently) accurate. Using erroneous information can lead to barrier failure. Furthermore, issues regarding quality and validity can occur in any stage of the information lifecycle, namely, in the generation, manipulation, presentation, conveyance, storage, or recall phases.
- **Availability:** for various reasons, important information may not be available when and where needed. This may be more common with emergency response barriers that require time-constrained decisions based on incomplete or conflicting information.
- **Latency:** the accessed information may have varying degrees of latency, for example, if may have been captured earlier but is no longer valid. Using outdated information can lead to barrier failure.
- **Time-sensitive:** some information may be acutely time-sensitive and requires an immediate response to prevent or limit the potential for a dangerous event or condition, hazard escalation, or materially degrade the performance of a barrier team.
- **Value over time:** information may have a shelf life. The value thereof may degrade over time or be conditional; for example, it remains high if the information is a request that requires an action that is delayed.

Table K.7 examines and normalizes information with respect to its type and attributes, aspects considered and addressed in the lifecycle model. It is important to consider this information in the information workspace design process because it may affect how the information should be acquired, conveyed, and presented, and if required, how it is stored for later recall and access. The table presents the information types addressed in the lifecycle model. To this, Hollnagel (2003, p. 560) adds the following additional types that may also be appropriate: "Acknowledge (communication received), Check (belief), Confirm, Propose (action), Agree, Disagree, and Correct (i.e., amend)."

The amount and flow of information (throughput, change rates, channels, medium, formats, and locations) in an activated barrier can be considerable, dynamic, and situational. This presents an information management challenge to the person assigned to provide, monitor, capture, comprehend, or respond to the information. At times, this flood of information will likely exceed the intended receiver's capabilities. When this occurs, the person (HE) may revert to an approach that selects and focuses attention on one item while purposely ignoring all others, a behavior that is not necessarily rational or consistent with the barrier or task goals and priorities. Plausible instances of dynamic information overload (chronic or excessive) should be identified. Its occurrence indicates a possible error in the barrier system design or management.

For barrier designers, the situational and dynamic aspects of the information flow and management challenge should be assessed and understood (sender, receiver and acquisitioner perspective). Identify situations when information flow is likely to exceed the capabilities of the intended sender or receiver(s). In such cases, solutions may be warranted. The problem (and the potential solutions) may be identified in the performance influencing factor assessment defined in Appendix H.2 (Task Complexity and Other Load PIFs in Table H.1), the barrier system acceptance test or B-SAT (defined in Appendix G.4), or the cognitive assessment process in Appendix I. One or more may contribute to identifying the peak information change/demand rates or information that is not needed or should be made available for later on-demand access. Once identified, corrective solutions may require changes to the communication workspace, task design, displays, or training and drills.

TABLE K.7
Barrier Information Types and Attributes

Type	Sensory Form	Info Function	Info Source (Examples)	Supports ad hoc access later?	Time-Sensitive	Info. Value over Time	Relies on WM/STM?
Notification barrier activation	Audible Visual Verbal	Barrier activator	Area audible / visual alarm. (May include voice message.) Audible / visual alarm at control panel. Barrier leader	**Yes**, for common barriers. If verbally conveyed, the answer is **Maybe** (*see status below*).		**High** if activation delayed. **Moderate** if promptly acted on. (Maintain awareness of activated barriers.)	Yes, if barrier leader issues verbal-only command to activate barrier. (PE/technical system not available to provide duplicate information.)
Notification Instruct/ command	Audible Visual Verbal Text	Barrier leader instructs HE (or others) to change task: start, wait, modify, or stop	Communication exchange (with barrier leader)	**Yes**, if a PE/technical system is the initial source, it retains the information for recall. If verbally conveyed, the answer is **Maybe.** *This qualified response assumes the person needing the information remembers to request it in the future, and the provider has time to provide it, AND both occur soon enough to achieve the task/barrier safety times.*	**High** Barrier/task may be time-of-the essence, so any delay is a risk.	**High** if instruction or command is delayed. **Moderate** if promptly acted on. (Maintains awareness of change action.)	Yes, if a verbal-only notification. (PE/technical system not available to provide duplicate information.)

Status Inform Coordinate Support	Audible Visual Verbal Text/ Symbol	Monitor PE/HE status, fitness, morale, etc. Team coordination Provide leadership	Barrier leader, others VDU and non-VDU displays Communication exchange Incident scene		Moderate to **High**	**Low to High** *Low to moderate* with frequent status updates. *Moderate to high* with less frequent updates, high-consequence information	Yes, if the PE/technical system is not the source of that information
Status Event Feedback		Feedback to verify response/ result to an action	HE (others) sensory access to act phase action result VDU and non-VDU displays	See above	**Moderate** to **High** Depends on the action	**Moderate** to **high** Confirmed failure of action requires additional actions	Yes, if the PE/technical system does not confirm action success/ failure
Request	Verbal Text	HE (or others) request information to support task	Barrier leader, HEs, others Communication exchange		Typically, **yes**, i.e., indicates current need	**Moderate** to **High** Delayed response may lead to event escalation, degraded team response	Yes, if a verbal-only request. (PE/technical system not available to provide duplicate information.)
Tracking hazard escalation	Audible Visual Verbal Text	Monitor/maintain change awareness (e.g., a change in the hazard scale or rate of change or enables a new hazard)	Visually identifiable changes New safety alarm activations New barrier activations	**Yes**, if the source of the tracked information is from a technical system or allows access on demand (e.g., a CCTV view is sufficient).	Moderate to **High**	**Moderate** to **High** Barrier failure may be likely if an escalation (or potential pathway) goes unnoticed	Yes, if tracking is not performed by PE/ technical system

(Continued)

TABLE K.7
(Continued)

Type	Sensory Form	Info Function	Info Source (Examples)	Supports ad hoc access later?	Time-Sensitive	Info. Value over Time	Relies on WM/STM?
Tracking Resource capacity or capability	Audible Visual Verbal Text	Track PE and HE for depletion, degraded capability, reliability over time	Barrier leader tracks team fatigue, morale, attitude, emotional state, etc. HE or technical system tracks HE/personnel availability PE displays: trends, remaining capacity values, etc.	**Yes**, if the information is tracked in a PE, technical system, or on paper. If verbally conveyed, the answer is **Maybe** (see status above). Otherwise, **no**	**Moderate** if the info is proactively used in planning **High** when approaching full depletion	**Moderate** to **High**	See above
Tracking Time	Audible Visual Verbal Text	Monitor and maintain awareness of remaining, available, or elapsed clock time, and time-critical events and actions	Clock (PE) Count up/down timers (PE) Clock-time alerts (PE)		**Moderate** with this info used in ongoing planning **High** when time is about to run out	**Moderate** to **High**	

Appendix L
Remote Barrier Support, Remote Operations Centers

This appendix is organized as follows:

- L.1 Introduction
- L.2 Remote Barrier Support (RBS) and Remote Operations Center (ROC) Design and Lifecycle Processes
- L.3 Potential RBS Opportunities
- L.4 Potential Contributors to Degradation

This appendix provides information and guidance on the remote barrier support (RBS) functions provided by one or more HEs who perform these functions from a remote location such as a Remote Operations Center (ROC).

TERMS USED

The term *remote barrier support* (RBS) refers to one or more barrier system tasks performed from a remote location by one or more HEs. Depending on how the tasks are defined, they may or may not be essential to achieving the barrier function.

The term *Remote Operations Center* (ROC) is used generically to refer to a room or facility where the RBS task is performed. Being remote, the ROC exists at a physically distant location that is typically not affected by a hazard event occurring at the protected facility. (Remote support for an offshore located O&G facility is typically provided from an onshore located ROC.)

The term *protected facility* refers to the facility (and the facility-located personnel) protected by the barrier function.

L.1 INTRODUCTION

Increasingly, owner/operators provide an integrated ROC to supply operational and supplemental support to facilities that process highly hazardous material. The ROC may provide the following:

- Basic, supplemental, or technical support to the protected facility. An example facility is an offshore floating drilling, production, or storage facility. (Taken from a different vantage point, the "facility" may be a regional or national transmission pipeline network.)

- Support to a facility that is not sufficiently staffed to perform highly specialized functions. Examples include a complex, multi-organization emergency response and advanced cyber security monitoring and response capabilities.
- A centralized location to manage an incident response in a highly distributed system such as a system of pipelines that transport highly hazardous fluids.

A barrier system may have one or more RBS tasks. The addition of the task introduces the following design and support activities:

- One or more RBS tasks are performed from a remote location by one or more barrier-assigned HEs.
- An ROC facility that provides the physical workspace, technical systems, external support systems, and environmental controls needed to support and enable the HE and RBS tasks performed in this workspace.
- Additions or modifications to technical systems (located in the ROC and the protected facility) and the additional infrastructure that enables the timely and reliable performance of RBS tasks from the ROC.
- Communication infrastructure that links the technical systems of the ROC to those located at the protected facility (e.g., the additional resources addressed in the preliminary and detailed design steps B20-4 and C14-5, respectively).
- Other resources, processes, and procedures needed to operate, maintain, and manage the ROC, the ROC located personnel, and the above-discussed additional resources over the full barrier lifecycle.

Appendix B.4 proposes a new document, the remote barrier support design basis, that defines the technical and functional basis for the facilities, equipment, capabilities and performance need to enable and support the performance of RBS tasks from the ROC. The initial draft (developed in preliminary design step B-1) provides input to early preliminary design phase activities, documents, and risk assessments. It is further developed and updated in steps B20-7 and detailed design process C-1.

Note: As is commonly true of design-basis–level documents, the early development, completion, and issue of such documents to all users is a critical contributor to timely and coordinated design development, improved inter-discipline coordination, and minimizing project team execution errors. The guidance in the proposed document affects many disciplines and organizations. Once issued, major changes may become increasingly difficult. Even "simple" changes can have unforeseen, knock-on effects. A change that is proposed late in the project cycle may require use of the change management process to ensure it does not create unintended consequences (technical, safety) and can be implemented by all affected parties.

The RBS enabling facilities, equipment, systems, and resources become integral parts of the barrier system. Section L.2 provides insight into how they may be integrated into the overall barrier system lifecycle. As such, they should be included (and addressed) in the barrier safety management plan and the safety requirements specification.

L.2 REMOTE BARRIER SUPPORT AND REMOTE OPERATIONS CENTER DESIGN AND LIFECYCLE PROCESSES

Adding a correctly defined, design and implemented RBS capability may create additional opportunities to enhance the barrier system capability, capacity, reliability, and resilience (robustness). However, the additions also increase the complexity of the barrier system. As such, they introduce the potential for new risks and increase the project scope and execution challenges. The following sections outline the activities and issues addressed in the prototype design and lifecycle model.

L.2.1 CONCEPTUAL DESIGN

The intention to add RBS tasks, an ROC, and the associated additional resources to the barrier system design should be identified and explained in the applicable conceptual phase documents listed in Table 3.1. Requirements of this type affect many technical and project disciplines and organizations in each lifecycle phase. The necessary documents should be developed and designed to adequately inform and guide others who may be required to integrate the RBS/ROC requirements into their technical discipline and organizational work scope, documents, and project interface activities.

L.2.2 PRELIMINARY DESIGN

If RBS and ROC elements are included or considered, the starred documents in Table 3.1 should provide that input to those charged with planning, scoping, budgeting, and preparing execution plans needed to prepare for and commence the preliminary design phase activities. The process of defining the RBS task, ROC, and additional resources requirements continues in the preliminary design phase. (See Figure 3.3 for an overview of the preliminary design phase steps.)

All preliminary design activities apply to the ROC and RBS additions to the barrier system. The following provides example guidance for the noted steps.

Step B-1, Early-Stage Design: If the additional of an RBS task is considered, participants in the risk assessment should be aware of the information in the RBSDB and the pros and cons Sections L.3 and L.4. The initial RBSDB draft should be available to support the risk assessments performed in step B-2, and the task analysis performed in step B-3. For additional considerations, see step B-25 (the note pertaining to staffing), Appendix C.4 (trust), Appendix J.5 (staffing), and Sections L.3 and L.4.

Step B-2, Risk Assessments: as applicable, the risk assessment process should consider design information applicable to the RBS, such as design guidance that may allow for or restrict RBS functions. Recommendations from this assessment are inputs to the step B3-1 task analysis.

Step B3-1, Task Analysis: this process identifies the tasks required to achieve the barrier function. One or more may be identified as RBS tasks. Participants in this analysis should be aware of the opportunities and potential concerns noted in Sections L.3 and L.4.

Step B-4, Base Barrier Performance Standards: as appropriate and required, the performance standards should include requirements that apply to the RBS task and the ROC facility and technical systems.

Step B-5, Shared Situation Awareness: defining an SSA requirement for an RBS task should consider the effects of the remote location. The RBS HE does not have access to the ambient information readily available to HEs located at the protected facility. If not addressed, this missing information may contribute to a deficiency in SSA that degrades the performance of the RBS HE. When defining the requirements for SSA, consider the information in Appendix C.1.2. It discusses the challenge of not having access to the rich sources of directly sensed information available to those located on the protected facility. Detected information of this type may include potentially important content in the ambient environment, including sounds, movements, vibrations, visibility, odors, or weather conditions. Depending on the task, closing this gap may require providing the RBS HE with access to live information from the protected facility, namely, information that creates the "being there" experience needed to achieve the desired RBS task performance. Example sources may include real-time access to live CCTV images, radio system communications, and teleconferencing systems. For additional information see Appendices K.3 – K.5.

Note: See Taber et al. (2012) for the potential opportunities for using SMART board technology for an Incident Command Board. Use of this technology offers one potential means to gain live (real-time) access to the Incident Command Board from the ROC and other remote locations. A continuous (live) CCTV view of the board may offer an additional option.

Step B-19, Reliability Assessments: the RBS HE, ROC, technical systems, communication channels, and infrastructure add new points of failure to consider in the reliability assessments.

If the communication bandwidth between the ROC and protected facility is limited, mutual agreements and procedures are needed to allocate the necessary bandwidth to support the required systems and equipment of the RBS. A requirement for maintain continuous access to live, high-resolution CCTV video from the protected facility may be a challenge if the communication bandwidth provided by the network infrastructure (additional resources) cannot fully support all network consumers at the required throughput and performance. The ROC link to the protected facility should be designed to meet all requirements under all plausible demand conditions. The assessment should consider the effects and consequences if the hazard event that activated a barrier can cause the loss or degradation of the communication infrastructure that enables this capability.

Step B20-3.1, Safety Requirements Specifications: this step adds the requirements and design information for the RBS task(s), ROC, and additional resources.

Step B20-3.2, Performance Standards: as applicable, this step adds the requirements from the RBS task(s), ROC, and additional resources to the base performance standard (Table 3.7), or develops a dedicated standard as required.

Step B20-4, Additional Resources: this step identifies the additional resources needed to enable the RBS capability from the identified remote location(s). (These resources should be identified in the RBSDB.)

Step B20-7, Preliminary Requirements for the ROC, and Remote Barrier Support: Update the RBSDB for issue. Once issued, a persistent effort may be needed to ensure others read, understand, and correctly applies the requirements. The addition of the RBS and ROC adds new staffing and facility requirements, technical systems, and additional resource requirements. The technical interface requirements may place additional design demands and constraints on technical systems located in the protected facility. If design teams have limited experience in this area, an interface coordinator may be needed to track and coordinate the needs, information exchanges, schedules, and conformance to the ROC and RBS requirements.

Note: The design and operation of the ROC may have many different stakeholders if non-barrier functions are performed from this workspace. The RBS functions, staffing, and performance requirements become one of the many inputs that guide the design and provisioning of this space. The design and implementation of the ROC may be managed by the owner/operator or a contracted third party. Those responsible for barrier design may need to coordinate with these stakeholders to ensure their design objectives and operational use plans do not conflict with the RBS goals and functions.

Step B-23, Control Center Conceptual Design Framework: as applicable, the activities in this step should be applied to the RBS task and ROC facility.

Step B-25, HE Staff – Roles and Organization: the scope of this step includes the RBS HEs and the personnel assigned to maintain the facilities and equipment that enables the RBS capabilities and functions performed in the ROC.

Step B-26, Fatigue Management (Fitness for Service): the scope of this step is expanded to include the personnel identified in the above step B-25.

L.2.3 DETAILED DESIGN AND IMPLEMENTATION

The information, requirements, and design documents developed in the preliminary design phase are essential inputs to the RBS, ROC, and additional resources processes and activities performed in the detailed design phase. External changes or constraints may develop (from others) that are inconsistent with one or more of these requirements. Issues of this type may require a return to preliminary design phase steps if required to find alternative paths and solutions. Examples include the following:

- A change in the ROC staffing policy, staff rotation plan, etc.
- A change in the RBS task or task assignments.
- A lower-than-planned bandwidth (capacity), reliability, or availability in the network/interfacing systems that connect the ROC-located technical systems to those located at the protected facility.
- Additional non-barrier activities are assigned to an RBS HE that may increase his/her workload during barrier activation. For example, the owner/operator may require the barrier team provide ongoing status updates to management or regulatory agencies when an emergency response barrier is active.

All detailed design and implementation phase processes apply to the ROC and RBS additions to the barrier system. The following provides additional guidance for selected processes.

Process C-2: the ROC, RBS tasks and additional resources should be included in the performance influencing factors assessment (Appendix H.2). RBS tasks increase barrier system complexity when they include communicating, coordinating, and other interactions with HEs located on the protected facility. They may also introduce new issues to consider such as trust, morale, motivation, other loads, safety culture, and rotation policies. The ROC also adds a complexity factor to facility operations, maintaining technical systems, and additional resources.

Process C-8: this process may apply to new HMI displays and other VDU-based displays needed to enable and support the RBS capability.

Processes C-9, C-10, and C-12: these processes may be applied to the ROC, control consoles, and the external support systems and protective barriers that enable, support, and maintain the RBS capability.

Process C-13: consider the possible changes to (or new) performance standards that may be needed to address the new RBS functions and systems (PE, OE, and HE) and the associated verification and validation activities.

Process C-14: process C14-2 may identify new technical system requirements when needed to enable RBS tasks and task functionality. Processes C14-3, C14-4, and C14-5 may do the same for external support and protective systems and additional resources.

Process C-15: this process identifies additional procedures to guide RBS tasks and ROC facility operations and maintenance.

Process C-17: this process updates the safety requirements specification with the added information attributed to the RBS task(s), task phase elements, ROC systems, additional resources, and other potential barrier system dependencies.

Process C-18: this process provides input to the procurement and contract documents as they apply to ROC systems and additional resources. The timing of these activities may significantly differ from the timing of the procurement and contracts at the protected facility.

Process C-19: this process adds the RBS tasks, ROC, and additional resources to the design verification process.

Process C-20: this process attempts to validate the elements of the RBS can adequately meet the needs of the RBS-assigned HEs and maintenance personnel. The information in Appendix C may provide additional guidance and considerations for assessing the "being there" sense of engagement in and the awareness of the HE situations and conditions at the protected facility.

Process C-30, C-31 and C-34: this series of procurement-related activities may apply to technical systems, the ROC facility, and the additional resources that enable and maintain the RBS capability and performance. The timing of these activities may significantly differ from the protected facility procurement and contract activities.

Process C-32: this series of steps applies to the RBS/ROC-applicable organization chart and staffing finalization, procedure development, and personnel training planning and development.

L.2.4 CONSTRUCTION, INSTALLATION, AND COMMISSIONING

All construction, installation, and commissioning phase activities apply to the ROC and RBS additions to the barrier system.

L.2.5 OPERATE AND MAINTAIN, MODIFICATION, AND DECOMMISSIONING

The phase E, F, and G processes apply to the ROC and RBS additions to the barrier system.

L.3 POTENTIAL RBS OPPORTUNITIES

Staffing on the protected facility may be marginal or insufficient at times given the potential for simultaneous barrier activations, blocked HE access to assigned barrier stations, potential injury/loss of barrier personnel, and unforeseen situations and events. This section provides example options for leveraging and supplementing the capabilities of HEs located on the protected facility (e.g., increasing the barrier team's capacity or capabilities).

L.3.1 SUPPORT FOR HIGH WORKLOAD ACTIVITIES (SEARCH, MONITOR, CHANGE DETECTION)

As a suggested pre-read, review the applicable information in Appendices C, D, F, J.6, and K.3–K.5.

This section suggests possible RBS tasks that address problematic workload, capability, flexibility, or reliability issues that are not adequately addressed by the HEs located on the protected facility. Examples of appropriate RBS tasks may include the following:

- Monitor an unsafe area (e.g., the location of a known present hazard) to identify when unauthorized personnel are in or entering the area.
- Assist with a search for personnel who failed to report to their assigned muster station (e.g., the person is missing, reported to a different muster location, or arrived but failed to report in).
- Monitor the information on the Incident Command Board for completeness, accuracy, and readability. (For background, see the preliminary design step B-6 example, and the detailed design step C7-8. Enabling this capability may require a smartboard or a live CCTV (real-time) view of a non-intelligent whiteboard.)
- Assist in identifying the source of an incident (e.g., a toxic of flammable gas release location) and its potential sources of escalation (e.g., an increased release rate, new release locations, and potential ignition sources).
- Assess if, how, and where the hazard event or its consequences (e.g., smoke) may block or impede progress by those attempting to perform a control and recovery activity, transit an egress/escape route, or conduct a search and rescue effort.

- Monitor the status of a barrier-dependent external support system, external protective barrier, additional resource, and other barrier-dependent elements.
- In real-time, monitor HE task performance for material delays and errors that place the barrier at risk. (This may be limited to specific roles or situations.)
- Provide support for complex situations that require integrating information (real-time and static) from many sources.
- Provide pre-determined assistance in complex (foreseeable) situations.
- Provide non-barrier communications (owner/operator required) to others, for example, provide incident and progress status updates to management or support a required regulatory notification.
- Monitor communications for errors (e.g., late, or incorrect/potentially misunderstood information) that may place the barrier at risk.
- Perform "what if" and preliminary planning to address plausible contingencies such as potential pathways for barrier escalation.
- Assist with a telemedicine response from an external resource.
- Assist with mobilization efforts of required external resources (e.g., evacuation vessels or helicopters).

Conceptual Exercise, Remote Barrier Support: Review information from the four case studies included in different sections within this book. (They include the Deepwater Horizon fire, explosion, and environmental release, the DuPont LaPorte toxic chemical release, the UK Buncefield fuel tank overflow and fire, and the PG&E natural gas transmission pipeline rupture and fire. Refer to the index for locations.) Review one or more for potential RBS opportunities. Once done, perform a validity screening by assessing the identified opportunity against the potential downside issues discussed in Section L.4. From that result, do any of the identified opportunities warrant further consideration or evaluation?

L.3.2 Decision Support

RBS opportunities may exist in the decision space for barriers that require complex decisions, often when full information is not available. A barrier leader may need to make a highly consequential and complex decision, often in a time-pressured situation. The information needed to make the decision may be incomplete, conflicting, or must be integrated (e.g., time-sensitive quantitative and qualitative information from several sources). The decider may be forced to choose between act phase responses that have different risks and possible outcomes. In such instances, the opportunity to receive timely input from a trusted peer may be beneficial, especially if the support is from a fellow HE who holds the same barrier role.

Note: Globally, ROC-provided support for a costly and complex deepwater production and drilling facility is relatively common. Barrier design and shift/rotation practices may swap the barrier leader (e.g., the Offshore Installation Manager) and other roles in scheduled onshore-offshore rotations. This approach should improve mutual trust, a requirement for accepting safety critical input from others. For additional information and background on trust, see Appendix C.4.

Note: One of the most complex decisions for an offshore O&G facility is deciding if and when to abandon a facility that has an uncontrollable active hazard (threat). A decision to do so sooner may provide the best overall protection for personnel. (The abandon process itself introduces some risks unique to this process.) Consequently, having all personnel leave a high value facility that is on fire or sinking increases the likelihood of losing that facility or a major environmental release. Table H.1 (Appendix H.1) includes "Societal Influencer" in the performance influence factors assessment performed in the detailed design phase and again in the operate and maintain phase. When attended by the barrier leaders, the assessment may reveal influencers that affect the decision-making process. The decider's beliefs become influencers. Beliefs may develop from a perceived (or misperceived) understanding of management expectations and the local safety culture (risk-adverse or risk takers). The decider may worry about making a wrong decision. How would that affect my career or how will my friends and peers view me? Beliefs and other influences can positively or negatively guide and skew the decision-making process. If thoughtfully included in barrier system procedures, taking a time-out to discuss a critical, pending decision with a trusted peer or manager may prove to be beneficial.

L.3.3 Monitor Cognitive Behaviors

Review Appendix F.2 for background on the behaviors and biases discussed in this section. For additional insight, review the Deepwater Horizon case study in Appendix M and other case study examples included throughout this book. (To locate, see the index.)

Undesirable cognitive behaviors and biases can occur in complex and time-pressured situations that may degrade or place a barrier at risk. This section considers the potential opportunities and viability of adding an RBS task to monitor HE behaviors that may be non-rational or placed the barrier at risk. The monitoring task defines the limits of the task; what to observe; when to take actions; and the permitted actions including who the information is communicated to, when, and how. An effort to validate the task should consider and confirm the action does not create new problems or degrade team trust. The acceptance and viability of this type of RBS task may depend on several performance influence factors such as task complexity, local safety culture, and individual morale and trust. Monitoring actions may look for the following behaviors and actions:

- Attention tunneling
- Confirmation bias
- Change blindness
- Inattention blindness
- Forget to remember (prospective memory failures)
- Losing place
- Decision framing error
- Losing track of time
- Task-switch errors (e.g., plan continuation error)
- Other forms of bias listed in Appendix F.2.5.

L.4   POTENTIAL CONTRIBUTORS TO DEGRADATION

RBS tasks should be guided by the underlying philosophy to *do no harm*. Adding HEs to a barrier increases the complexity of that barrier and opportunities for mistakes. Potential errors and miscues may be amplified when the HE performs the task from a remote location. Thus, care is needed when considering an RBS task. The following are example considerations:

1. RBS HEs performing barrier functions from a remote location do not experience the same physical and emotional effects as those performing their assigned functions at the protected facility. Those at the facility may physically experience the effects of an incident that activated the barrier, such as intense noise (a sonic velocity gas release), intense heat from a fire, choking and visually obstructing smoke, the violence of an explosion, or seeing a severely injured co-worker. *Those performing RBS tasks from the ROC do not equally share in the more visceral experience of "being there."* This may all be important input when determining whether an RBS task is appropriate. As such, the following should be considered:
   - Do not select RBS tasks if the HE needs to *be there* to adequately understand a situation and select and perform appropriate actions.
   - Conversely, consider RBS tasks that may measurably benefit from being physically separated from danger, such as those less likely to experience cognitive lockup, the negative effect of acute stress, or task-switch errors.
2. Consistent with a *do no harm* approach, select and design the RBS task following a process and mindset that prevents or minimizes its potential to negatively affect the efforts of the HEs located at the protected facility. For example:
   - Restrict verbally conveyed communications to pre-determined, disciplined communication channels. For example, communications may be limited to one designated HE in the ROC who is permitted to communicate with one designated HE at the protected facility. Procedures may define what communications are permitted (e.g., type, medium, and content) and their timing and duration.
   - Provide information using an approach that cannot divert attention at inopportune times, for example, use text or email to convey information that is not immediately needed but may be needed in the future. (See Meta SA, Appendix C.1.5)
   - Do not permit the RBS HE to change a display used by a protected facility HE (e.g., changing a CCTV camera view if not explicitly permitted to do so in the procedure).
   - Take no control actions unless explicitly requested by an HE at the protected facility.
   - Minimize or prevent RBS HE actions that may degrade the coordination, cohesion, morale, and trust of the barrier team.

Appendix M
Deepwater Horizon Accident: Active Barrier Causal Contributors

This appendix provides information on the Deepwater Horizon accident that occurred on April 20, 2010. Figure M.1 provides an event timeline indicating key events and active and active human barriers that were causal contributors to the accident. The extensive notes provide additional information to understand how each affected the accident escalation pathways, acceleration, and consequences. This information (and the examples provided throughout the book) is sourced from the accident investigation reports and books listed at the end of this appendix. See Skogdalen et al. (2011) for further insights into how emergency responders and others experienced these events.

NOTES

Note 1. Kick Detection (barrier activator): a detected kick is the activator for several active human barriers. On the day of the accident, simultaneous operations (SIMOPs) severely degraded the ability to accurately detect a kick and do so in a timely manner. Monitoring personnel (mud loggers) did not appear to be adequately trained (BOEM 2011, pp. 4, 99–102)

Note 1a: The kick progressed undetected for an estimated 49 minutes. Detection first occurred when one of the crew was idly viewing live CCTV camera views and noticed mud rising onto the drill floor (Boebert and Blossom 2016, pp. 177–178). At this point, it seems unlikely the crew knew how long the kick had been progressing, or when the mud/well fluid interface would move past the blowout preventer (BOP), enter the riser, and potentially progress to a blowout. When flammable gas from the well enters the riser, there are no inline valves to stop the rapidly expanding gas that can erupt violently. The interface movement speed accelerates as it moves up the riser.

Note 1b: A kick is detected when the measured fluid volume returned from the drill bore or well exceeds the injected mud fluid volume by a defined amount (e.g., 20 barrels). The increased volume in the returns occurs when well fluids enter the return stream (the well is flowing). To detect a kick, mud loggers access information from several systems and sources. Detection can be challenging under normal conditions. During this period, SIMOPs made detection more difficult if not infeasible. At one point, the detection effort stopped when the mud loggers misunderstood the operational events and activities (Boebert and Blossom 2016, p. 172).

TIME

ACTIVE & PASSIVE BARRIER STATUS

Event #

TIME	Event	#
20:51 (CSB 2014b, Fig 2.8)	Well becomes underbalanced. Kick begins.	1

Failed: Cement Barrier (BOEM 2011 Ch. V)
Severely degraded: Kick detection - Activator for several barriers. *Note 1.*

21:37 (CSB 2014b, Fig 2.8)	Well fluids enter the riser. (not detected)	2

"...the possibility of free gas getting into the riser in very deepwater locations is quite high and is probably the one events that is the most dangerous to rig floor personnel." (CSB 2016, p. 46)

21:40 (CSB 2014b, Fig 2.8)	Mud first seen on drill floor. (BOEM 2011, p.105) Mud shoots up derrick at 21:41. (BP 2010 p. 28)	3

21:43 (CSB 2014b, Fig 2.8)	Manually close BOP upper annular (CSB 201b, p. 29)	4

Failed: Manual closure of BOP upper annular. This first attempt to stop flow from the well fails, the valve leaks through. (CSB 2014b p. 29-30) *Note 6, 8*

21:43 (CSB 2014b, p.30)	Activated divert well flow to Mud Gas Separator (MGS) located near drill Floor.	5

Failed: Diverter not lined up to divert well-fluids to overboard lines. *Note 2*

21:44 (BP 2010 p. 28)	Mud begins to flow from open MGS vents	6

"Immediately after shutting the annular, the rig crew also activated a diverter...preset to flow directly to the MGS... the MGS was overwhelmed moments after the diverter was activated, and hydrocarbons began blowing out of exit points onto the rig." (CSB 2014b, p. 30)

21:47 (CSB 2014b, Fig 2.8)	Manually close one or more BOP pipe rams	7

Succeeds: Manual closure of one or more BOP pipe rams stops *new* flow into riser (CSB 2014b, p. 30). *Gas already in riser continues flow to the rig. Note 6*

21:46-21-47 PM (BP 2010 p. 28)	Well flow/pressure overwhelms the MGS causing it to rapidly vent gas that quickly spreads to unclassified areas: • HVAC air intakes to engine and switchgear rooms • Combustion air inlet to Main Generators #3 and #6 causing uncontrolled overspeed. (USCG 2010 pp. x, 3-7, 24/5, 32)	8

Bypassed/Inhibited: Automatic, flammable gas alarm-activated barriers: general alarm (muster), engine room ESD (trips HVAC, electrical circuits.) (USCG 2010 p. xi, 14, 20, Williams 2010)

Failed: Manual activation of above-listed barriers. (USCG 2010, p. 27, BOEM 2011, p. 197). *Notes 3, 4*

Failed: Generator #3 overspeed protection systems fails. Gas ingress causes uncontrolled overspeed to destruction and possible ignition source. The same may be true for Generator #6. Auto-start of backup generator fails. (BOEM 2011 pp. 115-117, 125-126, 197, USCG 2010 p. 14/5, Williams 2010) *Note 5*

21:47 (BP 2010 p. 28)	First gas alarm occurs at the shale shaker house followed by alarm from Drill Floor. 20 additional alarms indicate high gas concentration. (BOEM 2011 p. 106)	9

Explosion location and ignition source: USCG 2010 pp. 3, 23, H-14, BOEM 2011, pp. 122/3, Boebert & Blossom (2016, P. 179). *Note 9.*

21:48/9 (USCG 2010 H-14, CSB 2014b, Fig. 2.8, p. 30)	**First explosion at or near the MGS** • 10 fatalities – On/near/below drill floor and MGS (USCG 2010, Figs. 2, 4, p.23) • 9 injuries – Living quarters galley, laundry, gym, shower, hall, stairway. (USCG 2010, Fig. 4, Table 1)	10

Reports differ on the explosion locations and the explosions/loss of power event sequence. From BP 2010, p. 29, first power loss, 1st explosion 5 seconds later, 2nd explosion 10 seconds after first. (It remains unknown if both generators failed from uncontrolled overspeed.)

21:48/9 (BP 2010 p. 29, CSB 2014b p. 30, Figure 2.8)	**Second explosion in/near engine room #3** • 1 fatality – Explosion caused crew member to fall from crane cab. (USCG 2010, Fig. 3, Table 1) • 7 injuries - Electrical Control Room, crane, offices. (USCG 2010, Fig. 4, Table 1). • Destroys electrical distribution system, BOP communication links (USCG 2010 p. 31, BOEM 2011 p. 206, item 3) • Loss of power to FW pumps, DP thrusters, lighting • Destroys escape routes, areas near engine room # 3 (USCG 2010 p. 5, 47, H-16, Williams 2010) • Automatically activates AMF/Deadman closing blind shear ram. (CSB 2014b p. 42)	11

Explosion location and ignition source:
* One of the injuries occurred when transiting to or boarding lifeboat.

Disabled: Power loss disables electric-driven fire water pumps (fire-fighting barrier), emergency lighting (muster and abandon barrier) and dynamic position thrusters (station keeping barrier). (USCG 2010 p. xiii, 23, 35, 40-43, 55, H-16)

Failed: AMF/Deadman causes the blind shear ram valve (BSR) to close. BSR fails to fully close and seal the well. Well flow to the rig continued (CSB 2010b p. 27, 42). DWH remains tethered to the BOP and well-head. *Note 6*

Design Limited: Firewalls not rated to protect personnel from an oil fire (heat or explosion). Personnel at the drill floor, drill shack, mud pits, mud logger room and living quarters were not protected from a drill floor or engine room fire/explosion. (USCG 2010 p. 39.) *Note 11.*

21:50-21:55 (USCG 2010 H-14)	Chief Mate in CCR manually activates general alarm to initiate muster barrier. (USCG 2010 p. H-14)	12

"Personnel on the bridge waited approximately 12 minutes after sounding of the initial gas alarms to sound the general alarm, even though they had been informed that a "well control problem" was occurring. During this period, there were approximately 20 alarms indicating the highest level of gas concentration in different areas on the rig." (BOEM, 2010, p. 4)

21:51 – 21:56 (USCG 2010 H-14/5, BOEM 2011 p. 23)	Chief mate advises Master to abandon. Without authorization, subsea supervisor manually activates BOP Emergency Disconnect system (EDS). Mayday issued 21:52. (BP 2010, p. 29)	13

Failed: Late manual EDS activation fails. Crew unaware of the earlier possible activation by AMF/Deadman. See event 11 above. *Notes 6, 11*

22:00 (USCG 2010 H-17)	OIM arrives CCR @ 21:55-2200 (USCG 2010 p.H-15) Over public address, on-watch DPO gives verbal instruction to report to lifeboats and emergency stations. (USCG 2010 H-16, BOEM 2011, p. 107)	14

Degraded 'Abandon Barrier': Inadequate training. Fear and panic driven actions. Lifeboats not loaded as assigned/planned. Confusion delays - who has authority to give the lifeboat launch command? Persons left behind forced to jump to sea. Life raft launch and escape threatened by excessive heat, attached paint rope. (USCG 2010 pp. 48-59, 64/5, 128, H-16/17)

22:00 (USCG 2010 H-17)	Damon Bankston (DB) prepares FRC for launch	15

22:12-23:32 (USCG 2010 H-17 to H-20)	DWH life boats (x2) and life-raft (x1) launched. DB initiates Search and Rescue using Fast Rescue Craft (FRC)	16

Successful barrier: DB's rapid activation of S&R barrier (FRC) recovers 15 people from the ocean preventing further loss of life. (USCG 2010 p. 67) All 115 survivors recovered to Damon Bankston. (USCG 2010 p. 128/9, H-17 to H-20)

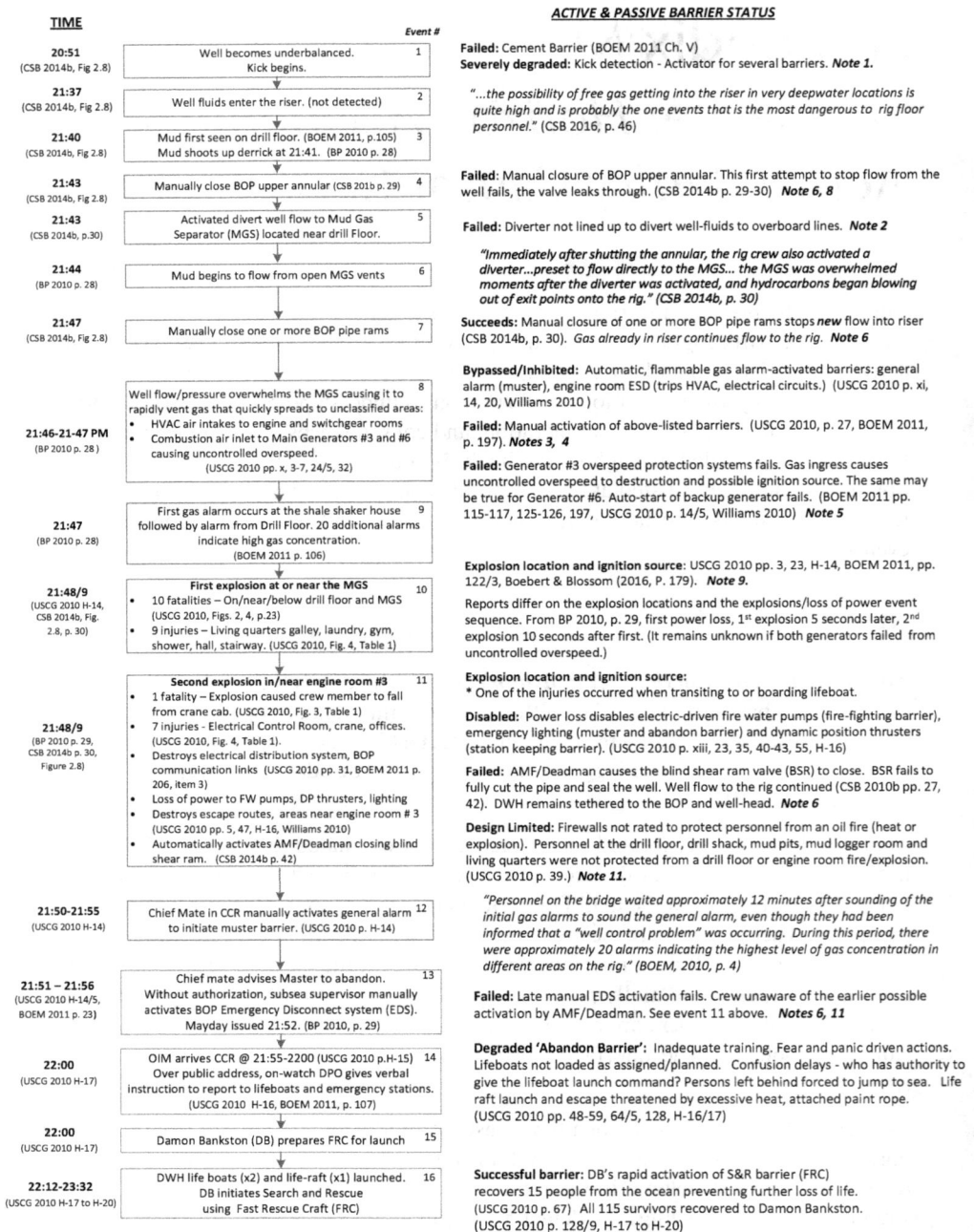

FIGURE M.1 Deepwater Horizon Accident: Barrier Failure Timeline; BOP – blowout preventer; CCR – Central Control Room, aka., the Bridge; DPO – Dynamic Position Officer; EDS – Emergency Disconnect System; ESD – emergency shutdown; HVAC – Heating, Ventilation, and Air Conditioning; OIM – Offshore Installation Manager; PA – public address system; S&R – search and recovery.

Note 1c: BOEM (2011, pp. 75–76) discussed an incident that had occurred six weeks earlier on the DWH. The crew missed a kick that went undetected for 30 minutes though did not progress to a blowout. The incident was not formally investigated. BP became concerned with the potential knowledge gaps among the mud loggers and DWH drillers. A BP well leader further believed that the DWH crew may be becoming overly comfortable and potentially complacent given their prior experience (and perceived successes) with difficult wells.

Note 1d: A well-control incident occurred at a different BP facility (different location and crew). BP procedures required that two independent barriers always be in place. Despite that requirement, a kick occurred during a transition phase when only a single barrier was in place. The single barrier failed to function as required. In response to the incident, BP issued a company-wide incident report that emphasized the two-barrier requirement. Unfortunately, DWH only first received the report the morning of the DWH accident.

Note 1e: On August 21, 2009, the Montara wellhead platform (located off the coast of Australia) experienced a blowout and long-term spill into the ocean. Similar to DWH, the incident started with a failed cement barrier. If BP received a report from the accident, it is not clear if it was distributed to the DWH facility or its onshore well support team.

Note 2. Divert Overboard System *(Divert Overboard Barrier):* on the day of the accident, the crew responded to the kick by manually lining the divert system to *the Mud Gas Separator (MGS).* BP's and Transocean's well-control manuals required diverting a kicking well to one of the two overboard lines "when flow rates are too high for the mud gas separator" (BOEM 2011, p. 104). As discussed in notes 1a and 1b, conditions on the day of the accident may have made this determination extremely difficult, if not impossible.

Once lined up, the MGS quickly filled and over pressured, forcing fluids into its two open vent lines. One vent was physically directed (goose-necked) downward toward the drill floor and the HVAC air inlets to several buildings. These buildings housed non-classified electrical equipment and the turbine-drive (diesel-fueled) generators providing normal and emergency power. The turbine combustion air inlets were located inside the buildings.

Note 2a: The MGS is a low-capacity, low pressure–rated process vessel (e.g., 60 PSI). It was designed to separate flammable gas from the returning mud under normal operating conditions. Procedures governing its use appear to present ambiguities.

> Transocean's 2008 well control handbook states it is essential to verify the [mud gas separator] system is capable of handling the maximum amount of fluid and gas that could be produced by the well in the case of a severe kick'.
>
> *(CSB 2016, p. 39)*

Earlier discussions addressed the conditions that required diverting the kick to the overboard lines. Doing so could dump oil into the sea, triggering a regulatory action.

It could also result in dumping recoverable and expensive mud into the sea, a possible conflict with corporate and local guidance to reduce costs.

Perhaps not considered, CSB (2016, p. 44) indicated ten steps (or more) were required to switch the divert system valve lineup from the MGS to the overboard lines. Doing so from memory while under time pressure warranted that this be learned as a skill-based competency. Achieving the skill would require repetitive exercises followed by testing to confirm the skill competency was met. Once learned, periodic competency testing would then be needed to verify the skill had not faded over time. (All complex skills fade over time if not used or refreshed.) Because that competency was not required and given the normal practice of setting the divert system lineup to the MGS, the actual reliability of the divert overboard barrier was likely much lower than might have been expected in the original design. (For information on skill fade, see Appendix J.7.1. The PIF assessment described in Appendix H.2 (e.g., the societal influencer PIF) may have identified this failure scenario as plausible and therefore a risk.)

Note 2b: This same procedure deviation occurred six weeks earlier in response to a kick that was also detected late (see Note 1c). It is plausible the crew perceived the deviation as safe because nothing bad happened with that incident. An inaccurate perception of this type could be countered by a post-event analysis that identified the event as a near-miss and possibly determine a gas-in-the-riser or blowout event was closer (timewise) than originally perceived. Further, had the change management process been used to update the MGS/divert system procedures, it may have revealed the unsafe ambiguities in the procedure and the risk posed by a major kick that quickly exceeds the MGS pressure ratings and capacity.

Note 2c: Consider the above procedure deviation as a potential drift scenario (individual and organizational drift). The default lineup to the MGS occurred at least twice, the prior incident discussed in Note 1c and the day of the accident. Diverting a kicking well to the overboard lines is the only remaining barrier when gas enters the riser. However, it appears that the operational use of the MGS in response to a kick was the default. CSB (2016, p. 41) identified this deviation as an example of organizational drift. (See Appendix J.7.2 for additional background on drift.)

Note 2d: Training and the divert system procedure ambiguities contributed to the accident (CSB 2016, pp. 38–42). (The operate and maintain phase in lifecycle model includes processes to monitor, detect, and address both conditions. See Chapter 7.)

Note 2e: The MGS had piping connections that directed well fluids in many different directions. This introduced new potential pathways that created additional loss-of-containment events and fires.

Note 2f: Weather conditions at the time of the blowout were a still night. The absence of the typical wind and air movements did not aid in diluting and dispersing the growing flammable gas cloud. Instead, the cloud expanded toward several building HVAC air-inlet stacks. The stacks to the two buildings containing the two "independent" electrical power sources and systems were spaced roughly 60 feet apart. The gas cloud encompassed both stacks 190 seconds after the gas first exited the MGS vents. The entering gas had already reached the lower flammable limits for methane and ethane (BP 2010, pp. 135–136). The cloud may have also migrated to other areas in the facility that had also exposed ignition sources.

Note 3. Flammable Gas Detector Alarms (activators for several active barriers): from witness testimony, personnel were not adequately trained on how to respond when the flammable gas alarms rapidly progressed from their lower to higher level alarm states (Fleytas 2010). (Also, some of the detectors were not operational, USCG 2010, pp. 3, 20). When queried on the delayed responses to critical actions, the Control Room Operator (a junior member of the marine crew) said the rapidly evolving accident event was overwhelming (Fleytas 2010).

Note 4. Bypassed/Inhibited Flammable Gas Detector Alarms (alarm-activated barriers): DWH personnel were not adequately trained on the added safety critical tasks that must be performed when the normally automatic active barriers were switched to a bypass/inhibit mode of operation (USCG 2010, p. 32; BOEM 2011, p. 107). In this mode, the barrier reverts to an active human barrier. Reports raised questions on who was responsible for monitoring and manually activating barriers in this mode (referring to the general alarm and building ESD function). Depending on the accident report, the responsibilities lie with the drillers (located in the drill shack) or with the dynamic position officers in the control room (bridge). Indicated as normal practice, the drillers seldom monitored the fire and gas displays located in front of the drill chairs (Boebert and Blossom 2016, p. 148).

Note: An activator bypass is designed and provided to support online testing and repair. In the Deepwater Horizon case, the bypasses were used for other purposes, e.g., preventing perceived nuisance/unnecessary barrier activations.

Note 5. Diesel-Drive Turbine Generators (provided electrical power to many active barriers.): The main power generators #3 and #6, located in different buildings, were operating the day of the accident. Regulations required two independent sources of electrical power. Failed active barriers and the consequence of the blowout resulted in the loss of both generators. (Those losses and the backup generator's failure to start resulted in the complete loss of all electrical power to the firewater pumps, the thrusters used to hold the DWH facility on station, and the facility's normal and emergency lighting.)

Note 5a: From note 2f, flammable gas entered both buildings and the combustion air inlets to generators #3 and #6. The gas ingress caused one (#3) or possibly both machines to overspeed uncontrollably to a catastrophic machine failure.

Note 5b: Each generator was protected by three different overspeed protection systems that were designed using diverse technologies and systems. A member of the crew testified that engine #3 progressed to an uncontrolled overspeed and failure (Williams 2010). Regulations at the time required an annual demonstration test for 20% of the overspeed protection systems. Given the triple failure on engine #3, it seems unlikely that all three systems were included in this rotational testing plan (BOEM 2011, pp. 116–117). Alternatively, it remains unknown if these systems were set to a bypass/inhibit state, disabling their safety function.

Note 5c: The backup generator failed to auto-start when generators #3 and #6 were lost. Manual attempts to start the generator also failed. The reason appears to be unknown. (Was it due to the nature of the damage to the electrical distribution system caused one of the initial explosions?)

Note 6. Blowout Preventer (BOP) Valves and Functions (act phase elements for many barriers):

The upper annular valve in the BOP was the first to be manually commanded closed in response to the kick. It failed to stop the flow. The slow-closing valve was a rubber, donut-shaped device that compresses around the pipe, blocking new well flow into the annular space. It was not designed to close under high flow conditions.

Note 6a: The upper annular valve was damaged four weeks earlier when a drill pipe (and a pipe joint) was accidentally moved so it was pulled through the tightly closed valve. At least some level of valve damage appeared to be confirmed when rubber pieces from the valve were discovered in the mud returns (Williams 2010). The incident was informally investigated. Though it appeared the primary focus of the investigation was discovering the cause of the unexpected pipe movement (e.g., a control system error or human error). No apparent action was taken to assess the potential valve damage or to repair it. The same valve was used in the failed negative pressure test on the day of the accident and was the first valve closed in response to the detected kick. Given the prior accidental pipe movement incident, it remains unclear if the crew considered if (and how) a potentially damaged valve could affect those uses and results.

One of the reports indicated the lower annular valve may have also been commanded closed in response to the kick. If so, it also did not stop the flow. The valve was previously modified to allow pipe movement when the valve was closed. With that modification, the valve no longer had tight shutoff capability.

Note 6b: BOP valves were monitored and manually controlled from subsea control panels located in the drill shack and the control room (bridge). The drillers typically controlled the annular valves. Other BOP valves were typically operated by a subsea specialist who was typically present in the drill shack during safety critical well operations. During the early efforts to control the kick, no specialists were in the drill shack. "This somewhat unusual arrangement may have led to delay or confusion in the last crucial minutes before the blowout, when no subsea specialist was in the drill shack" (Boebert and Blossom 2016, p. 149).

Note 6c: Blind Shear Ram Valve (BSR): when all other actions taken to stop the kick failed, the last remaining active barrier relied on the blind shear ram. The valve function was cutting through the pipe and sealing it in a way that blocks new flow from the well. (The BSR did not have the capability to cut through pipe joints.) CSB (2014b) determined the valve was commanded closed. Reports differ on the timing and the source of the closure command. The command may have resulted from the automatic activation of the AMF/Deadman function that resides within the BOP.

When the BSR closed, the pipe was not centered in the BSR assembly. Part of its diameter lay outside of the shear ram range. This created a worst-case scenario.

When the BSR closed, only part of the pipe diameter was cut through, opening a new flow path that allowed the well to flow unabated into the riser and toward the DWH facility. (One or more of the reports stated a pipe ram valve was closed stopping further flow from the well into the riser. If so, that action could not prevent the high-pressure gas already in the riser from offloading onto the DWH facility.)

Reports differ on what caused the pipe to move off-center in the BSR assembly. CSB (2014b, pp. 32–33) stated the closure of the pipe ram may have created differential a pressure within the pipe that caused it to buckle, an effect known as effective compression. Another report suggested the pipe movement occurred when the gas-filled riser became buoyant causing it to lift, hit the DWH facility, and then drop in a way that moved the pipe in the BSR assembly. (Several crew members stated they felt a jolt, a data point that may offer possible support for the buoyancy lift theory.)

Note 6d: Status feedback from the BOP (e.g., heath, activations, and valve positions) and the ability to control BOP valves was lost when the communication cables between the BOP and the subsea control panels were critically damaged by one of the initial explosions. This missing feedback and an unawareness that control from the panel was no longer possible were possible contributors to delayed decisions and actions, for example, the late decision to abandon the facility.

Note 6e: Activation of the EDS was designed to automatically close the blind shear ram and sever pins that held the top section of the BOP to its lower assembly. If these actions occur, DWH could use its thrusters to move away from the wellhead area towing the riser and BOP annular section behind it. Even if the severing of the pins had occurred, the incomplete cutting of the riser pipe ensured that the DWH facility remained tethered to the BOP and wellhead (CSB 2014b, Sections 2.2, 2.5, 3.2.3).

Note 6f: Permission to activate the EDS lies with the Offshore Installation Manager (OIM), who first reached the Central Control Room after the initial catastrophic explosions. (See Note 10.) Before his arrival, reports indicated the subsea supervisor attempted to manually activate the EDS, possibly without OIM authorization. By that time, damage to communication cables likely disabled the ability to transmit the EDS command to the BOP (Note 6d).

Note 6g: AMF/Deadman BOP Function – the BOP had self-contained logic that automatically activated the Emergency Disconnect System function if several simultaneous conditions occurred (CSB 2014b, Section 2.3.3). One or both initial explosions make it more likely that these conditions were met, causing the function to activate automatically.

Note 7. Cognitive functions and decision-making were likely affected by the speed, violence, and effects of the initial explosions and fires, the injuries and unknown status of co-workers, the major breakdowns in the emergency response, and the loss of emergency lighting and firefighting capability. Personnel at the lifeboat were unable to complete a simple count-off prior to loading. One person arrived at the lifeboats, became impatient, and then jumped into the sea, an act that could have resulted in a severe injury or a fatality (USCG 2010, 48, H-17; Skogdalen et al. 2011).

*Note 8. **The BOP was not adequately maintained*** or tested as required by regulations (USCG 2010, pp. xviii–xix, 96–97).

Note 8a: When the DWH arrived at the Macondo well site on January 21, 2010, the BOP was sitting on the deck. In preparation for its installation on the wellhead, Transocean personnel spent five days performing tests and maintenance identified by a condition-based maintenance system. The work was a subset of the maintenance tasks required by a past-due, regulatory-required, 5-year maintenance cycle. It did not appear the performed work included replacing several critical batteries that later failed or became fully discharged (Boebert and Blossom 2016, p. 58).

Note 8b: The BOP contained two independent pods that controlled its valves and other functions. By regulation, a failure in either pod required an immediate work stoppage and removal of the BOP for repairs. In the lead-up to the accident, one of the pods failed. The BOP was not removed or repaired.

Note 8c: A post-event analysis identified several wiring errors in the BOP and other changes that were not marked on the BOP drawings. Battery failures and wiring errors contributed to BOP malfunctions. The unmarked changes hindered post-blowout efforts to regain control of the well using the BOP. (The BOP assembly on the seafloor included a control panel that could be accessed by a remote-operated vehicle.)

Note 9. According to Boebert and Blossom (2016, p. 179),

> One explosion was centered in the area of the drill floor, probably at or near the mud-gas separator, and instantly killed the ten men on the floor and below it in the shaker room, pump room, and mud engineer's office. The explosion destroyed the entire area of cabins and offices, flinging the Transocean manager and senior toolpusher against the corridor walls and collapsing the OIM's shower around him. Others in that area were buried under the wreckage of partitions and ceilings, and many were hurt. The second explosion was aft, near one of the engines, and injured the members of the engineering group who were working in their space between the engine banks.

This information is consistent with the USCG (2010, p. 3) report.

*Note 10. **Barrier Leader Absence/Role Confusion/Communication Challenges:*** Throughout the accident, there was confusion among the crew on who was in charge, the captain or the OIM. A clerical error in the original facility registration placed the OIM in charge during drilling operations. The captain was in charge during emergency operations and when the Deepwater Horizon was physically moving. Uncertainties also existed as to whether the captain had the authority to activate the EDS.

Note 10a: Refer to Note 6f. The late OIM arrival to the control room may have contributed to the chaotic and confusing emergency response that preceded his arrival. (The ER team was missing its leader.) Personnel in the control room (bridge) could not visually see the urgent activities taking place in the drill shack, as live CCTV video feeds were not provided between these locations (Boebert and Blossom 2016, p. 149). Getting a status update required use of the radio or telephone, an attention capture task for both parties. Personnel in the drill shack were fully engaged in their efforts to regain control of the well.

Note 11. Delayed Muster Barrier Activation: The muster barrier, practiced weekly, required all hands to promptly transit to their assigned muster locations and emergency response stations when the general alarm sounds. According to USCG (2010, p. 61), the general alarm was designed to automatically activate when two or more gas sensors were in alarm. The same functionality applied when the automatic activation feature was placed in a bypass/inhibit mode. In this mode, the crew must continuously monitor alarms and, when defined alarm conditions were present, must then manually activate the appropriate barriers (safety functions).

From BOEM (2011, p. 4), the estimated interval between the initial flammable gas alarms and the general alarm activation was 12 minutes. From other sources, noted in Figure M.1, the alarm was activated 10–15 minutes after the kick was first noticed, 6–11 minutes after the initial kick response action failed, and 4–5 minutes after the initial two catastrophic explosions. If the latter number is correct, it appears plausible the delayed alarm might have contributed to one or more preventable injuries and injury severity. It seems less clear if the delay could have contributed to one or more fatalities. Further analysis seems prudent to more thoroughly assess and report the effects of the delayed alarm activation. From that information, it may be possible to assess the latest acceptable alarm activation. The accident demonstrated how quickly the loss-of-containment event placed personnel at risk. The suggested analysis may have revealed the need to place greater controls on the use of a bypass/inhibit feature that disables the automatic activation of the general alarm. (A safety function that relies on a human to detect an actionable alarm condition (the barrier activator) and complete all responding manual actions within ten minutes is generally not consistent with good engineering practice.)

The actual time needed to begin the muster response movement and complete the transit to the assigned emergency response station, the muster area of the lifeboat embarkation area, depends on many factors. A factor unique to each person, transit time, depends on several affecting factors present when the muster alarm sounds. Examples include current location (distance from the assigned muster station), what the person was doing (e.g., performing a task that must be made safe before leaving), if the person is injured or aiding another who may be injured, and the condition and availability of the egress/escape route (e.g., destroyed, blocked, or obstructed by smoke or loss of lighting). As such, the situations and transit times are different for each person. For those located near their assigned muster stations, the transit time may be a few minutes. For others, more time is needed. When the initial two explosions occurred, the transit time increased considerably. Personnel exposure to danger increased exponentially. The explosions destroyed the escape route to the aft lifeboats, caused the living quarter walls to collapse inward and ceiling tiles to fall. It also caused fire doors and computer floor tiles to blow out and the loss of all emergency lighting. Movements through many damaged spaces occurred in the dark. Similar (or worse) conditions occurred in the Electrical Control Room and the spaces at/near/below the drill floor. Movements were further hampered by dense fire and smoke, falling objects, flying shrapnel, low oxygen content, and the noticeable presence of methane gas.

From the included references, all injuries (with one possible exception) and all fatalities were attributed to the effects of these explosions and the subsequent fires. (One fatality may have resulted when the concussive effect of the explosion caused or contributed to a person's fall from a crane.)

Note 11a: A later book on the accident relied on testimony from the court case brought against BP. This source identified the first explosion as the likely event that caused 10 of the 11 fatalities (Boebert and Blossom 2016, p. 179). Two of the fatalities were maintenance personnel working on pumps in the space below the MGS. Perhaps an earlier general alarm activation might have caused one or both to leave the area before the fatal explosions or, at least, move farther from the explosion center and effects.

Note 11b: Not addressed in the various reports was: how did the absence of a general alarm affect personnel, contractors, and visitors who were unaware of the kick and the escalating danger? Did the alarm's absence contribute to a perception that the unusual sounds and a possible jolt from the now buoyant riser hitting the facility were not dangerous? To what extent did the absence of the alarm (in the presence of fire, explosions, and other violent events) degrade the crew's cognitive capabilities or contribute to the non-rational behaviors observed at the lifeboats?

APPENDIX M SOURCE REFERENCES

The information in Figure M.1 and the notes above were sourced or informed from the following documents and hearings:

- BP (2010), Deepwater Horizon Accident Investigation Report, BP, September 8, 2010
- Boebert, E., Blossom, J.M. (2016), *Deepwater Horizon, A Systems Analysis of the Macondo Disaster,* Harvard University, 1st Ed., 2016
- BOEM (2011), Report regarding the causes of the April 20, 2010, Macondo Well Blowout, BOEM, September 14, 2011
- CSB (2014a), Investigation report volume 1, Explosion and Fire at the Macondo well, Report No. 2010-10-I-OS, June 5, 2014
- CSB (2014b), Investigation report volume 2, Explosion and Fire at the Macondo well, Report No. 2010-10-I-OS, June 5, 2014
- CSB (2016), Investigation report volume 3, Drilling Rig Explosion and Fire at the Macondo well, Report No. 2010-10-I-OS, April 17, 2016
- Fleytas (2010), Andrea Fleytas live testimony @ joint USCG/BOEM inquiry
- Konrad, J., Shroder, T. (2011), *Fire on the Horizon, the Untold Story of the Gulf Oil Disaster,* HarperCollins Publishers, 2011
- Skogdalen, J.E., Khorsandi, J., Vinnen, J.E. (2011), Looking Back and Forward – Evacuation, Escape and Rescue (EER) from the Deepwater

Horizon Rig, Deepwater Horizon Study Group Working Paper – January 2011

- USCG (2010), Report of Investigation into the Circumstances Surrounding the Explosion, Fire, Sinking and Loss of Eleven Crew Members Aboard the MODU Deepwater Horizon in the Gulf of Mexico, April 20–22, 2010, Volume 1, United States Coast Guard
- Williams (2010), Mike Williams live testimony @ joint USCG/BOEM inquiry

References

Bea, R., Mitroff, I., Faber, D., Foster, H., Roberts, K.H. (2009), A new approach to risk: The implications of E3, *Risk Management*, 11, 30–43. doi:10.1057/rm.2008.12

Boebert, E., Blossom, J.M. (2016), *Deepwater Horizon a Systems Analysis of the Macondo Accident*, Harvard University Press, Cambridge, MA

BOEM (2011), Report regarding the causes of the April 20, 2010, Macondo Well Blowout, BOEM, September 14, 2011

Borders, J., Klein, G., Besuijen, R. (2024), Mental model matrix: Implications for system design and training, *Journal of Cognitive Engineering and Decision-making*. sagepub. com.

BP (2010), Deepwater horizon accident investigation report, BP, September 8

CAA (2021), Awareness of skill fade and suggested mitigations, UK Civil Aviation Authority, Safety Notice SN-2021/011

Carter, R., Aldridge, S., Page, M., Parker, S. (2014), *The Human Brain Book*, 2nd Ed, DK Publishing, New York

CCPS (2007a), *Human Factors Methods for Improving Performance in the Process Industries*, John Wiley & Sons, Inc., Hoboken, NJ

CCPS (2007b), *Guidelines for Safe and Reliable Instrumented Protective Systems*, AICHE, New York

CCPS (2015), *Guidelines for Initiating Events and Independent Protection Layers in Layer of Protection Analysis*, John Wiley & Sons Inc., Center for Chemical Process Safety (CCPS), New York

CCPS (2018), *Bow Ties in Risk Management, A Concept Book for Process Safety*, John Wiley & Sons Inc., Center for Chemical Process Safety (CCPS), Hoboken, NJ

CCPS (2022), *Human Factors Handbook for Process Plant Operations, Improving Safety and Systems Performance*, John Wiley & Sons Inc., Center for Chemical Process Safety (CCPS), New York

Chemuturi, M. (2013), *Requirements Engineering and Management of Software Development Projects*, 1st Ed, Springer, New York

Chiappe, D., Strybel, T., Vu, Kim-Phuong L. (2012), Mechanisms for the acquisition of shared SA in situated agents, *Theoretical Issues in Ergonomic Science*, 13(6), 625–647

CIEHF (2016), Human barriers in Barrier Management, a white paper by the Chartered Institute of Ergonomics and Human Factors, 12/2016, CIEHF

COS (2013), *COS-3-02, Skills and Knowledge Management System Guideline*, Center for Offshore Safety, 1st Ed, December 2013

COS (2020), *COS-3-06, Guidance for Developing and Managing Procedures*, Center for Offshore Safety, 1st Ed, January 2020

CSB (2014a), Explosion and fire at the Macondo well, Investigation report volume 1, Report no 2010-10-I-OS, U.S. Chemical Safety and Hazardous Investigation Board, Washington, DC.

CSB (2014b), Explosion and fire at the Macondo well, Investigation report volume 2, Report no 2010-10-I-OS, U.S. Chemical Safety and Hazardous Investigation Board, Washington, DC.

CSB (2016), Drilling rig explosion and fire at the Macondo well, Investigation report volume 3, Report no 2010-10-1-OS, U.S. Chemical Safety and Hazardous Investigation Board, Washington, DC.

Decker, S., (2011), *Drift into Failure, From Hunting Broken Components to Understanding Complex Systems*, Ashgate Publishing Ltd.

Durso, F.T., Hackworth, C.A., Truitt, T.R., Crutchfield, J., Nikolic, D., Manning, C.A. (1999), Situation awareness as a predictor of performance in en route air traffic controllers, Federal Aviation Administration (US Dept. of Transportation), DOT/FAA/AM-99/3, January 1999

Edmonds, J. (2016), *Human Factors in the Chemical and Process Industries, Making it Work in Practice*, Elsevier, Amsterdam

EI (2020a), *Guidance on Human Factors Safety Critical Task Analysis*, Energy Institute, London, 2nd Ed, January 2020

EI (2020b), *Guidelines for Management of Safety Critical Elements (SCE)*, Energy Institute London, 3rd Ed, January 2020

Endsley, M.R. (1995), Toward a theory of situational awareness in dynamic systems, *Human Factors*, 37(1), 32–64

Endsley, M.R., Garland, D.J. (2000), *Situation Awareness Analysis and Measurement*, CRC Press, Boca Raton, FL

Endsley, M.R., Jones, D.G. (2001), A Model of inter- and intrateam situation awareness: Implications for design, training, and measurement. In M. McNeese, E. Salas, M. Endsley (Eds.) *New Trends in Cooperative Activities: Understanding System Dynamics in Complex Environments* (pp. 46–67). Human Factors and Ergonomics Society, Santa Monica, CA.

Endsley, M.R., Jones, D.G. (2012), *Designing for Situation Awareness: An Approach to User-Centered Design*, 2nd Ed, CRC Press, Boca Raton, FL

Endsley, M.R. (2021), *Handbook of Distributed Cognition: Contemporary Research Models, Methodologies, and Measures in Distributed Team Cognition*, 1st Ed, M. McNeese, E. Salas, M.R. Endsley (Eds.), CRC Press, Boca Raton, FL

Fleytas (2010), Andrea Fleytas live testimony at joint USG/BOEM inquiry into deepwater horizon accident

Flin, R., O'Connor, P., Crichton, M. (2008), *Safety at the Sharp End: A Guide to Non-Technical Skills*, Ashgate Publishing, Aldershot

Gasaway, R.B. (2013), *Situation Awareness for Emergency Response*, PennWell Corporation, Tulsa, OK

Guastello, S.J. (2023), *Human Factors Engineering and Ergonomics, a Systems Approach*, 3rd Ed, CRC Press, Boca Raton, FL

Harris, Don (2011), *Human Performance on the Flight Deck*, CRC Press, Boca Raton, FL

Hendricks, J.W., Peres, S.C., Dumlao, S.V., Armstrong, C.A., Neville, T.J. (2021), The impact of hazard statement design elements in procedures: Counterintuitive findings and implications for standards, *Human Factors: The Journal of the Human Factors and Ergonomics Society* (Human Factors Society Journal), 65(7). https://doi.org/10.1177/00187208211050137

Hollnagel, E. (2003), *Handbook of Cognitive Task Design*, E. Hollnagel (Ed.), Lawrence Erlbaunm Associates Inc., Mahwah, NJ (Reprinted by CRC Press, 2010)

HPOG (2021), Best practice in procedure formatting, Human Performance Oil & Gas, 2/2021. www.HPOG.org

HSE (2007a), Managing competence for safety-related systems, Part 1, Key Guidance, UK Health and Safety Executive

HSE (2007b), Managing competence for safety-related systems, Part 2, Supplementary Material, UK Health and Safety Executive

IEC 61511-1 (2016), Functional safety – Safety instrumented systems for the process industry sector – Part 1: Framework, definitions, system, hardware and application programming requirements, Ed. 2, International Electrotechnical Commission

IEC 61511-2 (2003), Functional safety – Safety instrumented systems for the process industry sector – Part 2: Guidelines for the application of IEC 61511-1, 1st Ed, International Electrotechnical Commission

IFE (2022), *The Petro-HRA Guideline*, Rev. 1, Vol. 1, IFE/E-2022/001, ISBN 978-82-7017-937-4, Institute for Energy Technology

IOGP (2018), Introducing behavior markers of non-technical skills in oil and gas operations, IOGP Report No 503, International Association of Oil and Gas Producers

IOGP (2019a), Recommendations for enhancements to well control drills in the oil and gas industry, IOGP Report No 628, May 2019, International Association of Oil and Gas Producers, London

IOGP (2019b), Recommendations for enhancements to well control training, examination, and certification, IOGP Report No 476, 2nd Ed, International Association of Oil and Gas Producers, London, November 2019

ISO 11064-1:2000, Ergonomic design of control centres – Part 1: Principles for the design of control centres, International Organization for Standardization, 1st Ed, 2000

ISO 11064-3:1999, Ergonomic design of control centres – Part 3: Control room layout, International Organization for Standardization, 1st Ed, 1999

ISO 11064-4:2013, Ergonomic design of control centres – Part 4: Layout and dimensions of workstations, International Organization for Standardization, 2nd Ed, 2013

ISO 11064-5:2008, Ergonomic design of control centres – Part 5: Displays and controls, International Organization for Standardization, 1st Ed, 2008

ISO 11064-6:2005, Ergonomic design of control centres – Part 6: Environmental requirements for control centres, International Organization for Standardization, 1st Ed, 2005

ISO 13702:2015, Petroleum and natural gas industries, Control and mitigation of fires and explosions on offshore production installations – Requirements and guideline, International Organization for Standardization, 2nd Ed, 2015–08

Kahneman, D. (2011), *Thinking, Fast and Slow, Farrar, Straus, and Giroux* (paperback edition)

Karawowski, W., Szopa, A., Soares, M.M. (2021), *Handbook of Standards and Guidelines in Human Factors and Ergonomics*, 2nd Ed, CRC Press, Boca Raton, FL

Kelly, F.E., Frerk, C., Bailey, C.R., Cook, T.M., Ferguson, K., Flin, R., Fong, K., Groom, P., John, C., Lang, A.R., Meek, T., Miller, K.L., Richmond, L., Sevdalis, N., Stacey, M.R. (2023), Human factors in anesthesia: A narrative review, *Anesthesia*, 78, 479–490, Association of Anaesthetists.

Klein, G.A. (1993), A recognition-primed decision (RPD) model of rapid decision-making. In G.A. Klein, J. Orasuanu, R. Calderwood, C.E. Zsambok (Eds.), *Decision-Making in Action: Models and Methods* (pp. 138–147). Albex Publishing, Westport, CT

Klein, G.A. (2003), *The Power of Intuition*, Currency Doubleday

Konrad, J., Shroder, T. (2011), *Fire on the Horizon, the Untold Story of the Gulf Oil Disaster*, HarperCollins Publishers

Mlodinow, L. (2012), *Subliminal: How Your Unconscious Mind Rules Your Behavior*, 1st Ed, Vintage Books (Div. of Random House Inc.)

NATO (2023), Skill fade and competence retention: A contemporary review, STO Technical Report TR-HFM-292, North Atlantic Treaty Organization, Science and Technology Organization. Downloaded from www.sto.nato.ini on March 30, 2023

Norsok (2008), Technical Safety, NORSOK S-001, 4th Ed, Standards Norway

Norsok (2018), Working Environment, NORSOK S-002:2018 (en), Standards Norway

Norsok (2021), Technical Safety, NORSOK S-001:2020+AC:2021 (en), Standards Norway

NUREG (2016), Cognitive basis for human reliability analysis, NUREG-2114, Whaley, A.M., Xing, J., Boring, R.L., Hendrickson, S.M.L., Joe, J.C., LeBlanc, K.L., Morrow, S.L., Office of Nuclear Regulatory Research, U.S. Nuclear Regulatory Commission, Washington DC

NUREG (2020), Human-systems interface design review guidelines, NUREG-0700, O'Hara, J.M., Fleger, S., Rev. 3, Office of Nuclear Regulatory Research, U.S. Nuclear Regulatory Commission, Washington DC

OPITO (2014), Major Emergency Management Initial Response Training, Revision 1, OPITO Standard Code 7228, OPITO, March 13

Proctor, R.W., Van Zandt, T. (2018), Human Factors in Simple and Complex Systems, CRC Press, Boca Raton, FL

PSA (2013), Principles for barrier management in the petroleum industry, Petroleum Safety Authority, Norway, January 29

Radvansky, G.A. (2021), *Human Memory*, 4th Ed, Routledge, London

Rao, M.R, Mayer, A.R., Harrington, D.L. (2001), The evolution of brain activation during temporal processing, *Nature Neuroscience*, 4, 317–323

Reason, J. (1990), *Human Error*, Cambridge University Press, Cambridge

Reason, J. (2008), *The Human Contribution, Unsafe Acts, Accidents and Heroic Recoveries*, Ashgate Publishing Ltd.

Salmon, P.M., Stanton, N.A., Walker, G.H., Jenkins, D.P. (2009), *Distributed Situation Awareness, Theory Measurement and Application to Teamwork*, Ashgate Publishing Co., Aldershot

Salvendy, G. (Ed.) (2012), *Handbook of Human Factors and Ergonomics*, 4th Ed, Tsinghau University, John Wiley and Sons, New York

SINTEF (2011), CRIOP: A scenario method for crisis intervention and operability analysis, SINTEF Technology and Society, Report SINTEF A4312, 2011-03-07

Skogdalen, J.E., Khorsandi, J., Vinnen, J.E. (2011), Looking back and forward – evacuation, escape and rescue (EER) from the Deepwater Horizon Rig, Deepwater Horizon Study Group Working Paper – January 2011

Sneddon, A., Mearns, K., Flin, R. (2006), Situation awareness and safety in offshore drill crews. *Cognition, Technology & Work*, 8, 255–267.

SPE (2014), The human factor; process safety and culture, SPE Technical Report, Society of Petroleum Engineers, March 2014

Stanton, N.A., Salmon, P.M., N.A., Walker, G.H., Jenkins, D.P. (2010), *Human Factors in the Design and Evolution of Central Control Room Operations*, CRC Press, Boca Raton, FL

Sträter, O. (2005), *Cognition and Safety: An Integrated Approach to Systems Design and Assessment*, 1st Ed, Ashgate Publishing Ltd, Aldershot.

Sutcliffe, A. (2002), *User-Centred Requirements Engineering, Theory and Practice*, 1st Ed, Springer, London

Sylvestre, C. (2017), *Third Generation Safety: The Missing Piece*, ISBN 978-0-648 1200-0-1, National Library of Australia Cataloguing-in-Publication

Taber, M.J., McCabe, J., Klein, R.M., Pelot, R.P. (2012), Development and evaluation of an offshore oil and gas emergency response focus board, *International Journal of Industrial Ergonomics*, 43, 40–51

Tannenbaum, S., Salas, E. (2021), *Teams That Work, the Seven Drivers of Team Effectiveness*, Oxford University Press, New York

USCG (2010), Report of investigation into circumstances surrounding the explosion, fire, sinking and loss of eleven crew members aboard the MODU deepwater horizon in the Gulf of Mexico April 20–22, Vol. 1, Unites States Coast Guard

Wickens, C.D., Hollands, J.G., Banbury, S., Parasuraman, R. (2013), *Engineering Psychology and Human Performance*, 4th Ed, Pearson Education Inc., Upper Saddle River, NJ

Wickens, C.D., Gutzwiller, R.S. (2015), Discrete task switching in overload: A meta-analysis and a model, *International Journal of Human-Computer Studies*, 79, 79–84

Wickens, C.D., McCarly, J.S., Gutzwiller, R.S. (2023), *Applied Attention Theory*, 2nd Ed, CRC Press, Boca Raton, FL

Williams (2010), Mike Williams live testimony at joint USCG/BOEM inquiry into deepwater horizon accident

Index

Note: Page numbers followed by f, t and n indicate figures, tables, and notes, respectively. *Italic* page numbers indicate case studies and design criteria highlighted as *italicized* text.

For Product Safety Concerns and Information please contact our EU
representative GPSR@taylorandfrancis.com
Taylor & Francis Verlag GmbH, Kaufingerstraße 24, 80331 München, Germany